Springer Series in **Materials Science** 6

Edited by J. Peter Toennies

Springer Series in **Materials Science**

Volume 1 **Chemical Processing with Lasers**
By D. Bäuerle

Volume 2 **Laser-Beam Interactions with Materials**
Physical Principles and Applications
By M. von Allmen

Volume 3 **Laser Processing of Thin Films and Microstructures**
Oxidation, Deposition and Etching of Insulators
By I. W. Boyd

Volume 4 **Microclusters**
Editors: S. Sugano, Y. Nishina, and S. Ohnishi

Volume 5 **Graphite Fibers and Filaments**
By M. S. Dresselhaus, G. Dresselhaus,
K. Sugihara, I. L. Spain, and H. A. Goldberg

Volume 6 **Elemental and Molecular Clusters**
Editors: G. Benedek, T. P. Martin, and G. Pacchioni

Volume 7 **Beam Epitaxy-Crystallisation of Semiconductor Films from Atomic and Molecular Beams**
By H. Sitter, and M. A. Herman

Volume 8 **Physical Chemistry of, in and on Silicon**
By G. Cerofolini, and L. Meda

Volume 9 **Tritium and Helium-3 in Metals**
By R. Lässer

Volume 10 **Computer Simulation of Ion Solid Interactions**
By W. Eckstein

G. Benedek · T. P. Martin
G. Pacchioni (Eds.)

Elemental and Molecular Clusters

Proceedings of the 13th International School, Erice, Italy, July 1–15, 1987

With 218 Figures

Springer-Verlag Berlin Heidelberg New York
London Paris Tokyo

Professor Dr. Giorgio Benedek

Dipartimento di Fisica, Università degli Studi di Milano, I-20133 Milano, Italy

Professor Dr. Thomas Patrick Martin

Max-Planck-Institut für Festkörperforschung, Heisenbergstraße 1,
D-7000 Stuttgart 80, Fed. Rep. of Germany

Professor Dr. Gianfranco Pacchioni

Dipartimento di Chimica Inorganica e Metallorganica,
Università degli Studi di Milano, Via Venezian 21, I-20133 Milano, Italy

Guest Editor: Professor Dr. J. Peter Toennies

Max-Planck-Institut für Strömungsforschung, Böttingerstraße 6–8,
D-3400 Göttingen, Fed. Rep. of Germany

ISBN-13: 978-3-642-73503-5 e-ISBN-13: 978-3-642-73501-1
DOI: 10.1007/ 978-3-642-73501-1

Softcover reprint of the hardcover 1st edition 1988

2153/3150-543210

Preface

The 13th course of the International School of Materials Science and Technology was held in Erice, Sicily, at the Ettore Majorana Center for Scientific Culture, 1–15 July, 1987. At the suggestion of the European Physical Society, the course took the form of a summer school designed to present the fundamental concepts and methods needed to study elemental and molecular clusters. This volume is based on the lectures presented at the summer school. It has been organized into four parts: Electronic Properties, Structure, Fragmentation and Chemistry. Within each part the contributions have been ordered alphabetically by author; first the lectures followed by the seminars.

The summer school, held under the auspices of the European Physical Society, received generous support from a number of other organizations: The Italian Ministry of Education, The Italian Ministry of Scientific and Technological Research, The Sicilian Regional Government, IBM Italia, The National Research Council of Italy (CNR), The National Group for Structure of Matter, Italy (GNSM and CISM), and The Associazione Studi e Ricerche sull'Innovazione Tecnologica (ASRIT, Cameri, Italy).

Indispensable for the planning of the summer school was the support of the Director of the Ettore Majorana Center, Professor A. Zichichi, and of the Director of the International School of Materials Science and Technology, Professor M. Balkanski. We wish to express our sincere appreciation to the center staff, Dr. Alberto Gabriele, Dr. Pinola Savalli and Dr. Jerry Pilarsky for their expert assistance in all organizational matters and for their warm hospitality.

The success of a school is, in the last analysis, determined by the interest and commitment of the lecturers and participants. We are particularly grateful to the lecturers, not only for their carefully prepared formal contributions, but also for the spontaneous "evening sessions" and to all the participants for their inexhaustible enthusiasm.

Erice, August 1987 *The Editors*

Contents

Part I

Introductory Comments

Clusters: What Are They?

T.P. Martin

Max-Planck-Institut für Festkörperforschung, Heisenbergstr. 1,
D-7000 Stuttgart 80, Fed. Rep. of Germany

1. INTRODUCTION

How do the properties of a solid gradually evolve as atoms are brought together to form increasingly larger units? In order to answer this question one must study aggregates of atoms too large to be called molecules but still too small to have even the structure of a crystal. Recent advances in experimental methods have made possible some first explorations in this long neglected but fascinating field of study. Because these aggregates have attracted investigators /1-9/ from many branches of science and technology, a colorful but confusing array of terms has been used to describe them: small particles, microclusters, grains, oligomers, specks and microcrystals, to mention a few. Considering this state of affairs a few words of introduction might be useful addressing the questions: What are clusters? How do they differ from molecules? Why might they be of interest?

A ring of eight sulfur atoms or a tetrahedron of phosphorus atoms cannot with good conscience be called a cluster. Such stable units exist in the vapor, liquid and solid phases and have long been called molecules. Because molecules are readily available, their properties are usually well investigated. The term cluster might then be reserved

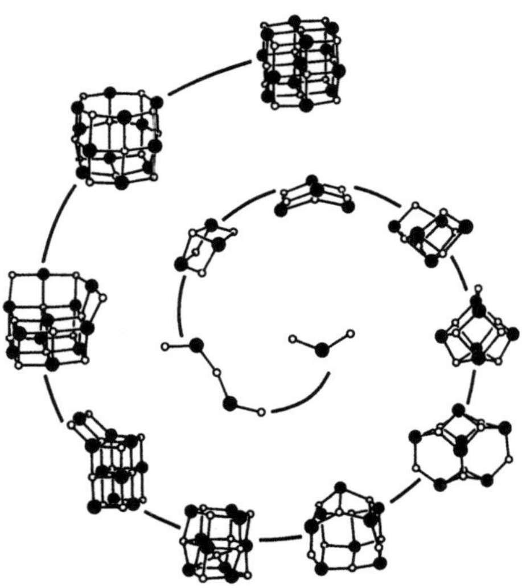

Fig.1. Calculated configurations for $Na^+(NaCl)_n$ clusters. Crystalline structures are not necessarily preferred during the first stage of growth of a solid.

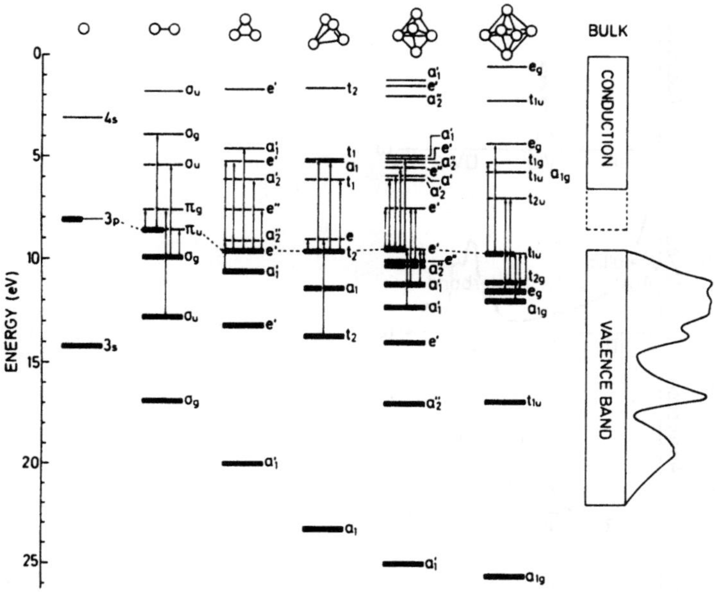

Fig.2. Calculated electronic energy levels of Si clusters having the indicated sizes and shapes.

for atom aggregates that are not found in appreciable numbers in an equilibrium vapor. Clusters, therefore, represent a new class of objects for investigation.

After a cluster reaches a critical size, it is no longer free to reconstruct each time a new atom is added. A final lattice structure becomes frozen into the cluster. This crystallization may occur for aggregates containing less than 100 atoms. Still it is convenient and meaningful to give these aggregates a new name, microcrystals, because deviations from the bulk structure can be described as a surface relaxation.

Why could the properties of clusters be of interest? For one thing we would like to understand the crystal growth process on a microscopic level. Everytime a new atom or molecule condenses onto a cluster, the atoms in the cluster completely rearrange themselves; the cluster reconstructs. We would like to know the sequence of structures a cluster assumes as it grows from a molecule into a crystal, Fig. 1.

Another reason for studying clusters is that we would like to know how the electronic band structure of a solid develops. This is represented in Fig. 2 by a simple Hückel calculation of the electronic energy levels of silicon clusters with increasing size. It is of interest to observe how the discrete levels coalesce to form bands and at what stage a distinct gap appears between filled and unfilled states.

A question often asked is, how many metal atoms are required to form a cluster that has metallic properties? One characteristic property of a metal is its optical response, the interaction with light. A single Li atom has a very simple absorption spectrum, Fig. 3. It consists of essentially one line in the visible. This absorption can be well described as a one electron transition from a 2s state to a 2p state. A lithium crystal, on the other hand, has a completely different absorption spectrum. The absorption is strong in the far-infrared, goes through a minimum in the visible and then rises again in the

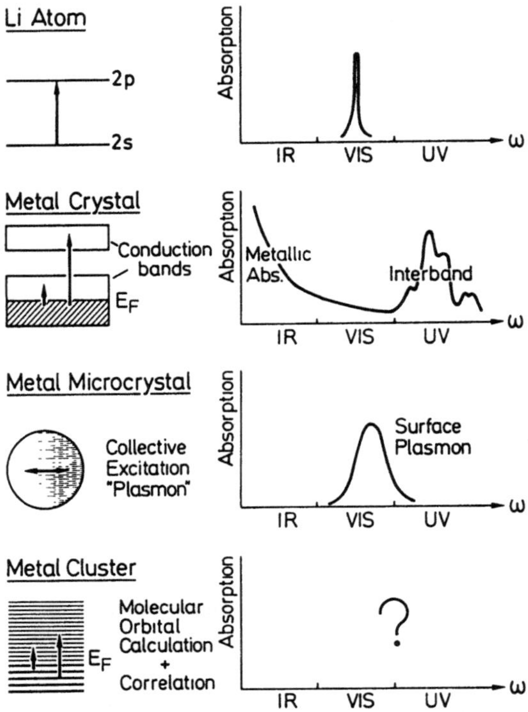

Fig.3. Absorption of light by Li in various stages of aggregation.

ultraviolet. The reason for this is that very low energy photons can excite electrons from the continuum of states just below the Fermi energy to states just above the Fermi energy. The strong uv absorption is caused by interband transitions. Suppose we chip off a small corner of the crystal, producing a microcrystal that contains only 50 atoms. Again the absorption spectrum changes completely. A relatively broad absorption maximum appears in the visible, Fig. 3. This absorption is due to a collective excitation of outer electrons, a plasmon excitation. The Li crystal also has plasmon excitations. However, in the crystal these excitations have a longitudinal character and do not couple to transverse light waves. In the microcrystal one can consider a plasmon as a collective sloshing motion of the electrons from one side of the microcrystal to the other side. It is clear that such a motion has a strong dipole moment, thus giving rise to the strong absorption band in the visible. The absorption of light by a Li atom is well described within the one electron picture. However, this picture cannot describe even qualitatively the plasmon excitations which dominate the optical response of a 50 atom cluster. Between a few atoms and 50 atoms we must completely change our way of looking at optical response. Clusters offer us the opportunity of studying this transition. Another reason for studying clusters is that the technique of cluster beam mass spectrometry offers new and convenient methods of allowing elements to react in a controlled manner and of determining the stability of the reaction products/9/.

2. CLUSTERS PRODUCED BY GAS EXPANSION

It has been known for several decades/10,11/ that when a gas under relatively high pressure (1 bar) expands into a vacuum through a small (0.1 mm) nozzle, clusters condense out of the cooled gas. The size distribution of the clusters can be controlled by varying the expansion parameters. Using this technique a large variety of gas and vapor clusters have been studied: hydrogen/12/, water/13-21/, sodium/22-27/, inert gases/28-33/ and tetracene/34,35/ to mention a few. For example, water clusters can be formed by expanding a mixture of water vapor and argon through a 0.22mm nozzle. The peak corresponding to $[H(H_20)_{21}]^+$ stands out particularly strongly in the mass spectrum. This mass spectrum is usually interpreted to indicate an enhanced stability for the 21-molecule water cluster. What does this cluster look like? Ice has a wurtzite structure where each oxygen atom is surrounded tetrahedrally by four other oxygen atoms. One, and only one, hydrogen atom is found on each bond. It is not possible to find a highly symmetric arrangement of 21 molecules within the wurtzite structure. However, using the tetrahedral bonding scheme it is also possible to construct a dodecahedral cage. In fact, such a cage has been identified in many clathrate compounds. The stability of the dodecahedron against both growth and decay could explain the strength of the n=21 peak in the mass spectrum.

3. CLUSTERS PRODUCED BY ION BOMBARDMENT

If an intense beam of high energy (5 KeV) ions Xe^+ is directed against a solid surface (CsI), secondary ions are ejected from the surface and can be detected in a mass spectrometer. These secondary ions include not only simple molecules but also large clusters containing hundreds of atoms/36-41/. A mass spectrum of $[Cs(CsI)_n]^+$ clusters produced in this way shows that the peak corresponding to the cluster $(Cs_{14}I_{13})^+$ is particularly strong. This cluster has been shown to have an enhanced stability not only for CsI but also for NaCl/41-43/ and CuBr/44/. Since this cluster is too small to be observed directly, we must rely on theoretical considerations in order to determine its structure. Stable cluster configurations can be calculated at widely varying levels of sophistication. The simplest method involves the packing of hard spheres. At the other end of the scale, stable structures can be determined by elaborate configuration interaction calculations of the total energy. Such calculations are appropriate for clusters with any type of bonding but are limited in practice to very small clusters of light atoms. The total energy of clusters with either purely ionic/45,47/ or van der Waals/48/ bonding can be determined much more simply. It is possible to define a size-independent, two-body interatomic potential. By summing this potential over all atom pairs a multidimensional total energy surface is obtained. Each minimum on this surface corresponds to a stable cluster configuration. The main computational difficulty is not to define the surface, but to find all true minima without getting trapped at a saddle point with a low curvature. Such total energy calculations indicate that $(Na_{14}Cl_{13})^+$ has more than the usual binding energy. The reason for this is quickly seen by looking at the structure of this cluster, Fig. 4. $(Na_{14}Cl_{13})^+$ is highly symmetric, resembling a portion of the rock salt structure. Calculations with parameters appropriate for the compounds CuBr and CsI indicate that the cubic cluster is responsible for the strong peaks in the mass spectra of these materials also. This is quite surprising because CuBr and CsI condense to form crystals of the zinc blende (coordination number, CN 4) and cesium chloride type (CN 8), respectively. It appears

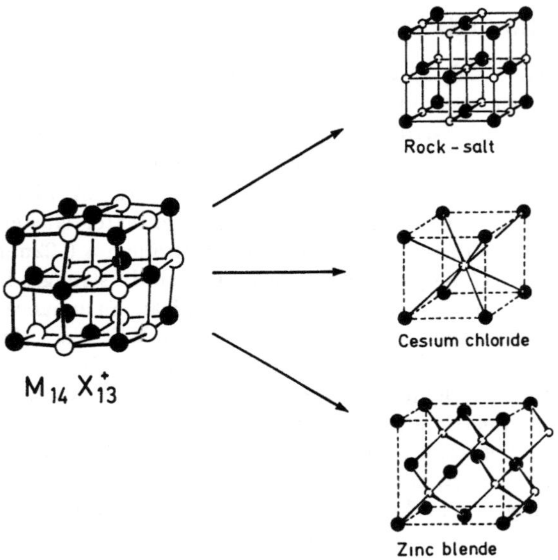

Fig.4. The most stable form of the metal halide $(M_{14}X_{13})^+$ and the crystal structures of NaCl, CsI and CuBr.

the rock salt form (CN 6) is energetically favored in the early stages of growth of not only NaCl but also CsI and CuBr.

4. VAPORS QUENCHED IN He GAS

Most materials when heated to a sufficiently high temperature in an inert gas atmosphere produce a dense smoke composed of microcrystals/49-52/. The size of the microcrystals varies from 10 to 1000 Å depending on experimental conditions. Recently, a time-of-flight mass spectrometric study has shown that the technique of inert gas evaporation produces not only microcrystals but also clusters containing fewer than 100 atoms/53,54/.

Not all elemental vapors produce a detectable number of clusters when quenched in He gas. Ge and Sn not only cluster easily/56-59/, but are particularly interesting because in the solid form they represent a transition from covalent to metallic bonding. Ge_n mass peaks are strong for n= 6,10,14,15 and 18, and particularly weak for n=13,17,20 and 24/60,61/. The meaning of these anomilies in the mass spectra is slowly emerging as a result of recent molecular dynamics computer simulations and ab initio calculations of the total energy /55/.

The dominant feature of the Sn mass spectrum is the weakness of a peak for n=14. In this respect Sn resembles Pb/53/. That Sn should more resemble Pb than Ge can be seen already in the solid state. Although Sn does have a low temperature semiconducting crystalline modification (α-Sn) with the diamond structure, at room temperature it is a metal with the β-Sn structure (CN 6). Once again we can only speculate as to the cause of the instability of the cluster corresponding to n=14. There are several close-packed structures corresponding to n=13 consisting of a central atom surrounded by a complete shell of outer atoms. An additional atom, which lies on the surface of such a highly symmetric compact core, could be expected to be weakly bonded.

5. CLUSTERS BY LASER EVAPORATION

A pulsed cluster source has been developed which uses a focused laser beam for vaporization/62-65/. The vaporization is timed to occur at the peak of a high pressure He pulse. Such a source is particularly appropriate for time-of-flight mass spectrometry. It has the advantage that essentially any material can be vaporized, even carbon. The mass spectrum of C_n^+ clusters has a bimodal size distribution. For n <30 both even and odd clusters were observed. The peaks corresponding to n=3,11,15 and 19 were particularly strong.

Under certain clustering conditions the C_{60} peak completely dominates the mass spectrum/66-69/. It seems unlikely that the special stability of such a cluster can be explained on the basis of a planar six-atom ring structure or on the basis of a tetrahedrally bonded diamond structure because such structures would possess too many dangling bonds. It has been suggested that C_{60} has the shape of a football (soccerball). It is spheroidal with icosahedral symmetry.

REFERENCES

1. Proceedings of the First International Meeting on Inorganic Clusters in: J. Phys. **38** Colloque C-2 (1977).
2. Proceedings of the Second International Meeting on Inorganic Clusters in :Surface Sci. **106** (1980).
3. Proceedings of the Third International Meeting on Inorganic Clusters in : Surface Sci. **156** (1985).
4. See papers in: Ber. Bunsenges. Phys. Chem. **88** (1984).
5. Proceedings of the International Symposium on Metal Clusters in : Z. Phys. D - Atoms, Molecules and Clusters **3** (1986); *Metal Clusters*, F. Träger and G. zu Putlitz, eds., (Springer-Verlag, Berlin, Heidelberg 1986).
6. *The Physics and Chemistry of Small Clusters*, NATO ASI Series, P. Jena, B.K. Rao and S.N. Khanna, eds. (Plenum, New York 1987).
7. *Microclusters*, S. Sugano, Y. Nishina and S. Onishi, eds., Springer Ser. Mat. Sci. (Springer, Berlin, Heidelberg 1987).
8. *Metal Clusters*, Martin Moskovits, ed. (John Wiley, New York, 1986).
9. A.W. Castleman, Jr. and R.G. Keesee: Chem. Rev. **86** ,589 (1986); J. Koutecky and P. Fantucci: Chem. Phys. **86** ,539 (1986); O. Echt: to be published in Hyperfine Interactions; T.P. Martin: Angew. Chem. Int. Ed. **25**, 197 (1986).
10. E.W. Becker, K. Bier and W. Henkes: Z. Physik **146**, 333 (1956).
11. O.F. Hagena and W. Obert: J. Chem. Phys. **56**, 1793 (1972).
12. J. Gspann and K. Körting: J. Chem. Phys. **59**, 4726 (1973).
13. S.-S. Lin: Rev. Sci. Instrum. **44**, 576 (1973).
14. J.Q. Searcy and J.B. Fenn: J. Chem. Phys. **61**, 5292 (1974).
15. G.D. Stein and J.A. Armstrong: J. Chem. Phys. **58**, 1999 (1973).
16. D. Dreyfuss and H.Y. Wachmann: J. Chem. Phys. **76**, 2031 (1982).
17. R.J. Beuhler and L. Friedmann: J. Chem. Phys. **77**, 2549 (1982).

18. A.J. Stace and C. Moore: Chem. Phys. Lett. **96**, 80 (1983).
19. G. Torchet, P. Schwartz, J. Farges, M.F. de Feraudy and B. Raoult: J. Chem. Phys.**79**, 6196 (1983).
20. H. Haberland, H.-G. Schindler and D.R. Worsnop: Ber. Bunsenges. Phys. Chem. **88**, (1984).
21. O. Echt, D. Kreisle, M. Knapp and E. Recknagel: Chem. Phys. Lett. **108**, 401 (1984).
22. A. Herrmann, S. Leutwyler, E. Schumacher and L. Wöste: Helv. Chim. Acta **61**, 453 (1978).
23. M.M. Kappes, R.W. Kunz and E. Schumacher: Chem. Phys. Lett. **91**, 413 (1982).
24. J.L. Gole, G.J. Green, S.A. Pace and D.R. Preuss: J. Chem. Phys. **76**, 2247 (1982).
25. K.I. Peterson, P.D. Dao, R.W. Farley and A.W. Castleman, Jr.: J. Chem. Phys.**80**, 1780 (1984).
26. W.D. Knight, K. Clemenger, W.A. de Heer, W.A. Saunders, M.Y. Chou and M.L. Cohen: Phys. Rev. Lett. **52**, 2141 (1984).
27. C. Brechignac, Ph. Cahuzac and J.Ph. Roux: Chem. Phys. Letters **127**, 445 (1986).
28. T.A. Milne, A.E. Vandegrift and F.T. Greene: J. Chem. Phys. **52**, 1552 (1970).
29. J. Farges, B. Raoult and G. Torchet: J. Chem. Phys. **59**, 3454 (1973).
30. O. Echt, K. Sattler and E. Recknagel: Phys. Rev. Lett. **47**, 1121 (1981).
31. A. Ding and J. Hesslich: Chem. Phys. Lett. **94**, 54 (1983).
32. L. Friedmann and R.J. Beuher: J. Chem. Phys. **78**, 4669 (1983).
33. A.J. Stace: Chem. Phys. Lett. **113**, 355 (1985).
34. A. Amirav, U. Even and J. Jortner: J. Chem. Phys.**75**, 2489 (1981); **86**, 3345 (1982).
35. U. Even and J. Jortner: J. Chem. Phys. **87**, 28 (1983).
36. F. Honda, G.M. Lancaster, Y. Fukuda and S.W. Rabelais: J. Chem. Phys. **69**, 4931 (1978).
37. J. Campana, T.M. Barlak, R.J. Colton, J.J. De Corpo, J.R. Wyatt and B.I. Dunlap: Phys. Rev. Lett. **47**, 1046 (1981).
38. T.M. Barlak, J.E. Campana, R.J. Colton, J.J. De Corpo and J.R. Wyatt: J. Chem. Phys. **85**, 3840 (1981).
39. W. Begemann, S. Dreihofer, K.H. Meiwes-Broer and H.O. Lutz: Z. Physik D **3**, 183 (1986).
40. T.M. Barlak, J.E. Campana, J.R. Wyatt and R.J. Colton: J. Chem. Phys. **87**, 3441 (1983).
41. W. Ens, R. Beavis and K.G. Standing: Phys. Rev. Lett. **50**, 27 (1983).
42. R. Pflaum, P. Pfau, K. Sattler and E. Recknagel: Surface Sci. **156**, 165 (1985).
43. T.P. Martin: Ber. Bunsenges, Phys. Chem. **88**, 300 (1984).
44. T.P. Martin and A. Kakizaki: J. Chem. Phys. **80**, 3956 (1984).
45. T.P. Martin: Physics Reports **95**, 167 (1983).
46. T.P. Martin: J Chem. Phys. **67**, 5207 (1977); **69**, 2036 (1980); **72**, 3506 (1980).
47. D.O. Welch, O.W. Lazareth, G.J. Dienes and R.D. Hatcher: J. Chem. Phys. **64**, 835 (1976); **68**, 2159 (1978).
48. M.R. Hoare: Adv. Chem. Phys. **40**, 49 (1979).
49. K. Kimoto, Y. Kamiya, M. Nonoyama and R. Uyeda: Japan. J. Appl. Phys. **2**, 702 (1963).
50. N. Wada: Japan. J. Appl. Phys. **7**, 1287 (1968).
51. C.G. Granqvist and R.A. Buhrman: J. Appl. Phys. **47**, 2200 (1976).
52. H. Abe, W. Schulze and B. Tesche: Chem. Phys. **47**, 95 (1980).

53. K. Sattler, J. Mühlbach and E. Recknagel: Phys. Rev. Lett. **45**, 821 (1980).
54. P. Pfau, K. Sattler, J. Mühlbach, R. Pflaum and E. Recknagel : J. Phys. F: Met. Phys. **12**, 2131 (1982).
55. E. Blaisten-Barojas and D. Levesque: Phys. Rev.B **34**, 3910 (1986); B. Feuston, R. Kalia and P. Vashishta: to be published in Phys. Rev.; K. Raghavacharia: J. Chem. Phys. **84**, 5672 (1986); S. Saito, S. Ohnishi, S. Sugano: Phys. Rev.B **33**, 7036 (1986); R. Car, M. Parrinello and W. Andreoni: in *Microclusters*, S. Sugano, Y. Nishina and S. Onishi eds. (Springer, Berlin, Heidelberg 1987); G. Pacchioni, D. Plavsic and J. Koutecky: Ber. Bunsenges. Phys. Chem. **87**, 503 (1983).
56. M. LeLeyter and P. Joyes: J. Phys. **36**, 343 (1975).
57. J.R. Heath, Y. Liu, S.C. O'Brien, Q.-L. Zhang, R.F. Curl, F.K. Tittel and R.E. Smalley: J.Chem. Phys. **83**, 5520 (1985).
58. J. Van de Walle and P. Joyes: Phys. Rev. B **32**, 8381 (1985).
59. J.C. Phillips: J. Chem. Phys. **85**, 5246 (1986).
60. T.P. Martin and H. Schaber: J. Chem. Phys. **83**, 855 (1985).
61. T.P. Martin: Z. Phys. D **3**, 211 (1986).
62. D.W. Powers, S.G. Hansen, M.E. Geusic, A.C. Pulu, J.B. Hopkins, T.G. Dietz, M.A. Duncan, P.R.R. Langridge-Schmith and R.E. Smalley: J. Phys. Chem. **86**, 2556 (1982).
63. E.A. Rohlfing, D.M. Cox and A. Kaldor: J. Chem. Phys. **8**, 3322 (1984).
64. V.E. Bondybey and J.H. Englisch: J. Chem. Phys. **76**, 2165 (1982).
65. S.J. Riley, E.K. Parks, G.C. Nieman, L.G. Pogo and S. Wexler: J. Chem. Phys. **80**, 1360 (1984).
66. H.W. Kroto, J.R. Heath, S.C. O'Brien, R.F. Curl and R.E. Smalley: Nature **318**, 162 (1985).
67. M.E. Geusic, M.F. Jarrold, T.J. McIlrath, L.A. Bloomfield, R.R. Freeman and W.L. Brown: Z. Phys. D **3**,309 (1986).
68. M.Y. Hahn, E.C. Honea, A.J. Paquia, K.E. Schriver, A.M. Camarena and R.L. Whetten: Chem. Phys. Letters **130**, 12 (1986).
69. A. O'Keefe, M.M. Ross and A.P. Baronavski: Chem. Phys. Letters **130**, 17 (1986).

Part II

Electronic Properties

Atomic Bonding in Clusters

J. Friedel

Laboratoire de Physique des Solides, Université Paris-Sud,
Faculté des Sciences, F-91405 Orsay, France

Framework.

The nature of bonding in clusters is directly related to their atomic structure and stability. The focus will therefore be here on the most stable electronic structures in the most stable arrangements. Electronic or atomic excitations will not be discussed in any detail ; they would require a whole set of lectures by themselves, and are anyway less well known in general.

General trends will be emphasised, without going into 'exact' quantitative estimates ; the most elementary description of bonding will therefore be used : the possibility of electron delocalisation will be described by using the simplest possible linear combination of atomic orbitals ; electron-electron interactions will be measured using atomic averages. Such a rough and ready approach has its inherent limitations. It cannot go as far as the elaborate computer studies of electronic structures ; these however mostly use a local (Slater, Kohn, Sham) correction to the one electron effective potentials which does not go far in the study of electron-electron correlations. Our way of treating these correlations, although better, does not go as far as the treatments using full blown configuration interactions ; these however only apply to small clusters of light atoms, while our range will be much wider.

After a reminder of what we know about very small clusters, and to start with about dimers, I want to discuss how the nature of bonding can change progressively when one goes through clusters of increasing sizes of a given element or through clusters of varying elements of the periodic table. Three points will be especially stressed :

- the role of electron-electron interactions, and therefore of electronegativity, in the choice between compact and molecular (dimerised) clusters.

- the concept of possible frustrations between the forces that keep atoms together in a small cluster and those that keep these clusters together in a larger aggregate. This choice dominates, at a macroscopic level, the choice between crystalline, quasicrystalline and amorphous structures.

- the concept of crossover between different regimes, when one considers clusters of increasing sizes of a given element or when one varies its chemical nature.

I - DIMERS.

We recall the case of the H_2 molecule and extend briefly its discussion to other dimers.

a - H_2 molecule.

When the two H atoms a, b are far from each other, they are neutral in their most stable state, with their electrons 1, 2 in 1s states of energy E_o ($|a\rangle$, $|b\rangle$, figure 1).

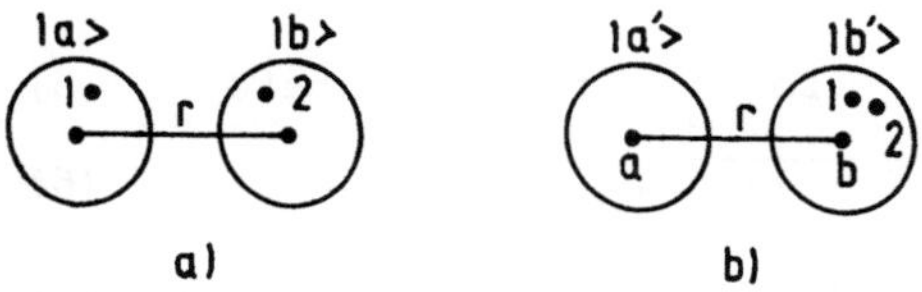

Fig. 1. H_2 molecule : a - H+H; b - $H^+ + H^-$

1 - Linear combination of 1s orbitals. The Hubbard model.

When the two atoms are brought to a smaller distance, their electrons can jump from one atom to the other, building temporarily H^+ and H^- ions. In its most stable state, a H^- ion has two electrons of opposite spins in 1s states, while for H+H, the two electrons can a priori have parallel or antiparallel spins. The simplest variational wave function $|\psi\rangle$ for the H_2 molecule, only using these 1s states, must be a linear combination of wave functions for H + H and H^++H^- configurations, which respect the symmetry of the molecule and the total spin symmetry. Starting with $|a(1)\rangle|b(2)\rangle, |b(1)\rangle|a(2)\rangle, |a'(1)\rangle|a'(2)\rangle$, $|b'(1)\rangle\ |b'(2)\rangle$, the four combinations are obviously

$$\alpha(|a(1)\rangle\ |b(2)\rangle - |b(1)\rangle\ |a(2)\rangle) \qquad (1)$$

for the triplet state,

$$\beta(|a(1)\rangle\ |b(2)\rangle + |b(1)|a(2)\rangle) + \gamma(|a'(1)\rangle|a'(2)\rangle + b'(1)\rangle|b'(2)\rangle)$$

$$\delta(|a'(1)\rangle|a'(2)\rangle - |b'(1)\rangle\ |b'(2)\rangle)$$

for the singlet states.

The corresponding energy **is** obtained from the Hamiltonian, with obvious notations,

$$\mathcal{H} \cong T_1 + T_2 + V_{ab} + V_{1a} + V_{2a} + V_{1b} + V_{2b} + V_{12} \ . \qquad (2)$$

As usual, the motion of the nuclei has been neglected (Born Oppenheimer approximation $T_a + T_b \cong 0$). We have also taken into account in (1) that the effective 1s states $|a'\rangle$, $|b'\rangle$ for the H^- ions are slightly more expanded than the 1s states $|a\rangle$, $|b\rangle$ of the H atoms. If we neglect these differences, an approximation often made, the functions (1) could be rewritten as linear combinations of products of one electron wave functions such as $\lambda|a\rangle + \mu|b\rangle$ with $|\lambda| \neq |\mu|$, where each electron is unequally distributed between the two atoms a, b. Thus (1) is a variational extension between the completely localised Heitler London approximation (λ or $\mu = 0$) and the completely delocalised Hund Mulliken approximation ($\lambda = \pm\mu$).

In the spirit of simplicity, we can concentrate on the obviously major factors of the problem, i.e. the intraatomic Coulomb interactions

$$U = \langle a'(1)\ a'(2) |\ V_{12} |\ a'(1)\ a'(2)\rangle, \tag{3}$$

and the one electron transfer integral, which reads

$$t = \langle a(i)\ |V_{ia}|\ b(i)\rangle = \langle b(i)\ |V_{ib}|\ a(i)\rangle, \tag{4}$$

when we neglect small differences such as between $|a\rangle$ and $|a'\rangle$. Because of the sign of V_{ia}, t is negative. The solution of $\mathcal{H}|\psi\rangle = E|\psi\rangle$ then reduces to

$$E_T \cong 2\ E_o \quad \text{for the triplet state.} \tag{5}$$

$$E_S \cong 2\ E_o + \frac{1}{2}\ (U - \sqrt{U^2 + 16\ t^2}), \tag{6}$$

$$E_{S'} \cong 2\ E_o + \frac{1}{2}\ (U + \sqrt{U^2 + 16\ t^2}), \tag{7}$$

$$E_{S''} \cong 2\ E_o + U, \quad \text{(for the singlet states.)} \tag{8}$$

This is equivalent to the Hubbard model for solids.

Figure 2 shows that, when decreasing $U/|t|$, one goes from the localised limit where the most stable state is H + H and the excited state is $H^+ + H^-$, to the completely delocalised one, where the most stable state is the completely bonding state (B + B) and the most unstable is the completely antibonding one (AB+AB).

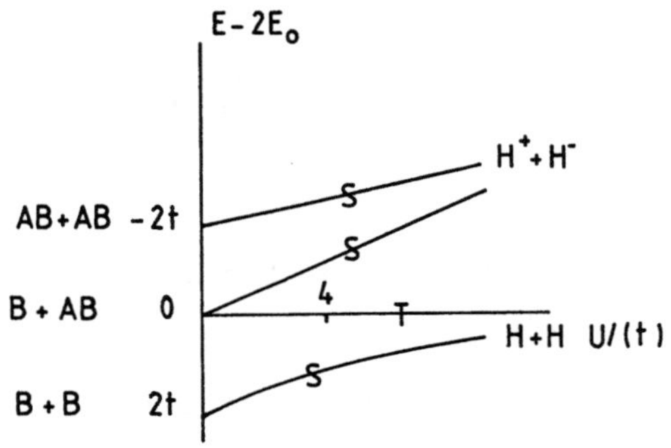

Fig. 2. Hubbard model for the H_2 dimer. S singlet state; T triplet state

The most stable state is the singlet state E_S. Development near the localised limit ($U \gg |t|$) gives

$$E_S \cong 2\ E_o - \frac{4t^2}{U} + O\ (\frac{t^3}{U^2}). \tag{9}$$

The term $-\frac{4t^2}{U}$ stabilises the singlet state with respect to the triplet one ; it is due to the fact that, in the singlet state, each localised electron can explore, in a perturbative way, the other atom, which is normally occupied by an electron of opposite sign. This is the Anderson term. Development near the delocalised limit ($U \ll |t|$) gives similarly

$$E_s \cong 2\,E_o + 2t + \frac{U}{2} + \frac{U^2}{16t} + O\,(\frac{U^3}{t^2}). \qquad (10)$$

Beyond the Hartree energy ($E_o + \frac{U}{4}$) of each electron and their Hartree Fock interaction $2\,\frac{U}{4} = \frac{U}{2}$, the first corrective term $U^2/16t$ is a lowering of energy due to electron correlations. The crossover between these two extremes occurs around $U \cong 4|t|$. It is smooth, as one expects for a finite system.

Finally let us note that U is independent of r while $|t|$ decreases exponentially with increasing r

$$t \cong t_o \exp(-\,p\,r)\,, \qquad (11)$$

where p^{-1} is of the order of the radius of the atomic orbits $|a\rangle$, $|b\rangle$. Figure (2) thus also gives the variation of E_s with r, where $U/|t|$ increases with increasing r.

2 - Corrections to the Hubbard model.

The corrections to this simple model keep within the 1s 1s variational scheme (1) or go beyond it.

- Short range repulsion. Within the scheme (1), the other matrix elements of the Hamiltonian $\mathcal{H}$ first include the electrostatic repulsion between the two neutral H atoms :

$$R = V_{ab} + \langle a(i)|V_{ib}|a(i)\rangle + \langle b(i)|V_{ia}|b(i)\rangle + \langle a(1)b(2)|V_{12}|a(1)b(2)\rangle \qquad (12)$$

This is a positive energy which decreases exponentially with increasing R :

$$R \cong R_o \exp(-q\,r), \qquad (13)$$

where

$$q \cong 2.5\,p. \qquad (14)$$

As $q >> p$, this short range repulsion stabilises the dimer at a distance which minimises the total energy $E_s + R$. In the delocalised limit (10), for instance, this gives

$$E_s \cong 2\,E_o + (1 - \frac{p}{q})\,2t + \frac{1}{2}\,U + (1 + \frac{p}{q})\,\frac{U^2}{16t} + \ldots\,, \qquad (15)$$

or

$$(q-p)\,r_2 \cong \ln \frac{q\,R_o}{2pt_o} + \frac{U^2}{32t^2}\,. \qquad (16)$$

The total binding energy of the dimer is the energy involved in the reaction

$$H_2 \rightarrow 2\,H\,. \qquad (17)$$

It is

$$2\,E_2 = 2\,E_o - (E_s + R)_{r_2}\,, \qquad (18)$$

where E_2 is the binding energy per atom.

- Spin dependent terms. In the localised limit $U >> |t|$, $E_s - 2\,E_o$ becomes small and might become comparable to other small spin dependent terms of scheme (1). There are two such terms :

α- the spin exchange term, which flips the spins of the two electrons without changing the atomic charges :

$$j = \langle a(1)\ b(2)\ |V_{12}|\ b(1)\ a(2)\rangle \qquad (19)$$

It is the Coulomb selfenergy of the charge of density $\langle a|b\rangle$: it is necessarily positive and stabilises the triplet state, destabilises the singlet one.

β- the non-orthogonality term

$$S = \langle a(i)|b(i)\rangle \neq 0 ,$$

which mixes the H + H and $H^+ + H^-$ configurations. In effect, the successive action of S on one electron from, say, a to b, and of t on the other electron from b to a produces the same final effect as j. The corresponding energy correction to the singlet state is

$$St + tS = 2\ St . \qquad (20)$$

As $S > 0$ and $t < o$, it is negative ; it stabilises the singlet state and destabilises the triplet one.

These two corrections (19) and (20) work, therefore, in opposite directions. It is easily seen that j should dominate all other binding contributions in the limit of large r's. But, at the relatively short equilibrium distances r_2, the Anderson and nonorthogonality terms outweigh the exchange one and stabilise the singlet state for H dimers as well as for most other dimers.

- Correlation terms using excited atomic states (2s, 2p ...). Coulomb interactions between the two electrons can produce virtual excitations to excited atomic states (2s, 2p ..). The corresponding energy corrections are better expressions for the electron correlations ; they necessarily lower the total energy of the system, thus increase bonding.

These terms take especially simple forms in the two extreme limits of figure 2.

α - In the localised limit, intraatomic correlations can be expressed as due to electrons being excited into orbitals with larger radii (2s ...) and lower symmetry (2p ...), so as to lower their Coulomb interaction. Within second order approximations, the leading (negative) correction to U is of the form

$$\frac{|\langle a(1)\ a(2)\ |V_{12}|\ a(1)\ a''(2)\rangle|}{E_a - E_{a''}} , \qquad (21)$$

where a" is, say, a 2s state. U is therefore reduced to an effective value U_e which can be experimentally deduced from properties of free atoms :

$$U_e = I - A , \qquad (22)$$

where I = 13.5 eV is the ionisation potential ($H \rightarrow H^+ + e$) and A = 0.7 eV is the electron affinity ($H + e \rightarrow H^-$). Thus, as expected figure 2 in this limit, U_e is the difference of energy between $H^+ + H^-$ and H + H. This gives U_e = 12.8 eV, to be compared with U = 17 eV.

Interatomic correlations lead to the van der Waals attraction potential in r^{-6}, which can be viewed as the Coulomb interaction of virtually excited and fluctuating dipoles on each H atom. Within second order perturbations, the (negative) terms are of the form

$$\frac{|\langle a(1)\; b(2)\rangle | V_{12} + V_{1a} + V_{2a} + V_{ab} |\; a''(1)\; b''(2)\rangle|^2}{E_a + E_o - E_{a''} - E_{b''}} .$$

With the z axis along the dimer axis, the potential energy $V_{12}+V_{1a}+V_{2a}+V_{ab}$ reduces to a dipole-dipole interaction at long range r $(x_{1a}\, x_{2b} + y_{1a}\, y_{2b} - 2\, z_{1a}\, z_{2b})\frac{e^2}{r^3}$.
Thus the leading terms in the perturbation are of the form

$$\frac{e^2}{r^6}\,\frac{|\langle 1s|\alpha|2p_\alpha\rangle|^4}{E_{1s}-E_{2p}} \quad (\alpha = x, y, z) \;, \qquad (23)$$

where $\alpha = z$ dominates.
β - In the strongly delocalised limit, the corresponding virtual excitations are not to individual (2s, 2p ...) atomic states but to excited bonding states which are built with these atomic states. As obviously the bonding of these states is larger than the bonding of the 1s states, one expects some lowering of U_e with respect to (22), but a disappearance of the van der Waals term.

γ - For intermediary configurations, the crossover regime is at distances r such that the transfer integral $|t|$ becomes comparable with the energy of atomic excitations (1s → 2s ...).

In fact, for the H_2 dimers, the excited states (2s, 2p) have energies well above that of the fundamental 1s state. As a result, the total reduction of U by correlations remains modest and near to (22). The ratio $U/|t|$ is thus sizeable, and the position of the dimer keeps to the left but near to the crossing point of figure 2. This means that the corrections from the Hubbard model such as (19),(20) are negligible.

b - Other elementary dimers.

We shall briefly extend these considerations to other elements of the periodic table.

1 - He_2 and other dimers of rare gases.

It is well known that, within the Hartree Fock limit and using only the most stable atomic orbitals, dimers of atoms or ions with closed shells can be equivalently analysed within the localised or delocalised limit.

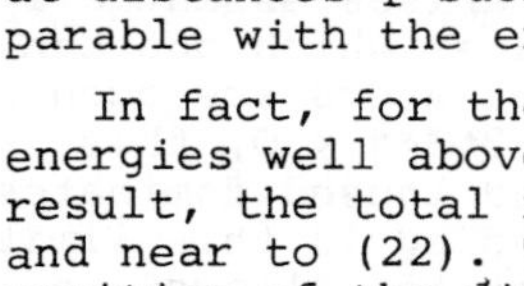

Fig. 3. H_2 dimer

Thus for a He_2 dimer (figure 3), with obvious notations, the orbital wave functions for the most stable singlet state in the localised and delocalised limits can be written

$$|\psi_L\rangle \propto \begin{vmatrix} |a(1),+\rangle & |a(2),+\rangle & |a(3),+\rangle & |a(4),+\rangle \\ |a(1),-\rangle & |a(2),-\rangle & |a(3),-\rangle & |a(4),-\rangle \\ |b(1)+\rangle & |b(2),+\rangle & |b(3),+\rangle & |b(4),+\rangle \\ |b(1)-\rangle & |b(2),-\rangle & |b(3),-\rangle & |b(4),-\rangle \end{vmatrix} \qquad (24)$$

$$|\psi_D\rangle \propto \begin{vmatrix} |a(1)+b(1),+\rangle & |a(2)+b(2),+\rangle & |a(3)+b(3),+\rangle & |a(4)+b(4),+\rangle \\ |a(1)+b(1),-\rangle & |a(2)+b(2),-\rangle & |a(3)+b(3),-\rangle & |a(4)+b(4),-\rangle \\ |a(1)-b(1),+\rangle & |a(2)-b(2),+\rangle & |a(3)-b(3),=\rangle & |a(4)-b(4),+\rangle \\ |a(1)-b(1),-\rangle & |a(2)-b(2),-\rangle & |a(3)-b(3),-\rangle & |a(4)-b(4),-\rangle \end{vmatrix} \qquad (25)$$

A well known theorem of determinants shows that $|\psi_L\rangle$ and $|\psi_D\rangle$ are equivalent.

Within this limit, one can then compute the energy from the Hamiltonian,

$$\mathcal{H} = \sum_{i=1}^{4} (T_i + V_{ia} + V_{ib}) + V_{ab} + \sum_{i>j} V_{ij} \ , \qquad (26)$$

using the localised approach $|\psi_L\rangle$.

Within the same approximation detailed for H_2, the total energy reduces to

$$\langle\psi_L|\mathcal{H}|\psi_L\rangle \cong 4\,E_o + 2U - 2(j+2tS) + R \ , \qquad (27)$$

where $2\,E_o + U$ is the energy of the He atom, R the short range electrostatic repulsion between the two He atoms and $-2(j+2tS)$ the exchange + non-orthogonality term which arises essentially because both the singlet and the triplet states which one can build with two electrons sitting on different atoms are occupied. Because, at the equilibrium distance, $2tS+j < 0$, this contribution to bonding is also repulsive and indeed dominates the electrostatic repulsion; there is no bonding in this limit.

Inclusion of excited (2s, 2p ..) atomic states allows a treatment of electron correlations. Now the equivalence between the localised and delocalised limits breaks down. As figure 4 shows, U_e is very large for He, because of the special stability of closed shells. Correlations must therefore necessarily be treated in the localised limit. As discussed for the H_2 dimer, one can consider the intraatomic correlations which reduce U to U_e and the interatomic correlations, which lead to the classical van der Waals attraction.

For all dimers of rare gases, a similar analysis leads to a bonding due to the weak Van der Waals attraction, and the equilibrium distance is fixed by the short range non-orthogonality repulsion.

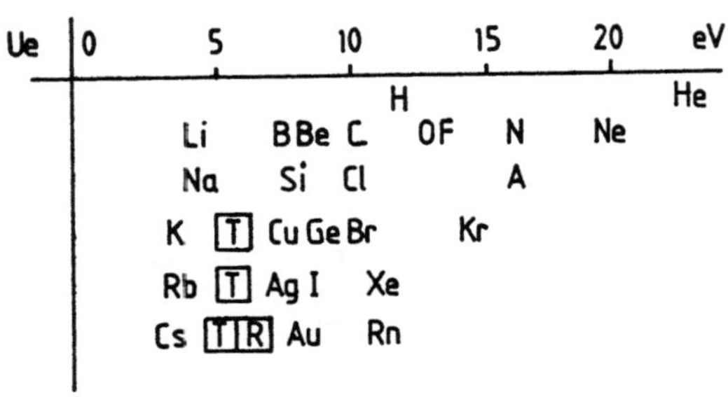

Fig. 4. Variation of $U_e = I - A$ in the periodic table. T transitional metals; R rare earth metals

2 - Alkaline dimers.

This is the most straightforward extension. In a qualitative way at least, the Hubbard model (6) can be still used, with a repulsive correction (13)(14) and an effective value of U given by (21). As figure 5 shows, U_e is small, essentially because the excited states are nearer in energy to the stable ns valence state. As $|t|$ is not expected to vary so much, the ratio $U_e/|t|$ is definitely smaller in the alkaline dimers than in the H_2 dimers, and the correlation terms less important in the bonding.

Fig. 5. 2p - 2p bonding: a - σ bond; b - π bond

3 - Dimers of halogens.

A direct extension is also possible in this case, with (np - np)σ bonding of the two electron holes of the np^5 shells (or equivalent bonding due to the non compensated σ bond of two np electrons, all the other s and p valence electrons occupying compensating bonding and antibonding σ and π bonds.

Figure 5 shows that here U_e is much larger, especially for light elements (because the excitation energy of n p states to (n+1)s is large). Correlation effects are much more important, and one is probably nearer the localised limit.

4 - Polyvalent elements. The case of C_2.

There are obviously intermediary cases as far as $U_e/|t|$, i.e. electron correlations, are concerned. Complications arise from the multiplicity of occupied (s and p) orbitals and from the multiplicity of electrons per atom.

Thus for divalent metals, the filling of the ns^2 shells would lead to a weak (van der Waals) bonding, as in rare gases, but to some sp hybridisation, or at least virtual excitation, by correlations which increase the bonding.

This sp hybridisation increases to a maximum about the middle of the sp series for the tetravalent elements, where the bonding can be described in a delocalised scheme by two electrons in a σ (sp sp)bonding state, two electrons in a σ(sp sp) non bonding state and 4 electrons in two orthogonal π(pp) bonding states (figure 6). The localised nature of the bond also increases with U_e (figure 4). If we nevertheless use to start with a Hartree delocalised scheme, the effective one electron Hamiltonian can be written for the σ bonds along the z axis,

$$\mathcal{H} = t_p(|p_z^a\rangle\langle p_z^b| + |p_z^b\rangle\langle p_z^a|) + t_s(|s^a\rangle\langle s^b| + |s^b\rangle\langle s^a|)$$
$$+ t_{sp}(|s^a\rangle\langle p_z^b| + |p_z^b\rangle\langle s^a| - |s^b\rangle\langle p_z^a| - |p_z^a\rangle\langle s^b|) + E_s(|s^a\rangle\langle s^a| + \quad (28)$$
$$|s^b\rangle\langle s^b|) + E_p(|p_z^a\rangle\langle p_z^a| + |p_z^b\rangle\langle p_z^b|) \text{ , where}$$

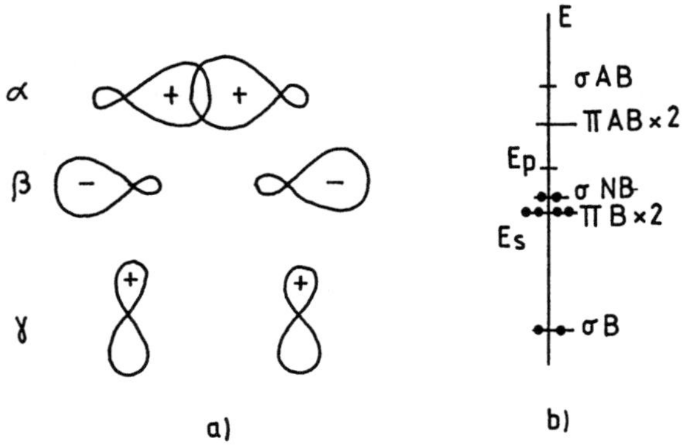

Fig. 6. C_2 bond; a - in space: α σ bonding state; β σ non bonding state; γ π bonding state; b - energy of one electron delocalised states

$$t_s = \frac{1}{2} \langle s^a | V_a + V_b | s_a \rangle \quad ,$$

$$t_p = \frac{1}{2} \langle p_z^a | V_a + V_p | p_z^b \rangle \quad , \tag{29}$$

$$t_{sp} = \frac{1}{2} \langle s^a | V_a + V_p | p_z^b \rangle \quad .$$

Hence

$$\begin{vmatrix} E_s - E & 0 & t_{sp} & t_s \\ P & E_p - E & t_p & -t_{sp} \\ t_{sp} & t_p & E_p - E & 0 \\ t_s & -t_{sp} & 0 & E_s - E \end{vmatrix} = 0 \quad .$$

This gives

$$(E_s-E)^2(E_p-E)^2 - 2(E_p-E)(E_s-E)t_{sp}^2 - (E_s-E)^2 t_p^2 - (E_p-E)t_s{}^2 \tag{30}$$

$$+ (t_{sp}^2 + t_s t_p)^2 = 0 \quad .$$

The result simplifies if we assume

$$t_s = - t_p = - t_{sp} = t \tag{31}$$

Hence

$$E = E_p + \frac{1}{2} \left[E_s + 2\ t \pm \sqrt{E_s^2 + 4\ t^2} \right] \quad . \tag{32}$$

Figure 7 schematises the variation of the bonding, non bonding and antibonding one electron states versus r. It is seen that, for small r, $|2t| >> E_p - E_s$. One has then two σ states of energies $\pm 2t + \frac{1}{2}(E_s+E_p)$, and two doubly degenerate π states of energies $E_p \pm t_\pi{}^2$; finally two non bonding σ states have energies near to $\frac{1}{2}(E_s+E_p)$. The six electrons in the bonding σ and π states build the triple bond C ≡ C ; the two electrons in the lower non bon-

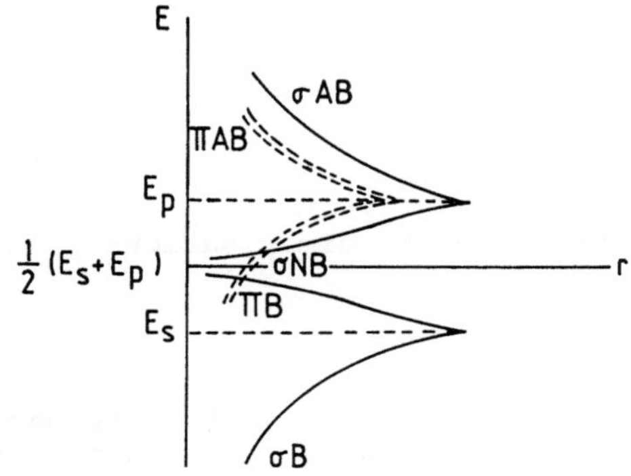

Fig. 7. Variation with distance r of the σ and π one electron state of the C_2 dimer

ding σ state correspond to the two non-saturated σ bonds on either side of the dimer.

Finally, instead of using the four s and p_z atomic orbitals, it is equivalent to use the 4 hybrid orbitals

$$\frac{1}{\sqrt{2}} \mid s^a \pm p_z^a\rangle \quad , \quad \frac{1}{\sqrt{2}} \mid s^p \pm p_z^b\rangle \ .$$

Approximation (29) then signifies that one neglects all transfer integrals between hybrid orbitals except the one schematised figure 6 a α:

$$\frac{1}{2} \langle s^a + p_z^a \mid \frac{1}{2}(V_a+V_b) \mid s^b - p_z^b\rangle = t_\sigma = -2t \ . \quad (33)$$

In fact the small difference in energy of the two non bonding states is bound to the fact that

$$\frac{1}{2} \langle s^a+p_z^a \mid \mathcal{H} \mid s^a - p_z^a\rangle = \frac{1}{2}(E_s - E_p) \neq 0 \ . \quad (34)$$

The hybrid orbitals can therefore only be considered as independent in the limit where

$$|t| \gg \frac{1}{2}(E_p - E_s) \ , \quad (35)$$

i.e. at relatively short distances r.

Finally dimers of di- and trivalent elements at the end of the sp series have sp bonds with decreasing weight of the s state when one goes towards the right and bottom of the periodic table, i.e. with increasing promotion energy $E_p - E_s$. The localisation character is, according to figure 4, especially marked for O_2 and N_2 ; O_2 being the only elemental dimer to have a stable triplet state.

II - EXTENSION TO SMALL CLUSTERS.

We wish now to discuss how these elementary bonding characters extend to small clusters. The case of trimers will already show most of the complications and characteristics of larger clusters.

a - Trimers.

Besides the relatively simple case of rare gases, we discuss the various problems that arise.

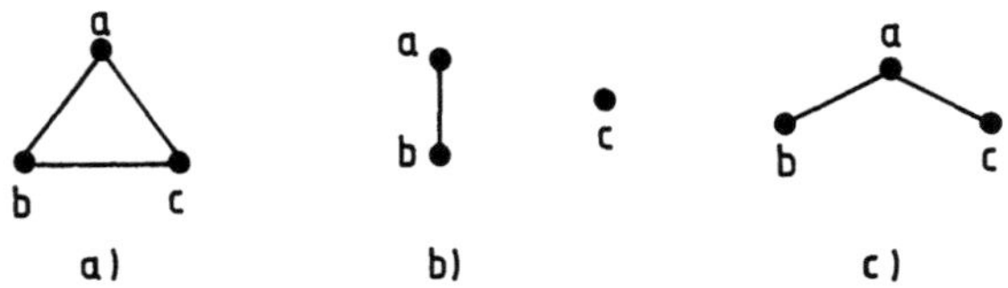

Fig. 8. Trimer. a) equilateral triangle; b) dimerisation; c) Jahn Teller distortion

1 - Rare gases.

It is customary to treat the interatomic forces in a trimer as additive, thus as the same central forces as for each of the three dimers involved (figure 8). Each is then the sum of a long range van der Waals attraction, and a short range repulsion. These three equivalent forces necessarily lead to the most stable state being an equilateral triangle (figure 8 a).

In fact, in the computation of the van der Waals interaction between a and b, the presence of c modifies the energy of excitation of the atomic a and b states. This reduction is small, however because the leading term involves p states which are not directed towards the third atom (cf (23)). Further small corrections involve three electron correlations and repulsions. But, for small clusters, these corrections are negligible. Thus trimers keep the same bond lengths as dimers.

2 - Closepacked trimers versus dimerised trimers. The role of correlations in open shells.

For open shells, the main problem is whether trimers keep fairly close packed (figure 8 a or c) or have a tendency to dimerise (figure 8 b).

It is clear that delocalisation favours close packing. Thus for monovalent elements with s bonding in the extreme delocalised limit, the energy spectra are given (figure 9) for the equilateral triangle, the three atoms in a row and the extreme dimerisation. The width of the energy band increases with the number of neighbours. As a result, the kinetic

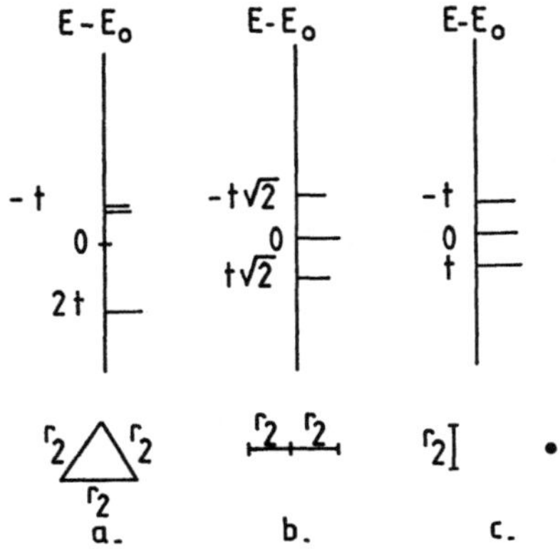

Fig. 9. One electron energy spectra for three extreme solutions of trimers in the delocalised limit

energy of the electrons increases but their total energy decreases ; this can be seen as a consequence of the virial theorem. More precisely, in this limit, with one electron per atom,

$$E_3 = 3\,E_o + 3\,t + 3\,R \quad \text{for equilateral triangle,} \tag{36}$$

$$E'_3 = 3\,E_o + 2\sqrt{2}t + 2\,R \quad \text{for the chain,} \tag{37}$$

$$E''_3 = 3\,E_o + 2\,t + R \quad \text{for the dimer.} \tag{38}$$

As R decreases much faster than t with increasing distance r, the first solution is the most stable.

It is also clear that <u>localisation favours dimerisation</u>. This can be seen in the extreme localised limit for the same monovalent elements; the only configuration that can be usefully compared to the singlet dimer plus an isolated paramagnetic atom (figure 9 c) is the linear chain (figure 9 b). For the dimerised triplet, the bonding energy is that of the dimer, thus in this limit (figure 10 c),

$$E''_3 \cong 3\,E_o + j + tS - \frac{4t^2}{U} + R\ . \tag{39}$$

For the chain, the only stable possibility is for the central atom to have a fixed moment antiparallel to those of the two end atoms. For each interacting pair of <u>fixed</u> antiparallel moments, the energy must be the average between those of the corresponding singlet $\frac{1}{}(|-\!>-|-+>)$, and triplet $\frac{1}{\sqrt{2}}(|+-\!>+|-+>)$, of energy $2E_o+j+tS-\frac{4t^2\sqrt{2}}{U}+R$ and $2E_o-|j+tS>+R$ respectively. The energy of the chain is thus (figure 10 b),

$$E'_3 = 3\,E_o - \frac{4t^2}{U} + 2R\ . \tag{40}$$

For the equilateral triangle, the solution of lowest energy is for three fixed moments at $\frac{2\pi}{3}$ to each other (Yafet and Kittel coupling). A similar argument gives (figure 10 a)

$$E_3 = 3\,E_o - \frac{3}{2}\frac{t^2}{U} - \frac{3}{8}\,(j+tS) + 3R\ . \tag{41}$$

It is then clear that the dimerised solution has the lowest energy E''_3.

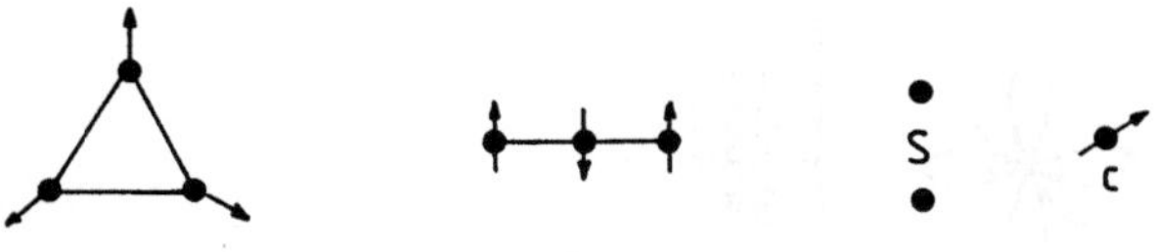

a. b. c.

Fig. 10. Solutions of lowest energy in the localised limit for triplets of monovalent elements

In actual fact, even in the rather delocalised limit, strong enough electron interactions stabilise the dimerised solution. Thus a straightforward extension of equation (10) gives for the triangle and the chain.

$$E_3 \cong 3\,E_O + 3t + \frac{2}{3}\,U + \frac{U^2}{18t} + 3R\ , \tag{42}$$

$$E'_3 \cong 3\,E_o + 2\sqrt{2}\,t + \frac{5}{8}\,U + \frac{U^2}{17.3t} + 2R\ . \tag{43}$$

Thus both terms in U and U^2 tend to stabilise the dimerised solution (10). Indeed for $q/p \cong 2.5$, the dimerised solution becomes more stable if

$$|U/t| \gtrsim 2 , \qquad (44)$$

a value for which developments such as (10), (42), (43) are still valid.

Following this discussion, we understand that alkali metals with small values of U_e (cf figure 4) have essentially close packed trimers, while H trimers, with definitely larger values of U_e, are dimerised.

The discussion can be extended directly to the positive holes of the trimers of halogens. The large values observed for U_e (figure 4) predict the stability of the dimerised solution. More generally, all elements with large values of U_e, thus large electronegativity, are expected to have stable dimerised trimers, while elements with smaller values of U_e are expected to have more close packed trimers.

3 - Dimensionality of close packing. The directionality of bonding.

In the limit of strong delocalisation, the exact nature of close packing is related to the directionality of bonding.

For metals, and especially alkali and noble metals with s isotropic bonding, one expects trimers not to deviate much from the equilateral triangle.

However if p bonding is involved, the one dimensional chain is obviously preferred to the equilateral triangle. This is the case for C_3 (figure 11) where s and p_z orbitals along the z axis of the trimer build bonding, non-bonding and antibonding sp bonds, while the p_x and p_y orbitals build two π bands with 3 doubly degenerate delocalised states. The 12 electrons occupy the bonding σ and π states and the non bonding σ states, which are nearly degenerate in the limit discussed for the dimer(figure 6).

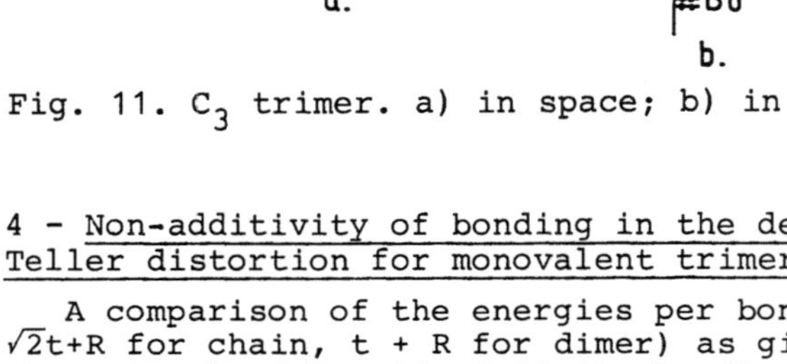

Fig. 11. C_3 trimer. a) in space; b) in energy

4 - Non-additivity of bonding in the delocalised limit. Jahn Teller distortion for monovalent trimers.

A comparison of the energies per bond (2t+R for triangle, $\sqrt{2}$t+R for chain, t + R for dimer) as given above clearly shows that, in the delocalised limit, bonding of open shells is not additive. Indeed the strength of each bond decreases with in-

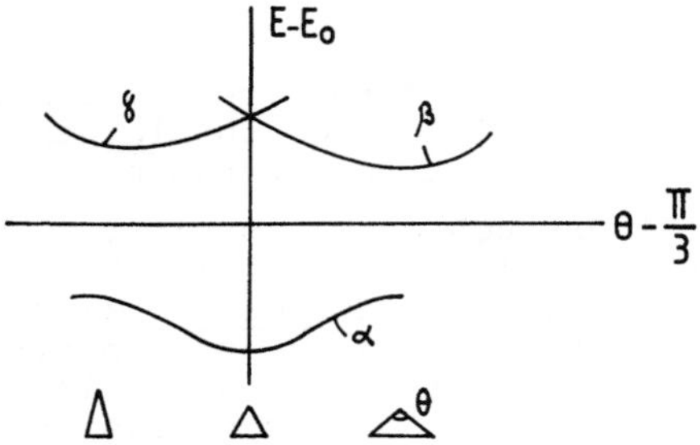

Fig. 12. Variation of the electron states in the delocalised limit with the Jahn Teller distortion $\theta - \frac{\pi}{3}$

creasing number of bonds ; and as a result the bond length increases with the number of bonds. This is a clear distinction of the s and p(π) bondings compared with van der Waals or (as we shall see later) sp(σ) bonds.

Another effect characteristic of the delocalisation is the (weak) Jahn-Teller distortion observed from equilateral triangle for monovalent trimers (figure 8 c) : starting from the equilateral triangle, with one electron in a doubly degenerate state (figure 9 a), one necessarily gains energy by distorting the triangle ; the energy of the one electron states will vary linearly with distortion while the elastic reaction will vary as the square of the distortion. A detailed analysis of this effect shows that the minimum of energy is reached for an isocele triangle with an angle larger than $\frac{\pi}{3}$ (figure 12). Thus in the delocalised limit

$$E_3 = 2E_\sigma + E_\beta + R' + 2R'' = 3E_o - \sqrt{t'^2 + 8t''^2} + R' + 2R'' \quad \text{for } \theta > \frac{\pi}{3}$$

$$E_3 = 2E_\sigma + E_\gamma + R' + 2R'' = 3E_o + \frac{3t'}{2} - \frac{1}{2}\sqrt{t'^2 + 8t''^2} + R' + 2R'' \quad \text{for } \theta < \frac{\pi}{3}$$

where t' and R' refer to the distance r' of the base and t" and R" refer to the distances r" of the sides of the triangle. Minimising E_3 with respect to r' and r" gives

$$(2p-q)(r''-r') = 2 \ln 2 > 0,$$

if as assumed here, $2p-q < 0$, r' is indeed larger than r".

This is indeed a characteristic of neutral trimers of monovalent metals. Also, as expected, positive trimers of the same elements, with only two electrons in the most stable state, are observed in equilateral triangles, with no Jahn-Teller distortion.

There is no Jahn-Teller distortion of C_3, i.e. the three C of the chain remain equidistant , as the non bonding π states which are primarily affected by the distortion are empty (figure 11).

b - Small clusters larger than trimers.

We shall only make some qualitative remarks in this field, which is far from completely explored.

1 - Rare gases.

Assuming additive central forces as in dimers, one has computed equilibrium configurations which, at least for the larger clusters, seem to fit with electron diffraction experiments.

The most stable arrangement of four such atoms is visibly the regular tetrahedron ; and for more than four atoms, the most stable arrangements can be visualised as a close packing of atoms which build locally regular tetrahedra so as to maximise the total number of bonds. We shall come back in the last lecture to the difficulties met in this way for large clusters.

2 - Metals.

The atomic structure of monovalent metals has been studied mostly from a theoretical point of view in this limit. The general tendency is expected to be towards rather close packed structures. But the stability does not increase linearly with the number of bonds and, for small enough aggregates, 2 dimensional geometries are preferred. This shows, for instance, for 4 atoms where a simple counting of one electron energies in the delocalised limit shows that the most stable configuration is not the tetrahedron but two equilateral triangles with a common base (Table I). There is, however, a tendency towards 3 dimensiona close packing for about 8 atoms or more, with, in some cases, the same difficulties as for rare gases.

3 - Tetravalent elements.

There are two limitations to the length of the chains of equidistant atoms which was discussed for C_3 (figure 11).

TABLE I. Comparison of various configurations for 4 monovalent elements.

Topology		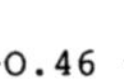	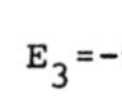	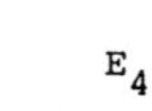	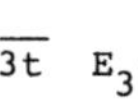
Energy spectrum	$E_1 = 1.57\ t$	$E_1 = 2.1\ t$	$E_1 = \sqrt{3}t$	$E_1 = \frac{1+\sqrt{17}}{2}t$	$E_1 = 3t$
	$E_2 = 0.46\ t$	$E_2 = 0.3\ t$	$E_2 = E_3 = 0$	$E_2 = 0$	$E_2 = E_3 = E_4 = -t$
	$E_3 = -0.46\ t$	$E_3 = -t$	$E_4 = -\sqrt{3}t$	$E_3 = -t$	
	$E_4 = -1.57\ t$	$E_4 = -1.4\ t$		$E_4 = \frac{1-\sqrt{17}}{2}t$	
Energy per bond $\lvert t \rvert$	1.35	1.2	1.15	1.0	0.65
Energy per atom $\lvert t \rvert$	1	1.2	0.8	1.3	1

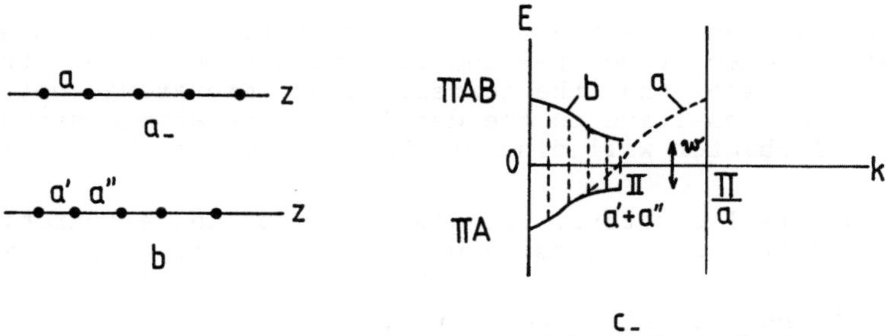

Fig. 13. Slight dimerisation of C_n chains. a) equidistant atoms; b) dimerised chains; c) energy diagram

If the chain becomes long enough, one expects it to dimerise slightly, i.e. the distances of the carbon pairs to alternate as short (a') and long (a") along the chain (figure 13). The reason is a kinetic energy effect analogous to the Jahn-Teller effect discussed for the trimers of monovalent metals. By doubling the period along the chain (a'+a" instead of a), one folds the dispersion curve E(k) of the π band in two (figure 13 c) and, by the usual argument, this opens an energy gap at the upper limit of the occupied half of the π band, with a value w proportional to dimerisation $|a" - a'|$. This lowers the energy of the quantized states (with $k \cong (n+1)\pi/Na$, n integral, N number of atoms) especially near the boundary $k = \pi/(a'+a")$ of the Brillouin zone. If N is small, the main effect is concentrated over a range of k near to $\pi/(a'+a")$ proportional to w, thus over a number of states proportional to $|a"-a'|$. Each being lowered in energy by a term proportional to w, the total effect is in $|a"-a'|^2$, as the elastic reaction of the σ bands of the chain. As σ bonding is stronger than π bonding, no dimerisation occurs. However, for N large enough, there are so many quantized states near $k = \pi/(a'+a")$ that one can treat their distribution as quasi-continuous. An integration over k, with a proper density of states per unit of k, leads to a total lowering of energy due to dimerisation in $w^2 \ln w$. In the limit of small distortions (a'-a"), thus of small w's, this is always dominant on the elastic reaction in w^2. Thus dimerisation should occur above a critical value of N, just as in $(CH)_x$ or carotenoids.

However for N larger than 8-10, it seems that other conformations are preferred, involving two dimensional sp^2 or three dimensional sp^3 bonding. This will be discussed below. Similar hybrid bonding is preferred for heavier Si, Ge and for heavy elements of the V and VI columns.

4 - Electronegative elements of the last columns.

Because of their large values of U_e (figure 4) halogens, O and N should occur mainly as molecular aggregates of dimers. This is certainly true for large aggregates.

III - ELECTRONIC STRUCTURE OF LARGE AGGREGATES.

We wish to discuss the way one goes from small clusters to large aggregates, up to macroscopic sizes. As usual, this poses the coupled questions of atomic and electronic structures, which have to be solved self-consistently. In this chapter, we shall assume some schematic atomic structures and concentrate on the electronic structure, to show that the corresponding interatomic potentials are not too sensitive to the details of the atomic structure. The final chapter will discuss the problems of the atomic structure.

As seen for small clusters, the discussion of the electronic structure has different aspects, depending on the type of element involved.

a - Rare gases and molecular aggregates.

For rare gases in a large aggregate, the (virtual) atomic excitations involved in the van der Waals attractions are easier than in vacuo. This is both because the surrounding atomic potentials attract the electron more in its excited than in its ground state and because the polarisability of the medium cuts down by an (electronic) dielectric constant ε_∞ the long range Coulomb field which binds the excited state. Indeed it is well known that the excitonic spectra in rare gases solids or liquids have lines of smaller energies than the corresponding optical spectra in gases. Because of this increase in the radius of the excited orbits, the matrix element in (24) should also decrease. These two effects compensate somewhat their effects. Which of these two effects dominates is not clear. Three body repulsions and correlations should also be added. It is clear however that, at least for atoms in the volume of the aggregate, the interatomic potentials are still central and not very different from those of dimers.

For molecular aggregates, either built from dimers (O_2, N_2, halogens) or from larger molecules (S_n ...), the weak multipolar and dispersion forces involved in their stability would also require a detailed analysis which has so far rarely been done. One does not expect, however, large deviations from additivity of two molecules interactions.

b - Metallic aggregates.

This is where deviation from additivity of two body potentials is the largest. We comment on this by comparing very large aggregates to those of reducing sizes.

1 - Large aggregates. Volume properties. Surface tension.

We restrict ourselves again to monovalent metals, studied in the strongly delocalised limit ($U/|t| \cong 0$).

We schematise an aggregate as a collection of N atoms in a close packed configuration, each with n nearest neighbours (n_s for the N_s atoms on the surface, n_v for the N_v other atoms). Each electronic state is delocalised on aggregate with a wave function

$$|\psi_\alpha> \cong \sum_{i=1}^{N} \lambda_{\alpha i}|i> , \qquad (45)$$

where the $|i>$ are the atomic (ns) wave functions. There are N such $|\psi_\alpha>$. Taking $E_o = 0$ as a reference of energy, the effective Hamiltonian can then be written

$$\mathcal{H} \cong t \sum_{\substack{i \\ j \text{ near to } i}} |i\rangle\langle j| \quad . \tag{46}$$

Writing $\mathcal{H}|\psi_\alpha\rangle = E_\alpha|\psi_\alpha\rangle$ and taking $\langle i|j\rangle \cong \delta_{ij}$, we obtain N linear equations

$$E_\alpha \lambda_{\alpha i} \cong t \sum_{j \text{ near to } i} \lambda_{\alpha j} \quad . \tag{47}$$

The <u>moment</u> M_ℓ of order ℓ of the distribution of E_α is related to the <u>number</u> S_ℓ of close circuits of ℓ interatomic jumps between near neighbours in the aggregate :

$$M_\ell = \sum_\alpha E_\alpha^\ell \quad . \tag{48}$$

But

$$M_\ell = \sum_\alpha \langle\psi_\alpha|\mathcal{H}^\ell|\psi_\alpha\rangle \quad .$$

In the approximation used here, the set of $|\psi_\alpha\rangle$ is quasi-complete and equivalent to the set of $|i\rangle$, and the operator $\sum_j |j\rangle\langle j|$ can be considered as equivalent to identity. Thus one can write

$$M_\ell = \sum_i \langle i|\mathcal{H}^\ell|i\rangle = \sum_i \langle i|\mathcal{H}| \sum_{i'} |i'\rangle\langle i'|\mathcal{H}| \sum_{i''} |i''\rangle\langle i''|\mathcal{H} \ldots |i\rangle .$$

With (46), this reduces to

$$M_\ell = S_\ell \, t^\ell \quad . \tag{49}$$

As a result, one has the following results

$$M_o = N , \tag{50}$$

$$M_1 = 0 , \tag{51}$$

$$M_2 = (N_v n_v + N_s n_s) t^2 \tag{52}$$

etc ... .

(50) expresses that the number of states in the band is equal to the number of atoms ; (51) expresses that E_o is the center of gravity of the band ; (52) gives an estimate $\left(\frac{M_2}{N}\right)^{1/2} = \left(\frac{N_v n_v + N_s n_s}{N_v + N_s}\right)^{1/2} |t|$ of the effective width of the band. M_3 (and more generally M_{2n+1}), which is <0, measures the asymmetry of the band with respect to E_o. For <u>alternate</u> structures such as figure 14 a, where one can define two interpenetrating subaggregates such that each atom of one has as neighbours only atoms of the other, $M_{2n+1} = 0$, thus the distribution of states is symmetrical in energy with respect to E_o (figure 14 b). This can also be seen from (47), where to each state $|\psi_\alpha\rangle$ of energy E_α one can associate a state $|\psi_{\alpha'}\rangle$ with $\lambda_{\alpha' i} = (-1)^n \lambda_i$ and $E_{\alpha'} = -E_\alpha$ (n = 0 for ion one subaggregate, and 1 on the other). No such symmetry exists for non alternate structures (figure 15). M_4 (and M_{2n}) gives another measurements of the effective width . One finds $\frac{M_4}{M_2^2}$ finite and not very large, showing that the distribution does not have very long tails. This is related to the fact that the overall width is of order $2\bar{n}|t|$, where $\bar{n}$ is an average of n_v and n_s. Indeed the lower limit of the band corresponds to the most 'bonding' state, of energy $\cong \bar{n}\, t$, where all $\lambda_{\alpha i}$ have the same sign and simi-

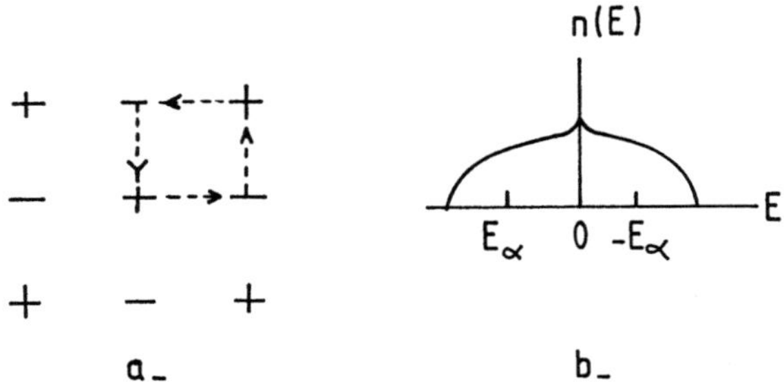

Fig. 14. Alternate structures. a) in space; b) density of states

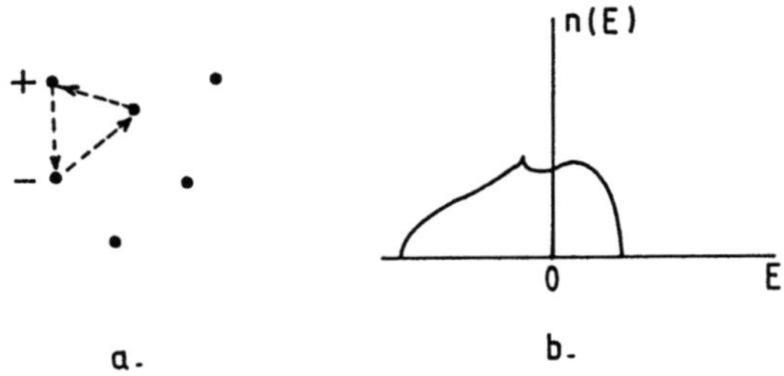

Fig. 15. Non alternate structure. a) in space; b) density of states

lar amplitudes ; the upper limit is symmetrically of energy $\cong -\bar{n}\, t$ and corresponds to the most 'antibonding' state in alternate structures (figure 14) ; in non alternate structure, such perfect antibonding is frustrated and, as a result, the upper limit is nearer to E_o.

Let us then schematise the real distribution of E_α by a continuous distribution of constant density $N\, n(E)$ between the energies $-\frac{w}{2}$ and $+\frac{w}{2}$. One must have

$$M_o = N = N\, n(E)\, w, \text{ hence } n(E) = 1/w \tag{53}$$

$$M_2 = N\, n(E)\, \frac{w^3}{12} = N\, \frac{w^2}{12} \tag{54}$$

The energy of the aggregate with one electron per atom (half filled band) is then

$$E_N = NE_o + 2 \int_{-\frac{w}{2}}^{0} N\, n\, E\, dE = N(E_o - \frac{w}{4}) = N\left[E_o - (\frac{3M_2}{4N})^{1/2}\right] \tag{55}$$

For large aggregates where $N_v \gg N_s$, one can write

$$E_N = E_\infty + \frac{N_s}{N}\, \gamma\,, \tag{56}$$

where

$$\frac{1}{N}\, E_\infty = (E_o - (\frac{3}{4}\, n_v)^{1/2}\, |t|) \tag{57}$$

is the <u>energy per atom of an infinite aggregate</u> and

$$\gamma = \frac{1}{2}\, \frac{(E_o - E_\infty)}{N}\, \frac{n_v - n_s}{n_v} \tag{58}$$

is the <u>surface tension per surface atom</u>.

In this approximation, the surface tension is proportional to the number of broken bonds on the surface ; but the total cohesive energy is not proportional to the number of bonds. However it increases with n_v, so that in this limit of large aggregates, the most compact aggregates are the most stable.

2 - Smaller aggregates. Orbital effects ; quantum alternations.

For decreasing sizes, two successive complications must be considered.

To begin with, for decreasing sizes of the aggregates, the fairly uniform distribution of states over the energy band regroups in successive families of quantized states.

This orbital effect is usually analysed assuming the valence electrons to be free, in a spherical aggregate of radius R. The electronic wave functions can then be analysed by spherical harmonics. For not too large angular quantum number ℓ, their asymptotic radial variation is, roughly, in $\sin(k\ r + \frac{\ell\pi}{2})$. Surface conditions usual in tight bonding would lead to

$$k(R + \frac{a}{2}) \cong (n + \frac{\ell}{2})\pi \qquad \ell,\ n \text{ integers} . \quad (59)$$

This regroups the successive quantized states at values of k equal to $2\,m\,\pi/(2\,P + a)$ or $(2m + 1)\ \pi/(2\,R+a)$, depending whether ℓ is even or odd. Each of such states has a multiplicity (spin included),

$$2\,[1 + 5 + 9 + \ldots(4m-3)] = 2m(2m-1) \text{ for } \ell \text{ even,}$$

$$2\,[3 + 7 + 11 + \ldots(4m-1)] = 2m\,(2m+1) \text{ for } \ell \text{ odd.}$$

A special stability should occur when all shells are either completely full or completely empty, thus for a total number of electrons (or atoms),

$$N = 2\,[1 + 6 + \ldots\ m(2m-1)] \qquad \text{for } \ell \text{ even,}$$

$$N = 2\,[3 + 10 + \quad m(2m+1)] \qquad \text{for } \ell \text{ odd.}$$

Table II compares these rough predictions with observations on Na aggregates. Deviations for large numbers of N can be explained by the fact that the asymptotic formula used for (59) is increasingly wrong for large values of ℓ, and that it is the states with large ℓ values (1g, 1h) which dominate at large values of N. The reasoning can be extended to the LCAO approximation used here if the form of the aggregate and its atomic structure at least roughly respects the symmetries involved in the degeneracies of the successive ℓ states. The agreement of Table II thus suggests the aggregates to be nearly spherical and not flat at least for $N \gtrsim 8$.

TABLE II. Orbital effect on stability of Na aggregates

	1p	2s,1d	2p, 1f	3s,2d,1g	3p,2f,1h
N obs	8	20	40	58	(92)
N comp.	8	20	40	70	112

For really small aggregates, another 'full shell' effect takes over. It corresponds to the fact that, for kinetic energy reasons, aggregates with an even number of electrons are more stable than those with an odd number : in the first case, each electronic state is either full with two electrons or empty ; in the second, the electronic state is occupied with only one electron (figure 16). This distinction has a meaning only for aggregates small enough for the energy difference at the Fermi level to be larger than k_BT, thus for N typically less than 20. It explains the alternation with size of stability, energy of ionisation, intensity of sputtering for metallic aggregates with an odd valency (alkali, noble metals, Al) in contrast with aggregates with an even valency (Mg, Pb) where these quantities vary smoothly with N.

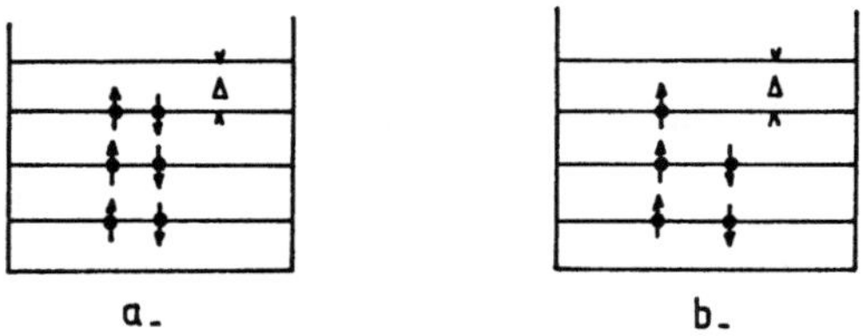

Fig. 16. Aggregates with even or odd parity of electronic population: a) even (diamagnetic); b) odd (paramagnetic)

c - Covalent aggregates.

If again one considers completely delocalised electrons, one expects large covalent aggregates of the tetravalent elements to build 3 dimensional bonds based on the diamond tetrahedron and the $\sigma(sp^3)$ hybridisation, because this gives the strongest bonding.

For reducing size, one could expect a critical value below which 2d aggregates based on $\sigma(sp^3)$ hybridisation + π bonding characteristic of graphite should be preferred.

And finally for very small sizes, one should switch to the linear chains already discussed, with $\sigma(sp)$ hybridisation $+2\pi$ bonding (indeed at least for the dimer).

The decrease in volume stability in these 'phase changes' is compensated each time by a decrease in surface tension, due to broken σ(or π) bonds : in the first case, this term is proportional to the surface, in the second case to the circumference of the graphitic planes, in the third, it is constant and due to two broken bonds.

These observations probably hold, at least roughly, for tetravalent elements.

For C_N chains, it is possible above a certain length to suppress the broken bonds at the ends of the chains, at the expense of some elastic bending energy, by forming a closed loop. This seems already predicted for N = 6.

For larger aggregates of C, Si, Ge, it is possible in the same way to imagine that a graphitic structure is curved along a full sphere to suppress the broken bonds on the edge of the graphitic plane. In the smallest such 'shell structure' possible, the curvature is made possible by replacing the six fold rings of the normal graphitic structure by five fold rings : all the atoms are then inscribed on a sphere and form a regular dodecahedron.

Comparison between C and Si or Ge aggregates show a special stability of C_N chains and a general **tendency** for C to prefer π bonding structures which will be attributed to correlation effects not discussed at the moment. For rather small Si aggregates, the presently computed structures show a no sign of chain structures, but shell structures. These have a mixture of bond angles representative of sp^3 bonding (109°) and 60° bond angles representative of a non hybridised and compact s and p bonding (as in metals). There is no sign even at high temperatures of the 90° bond angle seemingly correlated with 6 neighbours in liquid Si or Ge, perhaps because of a surface tension effect which tends to make small aggregates more compact than liquid droplets.

There is, however, a considerable gap between the small sizes of the clusters whose structures are studied in detail and larger aggregates to which this section is mainly devoted.

It is worth stressing finally that in aggregates respecting (at least approximately) their local directionality, <u>the hybrid bonds</u> are <u>roughly additive</u>.

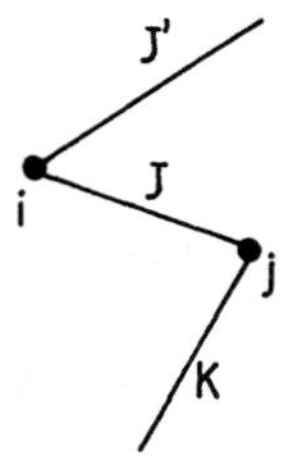

Fig. 17. Notation of sites i and bonds J

This can be seen for instance for sp^3 bonds. But very similar results obtain for sp^2 and sp bonds.

Let us consider a large aggregate with atomic sites i and bonds J (figure 17). Let $|iJ\rangle$ be one of the four sp^3 hybrids centered on i. In the spirit of our treatment, the $|iJ\rangle$'s build a quasi-complete and orthonormalised set to develop the delocalised electronic states

$$|\psi\rangle = \sum_{\substack{i \\ j \text{ near to } i}} a_{iJ} \, |iJ\rangle \quad . \tag{60}$$

We wish to solve $\mathcal{H}|\psi\rangle = E|\psi\rangle$ (61)

with

$$\mathcal{H} = T + \sum_i V_i \quad . \tag{62}$$

It is easy to see that

$$\langle i\ J|\mathcal{H}|i\ J\rangle = \frac{E_s+3\ E_p}{4} = E_o ,$$

$$\langle i\ J|\mathcal{H}|j\ J\rangle = t(<0) , \quad (63)$$

$$\langle i\ J|\mathcal{H}|i\ J'\rangle = \frac{E_s-E_p}{4} = \Delta(<0) .$$

Thus from (60), (61), (62),

$$(E_o-\Delta-E)\ a_{iJ} + \Delta \sum_{J'} a_{iJ'} + t\ a_{jJ} = 0 . \quad (64)$$

Similarly,

$$(E_o-\Delta-E)\ a_{jJ}+\Delta \sum_{K} a_{jK}+ t_{aiJ} = 0 . \quad (65)$$

Their combination gives

$$[(E_o-\Delta-E)^2 - t^2]\ a_{iJ} + \Delta[(E_o-\Delta-E) \sum_{j'} a_{iJ'}-t \sum_{K} a_{jK}] = 0. \quad (66)$$

Let

$$x_i = \sum_{J'} a_{iJ'} . \quad (67)$$

There are two sets of solutions :

1) $x_i \equiv 0$ for all sites i. Then (66) has nonvanishing $a_{iJ'}$s only if

$$E = E_o - \Delta \pm t = E_p \pm t. \quad (68)$$

These bands are obviously purely p in character.

2) $x_i \not\equiv 0$. Summation of (66) over J gives

$$[(E_o-\Delta-E)^2-t^2 + 4\ \Delta(E_o-\Delta-E)]x_i = t\Delta \sum_{j \text{ near } i} x_j \quad (69)$$

To this equation, one can associate a problem of s band <u>on the same sites</u>

$$\mathcal{H}'\ |\psi'\rangle = E'\ |\psi'\rangle \quad (70)$$

where

$$|\psi'\rangle = \sum_i x_i\ |i\rangle . \quad (71)$$

$$\mathcal{H}' = -\ t\Delta \sum_{i,j \text{ near } i} |i\rangle\langle j| \quad (72)$$

$$E' = -\ (E_o-\Delta-E)^2 + t^2 - 4\Delta\ (E_o-\Delta-E) . \quad (73)$$

Because each site i has n = four neighbours, the discussion above shows that

$$-\ 4t\Delta \leqslant E' < 4t\Delta \quad (74)$$

Figure 18 schematises the allowed energy bands when E' is replaced by E according to (73). The limits are

$$E - E_o + \Delta = \pm\ (t + 2\Delta), \quad (75)$$

$$E - E_o + \Delta = \pm\ (t - 2\Delta); \quad (76)$$

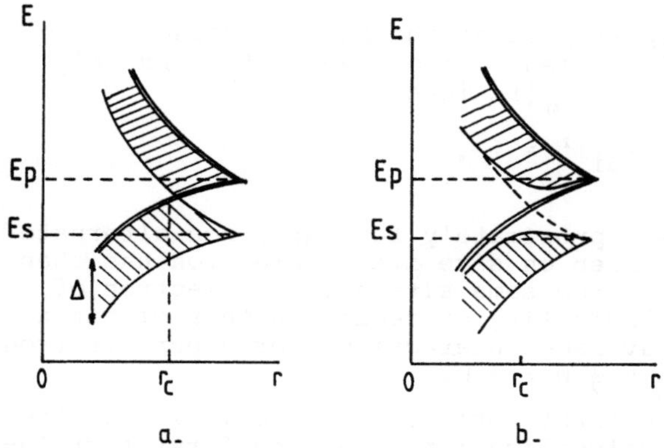

Fig. 18. Band structure of macroscopic aggregates with sp^3 bonds. a) alternate structures; b) non alternate structures

the exponential decrease of t with length r has been taken into account. In non-alternate structures, where the upper limit (74) is not reached for E', there is no exact crossing point between the conditions of s and p bands ($r > r_c$) and those for sp^3 bonding and antibonding bands ($r < r_c$). For $r \ll r_c$, which applies to C and, to a lesser extend to Si and Ge, the width Δ of the valence (bonding sp^3) band is not too large with respect to the band opening (2t) responsible for the gap. The average energy per atom can therefore be taken as

$$E \cong 4(E_o+t) \; E_s + 3E_p + 4t = 2(E_s + E_p) + 4(t-\Delta) \, .$$

In the same limit, each broken bond introduces an energy state of energy E_o instead of $E_o + t$: sp^3 bonding is at least roughly additive.

d - Electron-electron interaction effects in metallic and covalent aggregates.

In large metallic monovalent aggregates, a straightforward extension of (55) along the same line as (10) gives

$$E \cong N \; [E_o - \frac{w}{4} + \frac{U}{4} - \frac{U^2}{16w} + \frac{R}{2}] \qquad (77)$$

$$w \cong (\frac{12M_2}{N})^{1/2} = (12n)^{1/2} \; |t| \qquad (78)$$

The term in U is the Hartree-Fock correction for completely delocalised electrons. It is obviously positive. It can be understood as follows. In an s band, the total electronic density per site i for a fully occupied band is necessarily equal to the density due to $|i\rangle$, i.e. 2 electrons per site, as the equivalence noted for full shells between molecular and atomic orbitals extends naturally to the full band of any aggregate. Now in an alternate structure, the electronic density per site i, $2 \sum_{\alpha_{occ}} |\lambda_{\alpha i}|^2$, equals the electronic density for the same site i, of the holes of the upper part of the band, because of the symmetry property noted between $|\psi_\alpha\rangle$ of energy E_α and $|\psi'_\alpha\rangle$ of energy $-E_\alpha$ ($|\lambda'_{\alpha i}| = |\lambda_{\alpha i}|$). Thus

$$2 \sum_{\alpha \, occ} |\lambda_{\alpha i}|^2 = 1 \; .$$

The same property holds approximately for large non-alternate close packed structures. When on site i, each electron has then a probability $\frac{1}{2}$ to see on the same site another electron of opposite spin (and no possibility of seeing another electron of parallel spin). The average interaction energy per electron (or per atom) is then $\frac{1}{2} \times \frac{1}{2} U = \frac{1}{4} U$.

The term in U^2 is the first term due to a deviation of the electrons from perfect delocalisation, i.e. the first correlation correction . Because it is computed in a perturbative way, it contains in the numerator the square of the matrix element U of interaction, and in the denominator a term proportional to the energy of virtual excitation of the electrons from the occupied part to the unoccupied part of the s band. This is proportional to w. The term is necessarily negative, as these correlations stabilise the system. The numerical constant 16 given in (77) is deduced from a count of the probabilities for pairs of electrons of opposite spin colliding in this way. This very simple formula can also be obtained by developing to second order in U the Gutzwiller variational method which is especially good for U small ; it has also be checked numerically by exact developments for large closepacked crystals.

Minimising E with respect to r with $w \ll t$ gives

$$E_N = N \left[E_o - \left(1 - \frac{p}{q}\right) \frac{w}{4} + \frac{U}{4} - \left(1 + \frac{p}{q}\right) \frac{U^2}{16\,w} \right], \quad (79)$$

with length of bond r_N such that

$$(p-q)\; r_N = n \frac{q}{p} \left(\frac{n}{3}\right)^{-1/2} \frac{R_o}{t_o} + \frac{U^2}{48\; n\; t^2} \; . \quad (80)$$

This shows that the bond length decreases slightly with increasing number n of neighbours, and that electron-electron repulsions increase the bond length.

Furthermore, comparison with (10) shows that <u>the most close packed aggregates become unstable with respect to a collection of</u> dimers if

$$\frac{U^2}{t^2} > 16 \frac{q-p}{q+p} \frac{\frac{W}{4} - |t|}{W - |t|} \frac{W}{|t|} \; . \quad (81)$$

With (78), and $n \cong 12$, this gives again $U > 2\,|t|$, as for the trimer. The same conclusions hold for the stability of dimers for O_2, N_2, halogens.

For <u>covalent</u> aggregates, a development analogous to (77) reads for the $\sigma(sp^3)$ bonding

$$E = N \left[E_o + 4t - U + \frac{7}{16} \frac{U^2}{t} + \frac{R}{2} \right] \quad (82)$$

The correlation term in U^2 is expected to be larger when π bonding is involved because it has a smaller band width. This term is therefore expected to stabilise π bonding with respect to σ bonding, especially in C with its larger value of U_e (figure 4). This would help and explain why graphite is more stable than diamond, and why organic unsaturated molecules are much more frequent for C than for Si or Ge (C_N chains included). π electrons should be much more strongly correlated than σ electrons, a view prevalent among chemists.

IV - ATOMIC STRUCTURE OF LARGE AGGREGATES.

We now come back to the problem of atomic structure, mentioning aggregates of increasing sizes and introducing concepts which, while met in small aggregates, are useful to compare macroscopic structures.

a - Small aggregates. The concept of frustration.

As discussed above, in rare gases, the interatomic interactions are fairly well known and rather simple : central, additive and with a single minimum. In a dimer, a trimer or a tetramer, the equilibrium distance r_2 is the minimum of $V_{ij}(r_{ij})$ (figure 19 a). In a larger aggregate, the distance to the nearest neighbour r_N will be somewhat smaller than r_2, so that bonds between nearest neighbours are in compression; this equilibrates the (weak) long range attractions with further neighbours at distance $r'_N > r_2$ (figure 19 b) : this quite general result can be seen as a compromise between short and long range interactions ; or, if one is pessimistic as seems to be the fashion nowadays, one will talk of frustration. In this case, the frustration is weak : r_N is near the optimum $V_{ij}(r_2)$, and the longer range interactions $V_{ij}(r_{N'})$ can often be neglected, except in some specific cases.

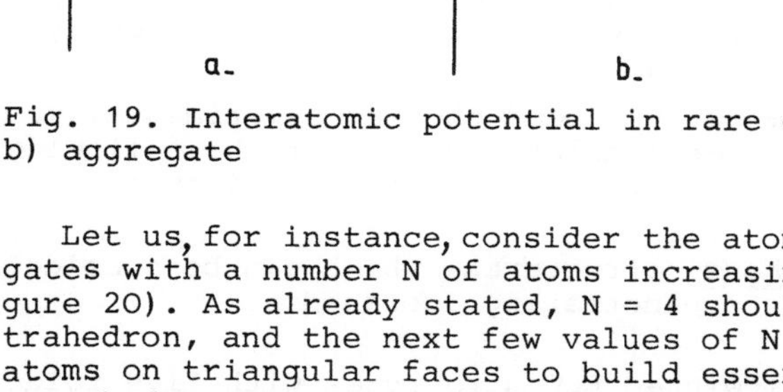

Fig. 19. Interatomic potential in rare gases: a) dimer; b) aggregate

Let us, for instance, consider the atomic structure of aggregates with a number N of atoms increasing from 4 upwards (figure 20). As already stated, N = 4 should give a regular tetrahedron, and the next few values of N (5,6) lead to adding atoms on triangular faces to build essentially new regular tetrahedra. However, for N = 7, the procedure leads to two equivalent configurations A, B (figure 20 d) which are near to each other ; position A is therefore not far from corner C and attracted by it. An obviously more stable configuration is to bring positions A and B together by extending slightly the bonds along A C D E F A. The most stable configuration for N = 7 is

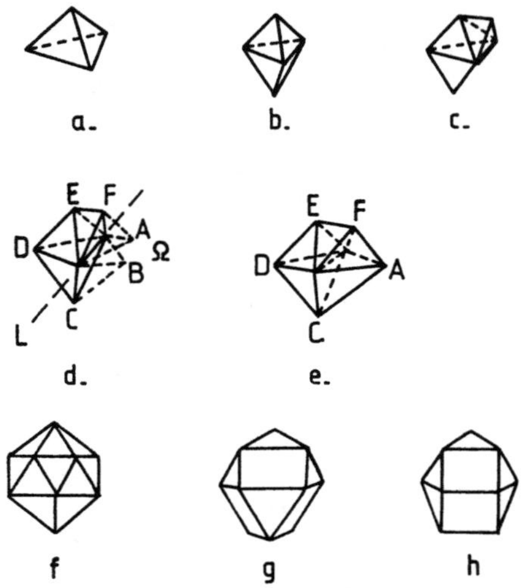

Fig. 20. Small aggregates of rare gases. a) N = 4; b) N = 5; c) N = 6; d) N = 7 unrelaxed; e) N = 7 relaxed; f) N = 13 icosahedron; g) N = 13 FCC; h) N = 13 HCP

therefore a regular pentagon A C D E F A, with two supplementary atoms above and below the axis L of the pentagon (figure 20 e).

This configuration is a regular pentagonal bipyramid, where the two summits have been squeezed towards each other to form a bond. It can also be considered as obtained by adding one basis atom on the regular square bipyramid which is actually the most stable configuration for N = 6. In a similar way, it is easy to show that with N = 13, the centred icosahedron with five fold symmetry axes (figure 20 f) is more stable than the nuclei of the crystalline FCC or HCP phases (figure 20 g, h). Again, the increase in the number of surface neighbours (here from four to five) is obtained at the expense of a slight elastic distortion, where the radial bond lengths are compressed in comparison with the surface bond lengths. The centred icosahedron is again obtained by close packing (and slightly distorting) regular tetrahedra.

Such predictions have not yet been checked experimentally. However for N > 13, computer simulations predict the following, which agrees with electron diffraction studies on jets :

15 < N < 50 : 'amorphous' structures, which can be considered as incomplete and interpenetrating icosahedra.

50 < W < 1000 : multitwinned FCC structures with five fold symmetry, which can be viewed as sectors of FCC crystals in spinels (111) twinning relative orientations, which are build coherently on a central icosahedron.

N > 1000 : irregularly twinned crystals.

All these structures can be viewed as due to a compromise (or frustration) effect between the maximum local bonding effects, which leads to the piling up of tetrahedra, and a topological constraint, due to the impossibility of paving the 3-D space continuously and perfectly in this way. One can say equivalently that, starting from a perfect FCC crystal, one can gain surface energy by multitwinning at the expense both of the usually small twinning energy and of an elastic distortion. The surface energy gain can surmount the volume elastic distortion only in relatively small aggregates.

Even in perfect crystals, the frustration effect mentioned in connection with figure 19 b shows up in that atoms at the surface of the crystal, having fewer neighbours, tend to have larger distances to first neighbours, nearer to r_2. As a result , one observes on most phases of rare gas crystals a normal (positive) relaxation (figure 21 a) and a tangential (positive relaxation) (figure 21 b) which both contribute to an increase of the atomic volume of the crystal inversely proportional to its size as well as to a slight change in form.

a- b-

Fig. 21. Surface relaxation in rare gases. a) Normal; b) tangential

Similar results hold for small aggregates of some metals (Au, Pt). There is however a surface contraction.

For aggregates of tetravalent aggregates, large enough to have sp^3 bonding, one predicts that the most stable aggregates are built on the dodecahedron, the figure inverse to the icosahedron, also with five fold symmetry axes. This corresponds, for this type of bonding, to a small number of broken bonds on the surface. This and other similar small aggregates have been invoked in describing aggregates formed in sputtering, C_n aggregates observed in space and Si_n aggregates deposited to form amorphous Si.

b - Analysis of the elastic distortions. Notion of disclination lines.

A way to analyse the elastic distortions in small aggregates such as in figure 20 e and f is by disclinations.

A disclination line L is a rotation dislocation, which can be introduced in solid matter by a Volterra process, i.e. removing or adding a sector of matter of angle Ω along the line (figure 22). Ω is the angle of unstressed matter added or removed, slightly different from the equilibrium angle ' of the same sector under the internal stresses developed

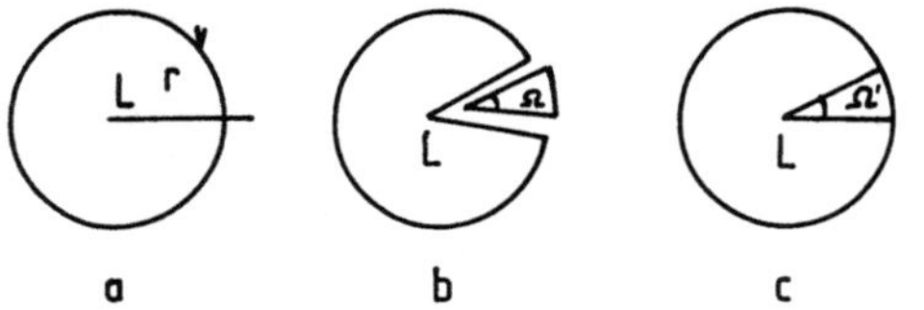

Fig. 22. Successive operations to create a line L of disclination of angle Ω in a cylinder

by the disclination. If, as in figure 22, matter is added, one talks of a negative disclination ; if matter is removed, of a positive one.

In figure 20 d and e, the Volterra process involved is to close down an initially empty sector of edge L and (small) angle Ω. This is a positive disclination.

In a similar way, the stresses developed in an icosahedron (figure 22 f) are those of positive disclination lines along its 6 five-fold axes. The angle $\Omega = \frac{\pi}{6}$, but the lines compensate most of their stresses.

In the multitwins with five-fold symmetry around an icosahedron, the stresses can again be described as due to 6 disclination lines along the 6 five fold axes of symmetry. A similar analysis holds for the dodecahedron.

The elastic energy per unit length of one disclination line such as in figure 22 c is

$$\tau \cong \frac{\mu \, \Omega^2 \, r^2}{16 \, \pi (1-\upsilon)} \, ,$$

where μ is the elastic (shear) constant of the solid and $\upsilon \cong \frac{1}{3}$ to $\frac{1}{2}$ the Poisson coefficient. The stresses developed by the disclination line of the pentagon of figure 20 e give an elastic energy of the order of $2 \, r \, \tau \cong \mu \, v \, \Omega^2/10$ where v is the atomic volume. As μv is the order of the cohesive energy per atom, and as $\Omega < 0.1$, this energy is indeed very small compared with the surface energy gained with respect to figures 20 d, a bit less than one broken bond, which is near to $\mu v/6$.

c - Projection from curved space. Extension to amorphous structures, crystals, quasi-crystals.

Another way of analysing the elastic stresses of the icosahedron is to use the projection from a curved space.

Indeed, the regular fivefold symmetry of the atom network on the outside sphere of an icosahedron is made possible because this surface is (positively) curved. If one tried to project this lattice on a plane, the lattice would become very distorted if one went too far from the center of projection. This is connected with the fact that one cannot have periodic two dimensional lattices with five-fold symmetry.

In a similar way, an icosahedral arrangement can be thought of as produced by the projection on the normal flat 3d space of a periodic lattice in a (positively) curved 3d space. This point of view has proved useful to predict other close packed aggregates by projecting different lattices from positively or negatively curved 3-D spaces. Again, the projection can only be local, as elastic distortions increase rapidly if one goes too far from the centre of projection.

There is obviously a connection between curved space and disclinations. Thus in two dimensions, the stress free figure produced by introducing a disclination in a plane is a cone, and the stressed flat figure obtained is the projection of the cone on the plane. A stress free sphere is obtained from a stress free plane by a continuous distribution of infinitesimal disclinations along a line of the plane. In the same way, a stress free curved 3-D space is produced from a stress free flat 3-D space by a continuous 2-D distribution of infinitesimal disclinations. The operation of projection from flat to curved space or vice versa is therefore equivalent to the introduction of such distributions of disclinations. For objects on an atomic scale like the icosahedron, stresses are only defined on an atomic scale, and a finite number of disclinations of finite size are involved.

One can then ask how to generalise these constructions to macroscopic matter. Two complementary points of view have been used.

1 - The first is to cover the space with small aggregates such as the icosahedra, as closely and perfectly packed as possible. When two such aggregates come into contact, one finds that bonding can be formed nearly perfectly locally, but cannot be extended over the whole surface of contact. The most stable arrangement is a bonding which becomes continuous and only slightly distorted everywhere except along lines of singularities which are situated at the edges of the polyhedra which each contain an aggregate (figure 23).

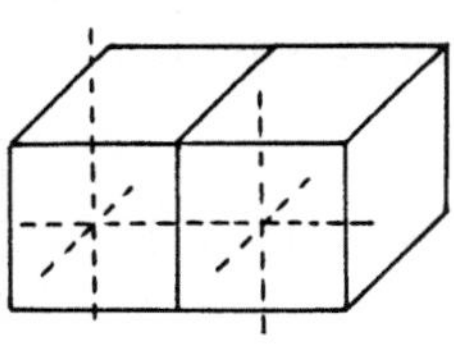

Fig. 23. Two interpenetrating networks of disclination lines in closepacked arrangements of say icosahedral aggregates

It is easy to understand that these lines are disclination lines of a sign opposite to those describing the stresses inside each aggregate. There are, therefore, two interpenetrating lattices of disclinations: those inside the aggregates, which join from aggregate to aggregate along their surfaces of contact ; those between the aggregates which run parallel to their surfaces of contact.

Such arrangements will be more stable than close packed regular crystals (say FCC/HCP in the case of icosahedral arrangements) if the free energy gained either by the local packing or by entropy is larger than the elastic energy developed by the disclination lines.

There are two cases where this is possible :

- liquid and amorphous structures, where the drive to stability comes from entropy, and the networks described must be irregular in space (and in time for liquids).

- crystals and quasi-crystals, where the drive to stability comes from local packing, usually of atoms of different sizes (alloys, compounds) which, for instance, favour an icosahedral arrangement of a large atom surrounded by 12 smaller ones.

In that case, the disclination lines build themselves regular lattices. In crystals, these lattices are necessarily commensurate with the atomic spacing (in curved space) of the aggregates. One talks of Frank and Kasper phases. Their large cells have sizes given by the spacing of the disclination lines, and their local arrangements are those of the aggregates. But one can also imagine that the spacing of the disclination lines, although regular, is incommensurate with the atomic spacing of the aggregates. The presence of the two network of disclinations essentially preserves the orientational long range order, but the incommensurability destroys the transitional order. This is one way of looking at quasi-crystals.

2 - The second point of view is to start from a perfect lattice in curved space, and to progressively decurve the space by introducing successive lattices of finite disclination lines all of the same strength. After each such operation, the curvature has been reduced but not suppressed ; one needs a further lattice of disclinations, with a longer lattice spacing, to suppress the major part of the remaining curvature. The process has no end if the matter is continuous or if the atomic grain of matter is incommensurate with the lattice spacing of the disclinations. This explains qualitatively the fundamental scaling property of quasi-crystals.

References

J. Friedel and M. Lannoo: J. Physique 34, 115, 483 (1973)

J. Friedel: Annales de Physique 1, 257 (1976).

J. Friedel: J. Physique 39, 651, 671 (1978).

J. Friedel: Physica 109 and 110 B, 1421 (1982).

J. Friedel and C. Noguera, Intern. J. Quantum Chemistry 23, 1209 (1983).

J. Friedel and M.C. Desjonquères: In La Catalyse par les Métaux, ed. B. Imelik, G.A. Martin and A.J. Renouprez, CNRS ed (1984).

J. Friedel: Helv. Phys. Acta 56, 507 (1983).

J. Friedel In Physics and Chemistry of Electrons and Ions in Condensed Matter; J.V. Acrivos, N.F. Mott and A.B. Joffé, ed. Nato ASI Series, Series C Mathematical and Physical Sciences Vol 130 (1984) D. Reidel Publ. Bordrecht.

J. Friedel: J. Phys. C Sol State Physics 17, 4259 (1984).

J. Friedel: J. Physique 48, 93 (1987).

J. Friedel (1987) in press

J.P. Gaspard, B. Mosseri and J.F. Sadoc: Phil. Mag. (1984)

A.M. Olès, F. Pfirsch, P. Fulde, M.C. Böhm; J. Chem. Phys. 85, 5183 (1986).

Electronic Properties of Metal Clusters

W.A. de Heer and W.D. Knight

Physics Department, University of California, Berkeley, CA 94720, USA

1. Introduction

Stimulated by the work of ROBBINS [1] and his collaborators on alkali clusters in molecular beams about 20 years ago, later workers [2,3] have extended the investigations to more elaborate experimental machines and to other metals. In these lectures we will mainly discuss our own experiments on alkali metal clusters and their properties, including shape, electronic structure, stability, magnetism, ionization potential, static and dynamic polarizability, plasma resonance, and photoabsorption cross section. We will include discussions of the experimental equipment and techniques, as well as applicable models relating to cluster behavior.

The results suggest that the clusters may be treated as a collection of free electrons embedded in uniform ellipsoidal distributions of positive charge, whose shapes give rise to the observed properties as a direct result of the symmetries imposed by the boundaries on the potentials. The uniform positive charge distribution and the delocalized electrons which are characteristic of the jellium of condensed matter physics, and the ellipsoidal harmonic oscillator [4] of nuclear physics combine to yield a satisfactory model which is a good match for the experimental results. Full application of the self-consistent ellipsoidal jellium model [5] validates and improves on the results.

2. Electric and Magnetic Dipole Moments

Following the methods of molecular beams the magnetic moments may be determined [6] for individual clusters identified by a quadrupole mass analyzer. Figure 1 shows schematically a supersonic jet cluster beam machine [7] including source, collimating slits, inhomogeneous field deflecting magnets, homogeneous magnet, and detector.

Magnetic moments. Magnetic deflection [8] is given, as in the Stern-Gerlach experiment, by (1):

$$z = K_m \mu_B (dB/dz)/Nmv^2, \tag{1}$$

where K_m is a constant depending mainly on apparatus configuration, μ_B is the Bohr magneton, dB/dz is the magnetic field gradient, N is the number of clustered atoms each of mass m, and the mean beam velocity is v, determined by time of flight measurements. Figure 2(a) shows beam deflection profiles for lithium, sodium, and potassium trimers. It is observed for Li_3 that the deflected peaks represent the m_s = +1 or -1 states for the odd electron. The small peak at zero deflection becomes more prominent in the lithium pentamer, whose profile resembles that for the sodium trimer. The Stern-Gerlach peaks are not observed for the sodium pentamer. The deflec-

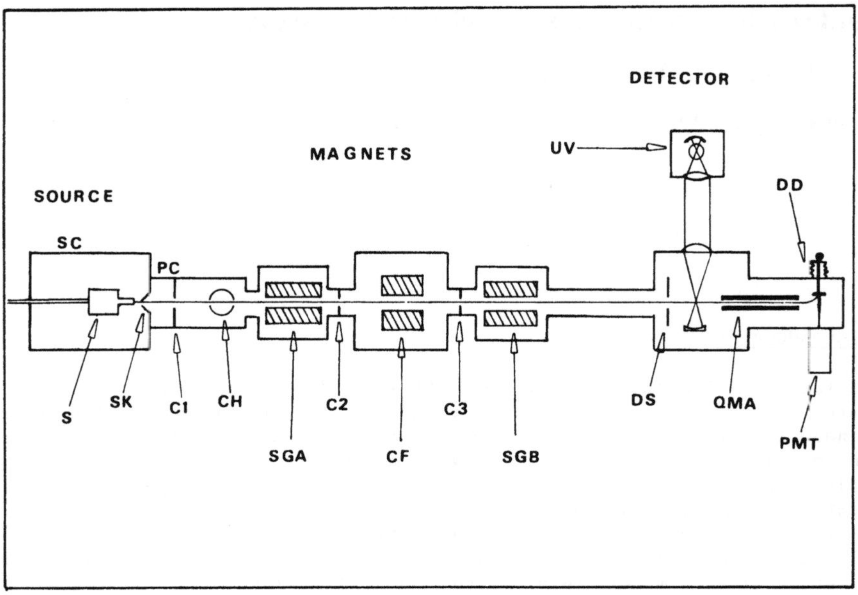

Fig. 1. Schematic diagram of cluster beam apparatus

tion profile for the potassium trimer is broadened slightly, but no deflected peaks are evident. In Fig. 2(b) we see field-on/field-off profiles for some heavier mixed clusters, none of which show Stern-Gerlach peaks. The undeflected beam profile is indicated for dimers.

It may be supposed, following (1), that clusters with even numbers of electrons will show zero deflection because of electron spin pairing. In fact deflections have not been observed for even clusters.

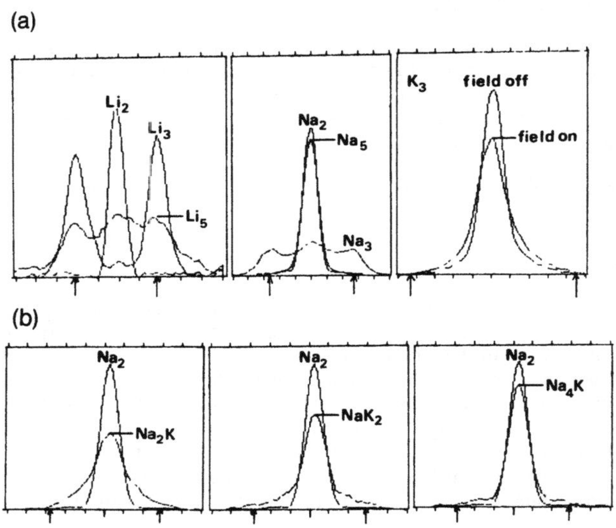

Fig. 2. (a) Magnetic deflection profiles for alkali trimers; (b) Field on/off profiles for some mixed alkali clusters. Arrows indicate expected positions for μ_B peaks

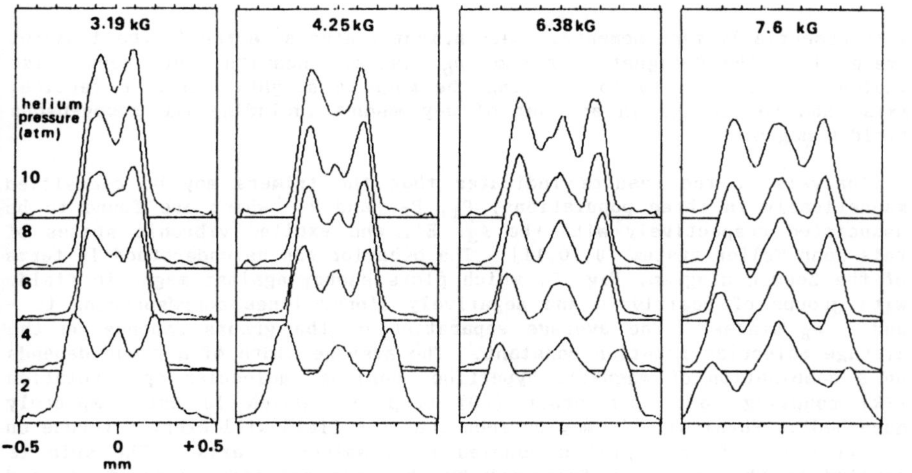

Fig. 3. Sodium trimer profiles for several magnetic fields and carrier gas pressures. Zero moment peak increases with increasing field and decreasing pressure

Further examination [6] of the behavior of the sodium trimer gives results as shown in Fig. 3, where the average magnetic moment decreases with increasing applied field B and with decreasing carrier gas pressure P. In Fig. 4 two deflecting magnets are employed as a state filter. Clusters selected in $m_s = +1$ (i.e. $+\mu_B$) states out of the A-magnet are, after a second deflection, observed to contain a $-\mu_B$ component, as well as some

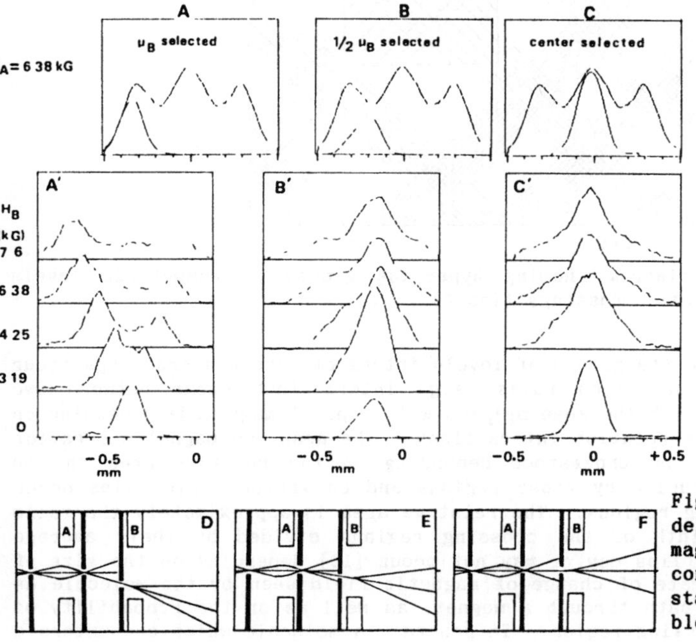

Fig. 4. Sodium trimer deflections by double magnet filters. Output contains selected input states plus other possible states

with essentially zero moments. Zero moment states selected in the A-magnet emerge from the B-magnet with some μ_B states. Results, not shown, also indicate a sensitivity to twisting the magnets slightly about a vertical axis, and to passage in and out of any magnet including the homogeneous field C magnet.

Analysis of the results indicates that the trimers may be classified magnetically in three populations, P_1, P_2, and P_3, which are found to be associated respectively with the A_2', E', and excited vibronic states of this Jahn-Teller system [9,10,11]. The behavior can be understood in terms of the Zeeman diagram, Fig. 5, which plots energy against magnetic field, with groups of positively and negatively sloped lines corresponding to + and - μ_B states. The average separation of the groups depends on the average molecular rotation constant. The average width of a group depends on a combination of magnetic hyperfine coupling, molecular spin rotation (SR) coupling, and spin orbit (SO) coupling (which is not completely quenched). Physically it may be seen that the field will exert a torque on the molecule if the spin is coupled to a molecular axis. The spin is coupled to the molecular framework by the spin rotation interaction, and also by the anisotropic g-tensor, which causes the electron in a magnetic field to have energies depending on the orientation of the spin with respect to the molecular axis. This torque mixes rotational states and will generate avoided crossings.

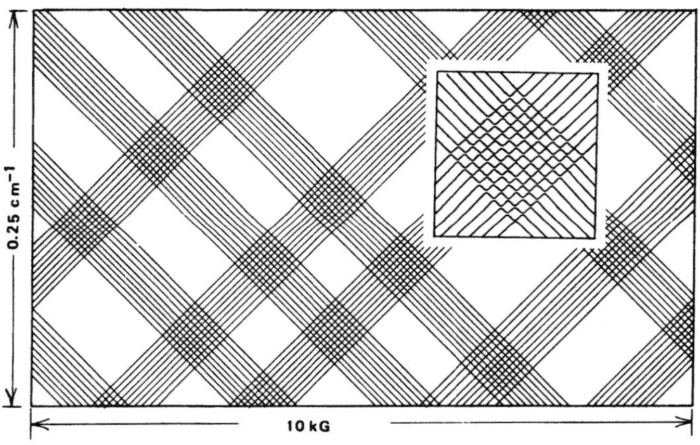

Fig. 5. Zeeman diagram showing hyperfine groups of compatible levels giving rise to avoided crossings (inset)

In regions where the groups of levels intersect, avoided crossings occur depending [7] on selection rules (e.g. interactions occur between two rotational levels with the same m_F), and with gaps of magnitude depending on the Zeeman and spin-rotation interactions. The observed magnetic behavior of the trimer can be understood depending on the relative area on the Zeeman diagram occupied by those regions and on whether spin flips occur within the crossing regions. The relative area is approximately the ratio of the average width of the crossing regions divided by their average separation. Spin flips may or may not occur [12] depending on the size of the gaps and the rate of change of magnetic field seen by the molecule as it negotiates the path through a magnet, as well as on the probability of encountering a crossing region. In general, a molecule which encounters a crossing region will emerge from a magnet showing a reduced effective magnetic moment.

The P_1 states are μ_B states exhibiting a high degree of spin memory which implies a low probability of spin reversal on entering or exciting magnetic fields. The spin reversals that do occur are estimated to be associated with crossing gaps >16 MHz and crossing separations of ~45 kG, implying that the average spacing between interacting rotational levels is ~2 cm^{-1}, which is ~20 times the average spacing of rotational levels. The last factor arises because of the symmetry of the A_2' states which is like that of the symmetric rotor [13] exhibiting many degeneracies. The selection rules determine that most crossings are not avoided. The P_2 states, on the other hand, show no spin memory when entering or exciting magnetic fields, and contain both a μ_B component and a low moment component. This latter population is associated with the E' states which resemble an asymmetric rotor with lifted degeneracies, resulting in a smaller average separation of rotational levels. Thus, the crossing regions will be more numerous and closely spaced at the average rotation constant (0.1 cm^{-1} for Na_3), accounting for the existence of both μ_B behavior and the observation of low moments.

The P_3 population, which is associated with excited vibronic states, has a moment distribution peaked at $\mu = 0$, varying as $-\ln|\mu/\mu_B|$, and decreasing with increased temperature. The logarithmic distribution indicates a strong coupling of the spin to a molecular axis. The temperature dependence is consistent with the assumption that this population is directly associated with excited states, and allows us to estimate the vibronic temperature to be ~80K. This results from the relative values of hyperfine couplings, SR and SO interactions, the latter of which dominate the effects for potassium. The experiments and theory are discussed completely in Ref. 6, which details some remarkable possibilities for magnetic spectroscopy of paramagnetic molecules.

Electric polarizability. The static electric dipole polarizability of metal clusters gives a direct measure of the metallic screening properties, and forms the physical basis for time-dependent phenomena such as ionization and plasma resonance. The polarizability may be measured directly [14] by deflection of clusters in an inhomogeneous electric field. The deflection is given by (2):

$$z = K_e \alpha E^2 / N m v^2 . \quad (2)$$

Here K_e is a constant depending on apparatus configuration, E is the value of the deflecting field at the cluster beam, α is the cluster polarizability, the E^2 term represents the product of E and dE/dz (which is proportional to E), and the other quantities are defined as in (1). While the deflection of a paramagnetic cluster varies as ~1/N, the electric deflection varies as α/N, which drops 40% between N = 1-20, and remains constant thereafter. It is therefore feasible to investigate the polarizability of clusters even to macroscopic sizes.

Experimental results [14] for the polarizability per atom are shown in the upper curve of Fig. 6. The average trend decreases from the atom at one end toward the bulk metal at the other. Variations of individual cluster polarizability from the average trend show minima at closed shell clusters. Since bulk metals exhibit screening of external fields, it would appear that the smaller clusters overscreen. Actually, a number of theoretical treatments [15,16,17] indicate that the size of the delocalized

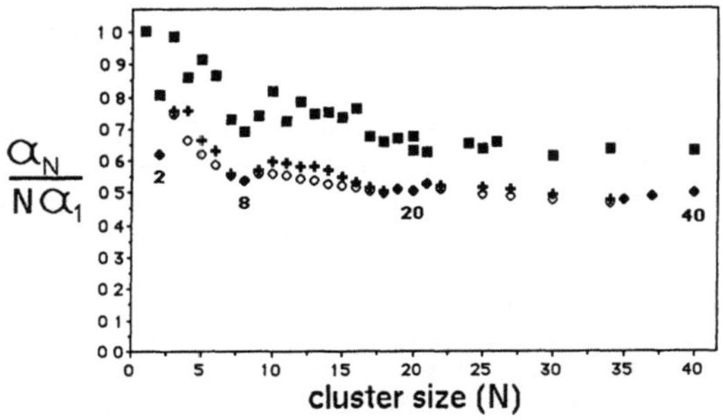

Fig. 6. Static electric polarizability per atom in units of atomic polarizability for sodium clusters, upper curve experiment, lower curve theory

electron distribution in clusters is larger than is estimated on the basis of the volume per atom in the bulk. In fact the polarizability turns out to be significantly larger than the classical value for a metal sphere, and is given to first order by

$$\alpha = (R + d)^3 , \tag{3}$$

where R is based on the bulk atomic volume and d is the electron spillout parameter. Since d ~ 1 a.u. [15], it is seen e.g. that for a cluster of 20 Na atoms where R = 11 a.u. the inclusion of the spillout parameter increases the polarizability by some 40%, in reasonable agreement with experiment.

A complete self-consistent calculation for the polarizabilities of clusters is not yet available. A recent calculation [18] using fractional occupation numbers allowing the use of spherical wavefunctions gives a good account of the polarizability when corrected for depolarizing factors [19] derived from known cluster shapes from Clemenger-Nilsson (CN) theory [20], see Sec. 5. The lower curve of Fig. 6 gives the calculated polarizabilities [18] for Na, and corrected for depolarizing factors as shown in the middle curve. The agreement between theory and experiment is rather good, although the theoretical values are all some 15% low. The reason for the discrepancy is not yet known, but it may result from the neglect of pseudopotentials and the associated crystal field splittings.

3. Ionization, Plasma Resonance, and Photoabsorption

Ionization. Following LANG and KOHN [21] the work function W for solid surfaces may be calculated in three parts which we write

$$W = E_{es} + E_{xc} - E_F , \tag{4}$$

where E_{es} is an electrostatic surface barrier term related to electron spillout beyond the jellium limit, E_{xc} accounts for exchange and correlation interactions among electrons, and E_F is the Fermi kinetic energy. According to (4) the calculated [21] work function for Na is W = 0.91 + 5.28 - 3.13 = 3.06 eV, compared to the experimental value 2.7 eV. For K, the calculation gives 2.74 eV compared to the experimental 2.39 eV. Inclusion of the ion pseudopotentials reduces the theoretical result by ~10%.

A corresponding self-consistent calculation of the ionization potential (IP) for Na clusters by EKARDT [22] gives the above asymptotic values for W and E_{es}, and for cluster N = 8 the IP = 4.8 eV compared to the experimental

[2] 4.1 eV. Like the calculations for the solid, the IPs for spherical clusters appear to be ~15% high. Similarly, inclusion of the pseudo-potentials may improve agreement with experiment. In any case the agreement is reasonably good and is expected to improve as the calculations develop.

Figure 7 shows the experimental values for the IP of potassium clusters N = 3 to 100. The IP decreases toward an asymptotic value of 2.4 eV in agreement with the bulk work function. In addition, abrupt drops in IP are observed following spherical shell closings. The theoretical predictions [23] give exaggerated values for these drops largely because they do not account for the non-spherical shapes of the open shell clusters. A recent ellipsoidal SCUBJ calculation [5] gives reduced IP drops as expected. Trends in IP for open as well as closed shell clusters are predicted in the ellipsoidal harmonic oscillator model, see Fig. 8, which shows groups of four, for example, for clusters 15-18, 23-26, and 41-44.

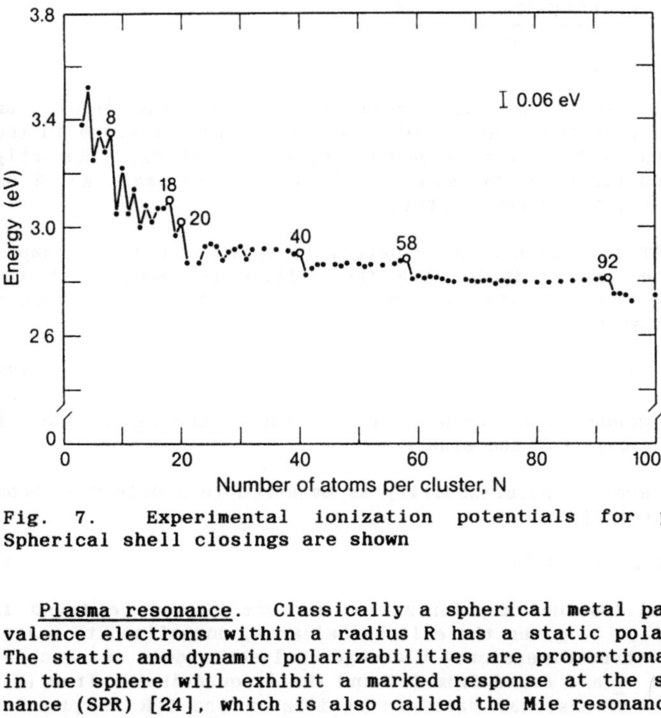

Fig. 7. Experimental ionization potentials for potassium clusters. Spherical shell closings are shown

<u>Plasma resonance</u>. Classically a spherical metal particle containing N valence electrons within a radius R has a static polarizability $\alpha_0 = R^3$. The static and dynamic polarizabilities are proportional and the electrons in the sphere will exhibit a marked response at the surface plasma resonance (SPR) [24], which is also called the Mie resonance. The bulk plasma resonance frequency is given by

$$\omega_p^2 = 4\pi ne^2/m = 3Ne^2/mR^3 , \tag{5}$$

where n is the valence electron density. The surface plasma resonance is given by

$$\omega^2 = \omega_p^2/3 = Ne^2/mR^3 . \tag{6}$$

This resonance represents the collective motion of the valence electrons against the positive charge background of the ion cores. For small

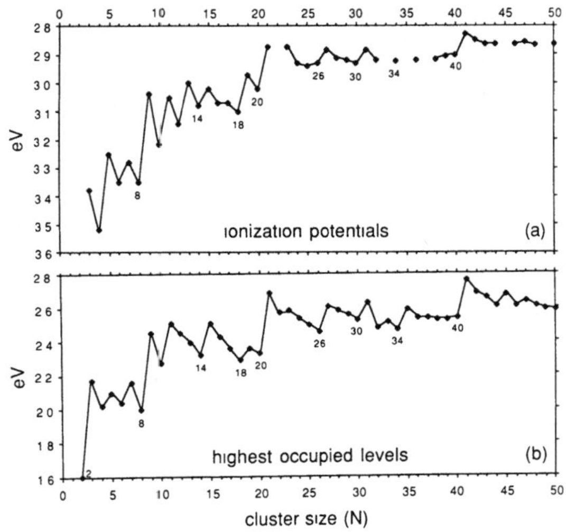

Fig. 8. Ionization potentials for potassium. (a) Experimental, (b) derived from Nilsson diagram. Note correspondences at spherical shell closings and ellipsoidal sequences 15-18, 23-26, and 41-44

particles in the classical limit this resonance exhausts the dipole sum rule for the valence electrons, and it will be the dominant feature in the optical spectrum. The SPR has been studied extensively [17,24]. Recently, experiments have been reported for noble metal clusters as small as ~2 nm diameter [24] in dielectric support matrix.

In general the SPR for ellipsoidal clusters is split into three component peaks associated respectively with the ellipsoidal axes. If the polarizability along the ith axis ($i = x,y,z$) is A_i, then the resonance associated with that axis is

$$\omega_i^2 = Ne^2/mA_i. \quad (7)$$

The polarizability components may be written in terms of the depolarization factors D_i, with $D_0 = 4\pi/3$ for the sphere.

Then, with $\bar{\alpha}$ the average polarizability as measured in a molecular beam deflection experiment [14],

$$A_i = 3\bar{\alpha}[D_i(1/D_x+1/D_y+1/D_z)]^{-1}. \quad (8)$$

The well known ellipsoidal depolarization factors D_i may be found in Ref. 19. Table 1, Sec. 5, gives the ellipsoidal axes, normalized to unity, derived [25] from a three-dimensional Clemenger-Nilsson model, see Sec. 5. The polarizabilities A_i are found from (8) and the frequencies ω_i from (7). The respective three-dimensional harmonic oscillator and SPR frequencies are related by appropriate scaling factors. Table 1 also gives the predicted wavelengths for the SPR according to the cluster shapes determined by Clemenger-Nilsson analysis and experimental values for $\bar{\alpha}$.

<u>Absorption cross sections</u>. Given the frequencies ω_i, we can determine the cluster absorption cross sections from the well known relation of classical electrodynamics [17,24].

$$\sigma = \Sigma_i(4\pi Ne^2/3mc)(\omega_i^2\omega\Gamma)/[(\omega_i^2 - \omega^2)^2 + \omega_i^2\omega^2\Gamma^2] , \quad (9)$$

where Γ is the damping factor.

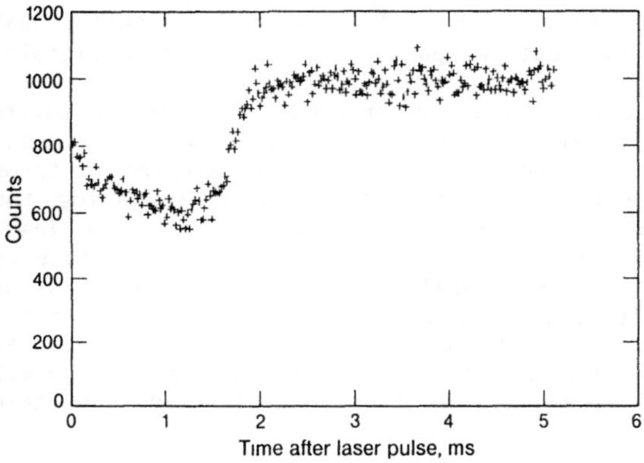

Fig. 9. Beam depletion for sodium N = 8 for time of flight (2 ms) after laser irradiation for 1 μs near t = 0

The above SPR analysis corresponds well with experiment. Longitudinal beam depletion spectroscopy (LBDS) [26] is used to determine absorption cross sections for sodium clusters between N = 2-40. We know [7] that a 2 eV photon will raise the temperature of an N = 20 cluster, for example, by ~400K, after which it evaporates an atom or dimer in order to cool. The radiation for longitudinal illumination of the beam is produced by 1 μs pulses from a flashlamp pumped dye laser. The transverse momentum produced during evaporation removes the cluster and daughter from the beam, thus reducing the signal of the selected cluster. Figure 9 shows a time evolution of the signal for cluster N = 8 following the laser pulse. The signal reduction lasts for the time of flight (~2 ms) of the beam. The ratio r is defined as the ratio of the light-on signal to the light-off signal, and the absorption cross section σ for a cluster N is

$$\sigma = -\ln r/\Phi \,, \tag{10}$$

where Φ is the number of photons/cm^2 per pulse. The cross section is determined by the slope of the curve of -ln r vs. light intensity, Fig. 10.

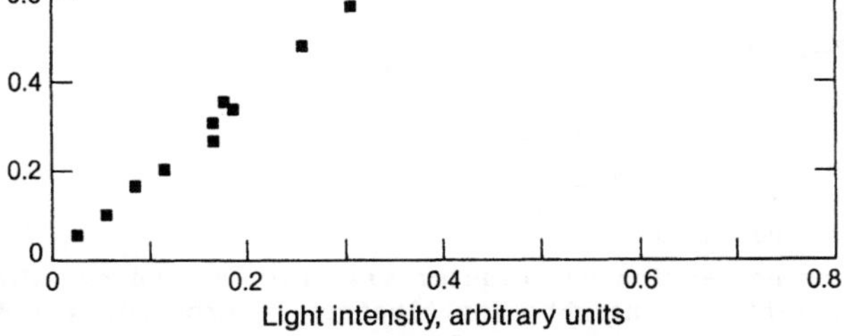

Fig. 10. Light on/off ratio vs. light intensity. Linear slope gives photoabsorption cross section

The response is linear for most of the range of the graph, indicating that the cross section is constant and that two-photon processes are negligible.

At this writing the experiments are in progress, and the complete absorption profiles for the clusters have yet to be determined. Nevertheless, the pattern of behavior is clear, on the basis of experiments taken at several wavelengths. The basic behavior is seen by comparing the experimental cross sections at 500 nm and 600 nm, Fig. 11. At the lower wavelength the closed shell clusters 8 and 20 show large cross sections by comparison with the open shell clusters, while at 600 nm the relative cross sections are reversed. This is consistent with the assumption that at the lower wavelength the closed shell clusters are closer to the predicted spherical resonances at ~475 nm, and that the neighboring open shell clusters are farther from their resonances at higher wavelengths, see Table 1, Sec. 5.

4. Mass Spectra of Neutral and Ionized Clusters

Shell structure in metal clusters was first recognized in the mass spectrum of neutral sodium, Fig. 12 [3], where certain clusters N = 2, 8, 20, 40,

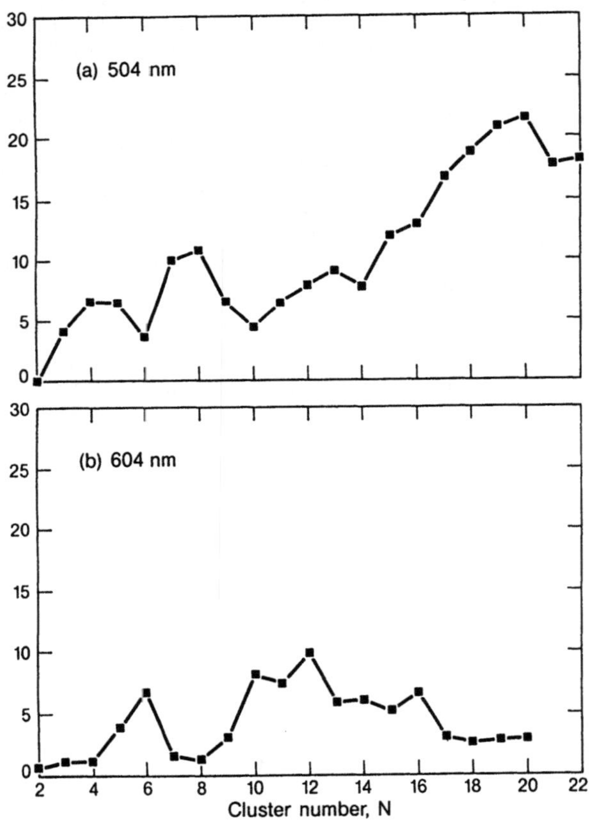

Fig. 11. Absorption cross section vs. cluster size for (a) 504 nm, (b) 604 nm. Cross sections relative to neighboring clusters reverse for N = 8 and 20. Vertical scale in square angstroms ± 15%.

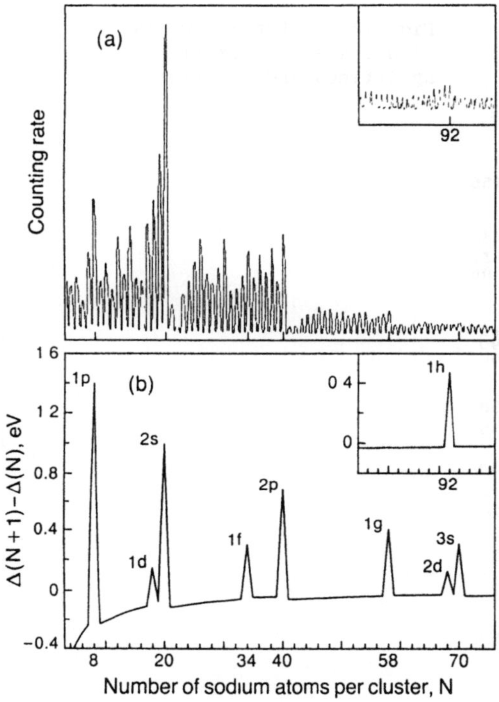

Fig. 12. (a) Mass abundance spectrum for sodium clusters showing spherical shell closings, (b) theoretical spectrum calculated with Woods-Saxon potential. Peaks near N = 70 disappear in SCUBJ calculation. Clusters N = 18 and 34 are ellipsoidal

58, and 92 are conspicuously more abundant than neighboring clusters. This is a direct consequence of the degeneracies in the energies of delocalized electrons in a suitable potential well [23]. It results [7] from the establishment of thermodynamic equilibrium in the nozzle of a supersonic jet cluster source. Similarly, positive and negative cluster ions [27] of e.g. silver show stabilities for the above number of electrons, i.e. at N = 21 (positive) and N = 19 (negative), both of which are closed shell systems with 20 electrons.

The electronic energy calculations have been done for spherical symmetry using the square well potential [28], Fig. 13, the intermediate rounded square well (Woods-Saxon) [3] potential, and in the self-consistent uniform background jellium SCUBJ approximation, Fig. 14, which improves on the former models and gives values for the IPs [22] and for the polarizabilities [17]. Calculations have also been done for the squared ellipsoidal harmonic oscillator well [20].

The thermodynamic treatment [7] assumes an equilibrium distribution involving all possible reactions among clusters. The relative cluster concentrations are found from the partition function and the law of mass action, assuming that a cluster has 3N - 6 vibrational modes described by an Einstein model with frequency ω_E, that they are approximately spherical, and that the ground state binding energy is approximated by

$$U_N = Nb + E_N , \qquad (11)$$

where E_N is the total electronic energy from the Clemenger-Nilsson model, and b is a smoothly varying function of N. The partition function is

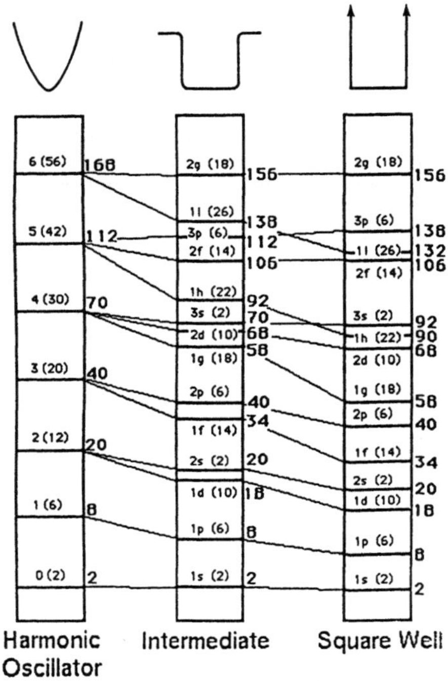

Fig. 13. Potential wells with respective level splittings and orderings

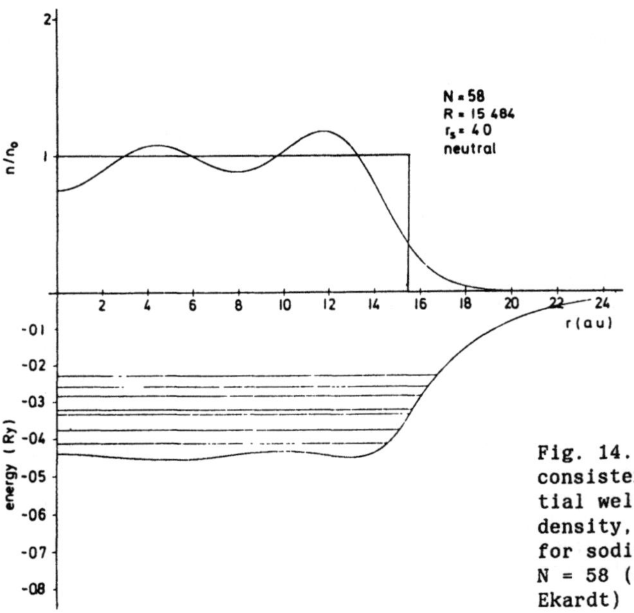

Fig. 14. Self-consistent potential well, charge density, and levels for sodium cluster N = 58 (after Ekardt)

factored into translational, vibrational, and rotational terms, and the cluster densities are found to follow the simple relation

$$\ln\,[\rho_N^2/\rho_{N+1}\rho_{N-1}] = [2E_N - E_{N+1} - E_{N-1}]/kT\,, \tag{12}$$

where ρ is the cluster density in the source. The detected cluster intensity I is proportional to ρ, and the RHS is the second difference of electronic energy with respect to N. Therefore (12) may be written [7] in the convenient form

$$\ln\,[I_N^2/I_{N+1}I_{N-1}] = \Delta_2(N)/kT. \tag{13}$$

Figure 15 shows values of the LHS derived from experimental abundance spectra for Na (above), and values for the RHS obtained from the Nilsson diagram (below). The Nilsson diagram is discussed in the following Sec. 5. The two plots are scaled for a best fit which occurs for a temperature T around 1000K, which agrees nicely with the experimental source temperature. The fact that the agreement is good suggests that the simple thermodynamic analysis is valid, and that cluster production in the jet source involves quasi-equilibrium processes. The detected intensities correspond to the above relations, provided the equilibrium distribution is frozen during the jet expansion. However, the large intensities of e.g. cluster N = 20 relative to N = 21 (or 22) suggests that following the expansion the latter have evaporated an atom (or dimer).

The normalized formation rate of species N from N-M and M (or the reverse) may be approximated by

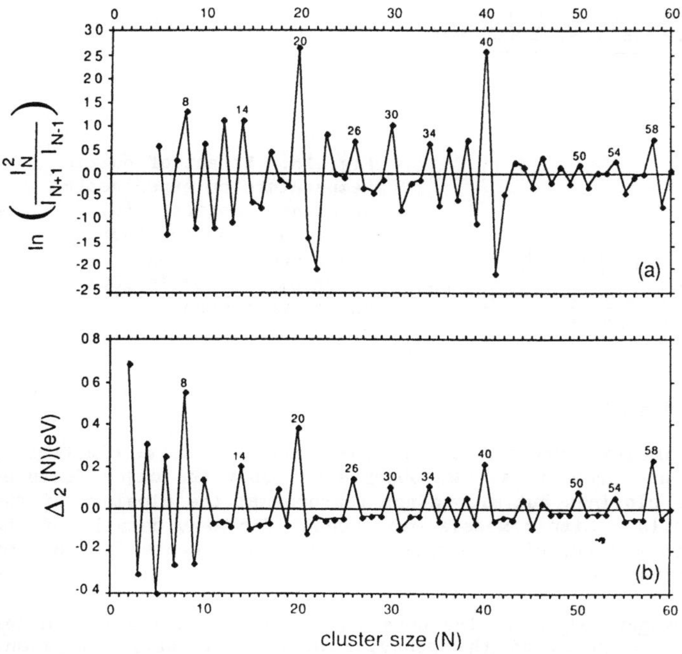

Fig. 15. (a) Relative measured cluster stabilities, (b) relative stabilities calculated on basis of levels from Nilsson diagram

$$W_{N-M} = \sigma_g v' (\rho_{N-M}\rho_M/\rho_N) , \quad (14)$$

where σ_g is the geometrical cross section, and v' is the relative velocity. The last term varies as the negative exponential of $[U_{N-M} + U_M - U_N]$. The probability for the process in (14) for $M > 2$ is small relative to the probabilities for atom and dimer. The rate of evaporation for $N = 20$ at 450K is ~5/s, while that of 22 is 4×10^7/s. In order to be stable for the beam flight time of ~1 ms, the temperature of cluster 22 cannot exceed 280K, compared to 500K for $N = 20$. Thus we conclude that under typical conditions the jet expansion is inadequate to cool clusters 21 and 22, and the probability is high for them to cool by evaporating atoms and dimers, respectively. This results in especially high detected intensities for cluster 20, and especially low intensities for 21 and 22 as is observed. If temperatures are somewhat higher, the probability for evaporation from even the more stable clusters becomes significant. Under these conditions the edges marking the closed shells are either blurred or moved to lower N. This effect may result from overheating the source nozzle, as is shown in Fig. 16.

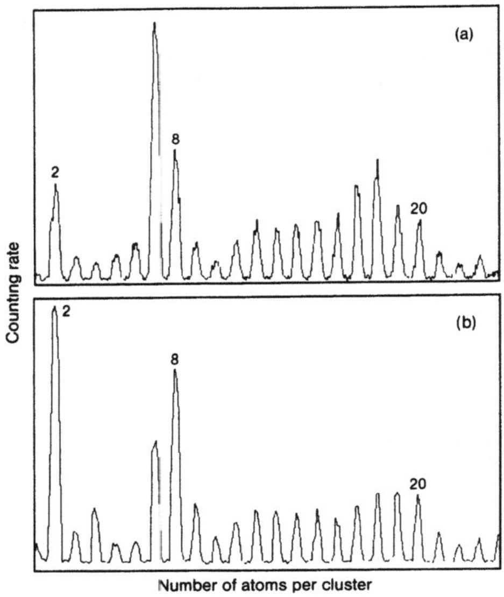

Fig. 16. Effect of nozzle temperature on experimental abundance spectra. (a) Nozzle 300K hotter than oven, shell closings blurred, (b) nozzle 150K hotter than oven

5. Why Shell Models Work

We have seen experimental evidence for the shell model of metal clusters in their stabilities and properties. We observe not only the major features at spherical shell closings but also minor features at the closings of the ellipsoidal subshells. Simple models can account for many details of the properties of atoms and nuclei, and these are good precedents for what we observe in metal clusters.

Electrons in a spherical box. The development of the ellipsoidal model for metal clusters is based on the behavior of non-interacting fermions under specified boundary conditions. The bounding potential arises from a uniform positive charge background. This idea is implicit in potential well models, stemming from the particle in a one-dimensional box, where the electron is free save for the constraints of the uniform potential. When the problem is extended to three dimensions, the uniform positive charge distribution is first assumed to be spherical [7]. The electrons are free, and collisions between them are suppressed by the Pauli principle. Solution of the single particle equation in spherical coordinates gives eigenstates with principal quantum number n, where n - 1 is the number of nodes in the radial wavefunction, and angular momentum quantum numbers are ℓ and m_ℓ. Then n = 1 for the lowest energy state of a given ℓ. Although the uniform potential differs sharply from the Coulomb potential, the conventional symbols are used: s, p, d, ... corresponding to angular momentum ℓ = 0, 1, 2, ... with the high degeneracy $2\ell(\ell + 1)$ which results from spherical symmetry. While for the Coulomb potential it is required that $\ell < n$, this restriction does not apply for the uniform potential, and high angular momenta are expected; see for example in Fig. 13 the spherical shell closing at N = 58, which is a 1g level. Notice that the patterns of level filling in spherical symmetry for the square, harmonic oscillator, and hybrid wells are similar. The hybrid or "intermediate" potential corresponds quite well to reality, and the exact shape governs the lifting of degeneracies and the ordering of levels.

<u>Electrons in an ellipsoidal box</u>. The extension to ellipsoidal symmetry is straightforward. Clusters undergo ellipsoidal deformations [20] at constant volume analogous to Jahn-Teller distortions in molecules. It is found that most non-spherical metal clusters are close to ellipsoids of revolution, with $x = y \neq z$, and major axes b and a, with a distortion parameter $\eta = 2(a-b)/(a+b)$. The ellipsoid of revolution may be treated as a three-dimensional harmonic oscillator, with two characteristic frequencies $\Omega_x = \Omega_y$, and Ω_z, where

$$\Omega_{x,y}/\Omega_o = [(2+\eta)/(2-\eta)]^{1/3}, \tag{15}$$

$$\Omega_z/\Omega_o = [(2+\eta)/(2-\eta)]^{-2/3}, \tag{16}$$

$$\Omega_o \sim E_F N^{-1/3}. \tag{17}$$

The CN model [20] for metal clusters uses a "squared" harmonic oscillator potential with a shape resembling the hybrid potential of Fig. 13. The Hamiltonian reflects the ellipsoidal symmetry and contains a quartic term to accomplish the squaring. The minimum total electronic energy for all filled levels is calculated for each cluster as a function of the distortion parameter η. The total energy for spherical clusters occurs for $\eta = 0$ as expected. For the ellipsoidal clusters, Fig. 17, two minima are found respectively for prolate and for oblate clusters. The lower of the two minima locates the most stable cluster, and thus each cluster is characterized by the unique value of η at the minimum.

The Nilsson diagram, Fig. 18, displays energy levels as a function of η, and each cluster is located by a dot at the appropriate level at the equilibrium value of η. The set of occupied cluster levels thus determined for all clusters, see Fig. 8, forms the basis for predicting $\Delta_2(N)$, relative stabilities, and ionization potentials. The energy scale is given by (17). CN theory gives accurate cluster shapes measured in terms of η, as has recently been verified in an ellipsoidal SCUBJ [5] calculation. The energy scale is only approximate, and the frequencies differ from the theoretical plasma frequencies. The shapes are given truly by the CN analysis and they do not depend on the energy scale.

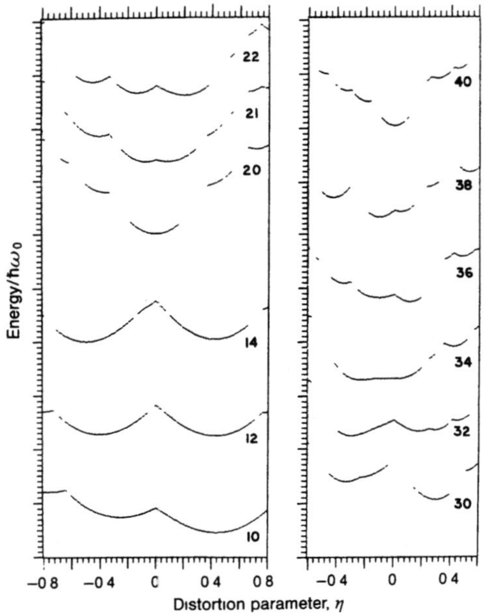

Fig. 17. Total electronic energy for several clusters as a function of the distortion parameter η. Spherical clusters show energy minimum at η = 0. Spheroidal clusters show two minima, the lower of which identifies equilibrium η

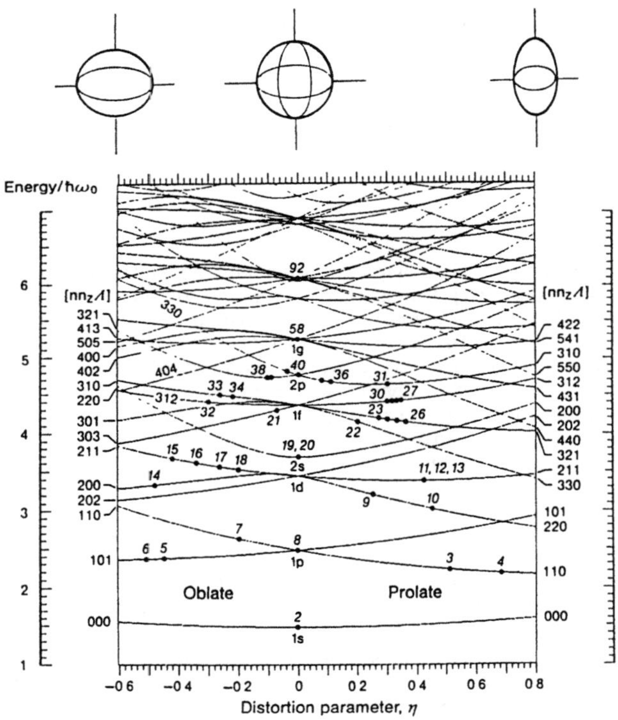

Fig. 18. Nilsson diagram for spheroidal clusters. Equilibrium values of η for numbered clusters identified by dots

As discussed in Sec. 3, the plasma resonance in ellipsoidal clusters splits, as occurs for the giant nuclear dipole resonance [29]. The peak corresponding to $\omega_{x,y}$ should have twice the oscillator strength of ω_z for an ellipsoid of revolution around the z axis. The photoabsorption data in Sec. 3 are consistent with this for shapes which are ellipsoids of revolution; for some other clusters three separate axes are needed, and correspondingly three peaks are expected. Table 1 is based on an explicit solution of a triaxial Nilsson Hamiltonian. Examination of the values in the table shows that half of the clusters N = 2-20 are ellipsoids of revolution. Experimental and theoretical (Γ = 0.15) cross sections for clusters N = 2-20 are compared in Fig. 19, and the agreement is seen to be very good. Spherical resonances for clusters N = 8 and 20 are predicted to be at ~475 nm. The wavelength dependence of the cross sections in this region is consistent with that prediction. Although the entire resonance curves are yet to be traced out experimentally, the agreement of the analyzed data with the ellipsoidal shapes derived from the CN model, and agreement with the predicted cross sections is sufficiently close to justify the application of the simple model as a first approximation. Since the locations of the peaks for open shell clusters depends only on the cluster shapes, the experimental data provide a direct measure of the shapes. The data agree both with Nilsson theory and with Ekardt's recent ellipsoidal jellium calculation [5].

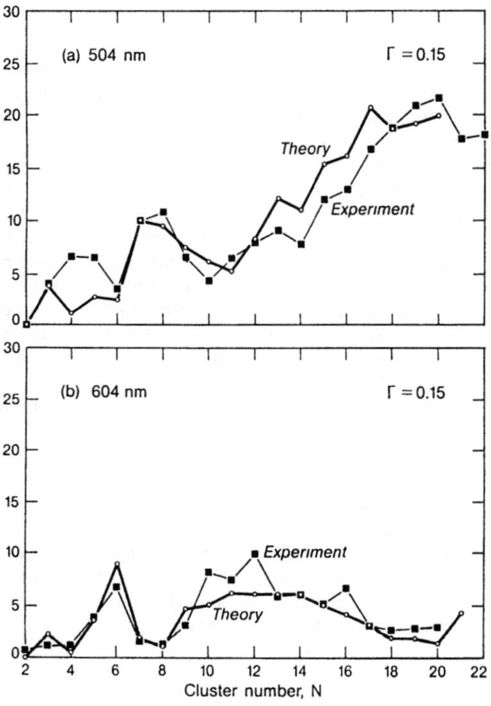

Fig. 19. Photoabsorption cross sections, theory and experiment for light at (a) 504 nm, (b) 604 nm. Surface plasma resonance for N = 8 and 20 calculated to be near 475 nm. Vertical scale in square angstroms ± 15%

Table 1. Ellipsoidal axes and resonance wavelengths

N	$\bar{\alpha}/N\alpha_{at}$	x_o	y_o	z_o	λ_x	λ_y	λ_z
2	0.80	1.00	1.00	1.00	520	520	520
3	0.99	0.84	0.84	1.41	500	500	690
4	0.86	0.79	0.79	1.59	440	440	680
5	0.91	0.74	1.03	1.32	450	540	640
6	0.86	1.19	1.19	0.71	580	580	430
7	0.73	1.07	1.07	0.88	510	510	450
8	0.69	1.00	1.00	1.00	480	480	480
9	0.74	0.91	0.91	1.20	460	460	550
10	0.81	0.86	0.86	1.35	460	460	610
11	0.74	0.81	0.92	1.35	430	460	580
12	0.78	0.77	0.96	1.35	430	480	600
13	0.74	0.74	1.08	1.26	410	510	560
14	0.75	1.19	1.19	0.71	540	540	400
15	0.73	0.76	1.12	1.19	410	520	540
16	0.76	0.79	1.06	1.19	430	510	550
17	0.67	0.83	1.07	1.13	420	490	510
18	0.65	1.08	1.08	0.85	490	490	420
19	0.67	1.04	1.04	0.93	480	480	450
20	0.67	1.00	1.00	1.00	470	470	470

6. Conclusions

We have showed that simple models give a remarkably good view of metal clusters and their properties. The SCUBJ model validates the basic symmetry patterns and improves quantitative agreement with experiment. The fourfold patterns evident in the abundance and ionization spectra follow directly from the ellipsoidal symmetry just as the major shell closings directly result from their spherical symmetry. The crystal field splittings associated with the pseudopotentials must be smaller than the ellipsoidal splittings, else the fourfold structures would not be evident. In fact the lack of fourfold structure in the cluster ion abundance spectra suggests that the ions may be spherical, witness the appearance of cluster N = 34, for example, in the Ag cluster ion abundance spectrum [27]. Similarly, the odd-even intensity pattern in Ag [27] appears to be regular up to large cluster sizes, suggesting that the spin pairing effects are more important than ellipsoidal distortions. The appearance of cluster N = 34 in the cluster ion spectrum for Rb [30] is also consistent with the above conjecture.

Observation of the properties of the alkali clusters involves several kinds of spectroscopy which may be developed further. The plasma resonance experiments have only just begun, and many determinations are yet to be made to find the exact resonance frequencies, the damping mechanisms, and their dependence on number of atoms, shape, temperature, and other variables. Fine structures will be observed. Both the single particle and collective electron behaviors will be elaborated and related.

7. Acknowledgments

This work was supported by the U.S. National Science Foundation under Grants DMR 84-17823 and DMR 86-15246. We have benefitted from ongoing discussions with Prof. Marvin L. Cohen and members of his theoretical group

M. Y. Chou, J. L. Martins, S. B. Zhang, Robert Dandrea, and David Tomanek from the group of Prof. Steven Louie. We are grateful to Prof. Eugene Commins for discussions, advice, and the loan of the pulsed dye laser and some accessories. We acknowledge useful discussions concerning laser operation with Jeff Hunt. We acknowledge with pleasure the collaboration of Kathy Selby, Vitaly Kresin, Jun Masui, Prof. Andre Châtelain, and Dr. Michael Vollmer in the plasma resonance work, which will be submitted for publication following completion of further experiments.

1. E. J. Robbins, R. E. Leckenby, and P. Willis, Adv. Phys. 16, 739 (1967).
2. A. Herrmann, E. Schumacher, and L. Wöste, J. Chem. Phys. 68, 2327 (1978).
3. W. D. Knight, K. Clemenger, W. A. de Heer, W. A. Saunders, M. Y. Chou, and M. L. Cohen, Phys. Rev. Lett. 52, 2141 (1984).
4. S. G. Nilsson, K. Dan. Vidensk Selskab. Mat. Medd. 29, No. 16 (1955).
5. W. Ekardt, private communication, submitted to Phys. Rev. B.
6. Walt A. de Heer, Ph.D. thesis, University of California, Berkeley, 1985.
7. Walt A. de Heer, W. D. Knight, M. Y. Chou, and Marvin L. Cohen, Solid State Physics, Vol. 40 (Academic Press, 1987), in press.
8. N. F. Ramsey, Molecular Beams, (Oxford, 1956).
9. G. Herzberg, Electronic Spectra of Polyatomic Molecules, (Van Nostrand-Reinhold, 1966).
10. G. Delacrétaz, E. Grant, R. Whetten, L. Wöste, and W. Zwanziger, Phys. Rev. Lett. 56, 2598 (1986).
11. W. Gerber and E. Schumacher, J. Chem. Phys. 69, 1692 (1978).
12. L. Landau, Phys. Z. Sowjetunion 2, 46 (1932); C. Zener, Proc. Roy. Soc. (London) A137, 696 (1932).
13. C. H. Townes and A. L. Schawlow, Microwave Spectroscopy (McGraw-Hill, New York, 1955).
14. W. D. Knight, Keith Clemenger, Walt A. de Heer, and Winston A. Saunders, Phys. Rev. B31, 2539 (1985).
15. D. R. Snider and R. S. Sorbello, Phys. Rev. B28, 5702 (1983).
16. D. E. Beck, Solid State Commun. 49, 381 (1984).
17. W. Ekardt, Phys. Rev. B31, 6360 (1985).
18. M. Manninen, R. M. Nieminen, and M. J. Puska, Phys. Rev. B33, 4289 (1986).
19. J. A. Osborn, Phys. Rev. 67, 351 (1945).
20. Keith Clemenger, Phys. Rev. B32, 1359 (1985).
21. N. D. Lang and W. Kohn, Phys. Rev. B12, 4555 (1970).
22. W. Ekardt, Phys. Rev. B29, 1558 (1984).
23. M. Y. Chou, A. Cleland, and M. L. Cohen, Solid State Commun. 52, 645 (1984).
24. L. Genzel, T. P. Martin, and U. Kreibig, Z. Physik B21, 339 (1975).
25. Walt A. de Heer, unpublished.
26. Walt A. de Heer, Katherine Selby, Vitaly Kresin, Jun Masui, A. Châtelain, and W. D. Knight, Bull. Am. Phys. Soc. 32, 484 (1987).
27. I. Katakuse, I. Ichihara, Y. Fujita, T. Matsuo, T. Sakurai, and H. Matsuda, Int. J. Mass Spect. Ion Proc. 67, 229 (1985).
28. M. G. Mayer and J. H. D. Jensen, Elementary Theory of Nuclear Shell Structure (Wiley, New York, 1955).
29. G. F. Bertsch, P. F. Bortignon, and R. A. Broglia, Rev. Mod. Phys. 55, 287 (1983).
30. N. D. Bhaskar, private communication. We thank Dr. Bhaskar for sending a preprint of his work on Rb cluster ions.

Optical and Dynamic Properties of Metal Clusters

G. Delacrétaz[a], *P. Fayet*[b], *J.P. Wolf*[a], *and L. Wöste*[a]

Institut de physique expérimentale, Ecole Polytechnique Fédérale de Lausanne, PHB-Ecublens, CH-1015 Lausanne, Switzerland

I. INTRODUCTION

Metal clusters are commonly associated with the very fundamental role of bridging the gap from atoms and small molecules to the bulk. Moreover, they have extraordinary chemical properties, which are due to their distinct structure and special electronic behaviour. Over more than two decades detailed spectroscopic information on matrix isolated particles has provided much insight into catalytic processes occurring on metal surfaces [1]. There are, however, obvious attractions of working in the gas phase, since one can avoid complications arising from matrix effects. Molecular beam techniques combined with mass spectrometry have proved to be very powerful tools for investigating optical and chemical properties of metal clusters of a specific size. This is documented by a series of impressive results, which - combined with new computational methods - contributed to the evolution of cluster science from what was originally a purely empirical field to a discipline where quantum mechanical understanding is verified by well-designed experiments. The development of the shell model by Knight et al. [2] and related calculations by Ekhard [3] provided an excellent example. The same is true for the observation of the gradual transition of the behaviour of mercury clusters from that of van der Waals molecules to metallic by Bréchignac et al. [4], Rademann et al. [5] and related calculations by Tomanek et al. [6]. A bridge between spectroscopic properties of metal clusters and their corresponding chemical properties was established by Whetten et al. [7] who measured and correlated the ionization potentials of iron clusters and the corresponding hydrogen absorption affinities as a function of their size. These results prove that the investigation of optical and dynamic properties of metal clusters contributes to the understanding of these particles, as well as indicating a way to future applications.

II. PRODUCTION AND CHARACTERIZATION OF METAL CLUSTERS

A major requirement for the progress of the field was the development of powerful cluster sources. Fortunately there was an impressive amount of knowledge available from molecular beam research, and Ramsey's famous book "Molecular Beams" has become a most valuable tool for experimental cluster

*present adresses:

a) Centre d'Applications Laser
Ecole Polytechnique Fédérale de Lausanne
CH-Ecublens
CH-1015 Lausanne

b) Exxon Research Laboratories
Clinton Township
Route 22 East
Annandale New Jersey 08801

science as well [8]: For an **effusive beam** expansion, for example, where the mean free path of the expanding gas inside the oven chamber is larger than the diameter of the oven nozzle (Knudsen number $K_n>1$), the intensity I_o of the beam is calculated - by Ramsey - simply as

$$I_0 = 1.118 \times 10^{22} \frac{p' A_s A_d}{d^2 \sqrt{MT}}, \qquad \text{[molecules/second]} \qquad (1)$$

where p' is the oven pressure [Torr], T the oven temperature [K], M the molecular weight, A_s the surface of the oven nozzle [cm^2], A_d the surface of the beam detector [cm^2] and d the beam length between nozzle and detector [cm]. The corresponding velocity distribution I (v) is given by

$$I(v) = \frac{2 I_0}{\alpha} v^3 \exp\left(- \frac{v^2}{\alpha^2} \right), \qquad (2)$$

$$\alpha = \sqrt{kT/m},$$

where m is the particle mass and k the Boltzmann factor.

In order to generate high-intensity cluster beams, however, it is obviously preferable to operate in the **supersonic beam** regime ($K_n<<1$), because the resulting spatial focussing, velocity focussing and state focussing (cooling) leads to significantly higher intensities and narrower distributions for the kinetic and internal energies of the beam particles. Using (1), the intensity of a supersonic beam is calculated by Anderson [9] as

$$\frac{I_{\text{supersonic beam}}}{I_{\text{effusive beam}}} = \left(\pi \frac{c_p}{k} \right)^{1/2} \gamma M^2, \qquad (3)$$

where c_p is the specific heat, $\gamma = c_p/c_v$, the specific heat ratio, and M is the Mach number. The intensity of a supersonic beam of N_2, for example, is calculated at Mach 10 to be 470 times higher, and at Mach 20, 1860 times higher than an effusive beam. The most important feature of the supersonic beam is the enhanced cluster formation, which occurs mainly inside the collision zone in front of the nozzle, where the strong adiabatic expansion leads to a highly supersaturated system, which favours the aggregation of the colliding particles.

The result of a supersonic beam expansion of mercury is shown in the mass spectrum of Fig. 1. It was obtained at a Hg stagnation pressure of 1 bar inside the oven cartridge, expanded through a nozzle of 0.4 mm diameter.The nozzle had a conical shape with an opening angle of 30°, which significantly favours the formation of larger aggregates [10]. The neutral particle beam was ionized by means of electron impact ionization; the mass spectrum shows series of singly and of doubly charged mercury aggregates. Both series exhibit rather similar intensity distributions until about Hg_{50}.The result was obtained with a modified quadrupole mass spectrometer [11]; the instrument was used later for the observation of the transition from van der Waals molecules to metallic bonding

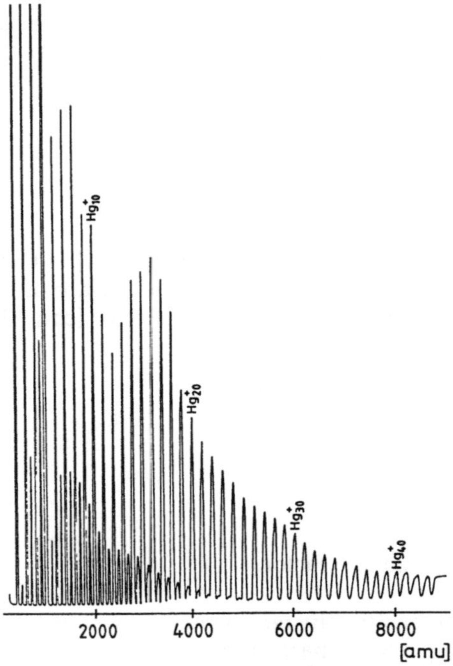

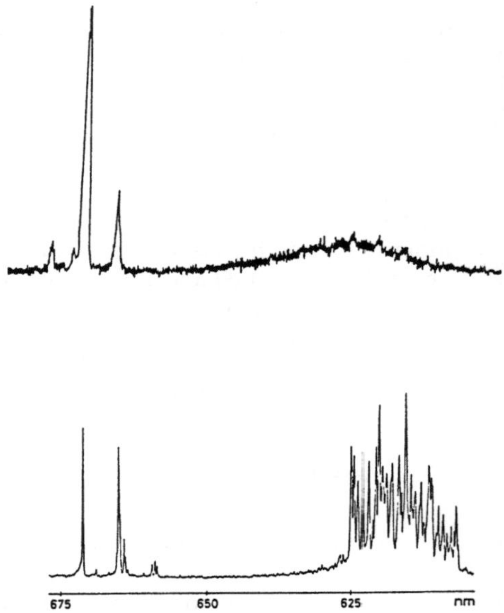

Fig. 1 Mass spectrum of mercury aggregates obtained from a pure-metal expansion. The stagnation pressure was 1 bar; a nozzle of conical shape and 0.4 mm diameter was used. The spectrum was recorded with a modified quadrupole mass spectrometer using electron-impact ionization [11].

Fig. 2 Comparison of Na_3 excitation spectra (a) recorded from a pure-metal expansion of 1 bar stagnation pressure [12], (b) recorded from a seeded beam expansion of 10 bar [13]. Surprising spectroscopic features arise from the spectrum, when the beam is cooled down; the warm spectrum, on the other hand exhibits pronounced hot band progressions.

in mercury clusters [4]. The advantage of **pure metal expansions** is the moderate requirement of active pumping speed , since cold traps can be used efficiently. The above-mentioned experiment was equipped with a 1500 l/s diffusion pump system.

Whenever cold particles are required, there are significant advantages to operating with **seeded beam expansions,** in which metal vapor is coexpanded with an inert carrier gas. In this way the heat of formation, which accumulates in a cluster growing inside the expansion zone, is effectively carried away by the inert gas collision partners. Figure 2 shows the TPI spectrum of Na_3 (Sect. IV) first obtained from a pure metal expansion of sodium at a stagnation pressure of 1 bar across a cylindrical nozzle of 0.4mm diameter [12]. The second curve shows the same spectral sequence recorded from a seeded beam expansion where 100 mbar sodium were coexpanded with 10 bar argon across a cylindrical nozzle of 50µm diameter. Based on an idea of Stein, the conditions were chosen such that only about 10 metal-metal collisions occurred in front of the nozzle. This favoured the formation of very small aggregates. The very high number of metal-gas collisions on the other hand provided very efficient cooling of the emerging beam paricles. The vibrational temperature of the Na_3 particles from the pure metal expansion is about 100 K, whereas the spectrum from the seeded beam expansion indicates a vibrational temperature of 30 K [13]. Since the supersonic beam expansion does not represent an equilibrium system, the rotational temperatures of the beam molecules differ significantly from these values. They were only determined for the dimer, due to insufficient spectral resolution. Under identical expansion conditions T_{rot} (Na_2) of the pure metal expansion is about 70 K, while the seeded beam expansion freezes down to 7 K [14].

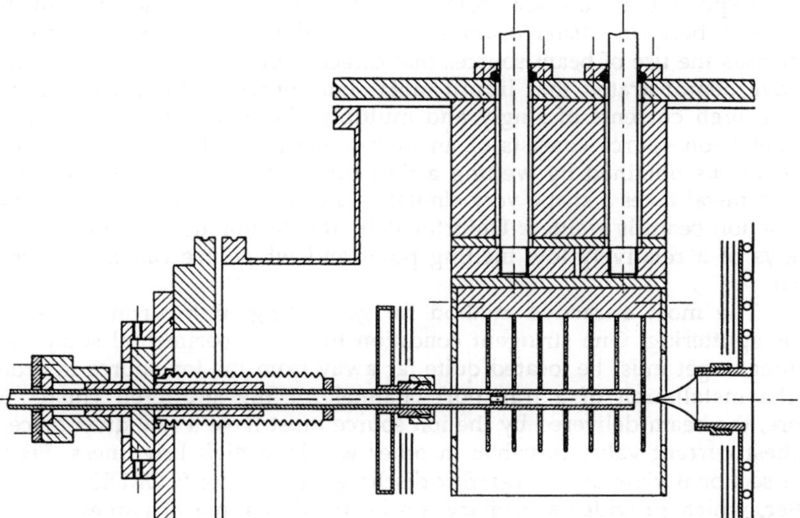

Fig. 3 Sectional view of the radiatively heated oven system (T_{max} = 2000K) for seeded-beam expansions (p_{max} = 10 bar). The oven cartridge is machined from TZM, the nozzle is Nd-YAG laser-drilled (diam. = 10μm).

The major limitation for pure or seeded metal beam expansions is the high vaporization temperature of most materials and their chemical aggressivity. Figure 3 shows a very advanced continuous flow molecular beam source, which allows the metal to be heated up to temperatures > 2000 K in coexpansion with carrier gas pressures > 10 bar. The interchangeable oven cartridge has the shape of a test tube. It is machined from a tungsten-zirconium-molybdenum alloy (TZM), which is resistant to most metal vapors even at high temperatures. The cartridge is fixed to the adjustable and water-cooled carrier gas inlet with a metal compression fitting. The oven nozzle is laser-drilled with a Nd-YAG laser system [LASAG], which allowed nozzle diameters < 10 μm to be obtained reproducibly. Oven and nozzle are radiatively heated by two independent sets of tungsten hairpins. The system is embedded in a three-layer radiative shield of tantalum foils and a water cooling jacket. It is pumped by a 3000 l/s diffusion pump. The nozzle is oriented, at a variable distance, towards a skimmer of 0.5mm diameter, which also serves as a differential pumping barrier to the next chamber, where the mass spectrometer is located. The molecular beam source allowed well-resolved spectra of Li_3 to be obtained for the first time (Sect. IV).

Whenever the experiment allows or requires pulsed beam operation, **laser vaporization sources,** developed by Smalley and co-workers [15], present an attractive alternative to hot oven systems. A metal rod or disk, which is located inside the carrier gas stream, is irradiated with high-power laser pulses such that a sufficient amount of metal is evaporated in order to form clusters. The laser vaporization source allowed impressive spectroscopic and chemical experiments to be performed in particular on transition metal clusters [7].

Experiments on size-selected clusters require the use of mass spectrometry, because cluster beam sources usually generate size distributions. This suggests the use of beam sources that directly produce cluster ions. Sudraud et al. have demonstrated that **liquid metal ion sources** (LMIS) are capable of delivering high currents of singly and multiply charged metal clusters [16]. A liquid metal ion source consists of an intense electric field applied to a liquid metal meniscus obtained by wetting a thin tungsten tip with the liquid metal. The liquid metal takes a shape very similar to a cone, and at the apex of the cone an intense ion beam is emitted. Unfortunately, this technique is limited to metals and alloys of a relatively low melting point and which are capable of wetting tungsten.

The most attractive method for generating high-current cluster ion beams is sputtering. One stringent condition for the experimental setup is that the sputter target must be located quite far away from the ion source in order to allow the installation of a transport channel for the sputtered cluster ions. Therefore, the beam delivered by the ion source must have a low divergence, but the highest current value possible; in other words, a high brightness. Figure 4 shows a sectional view of a cold-reflex discharge ion source (CORDIS) developed by Keller, which provides a primary ion beam of extreme brightness [17]. The primary electrons inside the CORDIS are generated by six resistively heated tantalum filaments suspended on a molybdenum rod near the source axis. A cylindrical anode surrounds the plasma, and because of 18 permanent magnets placed around the anode, plasma electrons can escape into the loss zones only,

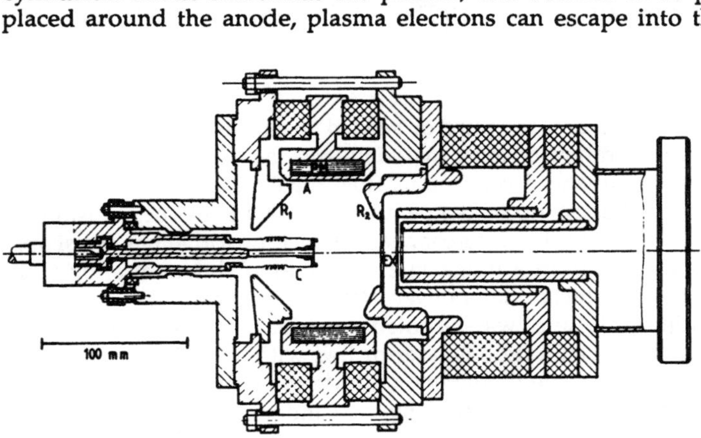

Fig. 4 Sectional view of the sputter ion source "CORDIS", developed by Keller [17]. Using argon, the source allows one to obtain at 20 kV extraction voltage typical primary ion currents on the target (diam. 8 mm) of 2 mA.

which are where the magnetic field lines of the cusps cross the anode wall. The source is operated with argon, at a pressure of about 4 Pa, as measured near the gas inlet tube. A single-aperture accel/decel extraction system is used to form the ion beam at 15 - 20 kV extraction voltage. Because the sputter target of 10 x 10 mm size is positioned 500 mm away from the source outlet plane, a maximum divergence half-angle of about 7mrad is allowed for the beam; the exact value depends slightly upon the aperture size. Up to 2mA beam current can be measured on the target when using a 6-mm diameter aperture. This allowed us to obtain particle beams of Ag_n^- clusters of up to 10^{12} particles per second [18] (see also Sect. V). A disadvantage of sputtering arrangements, however, is the fact that the sputtered molecules are rather hot [19]. This complicates spectroscopic applications, unless cooling can be provided.

The type of experiment should determine the kind of mass spectrometer system chosen: **Magnetic instruments** - in particular double focussing machines - provide excellent mass resolution ($M/\Delta M > 100\ 000$), and rather high mass values ($M/e > 10\ 000$) can be reached; the transmission values, however, are rather poor ($10^{-2} > T > 10^{-6}$). In addition to that, they are rather large and expensive. **Time of flight instruments** (TOF) are - in comparison - very simple. They have a practically unlimited mass range, and the a priori limited mass resolution ($M/\Delta M < 400$) can be substantially improved by using a reflectron assembly. The pulsed-mode requirement suggests that the TOF be used in combination with pulsed cluster sources. **Quadrupole mass spectrometers** have extremely high transmission values (typically 80%) and they reach decent resolution values ($M/\Delta M < 2000$). The non-linear mass discrimination and the rather limited mass range of commercial instruments (typically $M/e < 2000$) are, however, a handicap. As Fig. 1 indicates, this difficulty can be overcome, and mass values up to $M/e = 10\ 000$ can be reached by operating the system at a lower RF frequency,combined with computer control of the dc-RF voltage ratio [11]. A further advantage of quadrupole mass spectrometers is the rather low ion energy, which allows thermal ion-molecule reactions or soft-landing experiments without the necessity of slowing down the ion beam substantially (Sects. V and VI).

The mass spectrum presented in Fig. 1 was obtained by means of **electron impact ionization** (E_{el} = 50 eV). The very similar intensity distribution of singly and doubly charged aggregates, which does not change significantly when the electron energy is modified above the ionization threshold [20], suggests that the observed relative intensities reflect the size distribution of the neutral beam rather well. This assumption is confirmed by synchrotron light **photoionization** experiments ($h\nu$ = 50 eV), which yield rather similar intensity distributions [21]. In spite of this observation, fragmentation phenomena during ionization can never be neglected. Scattering experiments performed by Buck and Meyer have clearly shown that it is extremely dangerous to draw conclusions about intensity distributions of a neutral beam from mass spectrometry [22]. Fragmentation processes, however, can be reduced by choosing a "soft" ionization approach. Experiments on alkali aggregates, for example, have demonstrated that photoionization at photon energies barely above the ionization limit causes significantly less fragmentation than electron impact ionization [23]. **Penning ionization** experiments have equally proved to allow rather "soft" ionization [24]. The most elegant way to rule out fragmentation by means of spectroscopy will be discussed in Sect. IV.

III. SPECTROSCOPIC APPROACHES TO METAL CLUSTERS

As presented in the previous section, the photoionization of molecules offers several advantages compared to the electron-impact process. There is particular interest in combining absorption spectroscopy with photoionization, because it provides mass-selected spectra at high sensitivity [25, 26]. The combination is realized by **two-photon ionization spectroscopy** (TPI). In such a process (Fig. 5a), a first tunable laser (energy $h\nu_1$) excites the molecule into a real intermediate state, while a second laser at a fixed frequency ν_2 ionizes it ($h\nu_2$ < I.P., but $h\nu_1 + h\nu_2 \geq$ I.P.). By scanning the wavelength of the exciting photon and detecting the resulting ion current, one obtains the complete excitation spectrum of the molecule. If the energy of the first photon is larger than one-half of the ionization energy ($h\nu_1 \geq IP/2$), the same experiment can be performed with just one tunable laser (one-colour experiment). The experimental parameters of a one-colour experiment, however, are more difficult to control, as described later.

Since the excitation step is resonant, the related cross-section is rather large ($\sigma \sim 10^{-18}$ cm^2), and low laser irradiances are needed. The intermediate state, however, can be de-excited by fluorescence or stimulated emission. Fluorescence can lead to different vibrational ground-state levels, which cannot necessarily be re-excited again (optical pumping). Stimulated emission, on the other hand, represents the limit of excitation, in the sense that no more than 50% of the molecules can be excited simultaneously. This equilibrium situation is called "saturation". It occurs for the excitation step even at very low irradiances

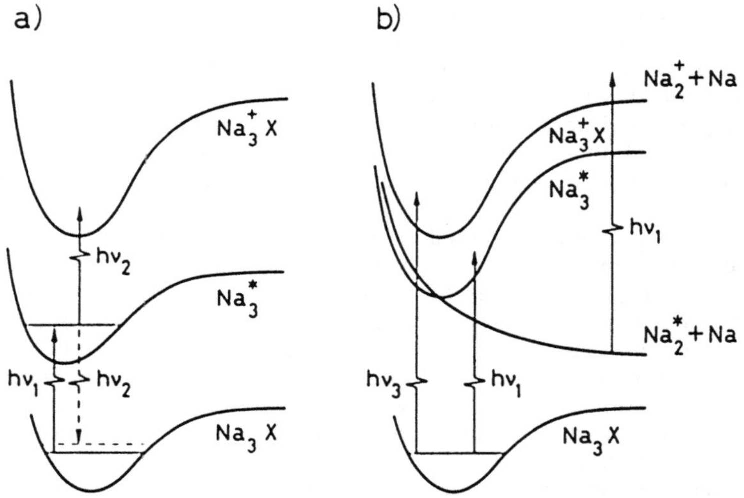

Fig. 5 (a) Two-photon ionization principle: a first laser $h\nu_1$ excites the molecule into the excited state of interest, while a second laser $h\nu_2$ ($h\nu_2 + h\nu_1$ > I.P) ionizes the molecule. When $h\nu_2$ is slightly smaller than $h\nu_1$, but still ionizes the molecule, stimulated emission pumping experiments (SEP) can be performed (see text).
(b) In the depletion spectroscopy scheme, a laser $h\nu_3$ continuously ionizes the effective Na_3 population, while another laser $h\nu_1$ depletes this population by exciting Na_3 molecules into the predissociated state of interest.

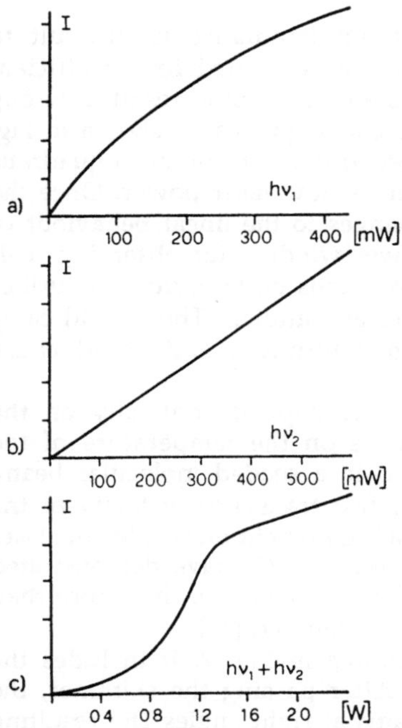

Fig. 6 Power dependency of the two-photon ionization signal: (a) as a function of the exciting laser $h\nu_1$,(b) as a function of the ionizing laser $h\nu_2$, (c) in a one-colour experiment when $h\nu_1 = h\nu_2$.

(Fig. 6a), due to the relatively large cross-section. In order to preserve real transition probabilities, the exciting laser should therefore be attenuated below saturation. In other words: the relation between ion signal and laser irradiance must remain in the linear regime.

The efficiency of the TPI process is strongly dependent on the interaction time between the ionizing laser and the excited particles, which is limited to approximately 10^{-8} s by the lifetime of the intermediate state. This short time window can be compensated to a certain extent by the laser power. However, since the ionization cross-section is rather small [27] , saturation cannot be achieved practically with cw lasers (Fig. 6b) [28] . The solution to the problem consists of using pulsed lasers, which deliver at high peak powers a pulse duration that corresponds to the lifetime of the intermediate state. The optimum power density for TPI spectroscopy is located around 100 kW/cm^2. Beyond 1 MW/cm^2, non-linear processes may occur, which drastically reduce the quality of the spectra obtained [29]. A typical example of the phenomenon is shown in Fig. 2: The upper spectrum was recorded with two cw lasers, and the A-state of Na_3 (around 675 nm) appears much more intense than the B-state (around 625 nm). The lower spectrum, recorded with two pulsed lasers, exhibits almost identical intensities for both states. This was understood later from the lifetime measurements of these excited states: the lifetime of the A-state (60 ns) is much longer than that of the B-state (14 nS) [30].

In one-colour experiments an optimum irradiance is difficult to determine, because the high power density, which is needed for an efficient ionizing step, should not saturate the excitation step. This would result in wrong line intensities. The power dependence of a one-colour process is shown in Fig. 6c. Since two photons from the same laser are needed to create an ion, a quadratic behaviour is observed in the unsaturated regime at low laser power. Once the excitation step saturates, the power function changes to the linear behaviour of the unsaturated ionization step. The usable power window for observing real-intensity spectra at a decent signal-to-noise ratio is thus quite narrow. A further complication can arise when highly excited states are studied. This would carry the total energy $2h\nu_1$ far into the ionization continuum and could cause fragmentation.

The sensitivity of the total TPI process depends critically on the molecular ground state population, and therefore on the temperature of the particles. The optimum conditions are produced in a seeded molecular beam, allowing temperatures as low as 7 K [14]. The fast transverse velocity of the particle beam, which is about 10^5 cm/s, implies a high repetition rate for the laser in order to irradiate a maximum number of molecules. We have demonstrated by using a high-repetition-rate copper vapour laser (f = 6kHz), that more than 20 % of a particular beam species could be selectively detected [31].

The TPI spectroscopy apparatus is shown in Fig. 7. It includes the supersonic beam source described in Sect. III. After passing the skimmer, the molecular beam interacts with the two laser beams at right angles.The resulting photo-ions are mass filtered with a quadrupole system and detected by an electron multiplier. The ion signal is averaged by a box-car integrator, which is synchronously triggered with the laser. The experiment allows a variable time delay to be introduced between the excitation and ionization steps. As a result, one can measure the lifetime of the intermediate level as a function of the decreasing ion signal. As mentioned before, we used this technique to determine the lifetimes of the excited states of Na_3 [30].

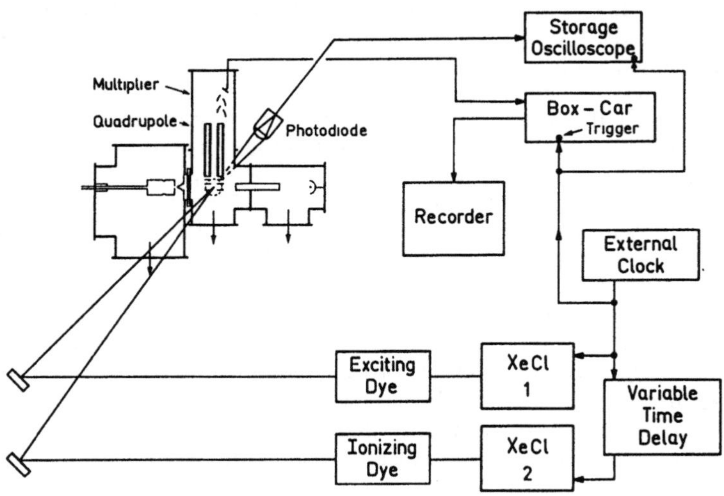

Fig. 7 Experimental setup for performing two-photon ionization spectroscopy on metal aggregates.

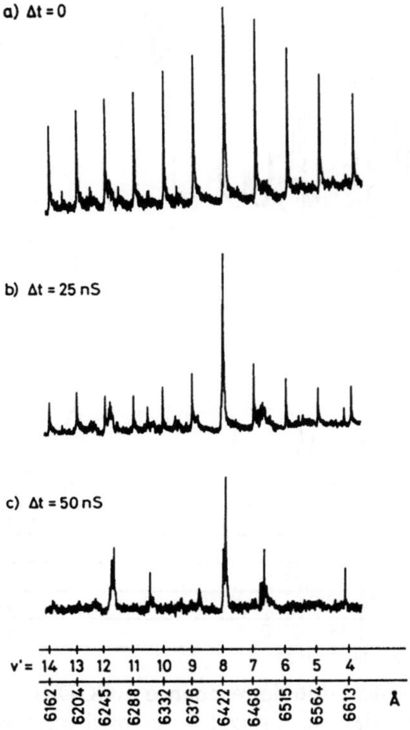

Fig. 8 Two-photon ionization spectra of the Na_2 A-state as a function of the time delay between the exciting and the ionization photon. At long delay times perturbed metastable levels clearly emerge (part c).

The lifetime measurements exhibit another possible perturbation of the TPI process: certain excited states can be crossed by long-living metastable states. This is, for example, the case for the Na_2 A-state, which is crossed by the triplet state $^3\Pi_u$ [32]. Figure 8a shows these triplet-induced perturbations as they appear in the normal TPI spectrum when no time delay is applied. Compared to their normal Franck-Condon envelope some transitions appear too weak, others too strong. By introducing the time delay, the short-lived states are progressively suppressed, and only the long-lived triplet states remain.

Another kind of perturbation can occur in TPI spectroscopy, when the total energy $h\nu_1 + h\nu_2$ reaches slightly above the ionization potential. In the Rydberg approximation, the electronic levels of a molecule are distributed as $R^*/(n-\Delta)^2$, where R^* is the Rydberg constant of the particle, n the principal quantum number, and Δ the quantum defect. With an additional rotational and vibrational energy, rovibrational levels belonging to highly excited Rydberg states can exceed the ionization threshold [33]. Molecules excited into these Rydberg states can autoionize. This will induce large intensity perturbations, because the coupling with a single level is much more favourable than the coupling with the continuum. The phenomenon is well observed in Fig. 9, where pronounced Rydberg series appear above the ionization potential. The same series are also

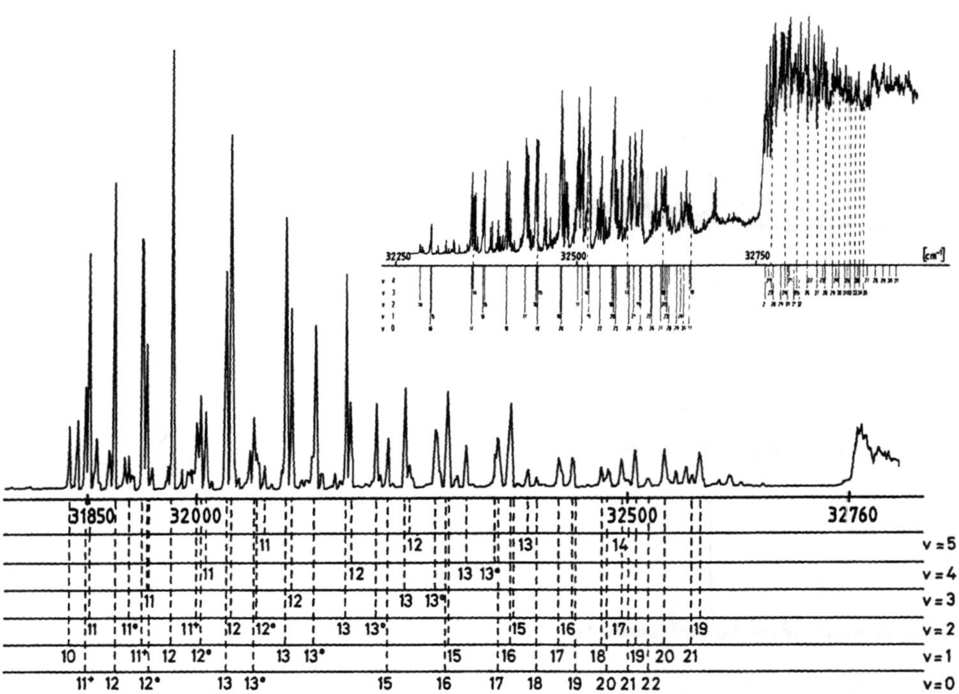

Fig.9 K_2 Rydberg states observed by optical-optical double-resonance (OODR). The levels located below the ionization potential are ionized by a third infrared laser (bottom) or by one of the two lasers used for the OODR (top). Autoionizing Rydberg states located above the ionization potential are generated in a two-photon process (top).

visible below the ionization potential, which, however, requires a three-photon process. In this case, the first photon is fixed on a level of an intermediate state, the second is tuned to the Rydberg level, which is then ionized by the third photon (Fig. 9a). The process is also called optical-optical double resonance (OODR). If the last photon again reaches a Rydberg series located above the threshold, perturbations appear in the original Rydberg spectrum (Fig. 9b). In order to prevent these Rydberg or multiphoton perturbations, the ionizing laser should be the more powerful source, and its wavelength should reach into the centre of the ionic ground state.

A process that competes efficiently with TPI is **fragmentation**. The excitation of a repulsive state may occur during the first or the second step. Figure 10 presents an example of fragmentation during excitation. The K_2 E-state is indeed crossed by a $^3\Sigma_u$ repulsive state, which induces partial dissociation of the dimer into excited and unexcited atoms (K + K*) [34]. The fragment K* can then be re-ionized by the same laser, exhibiting the same harmonic progression of the E-state also on the atomic channel. The comparison of both signal channels (K_2^+ and K^+) then allows the crossing point between the two potential curves to be determined precisely (Fig. 11a). One observes that the fragmentation begins at the vibrational level v'= 4, which means that the crossing point is located slightly

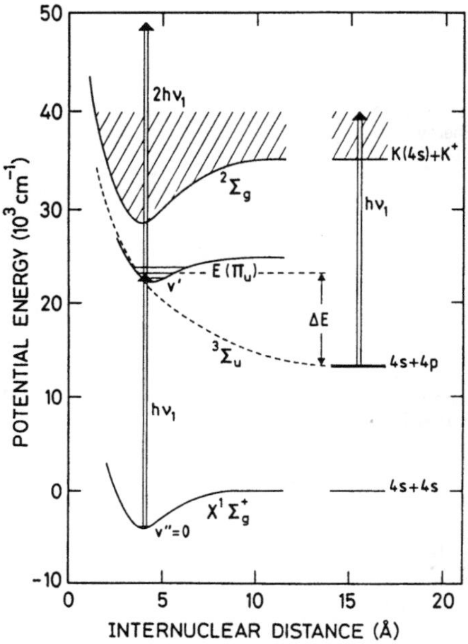

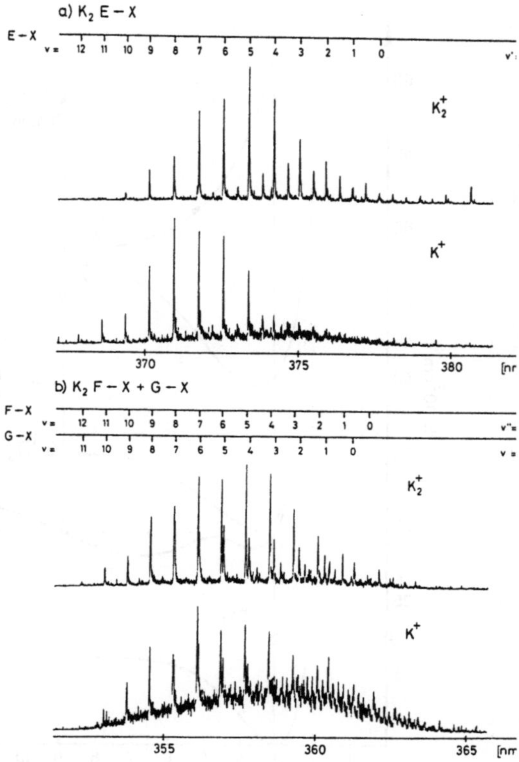

Fig. 10 (a) Two-photon ionization picture of the predissociated K_2 E-state. All the levels above the crossing with the repulsive $^3\Sigma_u$ state ($\upsilon' \geq 4$) are detected on the K^+ mass as well. (b) A similar situation is presented for the K_2 F- and G-states.

Fig. 11 Potential energy diagram of K_2 which illustrates the predissociation that occurs for the E-state, producing the signal on the K^+ channel.

below. In order to determine whether a crossing with a repulsive state occurs, one can either observe the ion signal also on the fragment channels (if re-ionization is energetically possible), or measure the lifetime of the intermediate state. If this lifetime is very short and differs for several vibrational bands, the state can be assumed to be predissociated. Figure 11b exhibits a similar behaviour which occurs for the K_2 E,- and G-states as well.

If the second photon reaches a repulsive state of the ion, a different kind of fragmentation can occur during ionization. The TPI spectrum of the well-known $C^1\Pi_u$-state and the double well (2) $^1\Sigma^+{}_u$ of Na_2 is particularly advantageous for observing this phenomenon [35]: In contrast to the C-state and levels below the apex of the double well, those vibrational levels which are located above the apex exhibit an ionic character with a large interatomic distance, and they converge into a region where a repulsive state of the ion opens up, as shown in the energy-level diagram of Fig. 12. By comparison of the spectra obtained on the Na^+ and the $Na_2{}^+$ channels (Fig. 13), one can observe that only

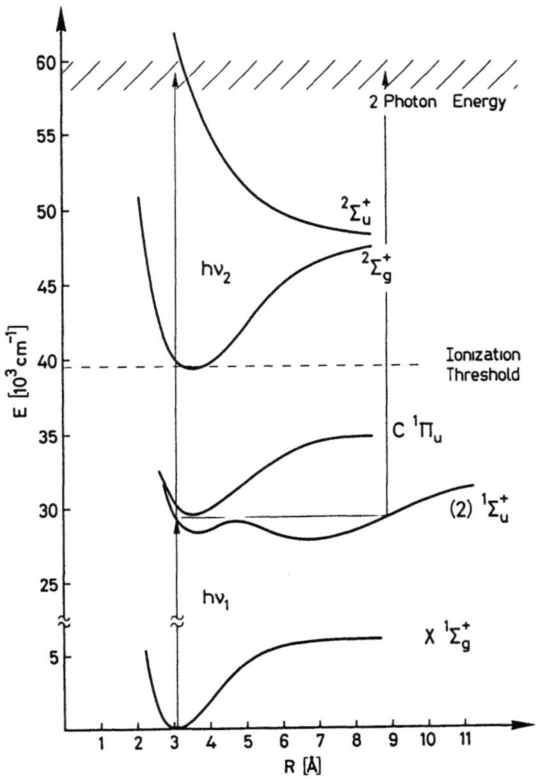

Fig. 12 Potential energy curves of the ground state of Na_2, its observed excited states and the first states of Na_2^+. The hatched area represents the total energy available from the two-photon process.

these levels are dissociated, when ionized. The others converge only to the ionic ground state, and do not appear in the fragment spectrum. In order to prevent this kind of fragmentation pertubation, one must again choose the energy of the ionizing photon very carefully, and again the pertubation can be avoided if the ionization energy is placed around the centre of the ionic potential well.

In order to observe the complete spectra of predissociated states, including the levels which conduct to fragmentation, we developed the **depletion spectroscopy** techniques [36]. In such a configuration (Fig. 5b), the laser with frequency ν_3 permanently ionizes the studied particles and monitors their quantity in the beam. Whenever another laser (frequency ν_1) is tuned on a predissociated transition, the induced dissociation reduces this quantity, resulting in a dip in the ion signal. This type of spectroscopy has been performed with success on Na_3 (Figs. 18 and 20).

Calculations on excited molecular states are difficult to perform, due to the large number of electrons and degrees of freedom. For this reason, most theoretical treatements are concerned only with the electronic ground state, except for Li_3 [37] and some recent approaches to Na_3 [38]. The spectroscopic observation of the ground state is, therefore, of particular interest. An ordinary

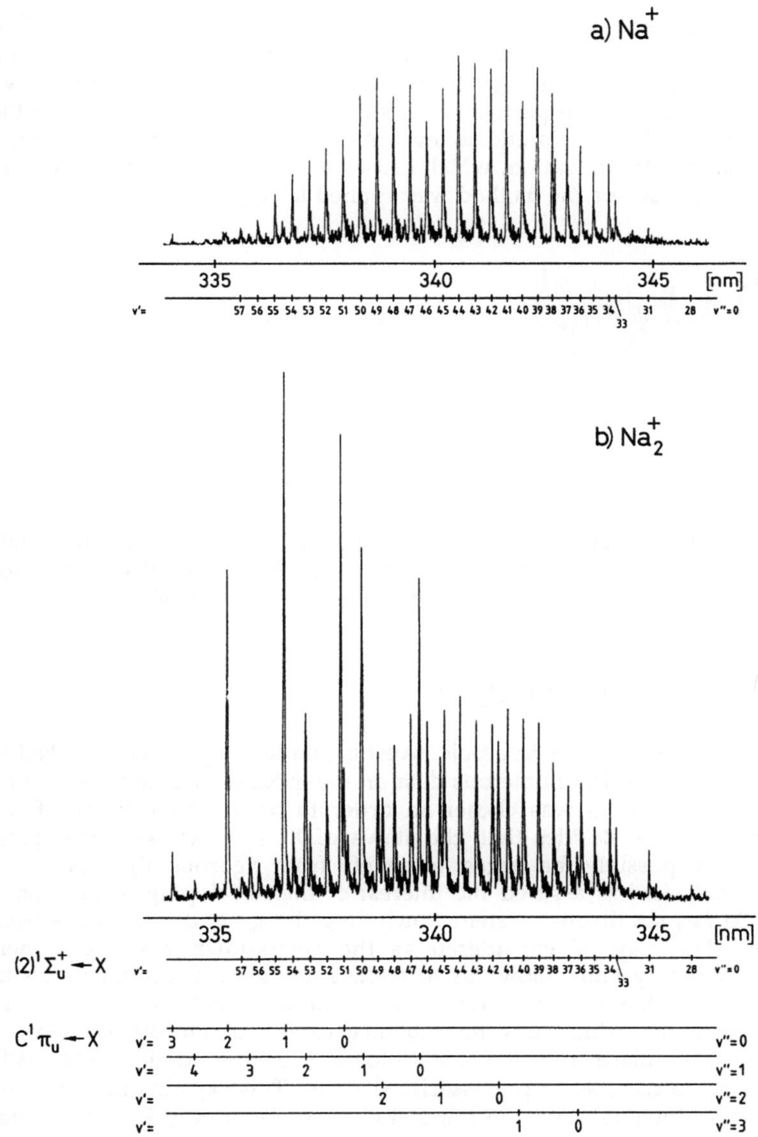

Fig. 13 TPI spectra of Na_2 between 333 nm and 345 nm showing the $C\,^1\Pi_u \leftarrow X\,^1\Sigma_g^+$ and (2) $^1\Sigma_u^+ \leftarrow X\,^1\Sigma_g^+$ transitions recorded on the non-dissociated Na_2^+ channel (b) and the dissociated Na^+ channel (a) [35].

technique is the analysis of hot bands as a function of the particle temperature [13]. This technique, however, only applies to rather low-lying ground-state levels, because a rise of the temperature lowers the sensitivity, the resolution, and introduces dissociation. **Stimulated Emission Pumping** (SEP) is an alternative technique [39,40] which provides much more information on the ground state. The scheme of the experiment is, as seen in Fig. 5a, identical to TPI; however, $h\nu_2$ is chosen quite close to $h\nu_1$. The first photon $h\nu_1$ is set to a transition of a fixed intermediate level, and $h\nu_2$ is scanned, so it can serve two functions: either it ionizes the particle, or it stimulates a transition down to the ground state, into a different vibrational level. As this latter process is resonant, it is more favourable than ionization, which causes a dip in the ion signal. Figure 14 shows an example where the method was applied to Na_3.

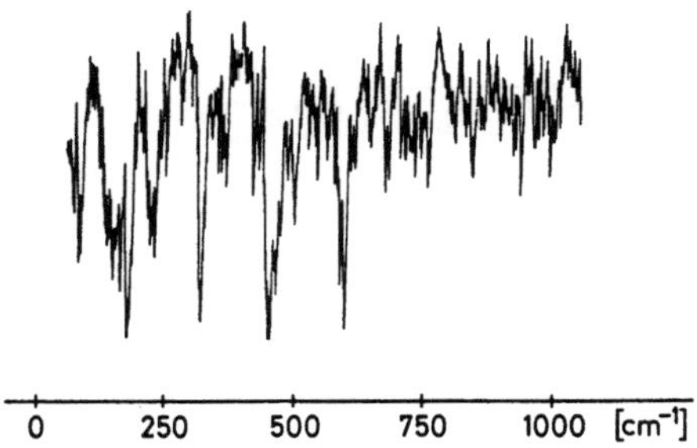

Fig. 14 Stimulated emission pumping (SEP) result of the Na_3 ground state originating from the 0 $(3/2^-)$ level of the C-state. The four main observed bands form a progession in the totally symmetric mode (ω_1 = 139 cm^{-1}) [40].

IV. SPECTROSCOPY OF Na_3 AND Li_3

The use of a wide palette of spectroscopic investigation techniques, as described in Sect. III, allowed us to collect numerous data on the ground state and the excited states of Na_3. This system was chosen in order to prove the validity of the spectroscopic approach. Besides that, the status of X_3 systems as the simplest metal clusters, and possibly the largest to benefit from rigorous high-resolution spectroscopic analysis, has assured the interest of the developing metal-atomic cluster field. So far, six distinct systems, including the ground state, have been observed, and Na_3 may be considered as the best-known metallic trimer. Contemplating these systems allows us to make some general comments about the salient features of the overall picture of Na_3 optical excitations. First, none of the five excited states exhibit any resemblance to any of the others, second, predissociation takes place from the C-state towards higher energies and leads above 22 000 cm^{-1} to completely predissociated states. This explains the observed Na_3^+ appearance potential, and the impossibility of seeing any Rydberg states with our optical-optical double resonance (OODR) photoionization technique.

1. The ground state of Na_3

Conventional hot band analysis of the A-, B- and C-states for different beam temperatures allowed us to determine some of the first vibrational levels of the Na_3 ground state [13]. The obtained vibrational intervals, ω_1 = 139 cm^{-1}, ω_2 = 49.5 cm^{-1} and ω_3 = 87 cm^{-1} correspond well to the vibrational frequencies: ω_1 = symmetric stretch- 142 cm^{-1}, ω_2 = bending mode- 58 cm^{-1}, ω_3 = antisymmetric stretch - 94 cm^{-1}, calculated from harmonic fits [41] of the very anharmonic Na_3

ground-state potential surface [42]. Unfortunately, higher ground-state levels cannot be seen with this technique as explained in Sect. III. They have been investigated by means of Stimulated Emission Pumping (SEP) [40]. It is important to mention that, because of resonant character of the SEP techniques, the observed intensities reflect the Franck-Condon factors between the intermediate state and the ground state, so intermediate-state-dependent information can be obtained. Energetically the C-state is located best to carry out the SEP experiment. Figure 14 shows the spectrum recorded from the second band of this state. Beside the bands already seen with the hot band analysis, a richly structured spectrum is observed up to 900 cm^{-1} above the ground-state origin band (Fig. 14). The most puzzling observation is the almost harmonic progression of nearly equally spaced bands at a distance of 139 cm^{-1}, which corresponds to the totally symmetric mode (Fig. 14). The series is accompanied by another weaker progression, which corresponds to the combination of $\omega_2 + n\omega_1$, a superposition of the bending and totally symmetric modes.

2. <u>The 675 nm system (A-state)</u>

The lowest excited state, first observed by Herrmann et al. [12] at a much higher temperature, is characterized by a short vibrational progression (Fig. 15). This might be explained either by predissociation processes, which could take place at higher vibrational levels, or by transition probabilities (Franck-Condon factors) between the A-state and the ground state. The recently measured A-state lifetime of 60 ns [30] and the lack of any substantial signal in depletion spectroscopy indicate that this state is not predissociated. Therefore, one can conclude that the shortness of the progression is due to the Franck-Condon factors only. Consequently the geometry of the A-state must be very similar to the ground state. This observation was recently confirmed by the theoretical CI interaction determination of the potential surface of the lowest states of Na_3 by Cocchini et al. [38]. Experimentally, the localisation of the molecule in a distorted shape (high pseudorotation barrier) is clearly evidenced by the first two observed vibrational intervals of 128.5 cm^{-1} and 47 cm^{-1}, which compare well with the corresponding

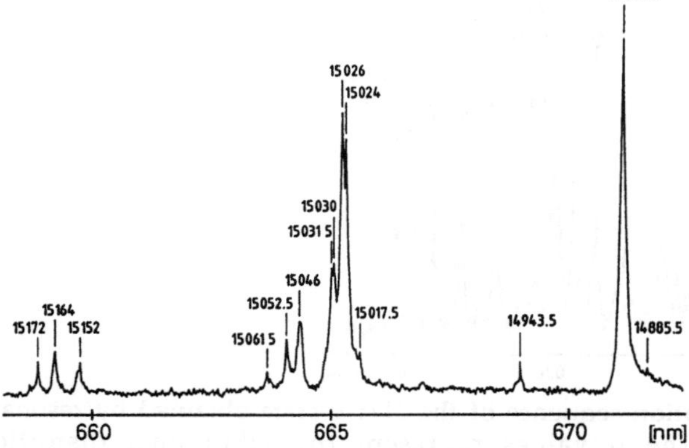

Fig. 15 Resonant two-photon ionization spectrum of the Na_3 A-state.

harmonic frequencies observed for the ground state [13,41]. A complete assignment of this state, however, has not been achieved, but its symmetry might be E' according to [38].

3. The 510 nm system (B-state)

The very characteristic progression observed for the B-system (Fig. 16) is the fingerprint of a strong Jahn-Teller effect, which is accompanied by a nearly free pseudorotation around the conical intersection. This result was the first Na_3 state to be interpreted with a full assignment of the twenty lowest vibrational levels [43]. The main feature of this band system is a long progression composed of nearly equally spaced bands ($\omega_0 \approx 128$ cm^{-1}), which appear to split into doublets and a series of closely spaced bands fanning out from each doublet of the main progression. This characteristic pattern was first predicted by Longuet-Higgins [44] in the analysis of large Jahn-Teller distortion cases accompanied by free pseudorotation; it is described by the simple formula

$$E(u, j) = (u + 1/2)\,\omega_0 + Aj^2 \quad , \tag{4}$$

where u = 0, 1, 2, ... corresponds to the distortion amplitude and $|j| = 1/2, 3/2, \ldots$ to the internal pseudorotation. However, this description cannot fully reproduce the actual spectrum and a detailed analysis using the dynamical Jahn-Teller model was required. The model consists of a description of the potential energy surfaces in the vicinity of the intersection, and a variational solution for the vibronic quantum states, including exactly the non-adiabatic coupling. The analysis fits the experimental data perfectly and reproduces in particular the splitting of the j = 3/2 level into two levels $3/2^-$ and $3/2^+$ because of the hindering of the free pseudorotation caused by a localization effect. Finally, the occurrence of both $3/2^-$ and $3/2^+$ levels together with the localization observed in the ground state, generally recognized as E' symmetry, allows one to deduce that the B-state symmetry is E' as well. Moreover, the potential surface solution of the

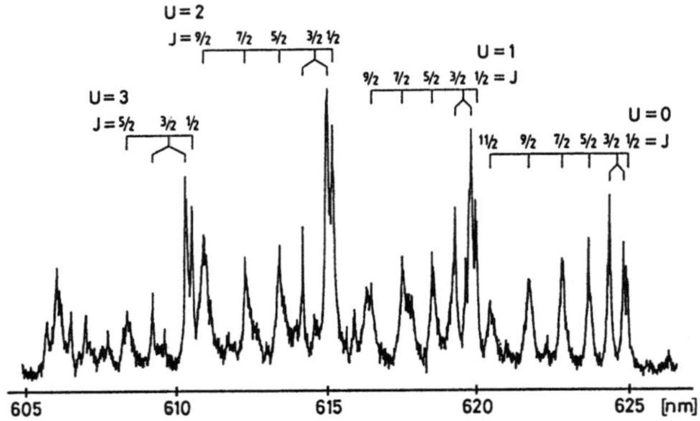

Fig. 16Vibronic pseudorotation sequence of the Na_3 B-state obtained by resonant two-photon ionization. The u values represent the vibrational distortion amplitude and the J values the non-integer pseudorotation quantum numbers [43].

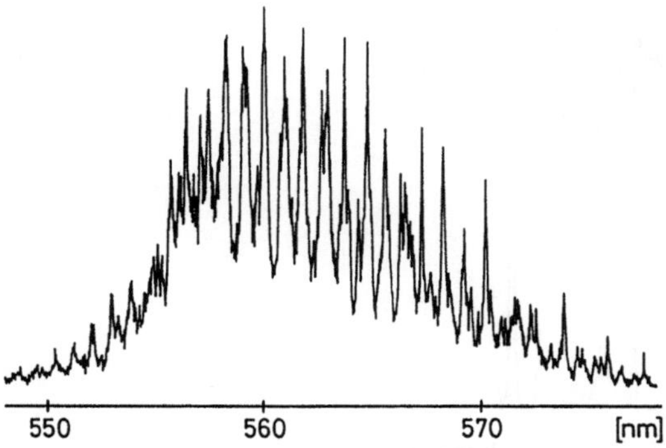

Fig. 17 Resonant two-photon ionization spectrum of the B'-state of Na_3.

dynamic Jahn-Teller problem gives a stabilisation energy of 1050 cm^{-1} and a pseudorotation barrier of 26 cm^{-1} [43]. The corresponding minimum energy structure is an obtuse triangle with an apex angle of 86°. Finally, the 14 ns lifetime of this state is substantially lower than that of the A-state [13].

4. The 560 nm system (B'-state)

The next higher system (B'-state) (Fig. 17) lies about 1 500 cm^{-1} above the B-state. Its very congested vibrational structure exhibits a rather complicated spectrum at a vibrational spacing of 20 to 30 cm^{-1}. The absence of a marked origin-band distinguishes this system from all the other observed systems. Moreover, the slow transition between the B- and B'-systems strongly suggests that the observed B'-system corresponds to transitions towards the upper Born-Oppenheimer surface associated with the split Jahn-Teller $^2E'$-state, of which the B-state is the lower surface. The observed lifetime of this state is 7 ns [13].

5. The 475 nm system (C-state)

The excitation spectrum of the C-state as seen by TPI spectroscopy (Fig. 18 bottom) shows another short progession. The photofragmentation spectrum, recently recorded by depletion of Na_3 from the beam [45], revealed however, that a long progression of bands had been invisible to the TPI method (Fig. 18 top). It appears, therefore, that ultrafast fragmentation occurs for all levels located more than 400 cm^{-1} above the origin-band. This was confirmed by the lifetime measurements, which give 7 ns for the first two levels, while the lifetime of the higher levels is substantially lower and can no longer be measured with our technique [13]. The vibronic structure of the C-state is complex at lower energies, and can only be explained by consideration of the dynamic Jahn-Teller effect. This is particulary true for the level that is located as low as 33 cm^{-1} above the origin-band, followed by another one, which lies 32 cm^{-1} above and appears only at very high laser fluence. Another characteristic feature of this state is the almost harmonic spacing of the band groups at higher energies.

A dynamic Jahn-Teller analysis [45] analogous to that carried out for the B-state demonstrates that the C-state has the electronic symmetry E'', and a total

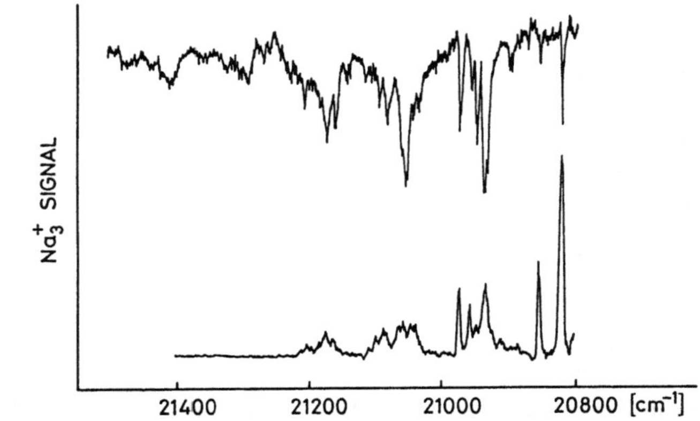

Fig. 18 C-state of Na_3 seen by resonant two-photon ionization (bottom) and by depletion of the Na_3^+ signal (top). The evidence of predissociation is well demonstrated by comparison of the two spectra [45].

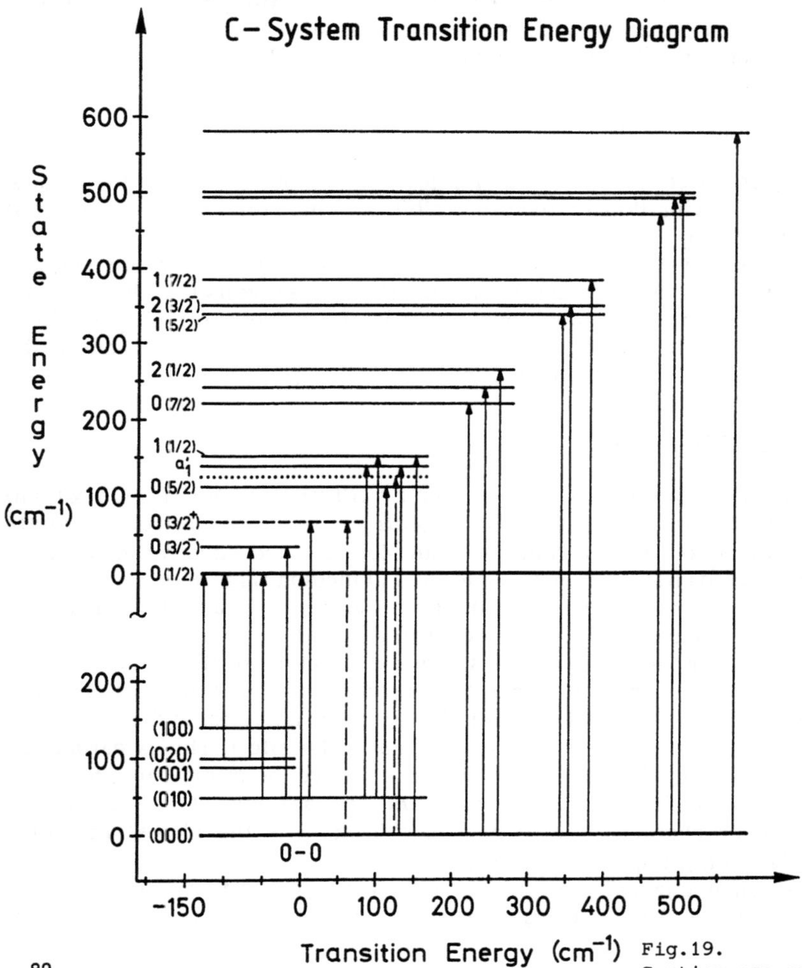

Fig.19.
Caption see opposite page

stabilization energy of 180 cm^{-1}. Indeed the first excitation to j = 3/2 is only accidentally degenerate in the Longuet-Higgins description, and it is expected to be split because of localization effects. Experimentally the 33 cm^{-1} band corresponds to $3/2^-$ excitation and the forbidden 65 cm^{-1} to $3/2^+$ (Fig. 19). The related minimum energy structure is an obtuse isoceles triangle of 65° apex angle, but the pseudorotation barrier of 40 cm^{-1} is small. Cocchini et al. [38] also predict an E'' state in that region, but the stabilization energy and the pseudorotation barrier they found seems too large according to the observed spectrum.

6. The predissociated 420 nm system (D-state)

TPI investigations from 450 nm up to 335 nm exhibited no further systems. During the systematic investigation, however, the Na and Na_2 fragmentation channels were always recorded as well. The 420 nm system observed on the Na_2 channel (Fig. 20 bottom) was rapidly suspected to originate from higher mass clusters, because of its symmetric band shapes and the lack of any predicted Na_2 excited state in the same region [46]. Consequently a depletion spectroscopy of the next higher masses (Na_3, Na_4, Na_5) was performed [46].

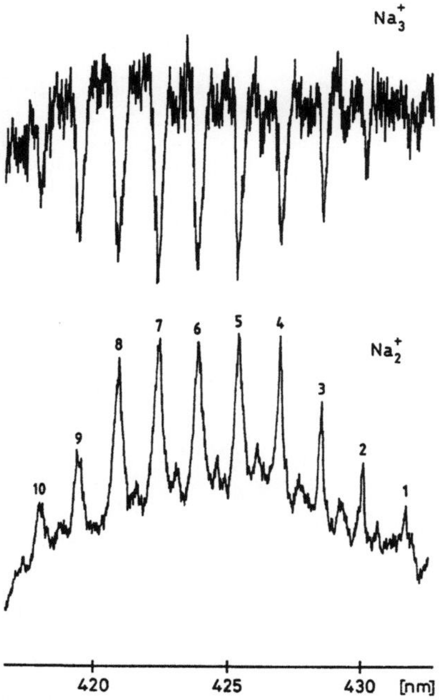

Fig.20 Simultaneous representation of the Na_3 depletion spectrum (top) and the Na_2 fragment spectrum (bottom) of the Na_3 D-state.

Fig. 19 C-state transition energy diagram showing the assigned levels of the C-state. The 0 ($3/2^+$) level appears as a dashed line, because it corresponds to a forbidden transition. Ground-state levels deduced from hot band analysis are also labelled (see text).

The use of the above-mentioned depletion technique allowed us to identify Na_3 as the parent cluster (Fig. 20 top). The vibrational structure of this last-observed Na_3 system is significantly different from the others. It has a harmonic band spacing, no recognizable origin-band and a symmetric band shape. The very small transition factors for the origin-band and the lowest level of the ground state indicate that the D-state has a substantially different geometry from the other observed states. The question of whether it corresponds to a Rydberg state, whose geometry is close to the equilateral triangle of the Na_3^+ ground state, remains unanswered. It might well be more open, approaching a linear configuration.

7. The Na_3-Ionization potential

TPI also allows the measurement of ionization potentials when the first laser is set on a resonant absorption line while the second laser is tuned across the ionization potential. In the case of sodium and potassium dimers [47,33] very abrupt ionization thresholds were obtained, because of the onset of vibrational autoionization from excited Rydberg states. These results allowed the characterization of the corresponding ion ground state [33,48] by identification of the Rydberg series.

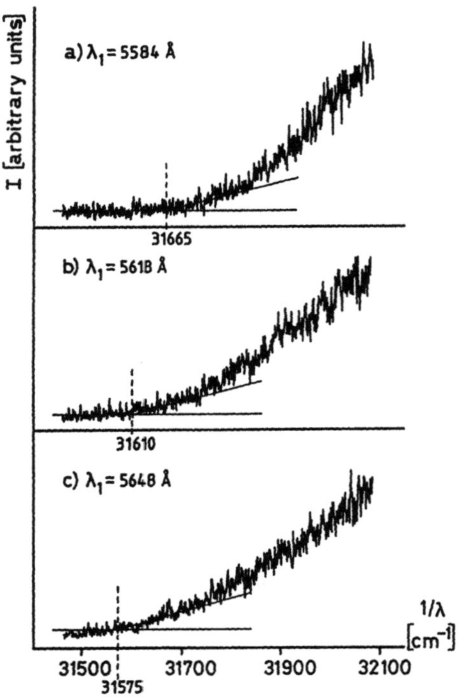

Fig.21 Na_3 appearance potential measurements obtained by resonant two-photon ionization from different intermediate levels of the B'-state.

Unfortunately the Na_3 ionization thresholds obtained by this method (Fig. 21) showed a gradual increase in the photoionization efficency, which indicates that the observed curves correspond to a direct photoionization process [49]. This is because the predissociation processes are obviously faster than the autoionization of Na_3 Rydberg states.

The slightly shifted thresholds measured as a function of the vibrational intermediate level reflect the Franck-Condon factors between the intermediate levels and the Na_3^+ ground-state levels. This explains the difficulty of obtaining precise values of the ionization potential, even when the TPI method is applied. Appearence potentials determined from six different intermediate states of the B- and B'-systems are all scattered between 3.915 eV and 3.935 eV, which is in agreement with the previously measured one-photon ionization thresholds [23,49,50].

8. The first observed excited state of Li_3

The Li_3 molecule, which has a total of only nine electrons, is the simplest metallic cluster. This fact allowed a theoretical ab initio determination of the ground state and the first excited states [37]. Li_3 was until recently [38] the only metallic trimer with ab initio calculated excited states. The experimental approach for Li_3, however, is much more difficult than for Na_3, mainly due to the very high boiling point and chemical agressivity of lithium. As a result, various attempts to observe the excited states of Li_3 have failed until recently.The oven design that we presented in Sect. II allowed us to obtain high enough cluster beam intensities in order to perform TPI spectroscopy on lithium [51]. The first observed excited state of Li_3 is shown in Fig. 22. It is located about 14 500 cm^{-1} (1.8eV) above the ground state and shows a quite short progression. It might correspond to the $^2E'$-state predicted by Gerber [37] at an energy 1000 cm^{-1} below. A vibrational analysis which uses the dynamical Jahn-Teller approximation as applied already for the B- and C-states analysis is in progress [51]. No further electronically excited state in the region 440 - 700 nm has been seen, so far.

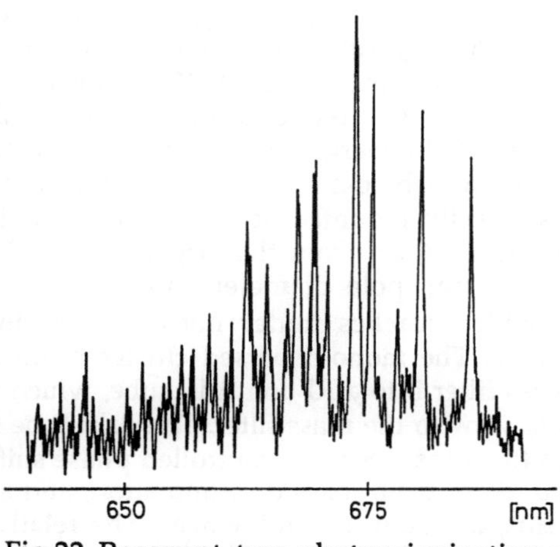

Fig.22 Resonant two-photon ionization spectrum of the 14 500 cm^{-1} Li_3 excited state.

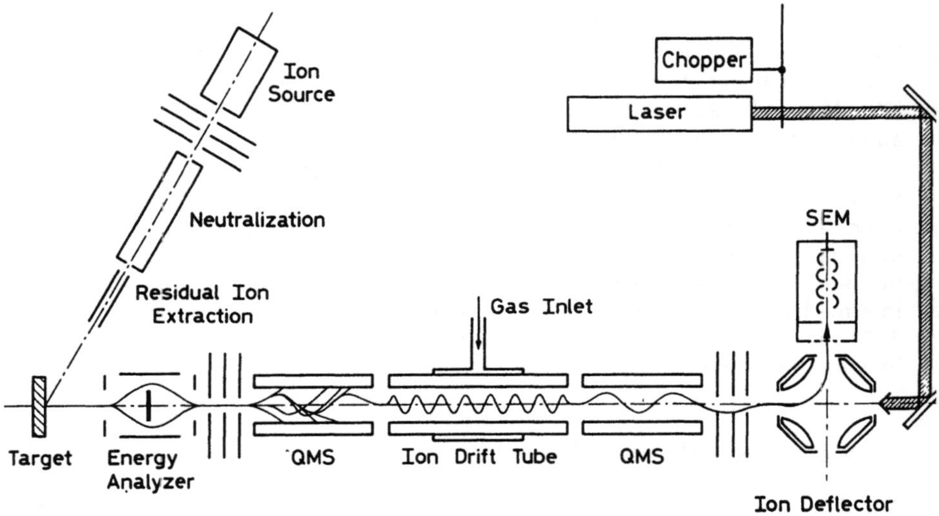

Fig.23 Triple quadrupole arrangement for performing experiments on size-selected metal clusters: The particles are produced by sputtering, energy-analyzed, mass-filtered and introduced into the ion drift tube. All inelastic scattering events are recorded with a subsequent quadru-pole mass analyzer.

V. CHEMICAL REACTIONS OF METAL CLUSTERS

An impressive number of chemical reactions of metal clusters have been performed by injecting reactive gas into the expansion of a laser vaporization source [7]. The experimental arrangement leaves, however, some doubt about fragmentation phenomena, which might occur during the reaction itself or as a result of the subsequent ionization process. For this reason, we have preferred to perform experiments on size-selected cluster ions, which - in a way - is a limitation. The approach, however, allows one clearly to distinguish parent-ion reactions from fragmentation products . The experimental scheme is shown in Fig. 23: The cluster ions are produced, as explained in Sect. II, by means of sputtering. The kinetic energy distribution of sputtered particles is typically between 0 and 10 eV. A cylindrical retarding field energy analyzer cuts out of this range an energy band of ± 2 eV. In addition, it blocks all fast primary ions and sputtered neutrals. The energy filter is directly mounted at the entrance of the quadrupole mass filter at a distance of about 4 cm from the target. We used a commercial quadrupole rod system of 9.5 mm pole diameter and a length of 20 cm (Extranuclear Model 4-162-8), which reaches under normal operating conditons a maximum mass value of 1700. The monodispersed cluster beam is directly guided from the quadrupole mass filter into a RF ion drift tube, which is 60 cm long and has an identical sectional view to the mass filter. The drift tube is operated with the same RF frequency as the mass filter at a controlled phase shift. However, since no dc voltage is applied, there is no mass discrimination, and all ions, reaction products or photofragments, are kept in confinement. The relative electric potential of the rod system allows the confined ions to be slowed down to

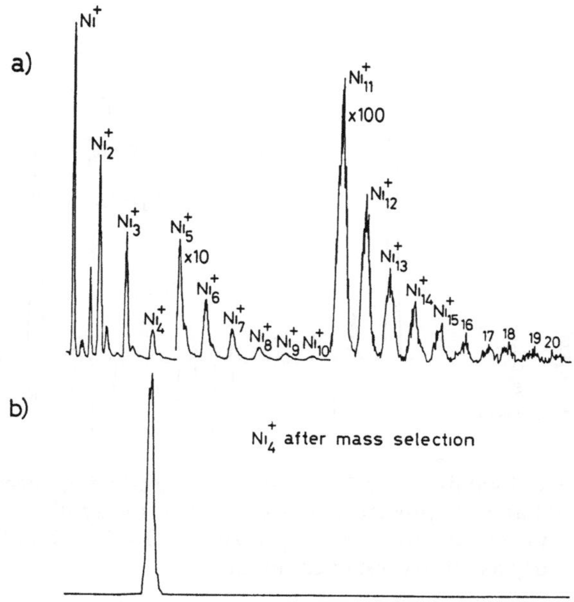

Fig.24 (a) Mass spectrum of sputtered nickel aggregates obtained by setting the first QMS and drift tube on RF only, while the last QMS was scanned. (b) Mass spectrum of a monodispersed cluster beam obtained by setting the first QMS on the mass of Ni_4^+ while the last QMS was scanned.

almost thermal velocities, which correponds to residence times of up to ~10 ms. The drift tube is surrounded by a scatter chamber of 10 cm length, where reactant gas can be introduced. The complete drift tube assembly is placed inside another differentially pumped ultra-high vacuum chamber, pumped by a 500 l/s turbomolecular pump. The background pressure under normal operation conditions is 10^{-8} mbar, which is sufficient to prevent undesired background collisions. It can be operated, however, at pressures up to 10^{-2} mbar, when reactant gas is introduced. At the exit of the drift tube the confined ion beam enters another quadrupole mass filter (Extranuclear Model 4-162-8). It is operated at the same frequency as the first mass filter and the drift tube at a controlled phase shift, and is placed inside a differentially pumped ultra-high-vacuum chamber pumped by a 500 l/s turbomolecular pump. The same chamber also contains a 21-stage Cu-Be secondary electron multiplier for detecting the ions. It is mounted on an electrostatic ion deflector, in order to allow a laser beam to enter the triple quadrupole system in order to interact with the confined particles.

Fig. 24a shows the mass spectrum of sputtered Ni^+ clusters that is obtained when the final quadrupole mass filter is scanned while the first quadrupole mass filter is kept - like the drift tube - in an ion-confinement mode at RF only. When the first quadrupole mass filter is set - by adding a dc voltage - into a discriminative mode as well, only clusters of a well-defined size can enter the drift tube and the final quadrupole mass filter detects only the signal of a monodispersed cluster beam, as shown in the example of Fig. 24b, where the first quadrupole was set to the mass of Ni_4^+. The introduction of a reactant gas into the ion drift tube allows one by this method to perform and trace ion-molecule reactions with cluster ions of any desired size and energy.

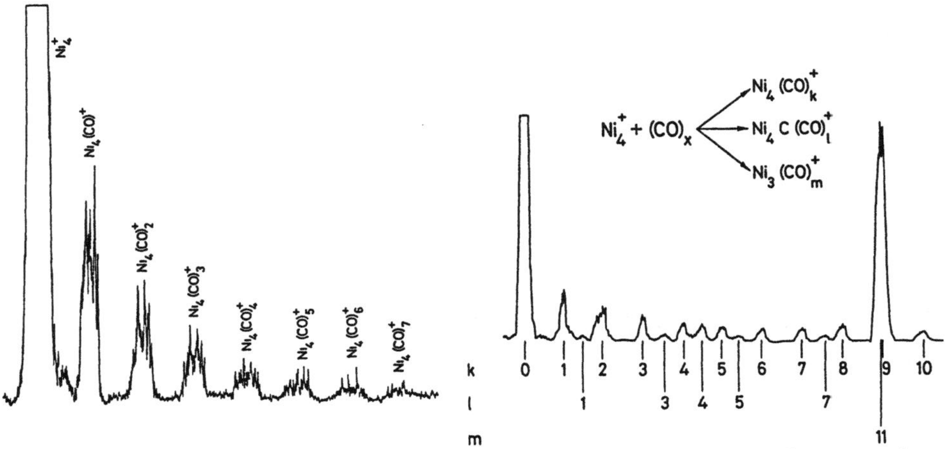

Fig.25 Reactive clustering of CO ligands to a Ni_4^+ cluster on introducing carbon monoxide into the drift tube: (a) at a CO presssure of 3 x 10^{-4} bar, the synthesis of saturated compounds is not yet achieved, (b) at a CO-pressure of 3 x 10^{-3} bar, the system $Ni_4(CO)^+{}_{10}$ appears clearly as the saturated compound.

When we introduced size-selected nickel clusters and carbon monoxide into the drift tube, we were able to perform the controlled synthesis of nickel carbonyls [$Ni_n(CO)_k{}^+$] and nickel carbides [$Ni_nC(CO)_l{}^+$]. A typical example obtained for Ni_4^+ at a kinetic energy of 2.0 eV and a CO pressure of 2 x 10^{-4} mbar is shown in the mass spectrum in Fig. 25a: The Ni_4^+ mother ions rapidly attract one or several CO ligands before leaving the interaction zone. The exponentially decreasing slope of the ligand distribution indicates an unfinished reaction. The pressure of CO was then gradually increased to the point at which the mass spectrum did not change any more; i.e., when saturation of the cluster with carbon monoxide ligands had occurred (pressure of CO: 2.4 x 10^{-3} mbar); this situation is depicted in Fig. 25b: We observe $Ni_4(CO)_{10}{}^{(+)}$ to be the saturated compound, which does not accept any further ligands. Applying the simple 18-electron rule to this situation, the fully occupied s-, p-, d-orbitals accommodate 4 x 18 = 72 electrons. From these, the four nickel atoms provide 40 electrons, the 10 CO ligands contribute another 20 electrons, which leaves 12 electrons to establish the metal skeleton. This corresponds to six metal-metal bonds, which corresponds perfectly with the most compact structure of a tetrahedron, having four terminal and six bridging carbonyl ligands (Fig. 26).

For larger aggregates the rather simple 18-electron rule loses sense, and it is now well-established that individual carbon monoxide ligands in multimetallic complexes rapidly interchange positions intramolecularly. Often, there is no one arrangement of ligands about the central metal core that is greatly preferred energetically over all others [52,53]. Indeed, it is readily apparent that individual ligands interact with the cluster as a whole and not just with individual metal atoms. The close analogies between this situation and the bonding of the so-called electron-deficient boranes, B_nH_m, have been recognized by Wade and by Mingos in a series of seminal papers [54,55], and there now exist simple electron-counting rules that enable us to correlate molecular geometries of metal cluster complexes with the total number of valence electrons [56,57].

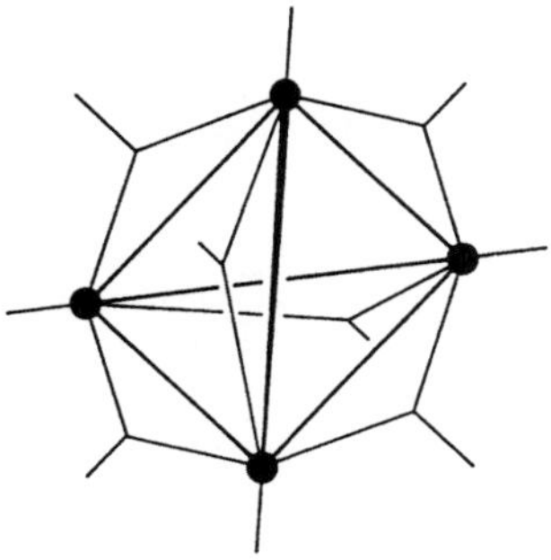

Fig.26 Structure of $Ni_4(CO)_{10}$ exhibiting four terminal and six bridging CO ligands.

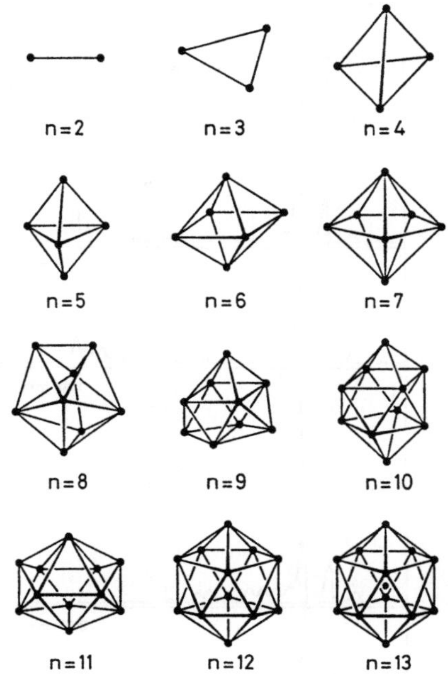

Fig.27 Shapes of the metallic skeleton of the saturated cluster compounds $Ni_n(CO)_m$ as predicted and measured according to Table 1 [61].

These ideas have also been extended to account for some aspects of the reactivity of organometallic clusters [58]. Of particular relevance to the results we have obtained is the work of Lauher [59] who not only calculated the most favourable molecular geometry for any given transition-metal cluster but also predicted the bonding capabilities of such clusters. The original calculations were based on clusters of rhodium atoms, but the results are readily transferable to most other transition metals. Since that time, many of Lauher's predictions have been experimentally verified [60].

The application of Lauher's framework to Ni_n^- carbonyls leads to the cluster geometries shown in Fig. 27 [61]. The corresponding number of ligands is given in Table 1. The predicted values are surprisingly close to results that we observed experimentally. The observation proves the full validity of Lauher's predictions. It shows furthermore that stoichiometrically reasonable results can be obtained with size-selected metal clusters in the proposed triple quadrupole arrangement.

Table 1 Predicted and observed CO bonding capabilities of Ni_n^+ clusters.

Cluster	Ni_2^+	Ni_3^+	Ni_4^+	Ni_5^+	Ni_6^+	Ni_7^+	Ni_8^+	Ni_9^+	Ni_{10}^+	Ni_{11}^+	Ni_{12}^+	Ni_{13}^+
maximum quantity of predicted CO ligands	7	9	10	11	13	15	17	19	21	23	25	20
maximum quantity of observed CO ligands	9	8	10	12	13	15	16	17	18	19	20	20*

*Very small peak appearing for 21 and 22

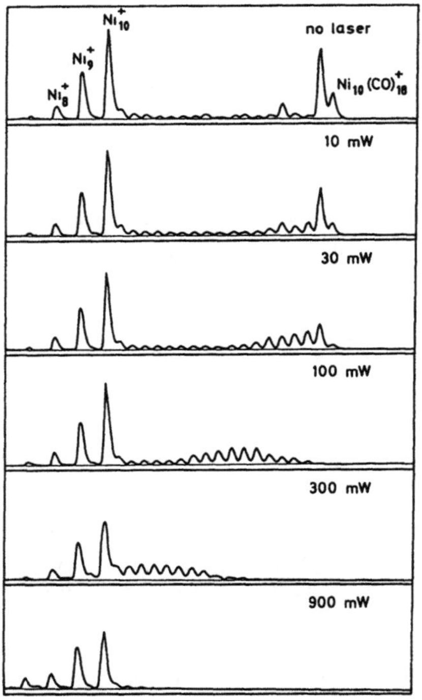

Fig.28 Progressive stripping of CO ligands from $Ni_4(CO)_m^+$ by means of photoexcitation of the metallic core (λ = 488nm) [62].

VI. PHOTODYNAMICS AND FUTURE PERSPECTIVES

Whenever a child has built something with his box of bricks, some time later he likes to destroy it. We were also tempted to destroy the nickel carbonyls, after having synthesized them in the ion drift tube (Sect. V): We passed a laser beam along the axis of the triple quadrupole and observed the $Ni_n (CO)_m^+$ signal as a function of laser wavelength and fluence. No changes were observed, neither when a 50 W CO_2 laser beam was applied (λ = 10.6 μ), nor when a 100 mW frequency-doubled output was used (λ = 5.3 μ), although we assumed this wavelength region to be in resonance with the stretching-mode of the CO ligands. Very surprising results were obtained, however, when visible or UV laser wavelengths were irradiated. An example for the ligand system of $Ni_{10} (CO)_x^+$ is shown in Fig. 28. With no laser irradiation the system accumulates CO ligands until $Ni_{10} (CO)_{18}^+$ is formed. The strongest peak occurs at $Ni_{10} (CO)_{17}^+$, which is due to the missing electron. When 10 mW laser power of the 488 nm line of an argon laser was applied, the mass peaks of the saturated complexes began to lose intensity. The phenomenon gradually increased at 30 mW, and at 100 mW no signal appeared for $Ni_{10} (CO)_{17}^+$ and $Ni_{10} (CO)_{18}^+$. Further ligands were stripped at 300 mW, and at a laser power of 900 mW no more ligands were seen and even the mother ion Ni_{10}^+ exhibited an increased fragmentation tendency towards Ni_9^+, Ni_8^+ and Ni_7^+. The result shows that very efficient photochemical reaction channels are opened. They are activated by an electronic excitation of the metallic core. This process is followed by a very rapid internal energy conversion into ligand vibrations, which successively shakes them off. The line shape of the excitation band indicates that this phenomenon takes place in about 10^{-14} s. Time-resolved measurements of the observed effects and a detailed analysis are given elsewhere [62]. A complication of the experiment was naturally the fact that the ion drift tube served two purposes: the synthesis and photodissociation of the attached CO ligands. It is obvious that future experiments should be carried out in a quintuple quadrupole: The first quadrupole filters out the mother ion for the metallic core, the second quadrupole (drift tube) serves as the ion-molecule reaction chamber, the third quadrupole filters a particular ligand system out, the fourth quadrupole (drift tube) serves to strip off the attached ligands, and the fifth quadrupole allows observation of the photon-induced dissociation events.

A system like the metal-carbonyls is of great interest for applications in catalysis, and the large electronic excitation cross-section of the metallic core raises hopes of photon-induced catalysis. Experiments in the gas phase are sensitivity limited, because the number of three-body collisions in the above-mentioned experimental scheme is low. This suggests depositing the size-selected particles on a surface. A classical system where metal clusters have extraordinary catalytic properties is the photographic system. We deposited therefore, as shown in Fig. 29, size-selected silver clusters on photographic plates [63]. The gelatin protection layer had been removed before the experiment. In addition, the plates were covered with an ITO layer, to prevent charging up during irradiation. The radiation dose was chosen such that about 10 clusters were deposited on each AgBr grain. Half of the irradiated area was covered with a conductive glass plate to compensate for a spurious light background which could come from the primary ion source. The irradiated plates were then taken out of the vacuum and developed by using a conventional developer. The results we obtained for

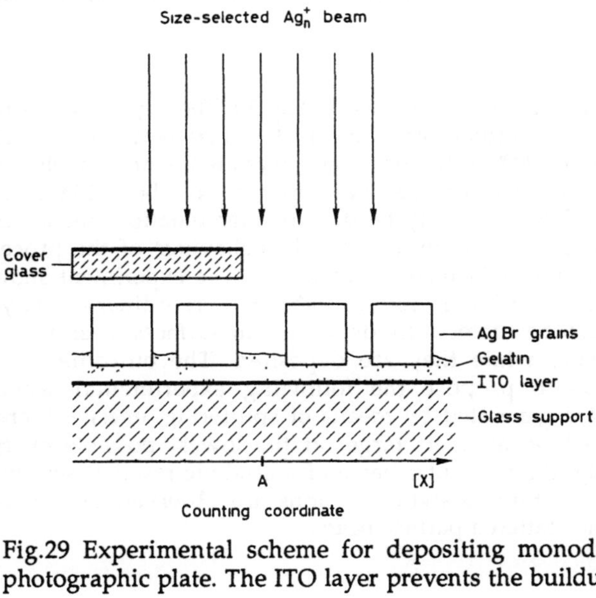

Fig.29 Experimental scheme for depositing monodisperse silver clusters on a photographic plate. The ITO layer prevents the buildup of repulsive charge.

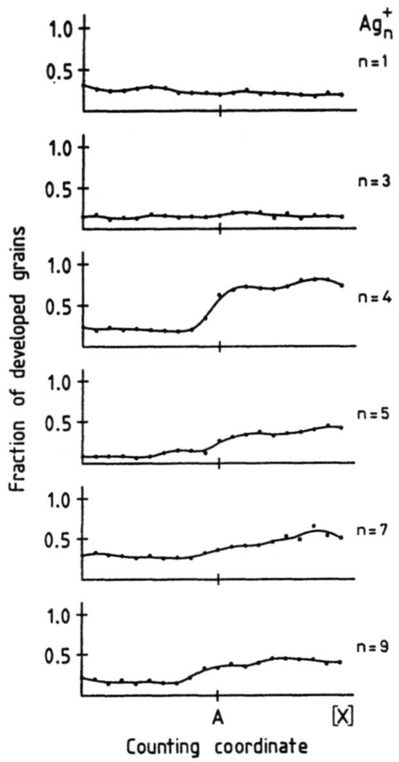

Fig.30 Fraction of developed grains vs. counting coordinate X. The result proves that Ag_n^+-clusters [$n \geq 4$] deposited on a photographic plate create latent image specks [63].

irradiation with different cluster sizes are shown in Fig. 30: The deposited clusters cause no change on the photographic plate for cluster sizes between Ag^+ and Ag_3^+. At Ag_4^+, however, over 90% of the irradiated crystals became developable within the cluster-treated area. An increased developability of about 40% is also observed at larger cluster sizes. This may be due to fragmentation phenomena during the deposition process or due to a threshold behaviour of the process, which at this point cannot be conclusively decided [63]. The experiment shows, however, that clusters of individual size do exhibit extraordinary catalytic properties, and that they can be successfully deposited on surfaces, even if some impact fragmentation occurs while they are deposited. The outcome of the experiment encourages us to propose the following experimental scheme (Fig. 31): Metal clusters are generated by means of sputtering, energy-filtered, mass-filtered and deposited on an inert target at the lowest possible energy (E_{kin}<2 eV). Simultaneously, the molecular beam of a catalytic reaction system is scattered on the target. Resulting catalytic reactions are observed by a mass-spectrometric analysis of the scattered particle beam.

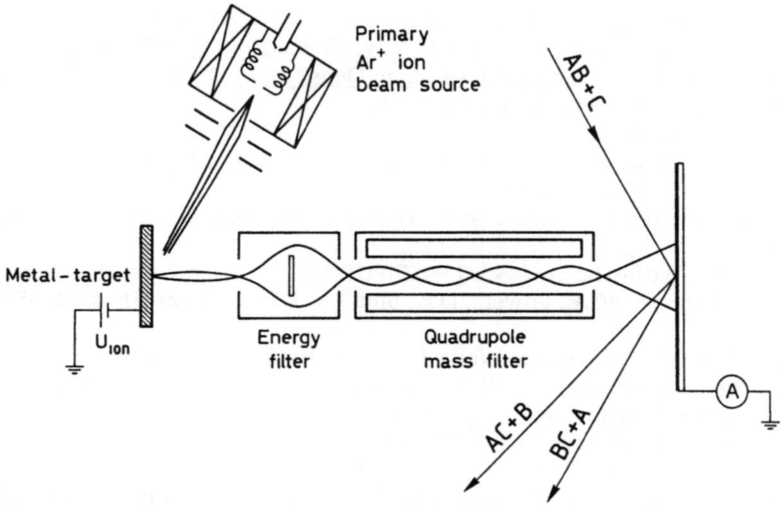

Fig.31 Experimental setup proposed for studying the catalytic properties of metal clusters of a specific size.

VII. ACKNOWLEDGEMENTS

The above experiments are the result of very close and stimulating collaborations with other research groups, and we would like to express our gratitude to several colleagues: M. Broyer and P. Labastie performed most of the spectroscopy experiments together with us; R.L. Whetten contributed fundamentally to the understanding of the Na_3 excited states; R. Keller provided the CORDIS design, M. McGlinchey established the organometallic understanding for the carbonyl systems; C. and Ph. Bréchignac and W. Saunders participated in the ligand-stripping experiment; E. Moisar, B. Pischel, G. Hegenbart and F. Granzer performed together with us the photographic experiment, and for one year we enjoyed the presence of Ni Guoquan in our laboratory. In particular, however, we do not want to forget that the idea of the cold molecular beam source was given to us by the late Gil Stein.

VIII. REFERENCES

1. Moskovits, M., Ozin, G.A.: In Cryochemistry (Wiley-Interscience, New York 1976)
2. Knight, W., Clemenger, K., de Heer, W., Saunders, W., Chou, M., Cohen, M.L.: Phys. Rev. Lett. 52, 2141 (1984)
3. Ekardt, W.: Surf. Sci. 152, 180 (1985)
4. Bréchignac, C., Broyer, M., Cahuzac, Ph., Delacrétaz, G., Labastie, P., Wolf, J.P., Wöste, L.: Phys. Rev. Lett. (in press 1988)
5. Rademann, K., Kaiser, B., Even, U., Hensel, F.: Phys. Rev. Lett. 59, 2319 (1987)
6. Tomanek, D., Mukherjie, S., Bennemann, K.H.: Phys. Rev. B 28, 665 (1983)
7. Whetten, R.L., Cox, D.M., Trevor, D.J., Kaldor, A.: Phys. Rev. Lett. 54, 1494 (1985)

8. Ramsey, N.F.: Molecular Beams (Oxford Univ. Press, Oxford 1956)
9. Anderson, J.B., Andres, R.P., Fenn, J.B.: Adv. Chem. Phys. 1, 275 (1966)
10. Hagena, O.F.: In Molecular Beams and Low Density Gas Dynamics ed. by P.P. Wegener (Marcel Dekker, New York 1974)
11. Labastie, P.: to be published
12. Herrmann, A., Hoffmann, M., Leutwyler, S., Schumacher, E., Wöste, L.: Chem. Phys. Lett. 62, 216 (1979)
13. Broyer, M., Delacrétaz, G., Labastie, P., Wolf, J.P., Wöste, L.: J. Phys. Chem. 91, 2626 (1987)
14. Delacrétaz, G., Wöste, L.: Surf. Sci. 156, 770 (1985)
15. Dietz, T.G., Duncan, M.A., Powers, D.E., Smalley, R.E.: J. Chem. Phys. 74, 6511 (1981)
16. Sudraud, P., Colliex, C., van de Walle, J.: J. de Phys. 40, L 107 (1979)
17. Keller, R.: Proc. Workshop on High-Current, High-Brightness, and High Duty-Factor Ion Injectors, San Diego (1985).
18. Fayet, P., Wöste, L.: Z. Phys. D 3, 177 (1986)
19. Fayet, P., Wolf, J.P., Wöste, L.: Phys. Rev. B 33, 6792 (1986)
20. Bréchignac, C., Broyer, M., Cahuzac, Ph., Delacrétaz, G., Labastie, P., Wöste, L.: Chem. Phys. Lett. 118, 174 (1985)
21. Bréchignac, C., Broyer, M., Cahuzac, Ph., Delacrétaz, G., Labastie, P., Wöste, L.: Chem. Phys. Lett. 133, 45 (1987)
22. Buck, U., Meyer, H.: Surf. Sci. 156, 275 (1985)
23. Herrmann, A., Schumacher, E., Wöste, L.: J. Chem. Phys. 68, 2327 (1978)
24. Haberland, H., Winterer. M., Worsnop, D.R.: Ber. Bunsenge. Phys. Chem. 88, 207 (1984)
25. Herrmann, A., Leutwyler, S., Schumacher, E., Wöste, L.: Chem. Phys. Lett. 52, 418 (1977)
26. Hurst, G.S., Nayfeh, M.H., Young, J.P.: Appl. Phys. Lett. 30, 229 (1977)
27. Bekov, G.I., Letokhov, V.S.: Appl. Phys. B 30, 1 (1983)
28. Wöste, L.: Laser und Optoelektronik 1, 9 (1983)
29. Delacrétaz, G., Ganière, J.-D., Monot, R., Wöste, L.: Appl. Phys. B 29, 55 (1982)
30. Broyer, M., Delacrétaz, G., Guoquan, N., Wolf, J.P., Wöste, L.: Chem. Phys. Lett. (in press 1988)
31. Delacrétaz, G., Wolf, J.P., Wöste, L., Broyer, M., Chevaleyre, J., Martin, S.: In Laser Spectroscopy VI, ed.by Weber, H.P., Lüthi, W., Springer Ser. Opt. Sci. 40 (Springer, Berlin, Heidelberg, 1983) p. 101
32. Kusch, P., Hessel, M.M.: I. Chem. Phys. 63, 4087 (1975) and references therein
33. Broyer, M., Chevaleyre, J., Delacrétaz, G., Martin, S., Wöste, L.: Chem. Phys. Lett. 99, 206 (1983)
34. Bréchignac, C., Cahuzac, Ph.: Chem. Phys. Lett. 112, 20 (1984)
Delacrétaz, G.: Thesis (Lausanne 1985)
35. Delacrétaz, G., Wöste, L.: Chem. Phys. Lett. 120, 342 (1985)
36. Broyer, M., Delacrétaz, G., Labastie, P., Wolf, J.P., Wöste, L.: Phys. Rev. Lett. 57, 1851 (1986)
37. Gerber, W.H., Schumacher, E.: J. Chem. Phys. 69, 1692 (1978)
Gerber, W.H.: Thesis (Bern 1981)
38. Cocchini, F., Upton, T.H., Andreoni, W.: to be published
39. Kittrell, C., Abramson, E., Kinsey, J.L., McDonald, S.A., Reisner, D. E., Field, R.W., Katayama, D.H.: J. Chem. Phys. 75, 2056 (1981)
40. Broyer, M., Delacrétaz, G., Guoquan, N., Wolf, J.P., Wöste, L.: to be published

41. Thompson, T.C., Izmirlian, G., Lemon, S.J., Truhlar, D.G., Mead, C.A.: J. Chem. Phys. 82, 5597 (1985)
42. Martins, J.L., Car, R., Buttet, J.: J. Chem. Phys. 78, 5646 (1983)
43. Delacrétaz, G., Grant, E.R., Whetten, R.L., Wöste, L., Zwanziger, J.W.: Phys. Rev. Lett. 56, 2598 (1986)
44. Longuet-Higgins, H.C.: Adv. Spectrosc. 2, 429 (1961)
45. Broyer, M., Delacrétaz, G., Guoquan, N., Whetten, R.L., Wolf, J.P., Wöste, L.: to be published
46. Jeung, G.: J. Phys. B 16, 4289 (1983)
47. Leutwyler, S., Herrmann, A., Wöste, L., Schumacher, E.: Chem. Phys. 48, 253 (1980)
48. Martin, S., Chevaleyre, J., Valignat, S., Perrot, J.P., Broyer, M.: Chem. Phys. Lett. 87, 235 (1982)
49. Broyer, M., Delacrétaz, G., Chevaleyre, J., Fayet, P., Wöste, L.: Chem. Phys. Lett. 114, 477 (1985)
50. Peterson, K., Das, P.D., Farley, R., Castleman, A.W.,Jr.: Chem. Phys. 80, 1780 (1984)
51. Delacrétaz, G., Wolf, J.P., Wöste, L.: to be published
52. Cotton, F.A., Jamerson, J.D.: J. Am. Chem. Soc. 98, 1273 (1976)
53. Benfield, R.E., Johnson, B.F.G.: J. Chem. Soc. Dalton Trans. 1743 (1980)
54. Wade, K.: Adv. Inorg. Chem. Radiochem. 18, 1 (1976)
55. Mingos, D.M.P.: J. Chem. Soc. Dalton Trans. 133 (1974)
56. Mingos, D.M.P.: Acc. Chem. Res. 17, 311 (1984)
57. Mingos, D.M.P.: Chem. Soc. Rev. 15, 31 (1986)
58. McGlinchey, M.J., Mlekuz, M., Bougeard, P., Sayer, B.G., Marinetti, A., Saillard, J.Y., Jaouen, G.: Can. J. Chem. 61, 1319 (1983)
59. Lauher, J.W.: J. Am. Chem. Soc. 100, 5305 (1978)
60. Johnson, B.F.G.: In Transition Metal Clusters, ed. by B.F.G. Johnson, (Wiley-Interscience, New York 1980)
61. Fayet, P., McGlinchey, M.J., Wöste, L.: J. Am. Chem. Soc. 109, 1733 (1987)
62. Bréchignac, C., Bréchignac, Ph., Fayet, P., Saunders, W.A., Wöste, L.: to be published
63. Fayet, P., Granzer, F., Hegenbart, G., Moisar, E., Pischel, B., Wöste, L.: Phys. Rev. Lett. 55, 3002 (1985)

Techniques for the Electronic and Structural Investigation of Cu Clusters on Graphite

M. De Crescenzi[1], M. Diociaiuti[2], L. Lozzi[2], P. Picozzi[2], and S. Santucci[2]

[1]Dipartimento di Fisica, Università di Roma II,
I-00173 Roma, Italy

[2]Dipartimento di Fisica, Università dell'Aquila,
I-67100 L'Aquila, Italy

1. Introduction

In the last few years the use of surface spectroscopic methods /1/ has greatly contributed to the knowledge of the electronic structure of clusters /2,3/, and the relaxation processes /4,5/ arising when the d-band and the core levels of the metal clusters evolve continuously towards those of the isolated atom. Moreover the surface effects due to the high surface/volume ratio seem to play a fundamental role in the cluster properties. For instance they are responsible for the high chemical reactivity in the oxidation process and in the heterogeneous catalysis /6/, and the lowering of the melting temperature due to the high mobility of the surface atoms /7/.

In this work we show that low energy electrons in the reflection mode can be very suitable to study many of the previous effects because of their low penetration depths. In particular we can obtain the following informations on a) the electronic structure, analyzing the energy losses features (ELS) /8/ close to the elastic peak E_p; b) the local structure in terms of the radial distribution function F(R), from the Fourier analysis of the oscillating structures which extend above the core edges (EELFS) /9/; and c) the changing of the valence band and of the core levels which cause shifts of the

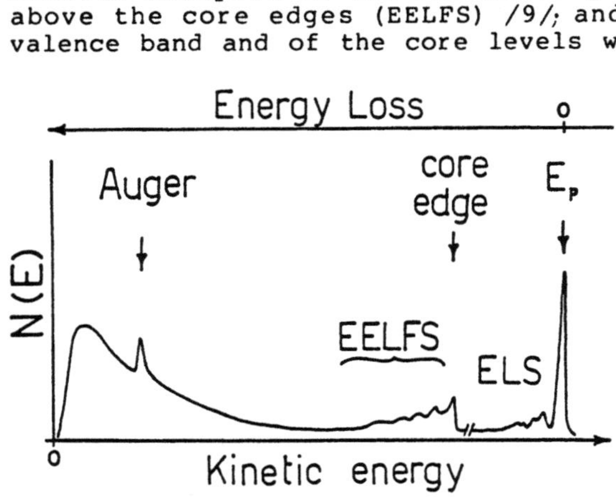

Fig. 1. Schematic picture of the electronic excitation processes occurring in the yield N(E) of the electrons reflected from a solid surface.

Auger peak positions /10/. The above three techniques are schematically shown in Fig. 1, where the electronic yield N(E) of the electrons reflected from a surface is displayed as a function of the kinetic energy, as well as of the energy loss from the elastic peak.

2. Experimental and Discussion

We applied different techniques to study the changes with size which occur in the structures of Cu clusters supported by polycrystalline graphite.
Copper clusters were deposited in UHV (10^{-10} torr) by evaporation from a tungsten wire on both a polycrystalline graphite and a grid for electron microscope. The grids were used to determine the filling factors and the clusters' size distribution by an image analyzer. Fig. 2 shows the distribution of cluster diameters for two typical evaporations, together with the evaluated mean diameter. The UHV chamber was equipped with a Cylindrical Mirror Analyzer (CMA) to detect the N(E) and the dN(E)/dE spectra in the different electron energy region of Fig. 1.

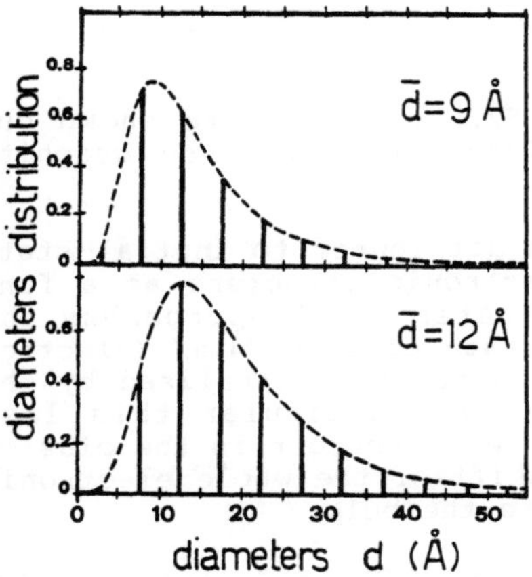

Fig. 2. Cluster size distribution of two granular Cu films deposited on graphite with mean diameters of 9 Å and 12 Å.

2.1 Auger lineshape

A quantitative analysis of the Auger lineshapes requires the knowledge of the peak energy position of the corresponding core levels and valence bands, because the LVV transition is located at a kinetic energy E_{kin} given by /1/:

$$E_{kin}(LVV) = E_c - E(V_1) - E(V_2) - U , \qquad (1)$$

where E_c is the core level energy, E(V) are the binding energies of the two independently created holes in the valence band and U the correlation energy between the two holes. One of the most remarkable effects of the electronic properties of clusters is the continuous shift of the binding energies of core levels and valence band centroid with respect to the Fermi

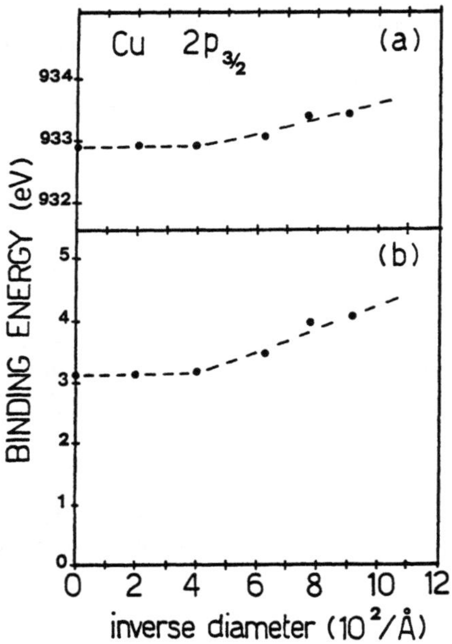

Fig. 3. (a) $2P_{3/2}$ core energies levels and (b) valence band centroid positions as a function of the inverse diameter of the Cu clusters on graphite /10/. The energies are referred to the Fermi level.

level, decreasing cluster size /2-5/. In Fig. 3 are shown our results for Cu clusters with different diameters on graphite /10/.

In the past, these shifts have been attributed to initial-state effects /3/ (variations of the electronic structure as a function of the size of the metal cluster which approaches the atom-like behaviour) and to final-state effects /4,5/ (electrostatic charging of the cluster which is not neutralized by the free electrons of the substrate). In particular this last effect should shift all energies by e^2/r, where r is the cluster radius, resulting in an apparent shift of the whole electronic structure of cluster with respect to the bulk.

An accurate analysis of the Auger lineshapes and energy variations of our Cu clusters shows an important broadening and shift towards low kinetic energies of about 2 eV for the smallest cluster ($\bar{d}$=3 Å).

We have explained /10/ the broadening and the shift of the Cu L_3VV lineshape following Cini approach /11/ and taking into account the measured broad distribution of the cluster diameters and weighting the different classes of diameters. The results are shown in Fig. 4 for the bulk and for 12 Å and 3 Å mean diameters.

The observed shift of the main peak can be understood, on the basis of Eq. 1, by considering shifts towards higher binding energies of both the core level and the valence band centroid. Our Auger analysis supports that the shifts reported in Fig. 3 are genuine and are caused by the electronic changes occurring in the clusters when their diameter decreases.

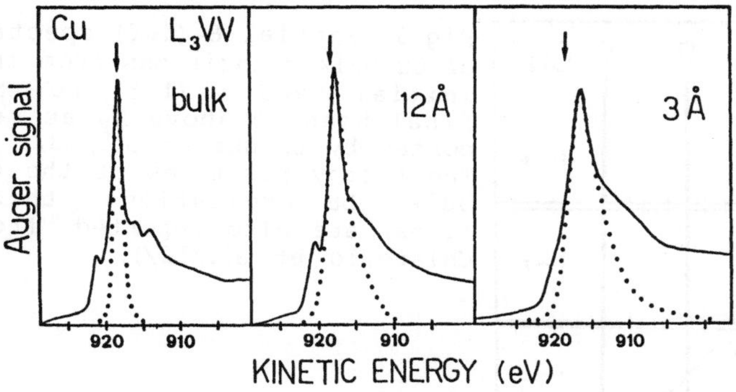

Fig. 4. Comparison between experimental and calculated L_3VV Auger lineshapes for continuous film and for 12 Å and 9 Å mean diameter Cu clusters on graphite. The arrows indicate the energy position of the Auger bulk main peak.

2.2 ELS results

Further support to the previous conclusions comes out also from the energy loss spectra close to the elastic peak reported in Fig. 5.

One can observe for graphite two main features located at about 8 eV and 26 eV. Increasing the Cu coverage, additional structures are observed /8/. For the 30 Å Cu nominal thickness (mean diameter $\bar{d}$=70 Å), these peaks, labelled A,B,C,D and E, coincide with those of the bulk /12/. Each feature observed in the N(E) spectrum is described, according to the dieletric theory, as the imaginary part of $1/\tilde{\varepsilon}$ where $\tilde{\varepsilon} = \varepsilon_1 + i\varepsilon_2$ /1/.

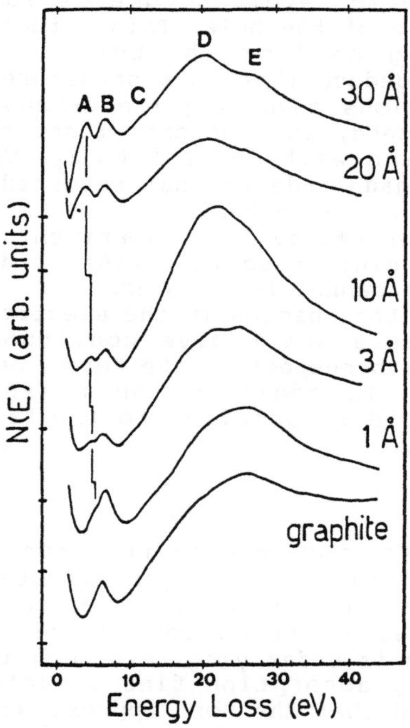

Fig. 5. Low energy loss spectra for clean graphite and different Cu nominal thicknesses. The primary beam energy was E_p=1000 eV.

As a consequence, each peak measured in N(E) should correspond to the minima of the imaginary part of $\tilde{\varepsilon}$ which gives the optical absorption. This close correspondence between energy loss and optics is evident in Fig. 6 for the Cu bulk if one sums the different contributions to the ε_2 of the interband transitions below E_F (levels i) /13/.

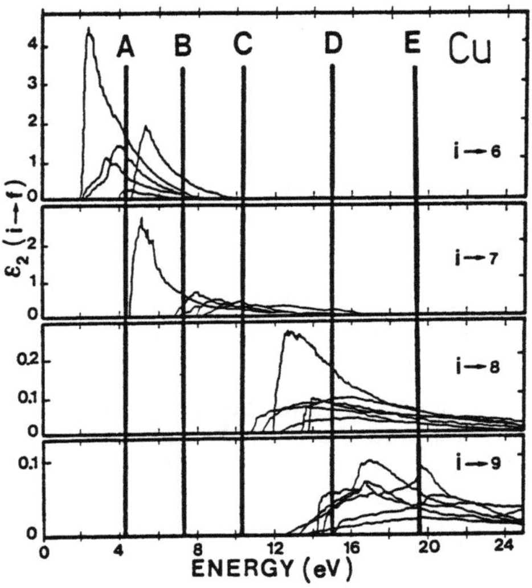

Fig.6. Partial $\varepsilon_2(i \to f)$ spectra of Cu bulk transitions from the initial band, i (1-5), to the final bands f above E_F as reported by Lasser et al./13/. The energy positions of the Cu bulk ELS excitations (thick lines) are also reported (from Chiarello et al./12/).

We note from the results of Fig. 5 an increasing structure A not present in clean graphite, which evolves towards lower energy losses up to the A structure of the bulk. This structure shifts of about 1.5 eV on going on from the bulk to the smallest cluster. From Fig. 6 we deduce that this structure is originated from interband transitions involving the d-band up to E_F. As a result an increased energy loss of this transition indicates a lowering of the d-band with respect to E_F. This lowering is of the same order of magnitude of that reported in Fig. 3 as obtained from valence band XPS data.
It is important to note that in our energy loss experiment the electrons do not escape from the cluster so that final states and electrostatic charging effects cannot be invoked.
In conclusion we can confirm that the changes in the electronic structure of clusters are sizeable and evolve continuosly towards higher binding energies with respect to the Fermi level for a decreasing cluster size. In addition the measured broadening of the lineshape should be ascribed to the broad distribution of cluster diameters.

2.3 Structural results

Structural information on clusters may be obtained from the EELFS spectroscopy /14/. From the study of the fine structures observed above the core levels in the energy loss spectra, it is possible to obtain the radial distribution around the excited atom. The structural analysis underlying this technique is the same of the extended x-ray absorption fine structure (EXAFS) spectroscopy /15/, provided that the Born approximation as well as the dipole selection rules are valid. This analogy is based on the assumption that the oscillating features in the electron yield N(E) may be interpreted as a modulation of the excited core electron final state given by /15/:

$$\chi(K) = \sum_j \frac{A(K)}{K\, R_j^2} N_j\, e^{-2\sigma_j^2 K^2} e^{-2R_j/\lambda} \sin(2KR_j + \Phi_j(K)), \quad (2)$$

where N_j is the coordination number of the backscattering atoms located at a shell of radius R_j around the absorbing atom, K is the wavevector, $A(K)$ is the backscattering amplitude and $\Phi_j(K)$ the phase shift. σ_j^2 describes the mean square displacement of the atoms and λ is the mean free path of the excited electron above the edge.

The Fourier transforms of the EELFS spectra, shown as a function of the interatomic distance, are reported in Fig. 7 for different cluster diameters. The different structures correspond to the distances of the excited atom from the surrounding atoms.

The important result of our measurements is the continuous shift in the peak position of the first neighbours with decreasing cluster size /9/.

We show in Fig. 8 the percent reduction of the nearest neighbours distances as a function of $1/\bar{d}$, in good agreement with the EXAFS results of Apai et al /16/.

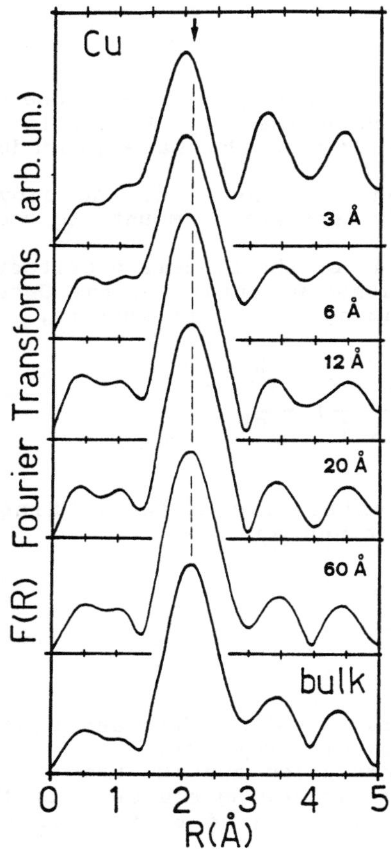

Fig. 7. Fourier transforms of the EELFS spectra of different Cu films deposited on clean graphite /9/. The nominal thickness is indicated. The dashed line and the arrow show the bulk position of the first nearest neighbours.

The observed linear dependence can be interpreted in the framework of the liquid drop model /7/. In this model the contraction ΔR due to the surface tension f is given by

$$\frac{\Delta R}{R} = - \frac{4}{3} \frac{k f}{\bar{d}} , \tag{3}$$

where k is the bulk compressibility, R is the metal bulk distance and $\bar{d}$ is the cluster mean diameter.

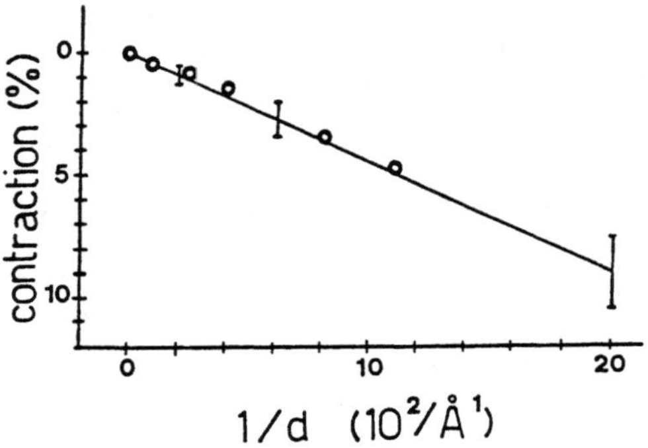

Fig. 8. Plot of percent decrease in nearest neighbour distance vs the reciprocal average diameter $\bar{d}$ of different Cu clusters (open dots) obtained with the EELFS F(R) analysis /9/. The solid line represents the EXAFS measurements of Apai et al. /16/.

From the backfourier filtering of the main peak of F(R) one can obtain information on the evolution of the Debye-Waller factor of the investigated cluster, following the same procedure described by Greegor and Lytle /17/.
The results indicate an important increase of the Debye-Waller factor for the smallest clusters which amounts to about twice that of the bulk.
Recently Solliard /18/, on the basis of the Lindemann criterion for melting point, found a relationship containing the Debye-Waller factors and some characterizing cluster parameters,

$$(4) \qquad \frac{T_m^{cl}}{T_m^{b}} = \left(1 - \frac{4\,k\,f}{3\,\bar{d}}\right) \frac{\beta^{b}}{\beta^{cl}} ,$$

where T_m^{cl} and T_m^{b} are the cluster and bulk melting temperature, respectively. β is defined as

$$(5) \qquad \beta = \frac{4\pi^2 \langle u^2 \rangle}{a^2\, T} ,$$

where a is the lattice parameter, $\langle u^2 \rangle$ is the mean square atomic vibration amplitude and T is the temperature.
A development of the Solliard formulation and the use of our experimental parameters lead to the following equation for the melting temperature of the cluster compared to that of the bulk:

$$(6) \qquad T_m^{cl} = T_m^{b} \left(\frac{a_b}{a_{cl}}\right)^3 \left(\frac{\sigma_b}{\sigma_{cl}}\right)^2 ,$$

where a_{cl} and a_b are the lattice parameters, and σ_{cl} and σ_b the Debye-Waller factors for cluster and bulk respectively /9/.
Our results indicate a sizeable increase in dynamic disorder of surface atoms and, as a consequence, a lowering of the melting point of the small clusters. In Fig. 9 the ratio T_m^{cl}/T_m^{b} is reported as a function of the cluster mean diameters. A very similar behaviour was observed by Buffat and Borel for gold clusters /19/.

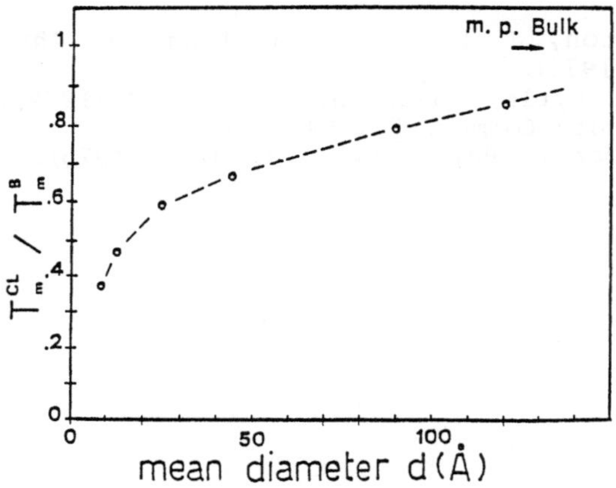

Fig. 9. Ratio between the melting temperature of small Cu clusters and the bulk, according to Ref. 18. The experimental lattice parameters and the mean relative displacements obtained from the EELFS analysis /9/ were used.

Aknowledgments

The authors are grateful to Osvaldo Consorte for his technical assistance.
This work was supported by the Consiglio Nazionale delle Ricerche through the Gruppo Nazionale di Struttura della Materia.

References

1. G. Ertl and J. Kuppers, Low Energy Electrons and Surface Chemistry Analysis 2nd ed. (Weinheim; Deerfield Beach, FL: VCH, 1985).
2. W.F. Egelhoff Jr. and G.G. Tibbets, Phys. Rev. B19, 5028 (1979).
3. M.G. Manson, Phys. Rev. B27, 748 (1983).
4. P.H. Citrin and G.K. Wertheim, Phys. Rev. B27, 3176 (1983).
5. G.K. Wertheim, S.B. Di Cenzo, D.N.E. Buchanan and P.A. Bennet, Solid State Communications 53, 377 (1985).
6. J. Sinfelt, Rev. Mod. Phys. 51, 569 (1979).
7. J.P. Borel, Surface Sci. 106, 1 (1981).
8. M. De Crescenzi, P. Picozzi, S. Santucci, C. Battistoni and G. Mattogno, Solid State Commun. 51, 811 (1984).
9. M. De Crescenzi, M. Diociaiuti, L. Lozzi, P. Picozzi and S. Santucci, Phys. Rev. B35, 5997 (1987).
10. M. De Crescenzi, M. Diociaiuti, L. Lozzi, P. Picozzi, S. Santucci, C. Battistoni and G. Mattogno, Surface Sci. 178, 282 (1986).
11. M. Cini, Solid State Commun., 24, 681 (1987); Phys. Rev. B17, 2788 (1978).
12. G. Chiarello, E. Colavita, M. De Crescenzi and. S. Nannarone, Phys. Rev. B29, 4878 (1984).
13. R. Lasser, N.V. Smith and R.L. Benbow, Phys. Rev. B24, 1895 (1981).
14. M. De Crescenzi and C. Chiarello, J. Phys. C18, 3595 (1985).
15. P.A. Lee, P.H. Citrin, P. Eisenberger and B.M. Kincaid, Rev. Mod. Phys. 53, 671 (1981).

16. G. Apai, J.F. Hamilton, J. Stohr and A. Thomson, Phys. Rev. Lett. 43, 165 (1979).
17. R.B. Greegor and G.W. Lytle, Phys. Rev. B20, 4902 (1979).
18. C. Solliard, Solid State Commun. 51, 947 (1984).
19. Ph. Buffat and J.P. Borel, Phys. Rev. A13, 3287 (1976).

Part III

Structure

Computer Simulation of Clusters

E. Blaisten-Barojas

Instituto de Fisica, Universidad Nacional Autónoma de México, Ap. Postal 20–364, 01000 México D.F., México

This contribution describes briefly the role of computer simulations in understanding physical properties of clusters. Clusters are ideal systems to be studied numerically because of their finite size. The calculations are useful to test potential models as well as to give the temperature and time description of physical phenomena with complete control of all the parameters entering in the experiment. Two computer experiments are presented as examples to explain some particular aspects pertaining to the calculations. The first example describes the structural changes of a Lennard-Jones 13-atom cluster when melting and freezing near a planar, structureless substrate. This system may be considered as a prototype for rare gas clusters physisorbed on a graphite surface. The second example emphasizes the effect of three-body, directional interactions in the description of the most stable structures of covalent clusters . These interactions are good to simulate clusters of semiconducting materials such as silicon and germanium.

1. Introduction

Computer simulations have added a new scope to scientific research in the rapidly growing field of cluster physics and chemistry. In the last few years, numerous novel innovations in experimental cluster measurements led to the discovery of a richness of cluster geometries built up in many cases by structures with a *magic number* of atoms. The possibility to observe clusters in the liquid state has also been raised. Heterogeneous catalysis, nucleation and physisorption, charge instabilities in electron microscopy studies are some of the phenomena that need theoretical support. As is common in scientific investigation, the validity of a comparison between theoretical predictions and experiments may be sometimes questioned because of the complexity of the experimental interpretation as compared to the simplicity of the theoretical model. Furthermore, the testing of a theoretical prediction may be restricted because of limitations in the experimental state-of-the-art. In other areas of physics [1,2], *computer experiments* have alleviated these bottlenecks to the understanding of phenomena, and, hopefully, this presentation will provide the reader with more tools to undertake the study of clusters. In Sec. 2 we give a brief description of numerical statistical mechanics, with emphasis on the two most popular methods, molecular dynamics and Monte Carlo.

We will consider two computer simulations as examples of cluster properties studies. The first example is presented in Sec. 3 and concerns the study of the structure, thermodynamics and phase stability of Lennard-Jones clusters adsorbed on a surface, with special emphasis on the cluster melting regime and surface wetting [3]. This investigation includes molecular dynamics simulations of the solid-like and liquid-like phases of the Lennard-Jones 13-atom cluster, of the melting and desorption process, evidence of surface wetting and mobility of the atoms within the cluster. The second example given in Sec. 4 describes the annealing procedure used to obtain the structural properties of neutral and charged silicon clusters of moderate size [4]. The molecular dynamics simulation used Stillinger-Weber model potential [5] which includes three-body directional interactions. The cluster equilibrium configurations resulting from quenches initiated at finite temperatures and ended at low temperatures established different growth sequences for neutral and charged clusters. The transition from liquid-like clusters to solid-like structures was also tackled. A model of an amorphous five-layer film was constructed using the 14-atom cluster as elemental building block. Sec. 5 closes this work with some comments.

2. Numerical Statistical Mechanics

2.1 Theoretical Concepts

The goal of Statistical Mechanics is to give theoretical prescriptions to evaluate observable macroscopic quantities in terms of the microscopic interactions between the N components of the model system. The *ansatz* consists of identifying the average value of a function of the dynamic variables $\mathbf{x}^N$, $\mathbf{p}^N$, with the observed quantity. At each time t this set of $6N$ variables represents a point in phase space defining an accessible state of the system,

$$\Gamma(t) = \{x_1(t), y_1(t), z_1(t), x_2(t), \ldots, z_N(t), p_{x_1}(t), \ldots, p_{z_N}(t)\}\,. \tag{1}$$

Macroscopically one is never concerned about the $6N$ variables defining the states of the system. For instance, in the laboratory, in order to know if a system is in thermal equilibrium at a temperature T in a volume Ω, we check if T is the same at different times and in different places inside Ω. But microscopically the usual classical definition of temperature is given in terms of the mean total kinetic energy of the system.

$$\frac{3}{2}NkT = \langle E_{\text{kin}}\rangle = \frac{1}{2}\sum_{i=1}^{N} m_i\langle \mathbf{v}_i(t)^2\rangle \tag{2}$$

where m_i and $\mathbf{v}_i(t)$ are the mass and velocity of each particle in the system and k is Boltzmann's constant. The brackets stand for an average over some equilibrium probability distribution function.

Therefore the problem is the calculation of these average values. Following Boltzmann's approach to Statistical Mechanics, a macroscopic quantity (A for example) is proportional to the *time average* over the sequence of instantaneous values $A(t)$ that the corresponding microscopic quantity acquires along the trajectory of $\Gamma(t)$ in phase space:

$$\bar{A} = \lim_{t\to\infty} \frac{1}{t}\int_{t_0}^{t} A(t')\,dt'\,. \tag{3}$$

But we can think that this average value is obtained differently. Suppose that we split the total time evolution from t_0 to t into a million time intervals of length such that the temperature variance is of the same order in all of them. Then, following Gibbs, each one of these intervals can be associated with a representative sample of the system, a replica. These million replicas are taken in an arbitrary order, without causal relation, and can be used to compute *ensemble averages* of the property A,

$$\langle A \rangle = \int A\, P(\mathbf{x}^N, \mathbf{p}^N)\, d\mathbf{x}^N\, d\mathbf{p}^N \,, \tag{4}$$

where $P(\mathbf{x}^N, \mathbf{p}^N)$ is the distribution function in equilibrium.

The two previous theoretical approaches correspond to two different computational methods. The molecular dynamics method generates the system time history by solving Newton's equations of motion for each particle [6]. The classical trajectory of the system in phase space is thus obtained and averages of any property according to Eq. 3 immediately follow. The Metropolis Monte Carlo method [7] produces at random the replicas of the system in equilibrium. Usually the algorithm proceeds by choosing an initial starting configuration $\mathbf{x}^N(t_0)$ with potential energy $V(\mathbf{x}_0^N)$ and making random steps $\Delta\mathbf{x}$. At each step, the change $\Delta V = V(\mathbf{x}^N + \Delta\mathbf{x}) - V(\mathbf{x}^N)$ is evaluated. If ΔV is negative, the step is accepted. If ΔV is positive, the step is accepted with probability $p = \exp[-\Delta V/kT]$. The series of accepted replicas generates a random walk wich explores the phase space, and which for many trials is governed by the probability distribution function

$$P(\mathbf{x}^N) = \frac{e^{-V(\mathbf{x}^N)/kT}}{\int d\mathbf{x}'^N e^{-V(\mathbf{x}'^N)/kT}} \,, \tag{5}$$

where the temperature T is fixed and can be controlled as an external parameter. Then, $P(\mathbf{x}^N)\, d\mathbf{x}^N$ is the probability that the walk will be in the volume $d\mathbf{x}^N$ on any given step after many trials. Averages according to Eq. 4 can therefore be calculated. Since the number of atoms N is kept fixed in the volume Ω and the temperature T is given, these averages correspond to the canonical ensemble.

2.2 Computational Aspects

The motion of a fixed number of atoms N in a fixed volume Ω is governed by their mutual interatomic interactions. Usually these interactions are assumed to be pair-additive such that V, the model potential energy of the system, is a summation of $N(N-1)/2$ pair potential functions.

Molecular dynamics proceeds as follows. A model is settled by assigning an analytical (or numerical) function to the potential energy of the system. Next, Newton's equations of motion are numerically integrated using some computational algorithm [6]:

$$\mathbf{F}_i = m_i \ddot{\mathbf{x}}_i(t) = -\nabla_i V(\mathbf{x}^N) \,. \tag{6}$$

The solution of these equations yields the trajectory of the system. One starts the calculation from an arbitrary, and initial, state $\Gamma(t_0)$ at time t_0. The initial value $A(t_0)$ of the quantity that we want to study is also arbitrary, but known. Since the initial geometrical configuration $\mathbf{x}^N(t_0)$ is given, the potential energy is known and from it the forces on each particle are calculated from Eq. 6. These forces move the system to a new state $\Gamma(t_0 + \Delta t)$, the quantity under study is recalculated $A(t_0 + \Delta t)$, and now we have a new configuration $\mathbf{x}^N(t_0 + \Delta t)$ with which the forces are calculated once more using Eq. 6. These new forces move the system again to a new state $\Gamma(t_0 + 2\Delta t)$; $A(t_0 + 2\Delta t)$ is recalculated and so on. This is the procedure to obtain a discrete number of A values along the trajectory of the system. The time average can now be calculated using Eq. 3, provided that the number of time steps is sufficiently large.

Since one is describing the evolution of N particles in a fixed volume Ω by classical Newtonian mechanics, the total energy E is a constant of the motion. Therefore, time averages of a a given property along the trajectory in phase space is an approximate measure of the microcanonical ensemble average of that property for a thermodynamic state N, Ω, E.

For a Monte Carlo approach to the calculation of averages the situation is very much like for molecular dynamics, except that the sequence of steps in time is replaced by the random generation of the configurations. Fig. 1 gives a block diagram for the strategy to follow.

The advantages of one method over the other depend upon the system to be studied, the computer facilities and the properties to be analyzed. Molecular dynamics is able to provide more information on dynamical properties calculated from time correlation functions. Monte Carlo method is best suited to calculate the free energy of the system.

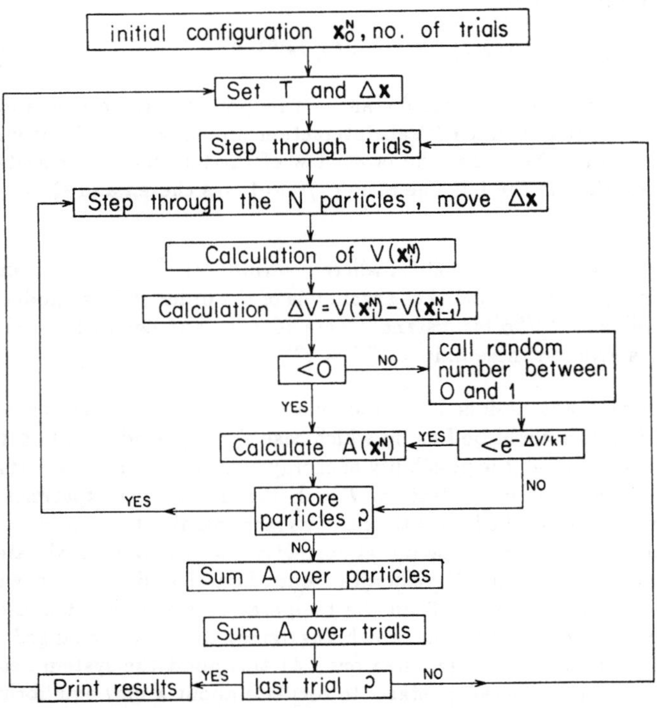

Fig. 1. Block diagram for the strategy of a Monte Carlo calculation.

3. Lennard-Jones Clusters Adsorbed on a Planar Substrate

3.1 Introduction

Both molecular dynamics [8-10] and Monte Carlo [11-13] have been used to describe the thermodynamic properties of Lennard-Jones clusters, with the observation that free clusters can exhibit phase changes analogous to the melting and freezing of bulk materials. The type of melting transition was under dispute because of apparently contradictory results giving evidence for a first-order [8-10] or second-order [11-13] transition. This contradiction has recently been reconciled [14] by recognizing that the results in the microcanonical [8-10] and canonical [11-13] ensembles need not coincide for finite systems. The study of clusters adsorbed on surfaces has received less theoretical attention [15]. We report in this section the results of the 13-atom cluster adsorbed on a planar surface obtained using molecular dynamics. The purpose of this study was to investigate the effect of the substrate on the melting-like transition and to describe the process of surface wetting by these small clusters.

3.2 Model and Methods

The Lennard-Jones (LJ) potential with depth ϵ and core σ was adopted to represent the pair interaction between atoms in the 13-atom cluster. Typical values for argon are $\epsilon/k = 121K$ and $\sigma = 3.4$ Å. The global minimum of this potential function corresponds to the icosahedron with potential energy -44.44ϵ [16]. Each cluster atom interacts with the surface by means of a model potential:

$$v(z) = 2.12\epsilon_s\{(\sigma_s/z)^{12} - (\sigma_s/z)^3\} \ , \tag{7}$$

where z is the perpendicular atom-substrate distance. This potential corresponds to a lateral averaging of the attractive van der Waals interaction similar to the LJ but with different parameters ϵ_s and σ_s. Note that the power-law of the first term represents a stiffer hard core repulsion than in the LJ. The ratio $\sigma_s/\sigma = 1$ was chosen and $\epsilon_s^* = \epsilon_s/\epsilon$ was changed from zero to 1.2.

Molecular dynamics with a time step of 0.01τ, where $\tau = (m\sigma^2/\epsilon)^{1/2}$ and m = atomic mass, was used in the integration of Eq. 6. For argon, τ is of the order of 2 ps. Reduced units of ϵ for energies and $r_0 = 2^{1/6}\sigma$ for distances were adopted and denoted by (*) in the text. Temperature is given in these units as $T^* = kT/\epsilon$.

The computer experiment was designed to simulate the process of heating and cooling the cluster. The cluster could be free, in which case $\epsilon_s^* = 0$, or adsorbed on the surface. If the cluster was adsorbed, the possibility of energy transfer towards the surface was excluded as to simulate a frozen substrate at $T = 0$. For the heating experiments, the initial configuration was an icosahedron with one of its faces parallel to the surface at a distance $-1.8r_0$. The temperature of this initial geometry was set to approximately $0.05\epsilon/k$ by choosing the initial set of $3N$ velocities from Maxwell's distribution with width equal to the chosen temperature. Then the temperature was calculated using Eq. 2, the system evolved for 70τ, time at which the temperature was changed by scaling the velocities with a factor α larger than one. At this point the system needs again a certain time until the temperature starts to oscillate about a new equilibrium value, 20τ in this example. Then we say that the system is equilibrated and start averaging the quantities of interest for the rest of the run, say 50τ more. Now we repeat the process: multiply the velocities by $\alpha > 1$, wait 20τ, obtain the new temperature and finally average the calculated quantities over 50τ. All heating experiments were done in this steplike manner [3,4,9] (in the cooling process $\alpha < 1$). Clusters were heated *slowly* at a rate of approximately $\Delta T^*/\Delta t = 1.7 \times 10^{-4}\tau^{-1}$, $\sim 10^9$ K/s for argon. Here we face one of the limitations of computer experiments. This rate of heating, that seems slow on the computer, is two or three orders of magnitude faster than in the real world.

3.3 The Melting Transition of the Free Cluster

For the free cluster the behavior of the equilibrium total energy as a function of temperature can be seen in Fig. 2a which is a graphical representation of the equation of state. At low temperatures the cluster is in the global minimum of the potential energy, the icosahedral structure. This is its *solid-like phase*. Furthermore, in the low temperature region the total energy is linear, with a slope of the order of the classical specific heat of a set of $3N$ harmonic oscillators. When the temperature is high, the cluster makes excusions over a set of potential minima with high potential energy. During the length of our computer runs, the icosahedral structure was not observed at high temperatures. The set of high energy structures visited in configuration space defines the *liquid-like phase* of the cluster. In the intermediate temperature regime the cluster can acquire

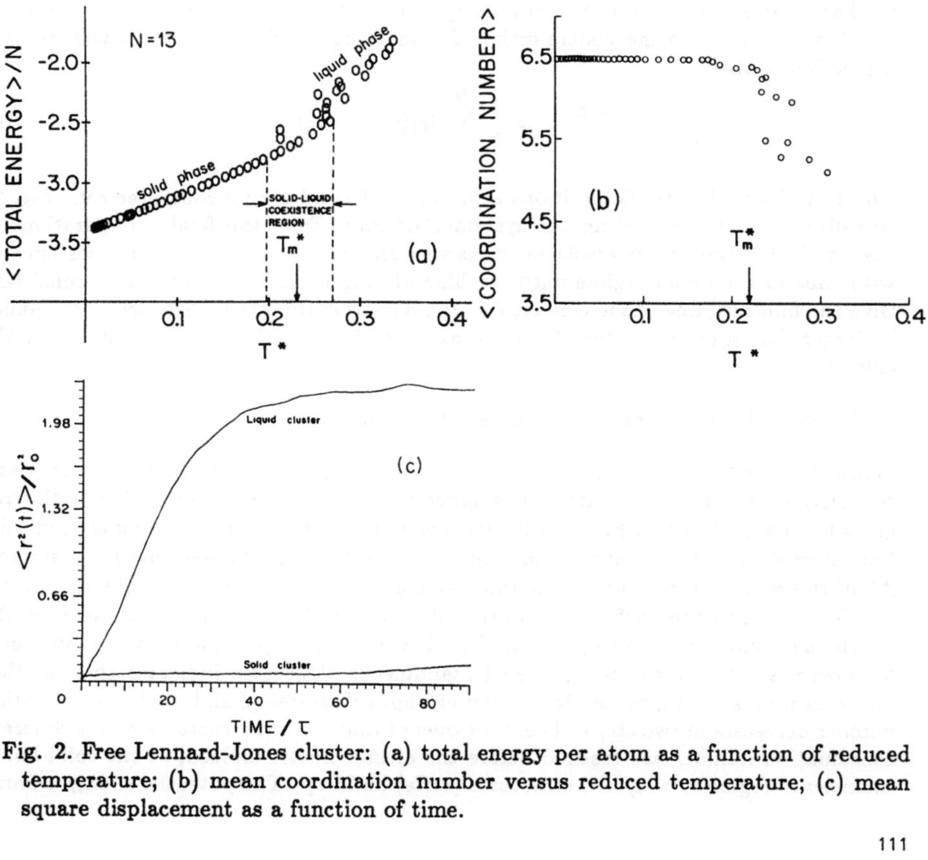

Fig. 2. Free Lennard-Jones cluster: (a) total energy per atom as a function of reduced temperature; (b) mean coordination number versus reduced temperature; (c) mean square displacement as a function of time.

one or two structures which correspond to local minima of high potential energy, but still visits the icosahedal configuration. This interpretation is consistent with the static analysis of configuration space [17] which shows the existence of an energy gap between the potential energy of the icosahedron and the potential energy of the remaining 987 local minima discovered for the 13-atom Lennard-Jones cluster. The solid and liquid phases coexist. When the liquid-like points are fitted to a straight line, it is seen that the slope is larger than in the solid-like region. All together, the behavior resembles a first-order phase transition of bulk materials. The melting temperature T_m is defined as the mid point of the coexistence region. Our results are in agreement with previous reports, both in the solid and liquid regions [10] as in the intermediate temperature regime [14].

Further evidence of the melting like transition in the cluster is given by plots of the mean coordination number n_c as a function of temperature (Fig. 2b) and of the mean square displacement $\langle r^2(t)\rangle$ as a function of time (Fig. 2c). To obtain n_c it is necessary to calculate the pair correlation function $g(r)$ [18],

$$n_c = \rho_0 \int_0^{r_{\mathrm{neig}}} g(r) 4\pi r^2 \, dr \ , \tag{8}$$

where $\rho_0 = N/\Omega$ is the reference density and r_{neig} is the distance between nearest neighbor atoms in the cluster. The coordination number exhibits a behavior very similar to that of bulk materials, it changes abruptly from a constant value in the solid phase to a lower value when the system melts. The mean square displacement of the atoms is a function of time:

$$\langle r^2(t)\rangle = \frac{1}{N} \sum_{i=1}^{N} \langle |\mathbf{r}_i(t) - \mathbf{r}_i(t_0)|^2 \rangle \ , \tag{9}$$

where $\mathbf{r}(t_0)$ are the atomic positions at an initial time during a computer run. Fig. 2c was obtained by propagating the system 100τ more from the final configurations of Fig. 2a. The mean square displacement is very small in the solid phase, it rises linearly with time in the liquid regime until reaching at long times a constant value consistent with the finite volume of the cluster, *i.e.*, the deviation from the mean positions cannot be larger than a certain value, if the atoms are to move in the volume occupied by the cluster.

3.4 Supported Clusters and the Wetting Transition

When the substrate is plugged into the calculations, qualitatively the total energy and coordination number as a function of temperature exhibit a behavior similar to the free cluster. This is shown in Fig. 3. All curves were generated during heating experiments that started with the cluster in the icosahedral structure adsorbed onto the surface. All of the results show that the melting like transition is taking place while the cluster is adsorbed onto the surface. However, the melting temperature, the width of the coexisting region and the slope of the liquid line depend upon the substrate strength. Moreover, a closer inspection of the instantaneous structures indicates that another phenomenom is taking place. For strong enough substrates (*c* and *d*) the coordination number decreases in two steps. There is evidence that a set of structures with a different coordination number are reached before the cluster melts. Indeed, in the solid-liquid coexistence region two-layered structures (*rafts*) build-up. The potential energy minima

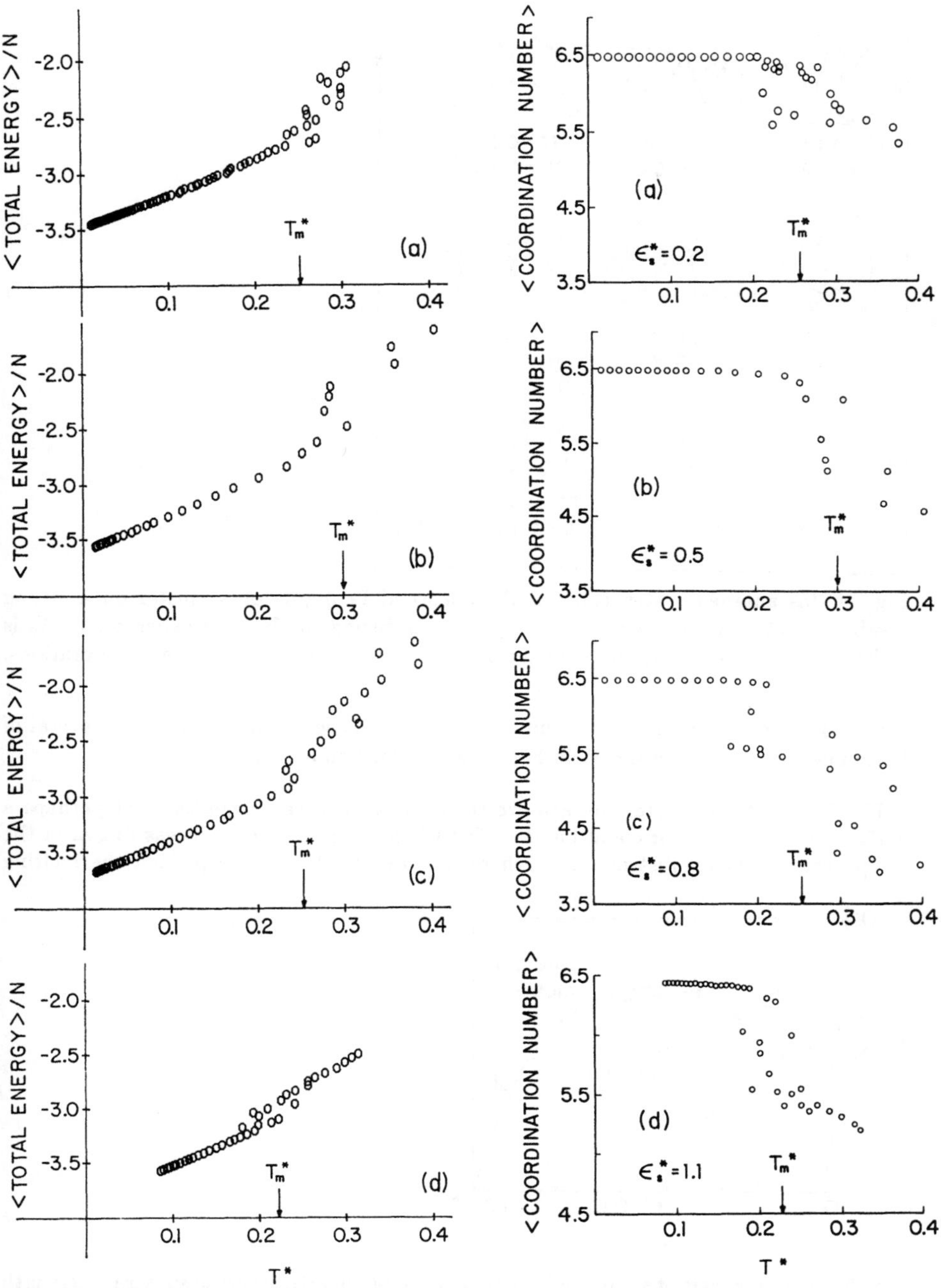

Fig. 3. Temperature dependence of the total energy per particle (left) and of the mean coordination number (right) for supported clusters: (a) $\epsilon_s^* = 0.2$, (b) $\epsilon_s^* = 0.5$, (c) $\epsilon_s^* = 0.8$, (d) $\epsilon_s^* = 1.1$. Total energy is in reduced units.

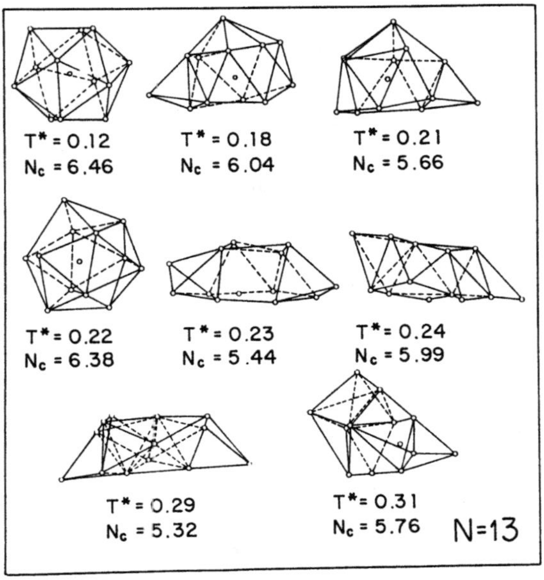

Fig. 4. Instantaneous structures of the adsorbed Lennard-Jones cluster on a strong substrate $\epsilon_s^* = 1.1$ at increasing temperatures during the heating experiment. N_c is the coordination number calculated from the atomic coordinates of the configurations.

corresponding to these rafts were not present in the case of the free cluster. Several of these new structures are shown in Fig. 4 for a strong substrate.

The fact that the cluster can acquire these raft structures suggested that perhaps a surface wetting behavior could be identified when the phase portrait was drawn in the temperature versus substrate strength plane. In Fig. 5 we have plotted the melting

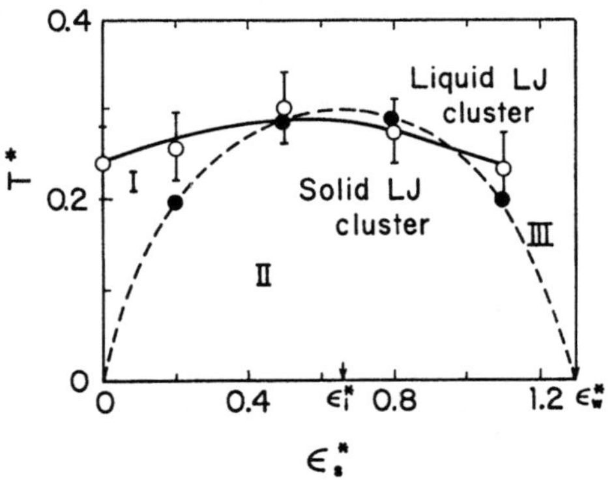

Fig. 5. Phase portrait of supported clusters in the temperature versus substrate strength plane. The full line joints loci of the melting temperature dividing the plane into liquid clusters (upper part) and solid clusters (lower part). Region I is a dry surface, region II corresponds to icosahedral clusters adsorbed on the surface and region III is a wet surface.

temperature as a function of substrate strength with open dots and bars indicating the width of the solid-liquid coexistence region. Above the line joining the open dots the cluster is liquid; below the line the cluster is solid. A second curve (dotted line) can be drawn through four points generated when a liquid cluster was cooled down to low temperatures. The point at $\epsilon_s^* = 0.2$ indicates the temperature at which the cluster ran away from the surface while freezing. For this weak substrate there is a transfer of kinetic energy from the atoms to the center of mass, the cluster cools but abandons the surface in the icosahedral structure. The value at $\epsilon_s^* = 0.5$ corresponds to the temperature where a two-layered structure was detected. Upon further cooling the cluster ended adsorbed onto the surface in its icosahedral structure. The two remaining points at $\epsilon_s^* = 0.8$ and 1.1 indicate the temperature where no more rafts were seen, and the cluster was mainly in the icosahedral structure. Three regions are now clearly shown in Fig. 5. Region I corresponds to a *dry* surface; no clusters are adsorbed. In region III, $\epsilon_s^* > \epsilon_w^*$, the surface is *wet* and the clusters are adsorbed as rafts. In region II, $\epsilon_s^* < \epsilon_w^*$, there is *partial wetting* and the clusters are adsorbed in the icosahedral structure. In region I and region II below ϵ_i^*, the cluster desorbs as a whole. In region II above ϵ_i^*, and in region III the cluster desorbs atom by atom. When a cluster is cooled down in region III and $\epsilon_s^* \gg \epsilon_w^*$, it loses atoms that remain adsorbed on the surface, *i.e.*, side evaporation from the cluster takes place.

In these computer experiments we have shown the large variety of structures that a supported cluster can acquire depending upon the mechanism of adsorption and upon temperature. Hopefully, these parameters could be equally controlled in the laboratory during a real experiment looking for equilibrium data.

4. Neutral and Charged Silicon Clusters

4.1 Introduction

The microscopic mechanism underlying the processes of melting or freezing is a consequence of the classical interatomic force field used in the description of the system. Pair potentials alone are insufficient to study covalent materials [19] and to simulate correctly their thermodynamic behavior. Very few simulations have been carried up to now using n-body terms in the potential model, both in $3D$ or lower dimensions, and both in the thermodynamic limit or for finite systems. Also, except for the triple-dipole terms [9,16], almost no 3-body class of potentials have been analyzed in all of its details. The purpose of this study was to provide a computational probe of the Stillinger-Weber (SW) potential function [5] in the description of annealing experiments on moderate-size (up to 32 atoms) neutral and charged silicon clusters. The SW potential includes 2- and 3-body terms and its parameters were adjusted to fit the silicon properties of condensed phases. Silicon was chosen as a prototypical semiconductor for which several experiments have been reported [20,21]. Other than SW potential, a new generation of empirical potentials [22] for silicon is now available in the literature. More effort should be invested to analyze them all. The functional form of the SW potential for silicon has been used to simulate the bulk amorphous and crystalline phases of germanium [23] by giving a different value to the parameter measuring the strength of the 3-body term. This new parametrization has been applied very recently to generate germanium clusters [24].

4.2 Model Potential

In general the structural energy of an N-atom cluster is a sum of 1-, 2- and 3-body potentials,

$$V(r_1, r_2, \ldots, r_N) = \sum_i^N v_1(\mathbf{r}_i) + \sum_{i<j}^N v_2(\mathbf{r}_i, \mathbf{r}_j) + \sum_{i<j<k}^N v_3(\mathbf{r}_i, \mathbf{r}_j, \mathbf{r}_k) . \tag{10}$$

We have investigated two cases. In the first case the one-body terms are zero to represent neutral clusters. In the second case the $v_1(\mathbf{r}_i)$ terms model approximately the interaction between one ionized atom with all other atoms in the cluster. The ionization takes place on one atom creating a positive charge q and inducing an energy

$$U^{(\mathrm{ind})} = \sum_i^N v_1(\mathbf{r}_i) = -\tfrac{1}{2} \sum_{j=2}^N q^2 \alpha / \mathbf{r}_{1j}^4 , \tag{11}$$

where the atomic polarizability α is considered isotropic and the contributions of induced dipoles to the electric field were neglected. The values $q = -e$ and $\alpha = 7.2$ Å^3 were adopted.

In both cases the 2- plus 3-body terms were modeled using the SW potential. This potential [5] is a seven-parameter function, where four parameters (A, B, p, q) belong to the pairwise interaction

$$v_2(r) = \begin{cases} A(Br^{-p} - r^{-q})\, e^{1/(r-a)} & \text{if } r < a \\ 0 & \text{otherwise,} \end{cases} \tag{12}$$

where the cutoff distance a is common to both 2- and 3-body terms and where the 3-body contribution is represented as a function of two bond lengths l_1, l_2 and the included angle θ_{12},

$$f(l) = \begin{cases} \sqrt{\lambda}\, e^{\gamma/(l-a)} & \text{if } l < a \\ 0 & \text{if } l \geq a, \end{cases} \tag{13}$$

such that

$$v_3(l_1, l_2, l_3) = \sum_{s<t=1}^{3} f(l_s) f(l_t) (\cos\theta_{st} + 1/3)^2 . \tag{14}$$

For silicon, the 3-body potential was constrained to have a minimum at the tetrahedral angle (109.57°) and confined to within a 3.77 Å cutoff. This short range makes it very attractive for molecular dynamics simulations. The seven parameters were chosen equal to those in Ref. (5). This choice allows two minima of the SW for the $N = 3$ cluster: the equilateral triangle which is the more stable ($V = -1.4786$ eV) and the isosceles triangle with one tetrahedral angle follows in stability ($V = -1.4447$ eV).

4.3 Annealing of Clusters

Molecular dynamics was used throughout solving Newtonian equations of motion with a time step of 7.66×10^{-4} ps. The units of length and energy were chosen to be $\sigma = 2.0951$ Å and $\epsilon = 2.167$ eV.

The annealing and quenching experiments were carried out in the following way. First, an initial configuration was cut from a large bulk configuration which had been equilibrated at very high temperatures. Second, the clusters were cooled at rates $\sim$ 100 K/ps using the step-like procedure described in Sec. 3.2 , such that each intermediate step in the quench was an equilibrated state of the cluster. The change to each new temperature was obtained by scaling the final velocities of the prior equilibrated state. Third, the clusters were led through a cycle of reheating followed by cooling to $T = 0$ to assure that the system was not trapped in metastable states at low temperatures.

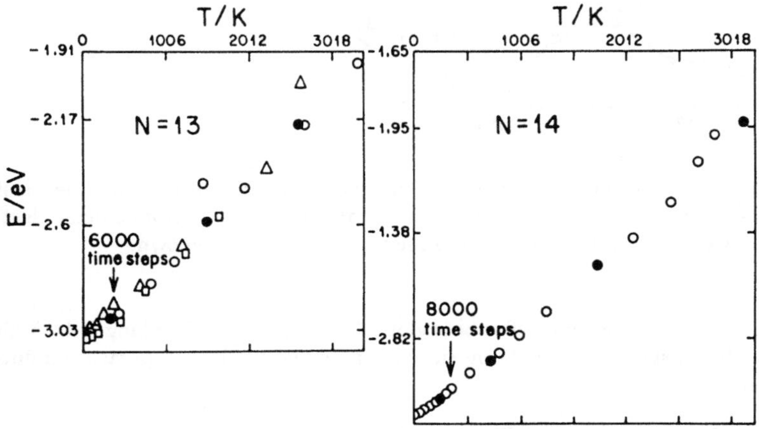

Fig. 6. Mean total energy per particle as a function of temperature for silicon clusters. The 13-atom cluster quenches ended in 3 energies: (a) open dots (E =-3.044 eV at T = 155K); (b) open squares (E =-3.0639 eV at T = 157K); (c) open triangles (E =-3.0225 eV at T = 161K). Black dots stand for MD runs at faster quenching rate.

The typical behavior of E, total energy per atom along a quench, is illustrated in Fig. 6 for the neutral 13- and 14-atom clusters (open symbols). Similar curves for other cluster sizes have been reported previously indicating extremely smooth changes of E between the low and high temperature limits. Each point is an average over 1.5 ps, but at T = 375K longer runs (7 ps) were performed before cooling to almost zero temperature. Black dots show a path followed by the system with a 5 time faster quenching rate. Clusters with odd number of atoms had a tendency to end the quench in different conformations of quasi-degenerate mean potential energy. Specifically, for the 13-atom cluster, three geometries were detected as indicated in Fig. 6, although more quenches ended in case (a).

All clusters changed shape with temperature because the atoms are mobile during a quench, as shown for the 14-atom cluster in the top part of Fig. 7. Eventually at high temperatures the cluster evaporated by losing one atom after certain *life time*, *i.e.*, the neutral 14-atom cluster lost one atom after 4.8 ps at T = 3168K. The mean coordination number n_c and the mean square distance of the atoms from the cluster center of mass $\langle R^2 \rangle$ are illustrated in Fig. 7. The bond distance $r_{\rm neig}$ is extracted from the position of $g(r)$ first minimum (2.94 Å). As it is seen in Fig. 7, a region of temperatures where the cluster has a coordination number larger than 3 without increasing its size is apparent,

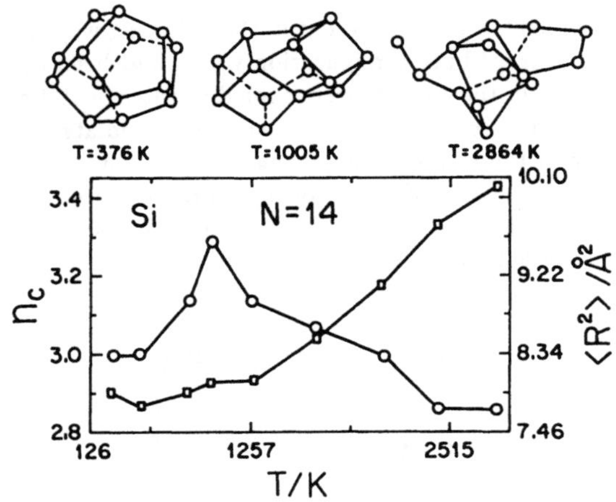

Fig. 7. Mean coordination number (dots) and mean square-radius from the cluster center of mass (squares) as a function of temperature for the 14-atom silicon cluster. Instantaneous cluster configurations at three temperatures are shown.

possibly indicating a more liquid-like behavior. The cluster is shrinking. At higher temperatures the cluster increases its mean size, melts and finally temperature induced fragmentation occurs.

It is not clear from an analysis of Fig. 6 and 7 at what temperature the cluster melts, although the structural characteristics are certainly different at low than at high temperatures. This is so because for this kind of clusters, the system can make excursions at low temperatures among several local minima. The situation is very different than for the Lennard-Jones 13-atom cluster. A discussion, on the definition of *solid* and *liquid* clusters was given [4] in terms of the temperature changes of structural quantities such as the pair correlation function or the density distribution of bond angles. Let us state again the arguments as a matter of completeness for the specific case of the neutral 15-atom cluster. Fig. 8 shows a selected heating experiment for the neutral 15-atom

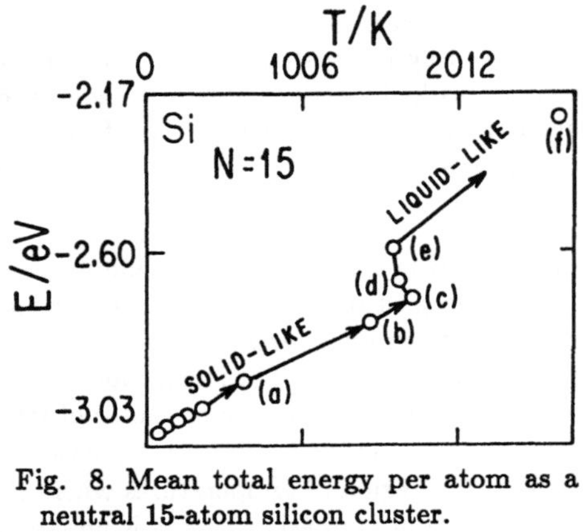

Fig. 8. Mean total energy per atom as a function of temperature while heating the neutral 15-atom silicon cluster.

cluster in which an abrupt change in the total energy is evident. In this diagram, the low temperature branch can be seen as a solid-like phase, the high temperature branch can be associated to a liquid-like phase and the region where the abrupt change takes place should correspond to a coexistence of phases. To see this effect more closely, in Fig. 9 is given the evolution through the coexistence region of the pair correlation function for states labeled (a) to (f) in Fig. 8. The first peak grows broader and lower, when going from low to high temperatures. The second coordination shell loses progressively its structure from state (a) to (e) and the gap between first and second coordination shells tends to disappear with increasing temperature. From this analysis it can be said that in state (a) the cluster has structural characteristics of a solid phase, whereas in state (e) the $g(r)$ has changed to be more like that of a liquid phase.

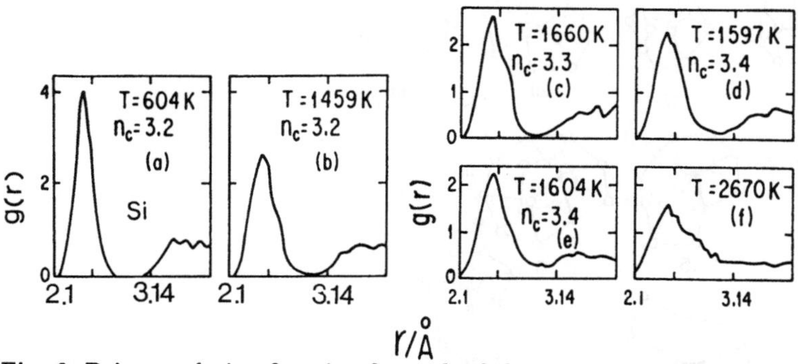

Fig. 9. Pair correlation function for each of the temperature steps labeled (a) to (f) in Fig. 8.

4.4 Cluster Configurations at Low Temperatures

The low temperature limit of the mean potential energy per atom reached after a quench is an interesting quantity. It is in fact the binding energy of the cluster. The values of this energy as a function of the cluster number of atoms were given previously [4,25] as well as the final temperatures reached in each simulation. The binding energy of the neutral clusters increases with increasing N, with a decrease in the slope at $N \sim 10$ suggesting that a sudden change in the way of aggregation takes place for that cluster size. An analysis of the derivative of the binding energy as a function of N shows a definite even-odd alternation in the cluster stability: clusters with even number of atoms are easier to be formed. The neutral 14-atom has the lower derivative and stands as a very stable cluster. For charged clusters the aggregation takes place in a steplike fashion with a slightly higher stability when $N = 13$ and 15.

The cluster geometries corresponding to the neutral sequence obtained at low temperatures at the end of the quenches are shown in Fig. 10. In certain cases more than one final structure was found with almost equal potential energy. The two geometries with lowest potential energy are labeled (a) and (b) in the figure. These geometries are hollow clusters with all atoms on the surface. The small neutral clusters tend to have a coordination of three for all of its atoms, until at $N = 15$ an incipient 8-member unit connecting two four-coordinated atoms builds up. When the cluster acquires a size of 17 atoms the tetrahedral seed is clearly present, although the dihedral angle is still distorted.

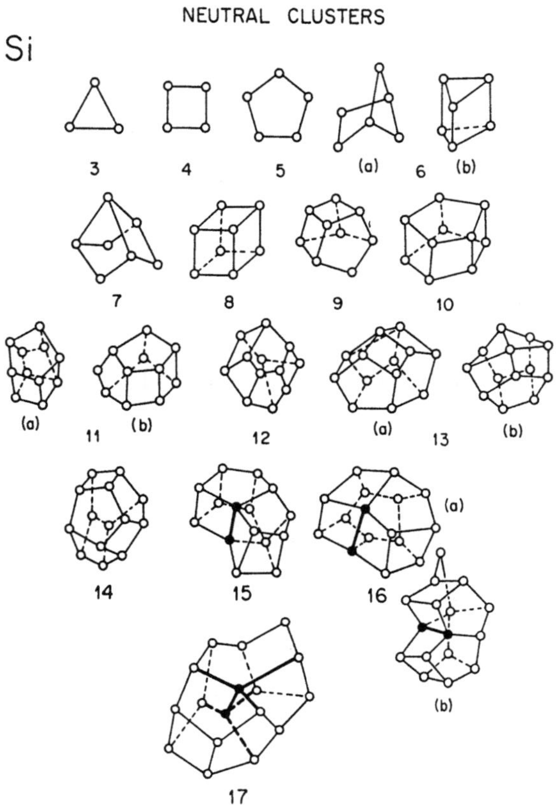

Fig. 10. Sequence of geometrical configurations of neutral silicon clusters at low temperatures as obtained from the molecular dynamics simulation. Bonded atoms are at distances < 2.94 Å.

Recently another molecular dynamics experiment [26] using SW potential led to the same structures of Fig. 10 by using a different method [27] than the one described in the previous paragraphs. Both classical results give rise to ground state geometries up to 10-atom clusters that do not agree with those obtained quantum mechanically, either by ab-initio calculations [28] or by local density approximations [29]. We should point out though, that all of these results, namely the more stable cluster geometries, do not contain seeds with tetrahedral symmetry —the preferred structures are in all cases hollow. Also, all calculations gave rise to several structures with quasi-degenerate binding energies, all lying within less than 0.5 eV from the more stable one.

In Fig. 11 is shown the evolution of $g(r)$ as a function of N at very low temperatures. It is interesting to point out that the cluster $g(r)$ carries over to the bulk *inherent* pair correlation function of the supercooled silicon melt (reproduced from Ref. 5 by dotted lines in Fig. 11c). This inherent $g(r)$ is the result of mapping the geometrical configurations of a system with 216 atoms onto nearby potential energy minima. The seeds to build up those minimum-energy configurations seem to be already present in the small clusters analyzed in this study.

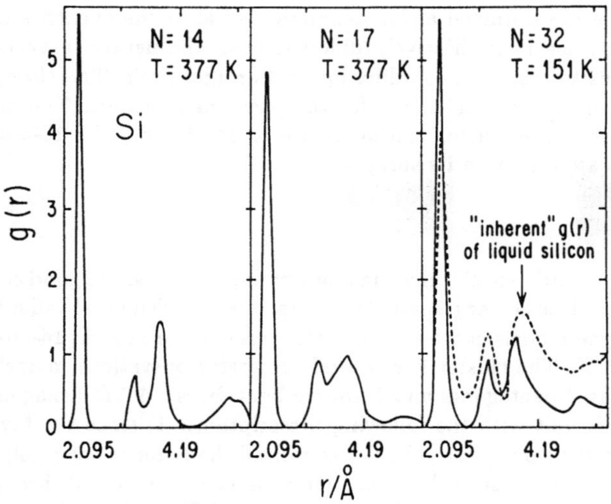

Fig. 11. Pair correlation function for the neutral 14-, 17-, and 32-atom silicon clusters.

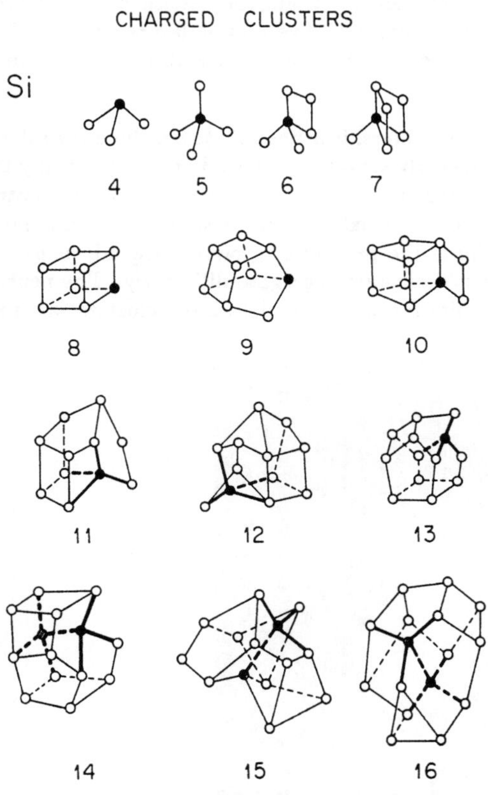

Fig. 12. Sequence of geometrical configurations of charged silicon clusters at low temperatures as obtained from the molecular dynamics simulation. Black dots indicate the charged atoms. Bonded atoms are at distances < 2.94 Å.

The charged cluster geometries obtained at low temperatures after the quenches are given in Fig. 12. This sequence is qualitatively different from the neutral sequence since the charged atom tends to assume a coordination of 4 around itself. Therefore, a distorted 5-member tetrahedral unit is built even for the very small clusters. The first 8-member unit is formed when the cluster reaches a size of 14 atoms. The 16-atom cluster presents already a 6-atom ring on its surface.

4.5 Larger Aggregates

The clusters generated in this work, specifically the more symmetric ones, are good candidates for the construction of larger aggregates made up of these elemental building blocks. Various noncrystalline models can be built. Lets consider the neutral 14-atom cluster in Fig. 7 at $T = 376$K. The cluster has several symmetry operations, namely: C_3, $3C_2$, $3\sigma_v$ and σ_h. It has 6 pentagonal and 3 square bent faces. A $2D$ hexagonal lattice can be built with this unit with the following prescriptions: 1) since the three C_2 axes of every unit are on the same plane, the C_3 axes of all the units are parallel; 2) every C_2 axis is common to two units such that neighbor clusters are connected through the atoms on the square faces. As a result, only pentagons are left on the surface and the cluster centers of mass form an hexagonal lattice, as shown in Fig. 13. This network has included hexagonal holes, which in turn form a triangular lattice by themselves. This gives a model of an amorphous silicon film with five atomic layers and a surface made up only by pentagons. The stability of the film increases with a deformation of square into rhomboidal faces.

In what concerns the annealing of silicon clusters, it is possible to conclude the following points. First, the clusters found in the early stages of growth do not have locally the tetrahedral symmetry of bulk silicon, the crossover starts to show at $N = 17$ for neutral clusters and $N = 14$ for charged clusters. For small clusters it seems that underlying stable structures more liquid-like dominate the process of growth. Second, using the same potential model, neutral and charged clusters aggregate differently. The neutral sequence shows a slight increase in stability for the 10- and 14-atom clusters, whereas

Fig. 13. Five-layer model built with the most stable 14-atom silicon cluster.

the charge sequence indicates the 13- and 15-atom clusters as more stable. Third, these stable cluster geometries can be used as elemental units to built larger aggregates of amorphous structures. Finally, since changes in the potential parameters can give rise to selected cluster geometries at zero temperature, *i.e.*, for $N = 3$ the equilateral triangle is a minimum depending on λ, it would be convenient to fit these parameters not only to bulk properties, but also to well established properties of small clusters. In this simulation we have tried to produce the clusters much in the same way as the bulk crystal is produced. However, this mechanism might not be the best to simulate the formation of amorphous clusters. Better simulations would be possible if more experimental knowledge of the parameters controlling a laboratory quench were available.

5. Final Comments

One purpose of this presentation was to familiarize the participants of the school with techniques that have already been used for many years. Another purpose was to insist on some technical details such that a simulation program would not be a black box producing million of numbers. Computer simulations can be very fruitful to study many open questions in cluster physiscs such as fragmentation, ionization process, infrared spectra, compound clusters, impurity effects, charge instabilities, among others. A challenge is to find a bridge between the classical methods and the quantum mechanical description of the clusters. Several groups are now working under this scope [30,31], as we will learn from other lectures during this school.

Acknowledgments

We are grateful to Prof. D. Levesque for his collaboration, continuous encouragement and interest in our work. The study on supported clusters was done in collaboration with Dr. I. Garzon and Dr. M. Avalos-Borja, we acknowledge their dedication and enthusiasm. We thank Dr. A. Robledo for illuminating conversations concerning the concept of wetting. This research was supported by Consejo Nacional de Ciencia y Tecnología, México, under grant No. PCCBBNA-022643 and joint contract with Centre National de la Recherche Scientifique, France.

References

1. J.P.Hansen and I.R.J.MacDonald, *Theory of Simple Liquids* (Academic Press, New York 1976, revised edition in press); F.F. Abraham, *Rep. Prog. Phys.* **45**, 1113 (1982).
2. K. Binder, (ed.) *Monte Carlo Methods in Statistical Physics. I* and *II*, (Springer, Berlin, Heidelberg 1979 and 1984).
3. E. Blaisten-Barojas, I. L. Garzon, and M. Avalos-Borja, to be published.
4. E. Blaisten-Barojas and D. Levesque, *Phys. Rev. B* **34**, 3910 (1986).
5. F.H. Stillinger and T.A. Weber, *Phys. Rev. B* **31**, 5262 (1985).
6. L. Verlet, *Phys. Rev.* **159**, 98 (1967).
7. N. Metropolis, A. Rosenbluth, M. Rosenbluth, A. Teller, and E. Teller, *J. Chem. Phys.* **21**, 1087 (1953).
8. C. L. Briant and J. J. Burton, *J. Chem. Phys.* **63**, 2045 (1975).
9. E. Blaisten-Barojas and H. C. Andersen, *Surf. Sci.* **156**, 548 (1985).

10. J. Jellinek, T. L. Beck, and R. S. Berry, *J. Chem. Phys.* **84**, 2783 (1986).
11. R. D. Etters and J. Kaelberg, *J. Chem. Phys.* **63**, 2045 (1975).
12. R. D. Etters and J. Kaelberg, *Phys. Rev. A* **11**, 1068 (1975).
13. J. B. Kaelberer and R. D. Etters, *J. Chem. Phys.* **66**, 3233 (1977).
14. J. D. Honeycutt and H. C. Andersen, to be published.
15. M. Weissmann and N.V. Cohan, *J. Chem. Phys.* **72**, 4562 (1980).
16. I. L. Garzon and E. Blaisten-Barojas, *Chem. Phys. Lett.* **124**, 84 (1986).
17. M.R. Hoare and J.A. McInnes, *Advances in Phys.* **32**, 791 (1983).
18. E. Blaisten-Barojas, *Kinam* **6A**, 71 (1984).
19. P.N. Keating, *Phys. Rev.* **145**, 637 (1966).
20. T.P. Martin and H. Schaber, *J. Chem. Phys.* **83**, 855 (1985).
21. L.A. Bloomfield, R.R. Freemen and W.L. Brown, *Phys. Rev. Lett.* **54**, 2246 (1985).
22. T. Takai, T. Halicioglu, and W.A. Tiller, *Scr. Matall.* **19**, 709 (1985); R. Biswas and D.R. Hamann, *Phys. Rev. Lett.* **55**, 2001 (1985); J. Tersoff, *ibid.* **56**, 632 (1986); S. Saito, S. Ohnishi, and S. Sugano, *Phys. Rev. B* **33**, 7036 (1986).
23. K. Ding and H.C. Andersen, *Phys. Rev. B* **34**, 6987 (1986).
24. G.A. Antonio, B.P. Feuston, R.K. Kalia and P. Vashishta, *Bull. Am. Phys. Soc.* **32**, 602 (1987).
25. E. Blaisten-Barojas and D. Levesque, in *Physics and Chemistry of Small Clusters*, ed. by P. Jena and S. N. Khanna, NATO ASI series, (Plenum Press, New York, 1987) p. 157.
26. B.P. Feuston, R.K. Kalia and P. Vashishta, *Phys. Rev. B* **35**, 6222 (1987).
27. F.H. Stillinger and T.A. Weber, *Kinam* **3A**, 159 (1981); *Phys. Rev. A* **28**, 2408 (1983); *Science* **225**, 983 (1984).
28. K. Raghavachari, V. Logovinsky, *Phys. Rev. Lett.* **55**, 2853 (1985); K. Raghavachari, *J. Chem. Phys.* **84**, 5672 (1986); G. Pacchioni, J. Koutecky, *J. Chem. Phys.* **84**, 3301 (1986).
29. D. Tomanek and H. Schluter, *Phys. Rev. Lett.* **56**, 1055 (1986); R. Car, M. Parrinello and W. Andreoni, in *Microclusters*, ed. by S. Sugano, Y. Nishina, and S. Ohnishi, Springer Series in Materials Science Vol. 4, (Springer, Berlin, Heidelberg, 1986) p. 134; W. Andreoni and P. Ballone, *Phys. Scripta*, in press.
30. U. Landman, D. Sharf, and J. Jortner, *Phys. Rev. Lett.* **54**, 1860 (1985).
31. R. Car and M. Parrinello, *Phys. Rev. Lett.* **55**, 2471 (1985).

The Electronic and Geometrical Structure of Small Elemental Clusters

P. Fantucci[1] and *J. Koutecký*[2]

[1]Department of Inorganic and Metallorganic Chemistry,
University of Milan, Via Venezian 21, I-20133 Milano, Italy
[2]Institute of Physical and Theoretical Chemistry,
Freie Universität Berlin, Takustr. 3, D-1000 Berlin 33, Germany

1. THEORETICAL AND COMPUTATIONAL PROBLEMS.

Recent theoretical investigations have shown that the quantum mechanical methods usually employed for molecular systems are also well suited for studying the electronic structure of small elemental clusters. The most commonly employed methods, aimed at establishing relationships between the electronic stucture and the geometrical shape of clusters, are derived from either the Hartree-Fock (HF) or the Density Functional (DF) theories. The DF methods are named Local (Spin) Density (LSD) when the functional dependency of the exchange (or exchange+correlation) potential upon the electron density is presumed to be only local. Characteristics, advantages and limitations of both methodologies have been discussed in the literature [1].

In this section, some theoretical and computational aspects of the quantum mechanical procedures will be considered to clarify their general applicability to the specific field of the cluster theory.

The first observation is that both HF and LSD methods are essentially one-electron theories, that is models describing each electron as moving in the averaged field created by all other electrons. Nevertheless, one-electron functions (orbitals or spin-orbitals when the electron spin state is also considered) and associated energies, even if not possessing any precise physical meaning, are neverthless useful tools for examining the leading bonding characteristics. For instance, all the symmetry restrictions imposed by a specific nuclear frame determine the nature of the one-electron functions, their nodal character, degeneracy and in some cases, their particular spatial orientation or localization. The fact that "orbitals" can also be defined within the framework of the DF theories (the so-called Kohn-Sham orbitals [2]) makes the HF and DF-LSD methods more similar than is usually assumed . In fact, the LCAO type methods based on LSD approximation differ from the HF only in the assumed form of the exchange potential. In fact, while the HF exchange energy U_{XHF} is defined in a completely natural way through the antisymmetric character of the wave function Ψ, the LD exchange potential is assumed to be dependent on $\rho^{4/3}$, where ρ is the total electron density:

$$U_{XLD}=c \int \rho(1)^{4/3}dv_1 .$$

U_{XLD} is an average potential which does not discriminate among the one-electron functions which determine ρ, thus ignoring all angular dependencies. On the contrary

$$U_{XHF}=-\sum_{ij}^{occ} K_{ij} = -\sum_{ij}^{occ} \langle\varphi_i(1)\varphi_j(1)|1/r_{12}|\varphi_i(2)\varphi_j(2)\rangle,$$

(where i and j run over the occupied spin orbitals with equal spin part), is formed by different contributions for different orbitals.

The LSD methods present the very appealing feature of requiring a much smaller computational effort (roughly proportional to m^2-m^3, m being the number of basis functions) than the HF method which is characterized by a m^4 dependency. Moreover, the minimization of the functional

$$\begin{aligned}\langle E\rangle &= \Sigma_i \langle\varphi_i|-1/2\,\nabla_i^2 - \Sigma_I Z_I/r_{iI}|\varphi_i\rangle \\ &\quad +1/2\Sigma_{ij}\langle\varphi_i\varphi_i|1/r_{ij}|\varphi_j\varphi_j\rangle + U_X + \Sigma_{I>J} Z_I Z_J/R_{IJ} \\ &= \Sigma_i\langle h_i\rangle + U_C + U_X + U_N,\end{aligned}$$

(the total energy of the system) for $U_X \equiv U_{XLD}$ may give results which mimic (for reasons not yet completely clear) those of calculations based on the HF model but which also include correlation corrections. However, shortcomings of the LSD methods (for instance a systematic overestimation of the binding energy, for instance) have been documented [3].

Since the inadequacies of the HF method are essentially connected with the one-electron approximation, the method can be systematically improved by considering the electron correlation effects at various levels of sophistication. In general, this can be done allowing the Slater determinants Ψ_i,

$$\Psi_i = \mathcal{A}\left|\varphi_1(1)\varphi_2(2)\varphi_3(3)\ldots.\varphi_n(n)\right|,$$

(or a spin-adapted combination of Ψ_i) to interact among themselves yielding a variationally improved representation of the n-electron system $\phi\,(1,2,\ldots n)= \Sigma_i\, C_i\, \Psi_i$. (Configuration Interaction, CI). It must be observed that the CI method is itself very demanding from a numerical point of view, even considering the high efficiency of algorithms and recently developed computer programs [4,5]. As a consequence, the applicability of the combined HF and CI procedures is necessarily limited to systems with a small number of electrons. This is the reason why studies on clusters of large nuclearity, especially if composed of transition metal atoms, have been mainly carried out using LSD or semiempirical approximate procedures [6].

The use of sophisticated quantum mechanical methods in studying small clusters of light atoms may be fruitful if simple laws and rules governing the electronic and geometrical structure can be established with confidence. Moreover, the results thus obtained for small systems may be used to check the reliability of more approximate procedures suitable for applications to large clusters.

As mentioned above, the HF method (and to a lesser extent the LSD ones) presents computational limitations essentially connected with the dimension of the basis set $\{\chi\}$ used to expand the one electron functions $\varphi_i = \Sigma_k V_{ki} \chi_k$.

The set $\{\chi\}$ must include functions adapted for describing the core and valence electrons of each atom plus the so called "polarization functions", of high angular momentum, useful (and sometime essential) to represent electron density distortions caused by bond formation. In choosing the size of the basis set a compromise must be accepted between large dimensions (which give high variational flexibility and accuracy) and applicability of the procedure to large systems. When a systematic study of a series of clusters has to be carried out, the size of the basis is necessarily limited by the nuclearity of the largest cluster considered. The following are examples of Gaussian basis sets of various dimensions employed in the case of HF (and CI) studies of Li_n clusters: Li_2(15s,3p,1d) [7]; Li_4(6s,2p) [8], (7s,2p) [7]; Li_6(5s,1p) [9]; Li_9(3s,1p) [10]; Li_2-Li_9(3s,2p) and Li_2-Li_{20} (2s,1p) [11]. HF calculations on large clusters are feasible only with relatively small basis, including, however, at least one p polarization component. The influence of the polarization functions is generally evident from the results obtained on Li_4 cluster [6]. Two sets of HF-CI calcutions have been carried out using a (10s) fully uncontracted and a (2s,1p) basis. The first basis gives for the lithium atom an energy very close to the HF limit but it is unable to describe a bound Li_4 tetramer, considered as built from two diatomics interacting along four different mutual approaches. On the contrary, for three of the examined geometries, the potential curves determined using the (2s,1p) basis show a minimum, even if the computed stability is smaller than that of the fully geometry optimized Li_4 cluster. It has been verified that the existence of minima is not a spurious effect due to the incompleteness of the basis (basis set superposition error).

For clusters composed of heavy atoms, an important simplification can be achieved using one of the several proposed Effective Core Potential ECP models [12]. The corresponding ECP-HF or ECP-LSD calculations are aimed at explicitly treating only the valence electrons, the innermost electrons being assumed to belong to atomic frozen cores. With this in mind, a modified one-electron operator,

$$\bar{h}_i = h_i + \Sigma_I W_{Ii},$$

(where I runs over the atoms with core electrons) is introduced. For W_{Ii} the following semilocal form is usually assumed

$$W_{Ii} = \Sigma_{\ell} A_{\ell I}(r_{Ii})^{n_{\ell I}} \exp(-\alpha_{\ell I} r_{Ii}^2)(\Sigma_{m=-\ell}^{\ell} |\ell m\rangle\langle \ell m|).$$

The operators W_{Ii} should represent a repulsive potential for the valence electrons i in the core region so as to avoid variational collapse and core-valence Coulomb-exchange potential. Extensive tables of the parameters of W_{Ii} suitable for HF [13] or LSD [14] applications have been reported.

Obviously, the valence—electron -only approach is a convenient and correct alternative to that of the "all electron" (AE) provided that the frozen-core approximation is valid; consideration of alkali metal clusters may clarify this point. HF-CI studies of the best geometry of lithium clusters can be carried out at the AE level without any particular difficulty, up to about Li_{20}. Already for Na_n clusters, however, difficulties may arise for n=10-12 when using a basis of type (4s,2p), and stronger limitations are expected in the case of potassium clusters. It would be natural to adopt an ECP procedure for large Na_n and K_n clusters. However, as the atomic number increases, the outermost core electrons become more and more polarizable in the molecular environment making a not negligible contribution to cluster stability, thus implicitly indicating a breakdown of the frozen core approximation. Theoretically well defined procedures have been outlined to include the core polarization effects in the W_{Ii} effective operators[15]. As an alternative, one can include the outermost core electrons in the valence space [16] but, of course, at expense of computational efficiency.

Precise identification of different stable isomers for a given cluster nuclearity is essential in order to deduce general features of cluster properties, for instance the preferred paths of cluster growth. From a computational point of view this is achieved by localizing the minima on the hypersurface of the nuclear configurations for a given electronic state but the procedure is evidently hindered by great difficulties connected with the surface dimensions (3N-6): there is no hope of carrying out a global search (the only way to surely locate all the stationary points) even for very small clusters. In addition, the spin state of the cluster is unknown in advance, and, in principle, the analysis of the surface should be repeated for different electronic states. This is worth mentioning because the requirement, that the geometry search has to be performed for electronic wavefunctions with well defined $\langle S^2\rangle$ values, is not fulfilled by the so called "unrestricted" self consistent procedures [17] which are often applied in connection with both HF and LSD approaches. The spin-unrestricted methods allow electrons with different spins to be described by different spatial functions. The resulting wavefunction for open-shell systems is not eigenfunction of $\langle S^2\rangle$ but is in general, a mixture of different spin states: its "spin composition" may vary during the optimal geometry search and thus lead to final structures associated with a not well defined spin wavefunction.

The geometry optimization procedure may be simplified by imposing symmetry constraints on the variable 3N-6 internal coordinates. This is equivalent to carrying out the geometry optimization only on given sections of the hypersurface but the corresponding minima may result in being saddle points on the full hypersurface. In order to avoid biased results, the characterization of the stationary points as true minima must be carried out by checking the positive character of the Hessian matrix built into the basis of all internal coordinates.

In terms of nuclear cartesian displacements, the gradient of the energy may be written as

$$g_i = -dE/dX_i = -\langle \Psi | \partial H/\partial X_i | \Psi \rangle - 2 \langle \partial \Psi/\partial X_i | H | \Psi \rangle .$$

The first term is the Hellman-Feynman force (HFF) while the second is the wavefunction force (WFF) which gives a vanishing contribution only for exact Ψ. As stated before, the quality of the HF wavefunction strongly depends on the adopted basis $\{\chi\}$ which, in the case of clusters, is usually reduced in dimension. This means that the computed Ψ is far from being an exact solution and that the consideration of the WFF term is mandatory. Unfortunately, this latter contribution is much more difficult to evaluate than the HFF one because it requires the computation of derivatives of 2-electron integrals with respect to all the nuclear displacements connected with the variable internal coordinates. Examples of cluster geometry optimizations carried out considering only HFF have been reported in connection with LSD methods [18].

Worthy of mention is the fact that the total cluster energy is usually a slow varying function of the nuclear displacement and its gradient components may be small even in points on the hypersurface far from local minima. This can give rise to numerical difficulties and eventually to meaningless results. For instance, starting from a given geometry for a M_n cluster, uncontrolled variations of the internal coordinates can produce a final geometry corresponding to a $M_{n-2}+M_2$ system or other combinations of dissociated fragments. This unpleasant feature is peculiar to weakly bounded systems like the metal clusters and does not occur in the geometry optimization of normal organic or inorganic molecules.

The limited accuracy of the HF method may be substantially improved by considering the electron correlation effects by means, for instance, of the CI method. The HF solution Ψ_o correspond to a given occupation of n of the m available spinorbitals, which can be conveniently classified as occupied $\Psi_i (i=1,2,\dots n)$ and empty (virtual) $\Psi_p (p=n+1,\dots m)$, respectively. Allowing single, double, triple,... excitations from Ψ_o, it is possible to construct a number (usually very large) of excited configurations, Ψ_i^p, Ψ_{ij}^{pq}, Ψ_{ijk}^{pqr}, ... The CI wavefunction is expressed as

$$\phi = C_o\Psi_o + \Sigma\, C_i^p\, \Psi_i^p + \Sigma\, C_{ij}^{pq}\, \Psi_{ij}^{pq} + \Sigma_{ijk}^{pqr}\, \Psi_{ijk}^{pqr} + \dots ,$$

and the highest excitation class included in ϕ determines the type and the accuracy of the CI expansion. The simplest one is based on single and double excited configurations only

(SD-CI). Triple (T), quadrupole (Q) and higher excitations usually play a minor role in the case of "normal" systems, characterized by a C_o coefficient largely dominant and already well described at HF level. Unfortunately, this is not the case for clusters (and in particular for metallic ones) which can be characterized by a C_o^2 contribution of the HF solution as small as 0.85, or even smaller in the case of large nuclearities.
If this occurs, Ψ_o can no longer be considered a good approximation to the true ground state and must be replaced by a more accurate wavefunction composed of Ψ_o and some important biexcited configurations,

$$\bar{\Phi} = \bar{C}_o \Psi_o + \Sigma \bar{C}_{ij}^{pq} \Psi_{ij}^{pq} .$$

If single and double excitations are generated from Ψ_o and all the Ψ_{ij}^{pq} included in $\bar{\Phi}$, the final CI wavefunction will contain up to quadruple excited configurations with respect to Ψ_o and a good choice of the additional reference configurations Ψ_{ij}^{pq} can correct the inadequacy of the single reference Ψ_o. These are the fundamental ideas on which the Multi Reference Double-CI (MRD-CI) method [5] is based.

It must be noted that all truncated CI expansions suffer from a severe limitation which is known as "size inconsistency error" [19]. Size inconsistency occurs because the correlation energy per electron given by a SD-CI (or, in general, by a truncated CI) is relatively high for systems with few electrons but decrease as the number of electrons increases. In

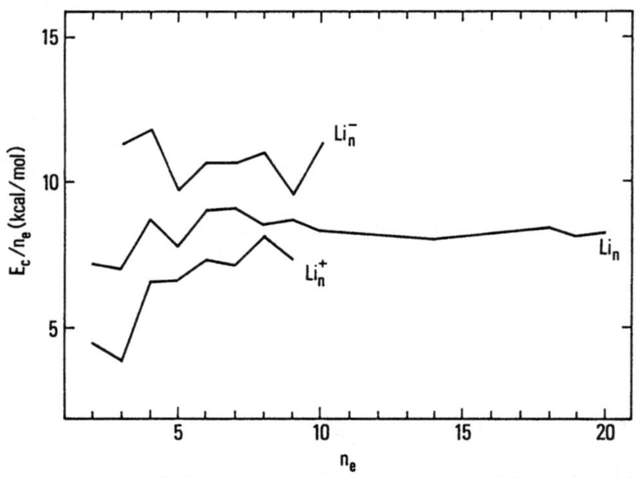

Fig.1. MRD-CI computed correlation energy per electron for neutral, cationic and anionic lithium clusters.

the case of lithium clusters, for instance, it is easy to completely recover the correlation energy (in a given basis) for trimers and tetramers, while the fraction of the correlation energy is necessarily smaller and smaller for higher clusters. Fig. 1 shows that the computed correlation energy per electron in the case of the most stable conformers of neutral, positively and negatively charged clusters increases for low nuclearity and then remains nearly constant for large nuclearity.

It is not excluded that the correlation energy per electron should further increase for larger nuclearity. Answer to this question can be obtained only by adopting "size-consistent" methods like the Linear Coupled Clusters Approximation (LCCA) [20,21].

2. ALKALI METAL CLUSTERS

2.1 Lithium Clusters

In this section the electronic and geometrical structure of neutral clusters up to Li_{20} and positively and negatively charged clusters up to Li_{10} will be analyzed on the basis of HF-CI results [11,22,23]. The adopted Gaussian basis is of type (2s,1p) (Basis A). A more flexible basis (3s,2p) (Basis B) has been adopted to check the reliablity of Basis A, which, being of reduced dimensions, allows to extend, in a homogenous and systematic way, the investigation to clusters even larger than Li_{20}.

As for the quality of the employed CI treatment, it has been proved that in the case of lithium clusters an MRD-CI approach is necessary in order to overcome the inadequacy of the HF solution as the ground state. Correspondingly, CI calculations have been carried out for all the considered clusters using more reference configurations (5-10) and including in the secular equation all the configurations which give an individual contribution larger than T (T = 0.01 - 0.3 μH) [5]. The geometry optimization carried out for this series of clusters is the most detailed and accurate performed up to now. It is based on a series of minimum searches starting from different guessed geometries subjected to few or no symmetry restrictions. Moreover, for final cluster geometries a complete vibrational analysis has been carried out (using the harmonic valence force field approximation) in order to prove that the stationary point is a true minimum, and to get information about the nature of the lowest frequency normal coordinate along which the cluster structure can easily be deformed. Such an analysis can give clues for designing new starting points for the search of different stable conformers. The minimum energy geometries determined at the HF level have been subjected to a refinement at CI level. In this case the optimization is not complete, being simply a one-dimension search along which the nuclear coordinates are uniformly scaled by a variable factor. Therefore, the best CI cluster geometries have their basic topologies in common with the corresponding HF minima. The shape of the lowest energy

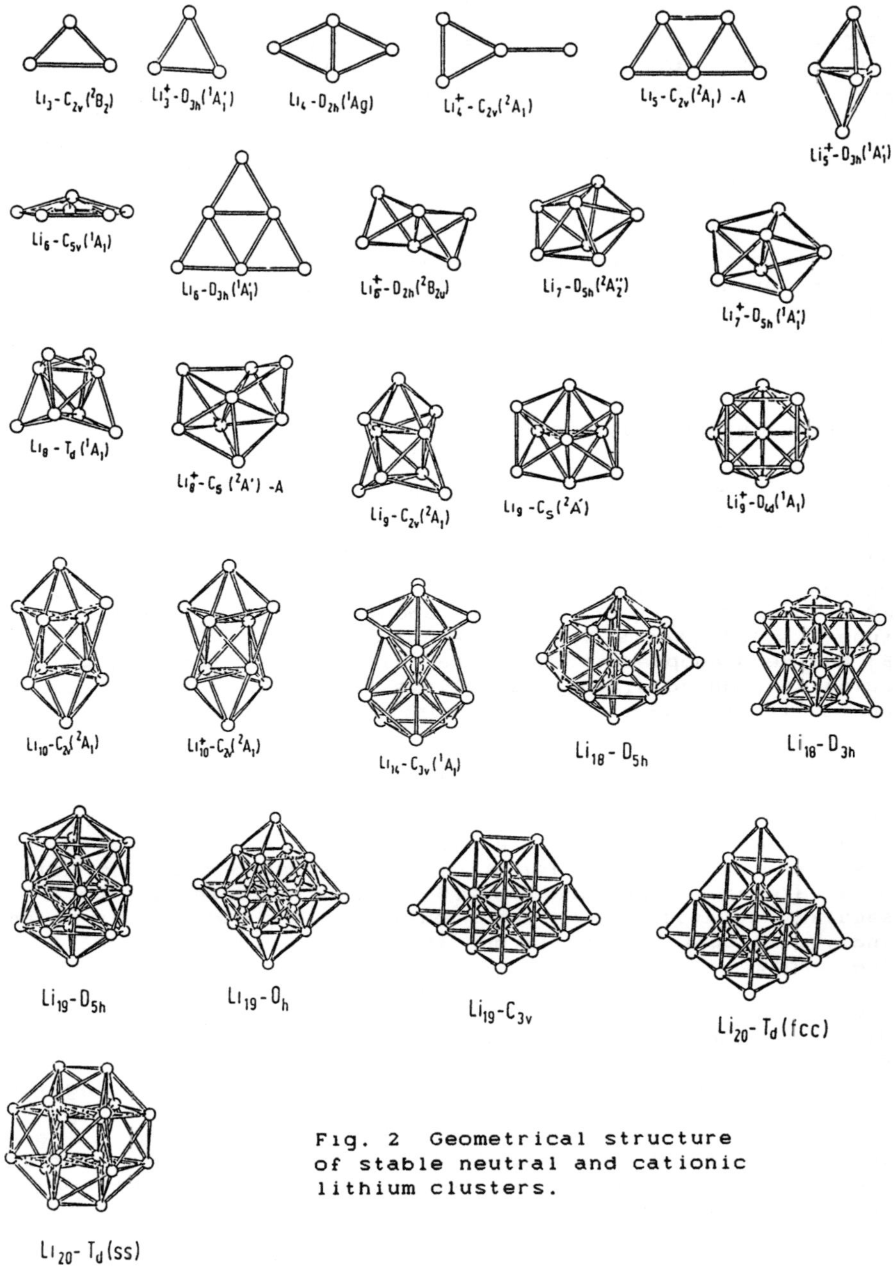

Fig. 2 Geometrical structure of stable neutral and cationic lithium clusters.

conformers are displayed in Fig. 2, while other forms of slightly lower stability are reported in Fig.3 of Ref.11. The binding energy per atom $E_b/n=(-E_n/n)+E_1$, where E_n is the total energy of the n-atom cluster (see Table 1 of Ref 11), has been computed using HF and MRD-CI energy values and according to Davidson's method [24] (CI-Q) which is frequently used to estimate the contribution to the correlation energy of the excitations higher than those explicitly considered in the variational treatment. It is apparent that the HF E_b/n values never exceed 50% of the corresponding MRD-CI values which are very close (90-95%) to the CI-Q estimates. The HF approach, in comparison with the CI trend, not only gives a poor description of the cluster stability but may also predict a wrong stability sequence for various conformers. This is the case, for instance, for Li_6 and Li_7 for which the best HF shapes are the bicapped tetrahedron and the bicapped trigonal bipyramid (Li_6-C_{2v}-B and Li_7-C_{3v} forms, respectively) instead of the planar form D_{3h} or the pentagonal pyramid C_{5v} for Li_6 and the pentagonal bipyramid D_{5h} for Li_7, which are preferred forms according to the CI energy values.

On examining the best cluster geometries some important rules can be deduced. Very often, the lowest energy conformer of a given nulearity does not exhibit the most compact and symmetrical structure. This is strikingly evident in the case of Li_3, Li_4, Li_5 and Li_6 for which a simple intuitive model of bonding would suggest structures like the equilateral triangle, tetrahedron, trigonal bipyramid and octahedron, respectively. Two different (and independent) mechanisms can be operative in determining the instability of the most symmetric cluster forms. The first is the Jahn-Teller (or pseudo Jahn-Teller, JT) instability which affects the structure of systems having an orbitally degenerate electronic state. All the above mentioned cases, characterized by E or T degenerate states in D_{3h}, T_d or O_h symmetry, are JT unstable. The distortions are usually very small but in the case of shallow energy surfaces the complete removal of the orbital degeneracy may require considerably large geometry variations. Two isosceles triangular forms (C_{2v}), a distorted tetrahedron (D_{2h}), a distorted trigonal bipyramid (C_{2v}) and an axially distorted octahedron (D_{4h}) are local minima characterized by a E_b/n value equal to 8.05, 8.98, 12.14 and 14.23 Kcal/mol, respectively. The distorted forms, still very compact, have a higher energy than other forms, particularly planar forms, which cannot be considered as derived from JT distortions. The particular stability of the planar forms (slightly deformed sections of the (111) fcc plane) which occur regularly for Li_4-Li_6 stems from the possibility of forming delocalized three center bonds, characterized by electronic charge accumulation at the center of triangular subunits. As will be discussed in Sect. 4, this particular topology of the electron density acts as the second important mechanism in stabilizing non compact structures.

Two conformers of the lithium hexamer (Li_6-D_{3h} and Li_6-C_{5v}) exhibit a nearly complete degeneracy, a fact which has been extensively checked by also performing "direct-CI" [4] calculations of large dimensions. In addition, these results have also been confirmed by the HF-CI calculations carried out with the Basis B. The equivalence of the two hexamers is an important result because it shows that Li_6 is the real transition point from planar to three dimensional structures which are the most favorable for all the clusters of higher nuclearity. Indeed, starting from Li_7, the most stable conformers are built from tetrahedral (or slightly distorted tetrahedral) subunits. Li_7 can be considerd as derived from Li_6-C_{2v}-A (only 2 Kcal/mol higher in energy than Li_6-D_{3h}): the addition of a lithium atom increases the number of tetrahedral units from three to five. Note that Li_6-C_{5v} and Li_7-D_{5h} are characterized by an exact five-fold symmetry which does not exist in space groups.

The cluster growth by condensing tetrahedra seems to be a very regular process for higher nuclearity clusters. Li_8 is composed of a central tetrahedron with all the faces capped. This form (5 tetrahedral units) is preferred to that of the pentagonal bipyramid capped on one face (6 tetrahedral units) just because of its high compactness and symmetry. The triply degenerate levels of Li_8-T_d are completely filled giving rise to a very stable closed shell structure.

The Li_9 cluster deserves special comment. Its more stable form Li_9-C_{2v}-A is slightly less stable than Li_8-T_d which, in turn, is more stable than Li_7-D_{5h}. This means that maximum stability occurs for Li_8-T_d: a fact of great importance in interpreting the experimentally observed cluster abundancies [25]. A large number of best geometry searches have shown that no stable Li_9 conformers exist with a E_b/n larger than that of tetrahedral octamer.

The cluster Li_{10}-C_{2v} and the largest clusters considered Li_{14}-Li_{20} are all built from tetrahedral subunits [22]. In particular, two very stable forms of Li_{20} have been found, both having a T_d global symmetry. The first is a section of the fcc lattice (T_d-fcc) while the second has a nearly spherical shape (T_d-ss). The CI results predict the T_d-ss form as the more stable by about 18 Kcal/mol, at variance with the HF treatment which favors the T_d-fcc conformer by about 6 Kcal/mol.

The geometry of Li_n^+ clusters up to n=10 has been optimized using the same procedure as for neutral clusters. [11].Table 2 of Ref.10 reports the E_b/n data for the most stable Li_n^+ cations. The first observation is that the ionization of the clusters with an odd number of atoms removes the possibility of the Jahn-Teller instability of the degenerate doublet states. This occurs for n=3 (D_{3h}) and n=5 (D_{3h}) which are now characterized by a stable singlet ground state and a symmetry higher than that of the corresponding neutral clusters. The ionization of Li_7 is not accompanied by an important cluster reconstruction which, on the contrary, occurs for Li_8^+ which would be characterized by a Jahn-Teller unstable state in T_d symmetry. Finally, an important change in cluster shape has been found for Li_9^+ which in the C_{2v} form (similar to the most stable one for the neutral species) has an energy of about 7 kcal/mol higher than the centered square antiprism (Li_9^+-D_{4d}).

The anionic clusters of low nuclearity (Li_2^- - Li_5^-) have linear arrangements in their most stable conformations [23]. This is due mainly to the possibility of minimizing the Coulomb repulsion between the excesses of electronic charge localized on the end atoms of the linear chains. Also in the case of planar or three dimensional clusters (Li_5^- - Li_{10}^-) the negative charge is always localized on the "peripheral" atoms having a low coordination number. The computed values of the adiabatic electron affinity (EA) are reported in Fig.3, where comparison is made with the corresponding ionization potentials (IP). Both IP and EA values exhibit regular oscillations for even- and odd-atom clusters. The IP values for Li_{2n+1} clusters are higher than for Li_{2n} and Li_{2n+2}. The opposite is true for EA values, which are higher for Li_{2n} than for Li_{2n-1} and Li_{2n+1}.

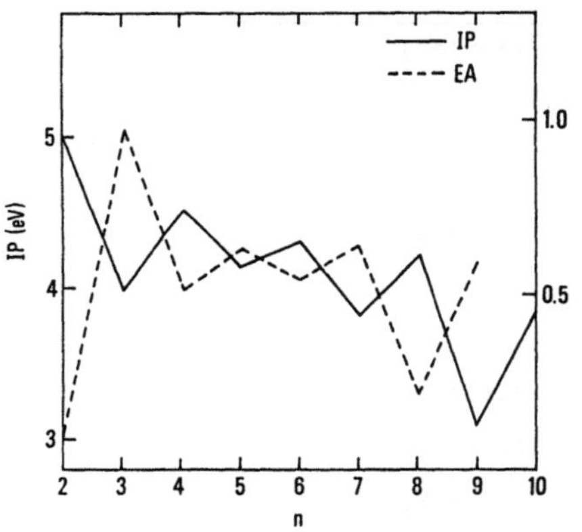

Fig. 3. Adiabatic ionization potentials (IP) and electron affinity (EA) for lithium clusters.

2.2 Sodium Clusters.

The Na_n (n=2,..9) clusters have been investigated with the aim of verifying that the basic cluster topologies and the regular trend in stability and ionization potentials are conserved passing from lithium to sodium clusters. This is an important point since the conservation of the basic cluster properties along the group Ia could bear out that the main bonding features are determined by the valence s electron only, and that the most simple lithium clusters can be considerd as model systems for a broad class of IA and IB Group metal clusters.

The study has been carried out [26] using the HF method in both the AE and ECP approaches and adopting a Gaussian basis of the same quality for the valence space and polarization components. The results reported in Table 1 show that the stability sequence found for Li_n is reproduced not only for the lowest energy conformer but also for conformers lying close in energy.

The Na_n clusters are characterized by E_b/n values smaller than the corresponding lithium species, confirming the stability trend of both the dimers and the bulk metals. The E_b/n data obtained by ECP-HF-CI treatment are parallel to the AE-HF-CI ones, but consistently smaller. This is an indication that core-polarization effects (considered in a natural way by the AE procedure but neglected in the present version of the ECP one) are of significant importance in determining the cluster stability. The IP-V values computed for Na_n are, as expected, smaller than those of Li_n clusters of the same topology. However, the characteristic oscillations for odd- and even-atom clusters is qualitatively reproduced.

Table 1. Stability of Na_n ($n \leq 8$) clusters

Cluster[a]	E_b/n(kcal/mol)[b] AE-CI	ECP-CI	IP-V(eV)[c]
$Na_2(D_{\infty h}, {}^1\Sigma_g^+)$	6.6	6.6	
$Na_3(C_{2v}, {}^2B_2)$	5.6	5.7	
$Na_4(D_{2h}, {}^1A_{1g})$	8.8	7.9	3.9(4.6)
$(C_{2v}, {}^1A_1)$	7.5	7.1	
$(C_{2v}, {}^3B_1)$	7.2	6.7	
$Na_5(C_{2v}, {}^2A_1\text{-A})$	9.5	8.3	3.8(4.5)
$(C_{2v}, {}^2B_1\text{-B})$	8.5	7.0	3.7
$Na_6(C_{5v}, {}^1A_1)$	10.5(10.6)	9.5	3.9(4.5)
$(D_{3h}, {}^1A_1')$	10.8(10.9)	9.7	4.0
$Na_7(D_{5h}, {}^2A_2'')$	11.0	10.0	3.4(3.9)
$Na_8(T_d, {}^1A_1)$	12.0	10.8	4.0(4.6)

a) See Fig. 2 and 3 of Ref. 11 for the identification of the cluster structure; b) Binding energy per atom from MRD-CI results, in parenthesis values from direct CI; c) AE-CI vertical ionization potentials, in parenthesis values for the lithium cluster of same topology.

The above discussed results of HF-CI investigations on lithium and sodium clusters may be compared with the results obtained using different theoretical approaches. A systematic GVB investigation [27] on Li_n ($n \leq 13$) and a LSD study on Na_n ($n \leq 13$)clusters [18] have been reported. The GVB theory predicts that the most stable forms for Li_3-Li_6 are planar and composed of almost regular equilateral triangles. The preferred three dimensional structures for higher nuclearities are realized by condensing the maximum possible number of tetrahedra. In this respect the GVB results substantially agree with the HF-CI ones. Similar agreement can also be noted between HF-CI and LSD results [18] as for sodium clusters. A few differences, however, have to be mentioned in this case. The LSD method for Na_8 gives an optimized geometry less compact than Na_8-T_d, and in general, the LSD E_b/n values are consistently higher and the nearest neighbour distances consistently smaller than the corresponding HF-CI ones. As mentioned in Sect.1, truncated CI expansions can give an underestimated binding energy. On the other hand, the local approximation applied to the exchange-correlation functional may yield a substantial overestimation of the cluster stabilities, especially if the "self-interaction" correction [28] is not taken into account. In addition, the non-perfect agreement between HF-CI and LSD results concerning the best cluster forms may be due to the different methodological procedures adopted in the geometry optimization (see Sect.1).

Despite some discrepancies, obviously expected from methods based on different theoretical models, a general qualitative agreement can be noted, at least for the basic cluster properties such as shape and stability trend. The satisfactory consistency of the results so far obtained gives confidence in the predicting power of the theory in the field of light metal clusters.

3. MAIN GROUP ELEMENTS AND TRANSITION METAL CLUSTERS.

In discussing the electronic structure of elemental (other than alkali metals) clusters, a general observation has to be made. The theoretical and computational problems met in studying Ia Group clusters are much more pronounced for the atoms considered here because of i) the larger number of electrons involved and the more critical role played by the associated correlation energy; ii) the need to include functions with high angular momentum in the basis set to properly describe polarization and angular correlation effects of sp and sd valence electrons; and iii) the need to accurately describe a manifold of low lying valence states which characterizes the transition metals. All facts considered, it is not surprising that a systematic investigation by quantum chemical methods of the electronic-geometric structure relationships are lacking for this class of elemental clusters.

IIa Group elements are a very interesting example of how important the sp hybridization can be in determining cluster stability. If this occurs, a cluster is expected to be stable only if the energy gained from the formation of metal-metal bond is higher than the sum of the atomic promotion energies. The Be clusters can offer a good example of such behaviour. The Be-Be bond strength is essentially determined by the participation of sp hybrids. The Be_2 and Be_3 species are unstable (or very weakly bound) while Be_n ($4 \leq n \leq 13$) clusters are characterized by a E_b/n value ranging from about 10 to 20 Kcal/mol (see papers quoted in Ref. [6]). The relative high stability of Be_4 has been confirmed by a variety of theoretical works carried out at different levels of approximation. In a non-systematic optimal geometry search, the Be_n clusters ($n>4$) have been found to be planar or three-dimensional sections of the hcp lattice, although recent ECP-HF-CI investigations have pointed out that $Be_{13}(D_{3h})$ derived from fcc lattice has a stability slightly higher than $Be_{13}(D_{3h})$ hcp. Worth noting is the fact that Be_4 and Be_5 assume a very compact and symmetrical shape: the number of valence electrons is high enough to completely fill all the degenerate orbitals of T_d or D_{3h} symmetry and no Jahn-Teller instability can occur.

In the case of Mg and Ca clusters, the importance of p and d polarization functions is even more evident. HF calculations adopting a double-zeta basis (6s,4p) give, for Mg_4, only a very shallow minimum for a Mg-Mg distance of 4.1 Å (see papers quoted in Ref. [6]), while sophisticated correlation treatments using a basis of type (5s,3p,1d) are needed in order to describe a Mg_4-T_d cluster stable by about 24 kcal/mol, with a Mg-Mg distance of 3.33 Å.

Results from ECP-HF-CI, AE-HF-CI and ECP-LSD calculations (see papers quoted in Ref. [6]), all show that d polarization functions are important in stabilizing the metal-metal bond in Mg_4 and Ca_4. Note that the approximate ECP-HF-CI study adopted a basis set of 15 contracted functions, and that the corresponding computational cost is at least as high as that required for Li_{12}-Li_{14} clusters. No theoretical works have been reported for Mg_n or Ca_n clusters for $n>4$.

Among the IIIa Group elemental clusters, the Al_n systems are the most widely studied. Al_3 is an interesting example because of its similarity to the alkali trimers. The equilateral form is a saddle point between two energetically nearly equivalent isosceles triangular forms (acute and obtuse, respectively) [29]. The energy barrier governing the fluxionality (and the pseudorotating behaviour) is no higher than 2.5 kcal/mol. From several theoretical studies [30,31] (all performed with correlated methods) evidence has been obtained for preferred planar structures for $Al_4(D_{2h})$ and $Al_5(C_{2v})$, similar to those found for small alkali metal clusters. However, the octahedral form is largely preferred with respect to planar ones, and close in stability to a conformer built from three condensed tetrahedra. The largest cluster Al_{13} has been studied only at the ECP-HF level [31] because the extended basis set used (3s,3p,2d) and the high number of electrons make a reliable CI treatment unfeasible. The icosahedral structure and compact forms derived from hcp and fcc lattices are characterized by a computed E_b/n value of 25.6, 24.9 and 23.5 Kcal/mol, respectively.

The electronic structure investigation of the IVa Group covalent clusters [32-35] has revealed important differences between carbon and silicon or germanium clusters. The bonding in C_n is dominated by strong π interactions giving rise to preferred linear or monocyclic forms. No evidence has been obtained [33] for the existence of fused-ring conformers with a graphite-type structure up to C_{10}. The clusters with odd numbers of atoms have linear structures and singlet ground states. Cyclic structures are more favorable for C_{2n} clusters, especially in the case when ring closure is not accompanied by steric strain. C_n (n=6,8,10) have ring forms distorted, however, with respect to the regular polygonal form, and characterized by a symmetry lower than D_{nh}. ECP-HF-CI [34] and AE-HF-CI [35] studies have been reported on Si_n clusters. Ref. [34] also reports comparative studies between Si_n and Ge_n clusters. The most stable structure for M_n is derived from that of M_{n-1} by edge or face capping. For instance, Si_n (n=7-10) are composed of a basic octahedral unit and with up to four faces capped. Si_{10} has the same global symmetry (T_d) as an adamantane-type structure but the two forms seem to be characterized by very different electronic structures and bond distances. Analysis of the wavefunction of Si_n and Ge_n has shown that the valence configuration assumed by the cluster atoms is intermediate between that of the free atoms (s^2p^2) and that in the bulk crystal (sp^3). All facts considered, it seems questionable to assume that 10-atom clusters can have features in common with crystalline solids.

It is superfluous to recall that the problem of understanding the electronic properties of the transition metal (TM) clusters is very challenging because of their possible implication in several other fields (catalysis, inorganic and metallorganic chemistry, material and surface science, etc.). However, up to now both experimental and theoretical research has been able to give only few systematic results from which a general knowledge of the TM clusters can be unambiguously derived.

Rigorous quantum mechanical studies of TM systems as small as diatomics have revealed the existence of such great difficulties that one can hardly hope for an early development of computational techniques routinely applicable to such complicated systems. The TM atoms are characterized by closely spaced excited states generated by intershell d-s and s-p excitations. Even a good description of the atomic ground state at HF level is difficult in the sense that the quality of the results seems to be much more basis set dependent than in the case of non-TM atoms. In particular, the d one-electron functions of the first transition atoms may require at least a double-zeta representation in order to compute properly bound d states. For example, the 3d one-electron energy for $Cu(^2S)$ is 25 eV higher than the corresponding numerical HF value. Moreover, it has been found that in molecular calculations the double-zeta representation of d orbitals must be augmented by (at least) one diffuse component. In addition, if electron correlation has to be considered, f (and g) functions are expected to play an important role. Finally, quantitative predictions of binding energies and bond lengths are possible provided that the leading terms of the relativistic corrections are taken into account. The numerous studies carried out on Cu_2 attest to the importance of all the mentioned aspects (see papers quoted in Ref. [6]).

The situation is even more complicated for other TM diatomics due to the presence of unpaired d electrons which, at molecular level, can give rise to a very large number of closely spaced states. For instance, in the case of Fe_2 it has been found [36] that 112 states arising from two interacting $3d^74s^1$ iron atoms lie within a range of only 0.5 eV.

Another source of difficulty is the fact that almost all the experimentally and/or theoretically determined ground states of the TM diatomics do not correlate with separated atoms in their ground state. One or both TM atoms need to be promoted from the $d^{n-2}s^2$ lowest configuration to the $d^{n-1}s^1$ one. This occurs for all the TM diatomics with the obvious exception of Cr_2 and Cu_2. $Cr_2(^1\Sigma_g^+)$ correlates with two atoms in their ^7S ground state but the Cr-Cr bond formation is accompanied by a complete reorganization of the electron spins. HF and multiconfiguration self-consistent calculations, followed by extensive CI treatment, up to now performed seem to be unable to describe neither this peculiar feature or a stable chromium dimer [37]. The Cr_2 molecule has also been the subject of several LSD studies which were able to nicely reproduce the experimental data concerning r_e and ω_e by means of a computed wavefunction in which spatial and spin symmetry restrictions were removed [38]. LSD studies pointed out another very important aspect of the interaction among d electrons which can be relevant also for other TM diatomics and clusters. As the Cr-Cr distance is slightly increased with respect to r_e, the d-d interaction leading to an hexuple strong bond rapidly decreases giving rise to only a weakly bounded system, thanks to an anti- ferromagnetic coupling of the d unpaired electrons.

While from Sc up to Cr the metal-metal bond in diatomics is dominated by the d-d interaction, only a weak antiferromagnetic coupling occurs in Mn_2 and for the right-side atoms the d-d interaction is too small (due to the spatial contraction of orbitals) to appreciably contribute to the bond. This general behaviour is confirmed by the fact that the (most probable) electronic ground state for Fe_2, Co_2 and Ni_2 is characterized by 6,4 and 2 unpaired electrons, respectively, and also by the theoretical results on TM clusters. For instance, ECP-HF studies of the $Ni_n(n \leq 6)$ systems [39] and HF-CI of Ni_3 [44] have shown that the nickel atoms assume a configuration very close to d^9s^1; the Ni-Ni bond is essentially due to the s electrons and the high spin state of the cluster is due to the presence of one unpaired d electron per nickel center (plus eventually s unpaired spins). These basic observations can be derived also from a systematic X_α-SW analysis of the spin density distribution in M_{13}(hcp or fcc) and M_{15}(bcc) clusters[40]. Sc_{13}, Ti_{13} and V_{13} are found to be essentially non magnetic, Cr_{15} and Mn_{13} have 12 and 11 unpaired spins, respectively, but are involved in a strong antiferromagnetic coupling. For Fe_{15}, Co_{13} and Ni_{13}, an average magnetic moment per atom equal to 2.7, 1.6 and 0.6 has been computed while the magnetic moment found for Cu_{13} is entirely due to an unpaired s electron. These theoretical findings correlate well with the experimental magnetic measurements and could suggest that the spin ordering in TM clusters of relatively low nuclearity is already close to that of the bulk metals. However, the conclusion must be considered with care because the spin density distribution in clusters may greatly depend upon the assumed cluster geometry and details of the adopted computational procedure. Moreover, the spin density on cluster atoms is markedly not homogeneous and the spin-up spin-down segregation seems to be controlled by a mechanism different from that of the bulk metal.

Attempts based on theoretical results of proving that the electronic structure of a relatively small TM cluster is in many respects similar to that of the bulk metal have been made by comparing the Density Of States (DOS) of band structure calculations with that derived by broadening the discrete one-electron energy spectrum of the cluster. X_α-SW and other LSD calculations have shown that a narrow "d band" is superimposed on a broad "s band", in agreement with the band structure results, in the case of relatively small clusters (see Ref.6 and references therein). In the case of Cu_n clusters an analogous analysis based on HF one-electron energies led to a completely different description of the DOS. The d states, being characterized by a much higher binding energy,are well separated from the s ones. Such contrasting predictions based on HF or LSD calculations gave rise to an intense debate about the physical meaning of the one-electron energies as a measure of ionization energies. In fact, when the electron relaxation caused by the ionization process is taken into account and the d hole is described as localized (requiring a symmetry broken solution), the HF ionization energies decrease and some overlap can occur with the corresponding values for ionizations of s electrons which do not seem appreciably affected by relaxation effects. This can bring the HF results in close agreement with the LSD ones. However, the similarity between DOS profiles for cluster and bulk metal still does not prove any physical similarity since the overlap between d and s ionization energies has also been found for clusters as small as Cu_2 [41] and Ni_4 [42], the electronic structure of which is certainly very dissimilar from the bulk one.

Beside trimers [43,44] and a few other examples of TM clusters of low nuclearity (see Ref [6] and references therein) no systematic investigations have been reported aimed at solving basic problems connected with the general dependency of the cluster properties (stability, ionization potential and geometrical shape) upon the cluster size. All these problems can in principle be addressed by employing LSD methods whereas the applicability of HF correlated methods seem to be outside present computational capabilities. The impossibility of comparing results given by different methodologies leaves many questions of basic importance completely open.

4. MANY-BODY INTERACTIONS IN METAL CLUSTERS

The understanding of the nature and topology of the interatomic interactions in metal clusters is essential for two kinds of reasoning. On one side, it can supply a general rationalization of the preferred cluster shapes and paths in cluster growth and, on the other side, it can help in developing analytical many-atom potentials for molecular mechanics (and dynamics) studies, the only possible approach of investigating systems of large nuclearity. A very appealing project would be that of outlining for elemental clusters computational techniques as fast as those currently used, for instance, for complex organic molecules. Methods of molecular mechanics [45] have been successfully developed for organic systems thanks to the existence of a very large collection of experimentally determined structures. On the contrary, for elemental clusters there are only few structural data available up to now, as at the moment they can be obtained only from cumbersome theoretical calculations.

A model interaction potential may have the general form

$$V(1,2,...N) = \Sigma'_{ij}\, E^{(2)}_{ij} + \Sigma'_{ijk}\, E^{(3)}_{ijk} + \ldots\, E^{(N)} = E^{(2)} + E^{(3)} + \ldots\, E^{(N)}, \qquad (1)$$

(where Σ' indicates summations restricted only to distinct n-ples). A potential of form (1) can be of practical use only if the expansion is very rapidly convergent so that the high order terms can be neglected. On the other hand, it is difficult to find a general analytical form even for $E^{(4)}$ terms.

The use of two-atom terms only leads obviously to very compact structures [46] also for small clusters; this does not seem to be case for IA and IIIA group elements. The inclusion in the potential of the $E^{(3)}$ term gives rise to severe problems because it is difficult to decide which kind of structure-stability data the resulting $E^{(2)}+E^{(3)}$ potential must reproduce. For instance, Stillinger and Weber (SW) [47] have parameterized a 2- and 3-body potential aimed at reproducing the bond distances and cohesive energy of the bulk silicon. However, applications of the SW potential to small Si_n clusters [48] gave preferred structures notably different from the HF-CI ones [32,34-35], a result simply confirming the deep differences occuring between clusters and crystalline solids.

An alternative approach would be that of deriving an $E^{(2)}$ $+E^{(3)}$ potential from theoretical data, since the energy surfaces of the dimer and trimer may be easily computed with high accuracy, at least for light atoms. For instance 2- and 3-body analytical potentials have been obtained for Li_2 and Li_3 [49]. However, the question of how important can be the neglected higher order terms remains open.

A detailed analysis of the relative magnitude of $E^{(n)}$ (n=2,3,...6) has been carried out for planar lithium clusters Li_4-D_{2h}, Li_5-C_{2v} and Li_6-D_{3h} [50]. The $E^{(n)}$ (n=4,5,6) contributions have been computed for all the relevant 4-, 5- and 6- atoms groups using MRD-CI energy values determined with the Basis A (see Sect.2). Two and three-center terms are evaluated from an analytical potential in the form of extended Rydberg-type functions [51] fitting numerous MRD-CI energy data (Basis A) for Li_2 and Li_3. The relative magnitude of the $E^{(n)}$ contributions has also been computed for different Li-Li nearest neighbour distances ranging from 5.5 to 10 au. The results clearly show that in this case an expansion of type (1) is extremely slowly convergent; the $E^{(6)}$ term in Li_6-D_{3h} is as large as -2.5 eV and accounts for more than 50% of the total cluster stability. The behaviour of the many-body expansion is essentially determined by the particularly low stability of the trimer which produces an $E^{(3)}$ term strongly repulsive with respect to the three 2-body contribution. The $E^{(2)}+E^{(3)}$ sum needs to be corrected by a relatively large attractive $E^{(4)}$ term, and so forth. The obvious conclusion is that 2- and 3-body potentials cannot be derived from stability data of the dimer and trimer and transferred to higher nuclearity systems. The same considerations are likely applicable to all the elements for which the binding energy of the trimer is significantly smaller than those of the dimer and tetramer.

Simplified model potentials $V=E^{(2)}+E^{(3)}$ can be assumed to have the Lennard-Jones (6-12) (LJ) form for $E^{(2)}$ and the Axilrod-Teller (AT) form for $E^{(3)}$[51]. In the case of Al_n clusters it has been found [31] that LJ+AT potential may be useful only in suggesting "possible" cluster shapes being, on the other hand, unable to correctly reproduce, for instance, the stability sequence I_h > hcp > fcc > planar predicted by ECP-HF calculations on Al_{13}. The sequence given by the LJ+AT potential is planar > hcp = fcc > I_h and the corresponding E_b/n values are about 70% larger than the HF ones. Application of LJ+AT potential to small lithium clusters [50] gave not significantly better results than for Al_n

The question can be raised why the expansion of type (1) for the many body interaction fails when applied to metal clusters. A plausible explanation can be offered by the analysis of the electron density in lithium clusters. Plots of the total density ρ and distortion density $\Delta\rho = \rho - \Sigma_i \rho_i$ (ρ_i being the density of the individual non-interacting atoms) reported in Ref [11], are characterized in the case of planar clusters by maxima at the center of triangles. The occurence of a maximum at the Li-Li midpoint (a situation similar to that of the dimer) is observed only for the shortest edge of the trapezium Li_5-C_{2v}. A quantitative analysis of ρ carried out according to Bader's theory [52] on planar Li_n (n=4,5,6) clusters [7] confirmed this qualitative picture. The results of Ref.[7] can be briefly summarized as follows. The maxima of ρ occuring at the center of the triangles and at the Li-Li midpoint are essentially responsible for the cluster stability in the sense that they act as "attractors". The bonding interaction existsonly among Li atoms and attractors and among attractors themselves. In other words, no direct Li-Li bond exist. On the basis of the bond-paths identified according to Bader's theory it is possible to define a formal valence both for real atoms and attractors. This formal valence is equal to one for Li atoms with one or two nearest neighbours and equal to two for atoms having three or four neighbours. The formal valence of the attractors associated with ρ maxima at a position other than nuclei can be as high as five due to their tendency of connecting the maximum possible number of attractors and real atoms. It is apparent that if a suitable form of a many-center potential has to be worked out, the above described particular topology of ρ must be taken into account. For instance, the $E^{(2)}$ term seems to have little significance in this context. In addition, in the expansion (1) all the ijk triples contribute to $E^{(3)}$ which has a repulsive character. On the contrary the analysis of ρ shows that the 3-center term is mostly responsible for the cluster stability and it is composed only by topologically well defined triples.

Extension of the analysis to larger clusters [53] will help to define the nature and localization of the attractors responsible for the stability of three dimensional clusters and to suggest a possible strategy in defining an interaction potential depending not only upon the spatial positions of the real atoms.

References

1. a) R. McWeeney, B.T. Sutcliffe; Methods of Molecular Quantum Mechanics, Academic Press, London (1969); b) H.F.Schaefer III, Ed., Methods of Electronic Structure Theory, Plenum Press, New York (1979); c) W.J.Here, L.Radom, P. v.R.Schleyer, J.A. Pople; Ab Initio Molecular Orbital Theory, John Wiley, New York (1986); d) A.K.Rajagopal; Adv.Chem.Phys. 41, 59 (1980); e) R.G.Parr; Annu.Rev.Phys.Chem. 34, 631 (1983); f) J.P. Dahl, J.Avery, Eds., Local Density Approximations in Quantum Chemistry and Solid State Physics, Plenum Press, New York (1984); g) R.M.Dreizler, J.da Providencia, Eds, Density Functional Methods in Physics, Plenum Press, New York (1985)
2. W.Kohn, C.J.Sham; Phys.Rev. B 140, 1133 (1965)
3. O. Gunnarson, R.O.Jones; Phys.Rev. B 31, 7588 (1985)
4. a) B.O. Roos; Chem.Phys.Lett. 15,153 (1972); b) See the contribution of B.O. Ross, P.M. Siegbahn in Ref.2b; c) M.F. Guest, J. Kendrick; Daresbury Laboratory Report No. CCP1/86, unpublished
5. a) R.J. Buenker, S.D. Peyerimhoff; Theoret.Chim.Acta 35,33 (1974); b) R.J. Buenker, S.D. Peyerimhoff, W.Butscher; Mol. Phys. 35,771 (1978); c) R.J. Buenker in Studies in Physical and Theoretical Chemistry. Current Aspects of Quantum Chemistry (R.Carbo Ed.), Elsevier, Amsterdam (1980); d) R.J. Buenker, R.A. Philips; J.Mol.Struct. Theochem 123,291 (1985)
6. J.Koutecky, P.Fantucci; Chem.Rev. 86,539(1986)
7. C.Gatti, P.Fantucci, G.Pacchioni; Theoret. Chim. Acta, in press
8. H.-O. Beckman, J.Koutecky, P.Botschwina, W.Meyer; Chem.Phys.Lett. 67,119(1980)
9. P.Fantucci, V. Bonacic-Koutecky, J. Koutecky; J.Comput. Chem. 6,642 (1985)
10. a) R.F. Marshall, R.J. Blint, A.B.Kunz; Phys.Rev. B 13,333 (1976); b) R.F. Marshall, R.J. Blint, A.B. Kunz; Solid State Commun. 18,731 (1976)
11. I. Boustani, W. Pewestorf, P.Fantucci, V. Bonacic-Koutecky, J. Koutecky; Phys. Rev. B, 35, 9437 (1987)
12. L. Szasz; Pseudopotential Theory of Atoms and Molecules, John Wiley, New York (1985)
13. P.J. Hay, W.R. Wadt; J.Chem.Phys. 82,270 (1985); ibidem, 82, 284 (1985)
14. G.B.Bachelet, D.R.Hammann, M. Schulten; Phys.Rev. B 26,4199 (1982)
15. a) G.-H. Jeung, J.P. Malrieu, J.P. Daudey; J.Chem.Phys., 77,3571 (1982); b) G.-H. Jeung, J.C. Barthelat; J.Chem.Phys. 78,2097 (1983); c) P.Fuentealba, H.Preuss, H.Stoll, L. von Szentpaly; Chem.Phys.Lett. 89,418 (1982)
16. P.J. Hay, W.R. Wadt; J.Chem.Phys. 82,299 (1985)
17. J.A. Pople, R.K. Nesbet; J.Chem.Phys. 22,571 (1954)
18. a) R. Car, J.L. Martins; Surf.Sci. 106,280 (1981); b) J.L. Martins, J. Buttet, R. Car; Phys.Rev.Lett. 53,655 (1984); c) J.L. Martins, J. Buttet, R. Car; Phys.Rev.B 31,1804 (1985); d) J.L. Martins, R. Car, J. Buttet: J.Chem.Phys. 78,5646 (1983); e) J.L. Martins, J. Buttet, R. Car; Ber. Bunsenges. Phys.Chem. 88,239 (1984)

19. A. Szabo, N.L. Ostlund; Modern Quantum Chemistry: Introduction to Advance Electronic Structure Theory, MacMillan, New York (1982)
20. J. Cizek, J. Paldus; Physica Scripta 21,251 (1980)
21. P.Scharf, S. Brode, R. Ahlrichs; Chem.Phys.Lett. 113,447 (1985)
22. I. Boustani, J. Koutecky, to be published
23. I. Boustani, J. Koutecky, to be published
24. S.R. Langhoff, E.R. Davidson; Int.J.Quantum Chem 8,61 (19874)
25. See the contribution of J.Koutecky, V.Bonacic-Koutecky, I.Boustani, P.Fantucci, this book
26. V.Bonacic-Koutecky, P.Fantucci, J. Koutecky, to be published
27. a) M.H. McAdon, W.A. Goddard III; Phys.Rev.Lett. 55,2563 (1985); b) J. Non-Crystalline Solids 75,149 (1985); c) J.Phys.Chem. 91,2607 (1987)
28. a) J.P. Perdew; Chem.Phys.Lett. 64,127 (1979); b) S.-W. Wang; J.Chem.Phys. 82,4633 (1985)
29. G.Pacchioni, P.Fantucci, J. Koutecky, to be published
30. a) G.Pacchioni, D.Plavsic, J. Koutecky; Ber.Bunsenges. Phys.Chem.86,503 (1983); b) G.Pacchioni, J. Koutecky; Ber. Bunsenges.Phys.Chem. 88,242 (1984); c) J. Koutecky, G. Pacchioni, G.-H. Jeung, E.C. Hass; Surf.Sci. 156,60 (1985)
31. a) W. Bauschlicher,Jr, L.G.M. Petterson; J.Chem.Phys., in press.; b) L.G.M. Petterson, W. Bauschlicher,Jr.; J.Chem. Phys., in press, and references therein.
32. W.L. Brown, R.R.Freeman, K. Raghavachari, M. Schluter; Science, 235,860 (1987)
33. K. Raghavachari, J.S. Binkley; J.Chem.Phys., in press
34. G. Pacchioni, J. Koutecky; J.Chem.Phys. 84,3301 (1986)
35. a) K. Raghavachari, V. Logovinsky; Phys.Rev.Lett. 55,2853 (1985); b) K. Raghavachari; J.Chem.Phys. 84,5672 (1986)
36. I. Shim, K.A. Gingerich; J.Chem.Phys. 77,2490 (1982)
37. a) A.D. McLean, B. Liu; Chem.Phys.Lett. 101,144 (1983); b) M.M. Goodgame, W.A. Goddard III; Phys.Rev.Lett. 54,661 (1985); c) S.P. Walch, C.W. Bauschlicher,Jr, B.O. Roos, C. Nelin; Chem.Phys.Lett. 103,175 (1983)
38. N.A. Baykara, B.N. McMaster, D.R. Salahub; Mol.Phys. 52,891 (1984)
39. H. Basch, M.D. Newton, J.W. Moskovits; J.Chem.Phys. 73,4492 (1980)
40. D.R. Salahub: In Contribution of Cluster Physics to Materials Science and Technology, J. Davenas, P.M. Rabette, Eds., Martinus Nijhoff, Dordrecht (1986)
41. P.A. Cox, M. Benard, A. Veillard; Chem.Phys.Lett. 87,159 (1982)
42. M.D. Newton; Chem.Phys.Lett. 90,291 (1982)
43. a)E. Miyoshi, H. Tatewaki, T. Nakamura; J.Chem.Phys. 78,815 (1983); Int.J.Quantum Chem. 23,1201 (1983); b) G.-H. Jeung, M. Pellissier, J.C. Barthelat; Chem.Phys. Lett. 97,369 (1983); c) G. del Conde, P.S. Bagus, O. Novaro; Phys.Rev. B 25,7843 (1982), G. del Conde, J. Garcia-Prieto, O. Novaro; Mol.Phys. 44,477 (1981)
44. (a) J.P. Walch; J.Chem.Phys. 86,5082 (1987). b) S.P.Walch, C.W.Bauschlicher, in: Quantum Chemistry: The Challenge of Transition Metals and Coordination Chemistry, A.Veillard Ed., D.Reidel, Dordrecht (1986)

45. N.L. Allinger; J.Amer.Chem.Soc. 99,8127 (1977)
46. M.R. Hoare; Adv.Chem.Phys. 40,49 (1979)
47. F.H. Stillinger, T.A. Weber; Phys.Rev.B 34,29 (1985)
48. E. Blainsten-Barojas, D. Levesque; Phys.Rev. B 34,3910 (1986)
49. A.J.C. Varandas, V.M.F. Morais; Mol.Phys. 47,124 (1982)
50. P.Fantucci, J.Koutecky, to be published
51. J.N. Murrel,, S. Carter, S.C. Farantos, P. Huxley, A.J.C. Varandas, Molecular Potential Energy Functions, John Wiley, New York (1984)
52. R.F.W. Bader; Acc.Chem.Res. 18,9 (1985), and references therein
53. C.Gatti, P.Fantucci, G.Pacchioni, to be published

Molecular Clusters

J. Jortner[1], *D. Scharf*[1], *and U. Landman*[2]

[1]School of Chemistry, Tel-Aviv University, 69978 Tel-Aviv, Israel
[2]School of Physics, Georgia Institute of Technology, Atlanta, GA 30332, USA

I. INTRODUCTION

Structural, electronic and chemical characteristics of materials depend primarily on the state (phase) and degree (size) of aggregation. Clusters, i.e., finite aggregates containing 2-10^4 particles, exhibit unique physical and chemical phenomena, which provide ways and means to explore the gradual transition from molecular to condensed matter systems [1-4]. As has been repeatedly emphasized [5,6], the elucidation of the level structure of elementary excitations, e.g., phonons, electrons and excitons, the dynamics of nuclear motion, the nature of electronic states, electronic and vibrational relaxation phenomena in clusters cannot be characterized in terms of a conceptual framework resting solely on one of the traditional disciplines of molecular physics or of condensed matter (Fig. 1). The area of structure, level structure and dynamics of clusters provides a bridge between molecular and condensed matter physics. Furthermore, the large surface to volume ratio in clusters allows for the exploration of microscopic surface chemical physics [7,8]. The progress in our understanding of the evolution of quantum, thermodynamic, dynamic and chemical size effects in clusters is expected to provide a central contribution to the merging between the microscopic and macroscopic points of view in the unified description of the interrelationship between molecular, surface and bulk phenomena in large finite systems.

Ten years ago when cluster physics and chemistry were in their embryonic stages, the study of clean and isolated clusters was viewed as the 'theoretician's dream' [5]. This dream has come true! The remarkable progress in the areas of supersonic jets and cluster beams [9,10] makes isolated clusters amenable to experimental interrogation, while the merging between the beam and laser techniques provides experimental information pertaining to the dynamics of nuclear motion, the nature of electronic states, electronic and vibrational relaxation as well as chemical processes in clusters. There are some compelling reasons for the study of isolated clusters.

(1) The exploration of a wealth of energetic, dynamic and chemical phenomena in large finite systems, whose size can be continuously varied, provides the ways and means for the exploration of size effects.

(2) These systems allow for a microscopic approach to macroscopic phenomena, such as nucleation, adsorption, desorption and catalysis.

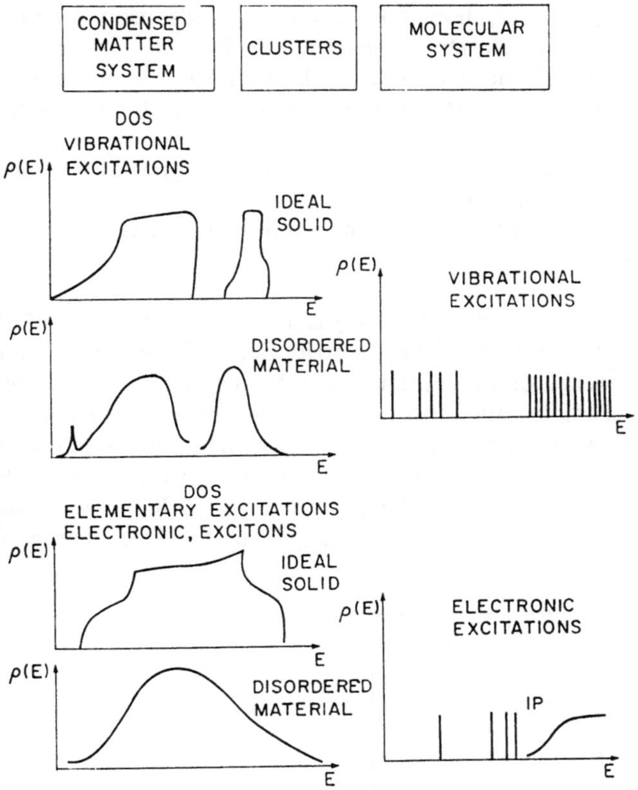

Fig. 1. Density of states (DOS) of vibrational and electronic excitation in condensed matter systems and in molecular systems. The DOS for electronic and vibrational excitation in condensed matter systems is continuous, being characterized by van Hove topological singularities for ordered structures and by exponential Mott tails for disordered materials. The DOS electronic excitation of molecular systems is discrete below the first ionization potential. In molecular systems the vibrational DOS is discrete, while in large molecules a quasicontinuum of vibrational states is manifested at high energies.

(3) A variety of astrophysical applications rest on cluster physics and chemistry. These studies include the formation mechanism and properties of cosmic dust [11]. Furthermore, clusters in the interstellar medium may be responsible for some fascinating phenomena in astrophysical spectroscopy [12,13], e.g., the ubiquity of undefined ultraviolet and visible absorption bands in the range 443-800 nm [13] may originate from inorganic clusters, while graphite clusters may give rise to large polycyclic aromatic hydrocarbons, which presumably are responsible for some infrared interstellar emission bands in the range 3.3-11 μ [12].

The first obvious question is, how does a cluster holds itself together? The energetic stability of a cluster is determined by the binding. A classification of the nature of the chemical bonds in clusters is presented in Table I. Clusters fall into two general categories according to the strength of the chemical binding:

(1) "Weak interactions" prevailing in van der Waals (vdW) type, as well as in molecular and hydrogen-bonded clusters.

(2) "Strong interactions", which encompass ionic and valence clusters, as well as many metallic clusters.

The atomic structure of the cluster (at 0 K) is a direct consequence of the binding. At finite temperatures the contribution of the vibrational excitations to the free energy of the cluster has to be incorporated. Figure 2 provides a typical example for the temperature dependence of the equilibrium configuration of alkali halide clusters, which were explored by MARTIN [14] and by us [15]. The cubic structure constitutes the stable equilibrium configuration of the $(NaC\ell)_4$ cluster at 0 K. At finite temperatures the vibrational entropy contribution, S, to the free energy, G,

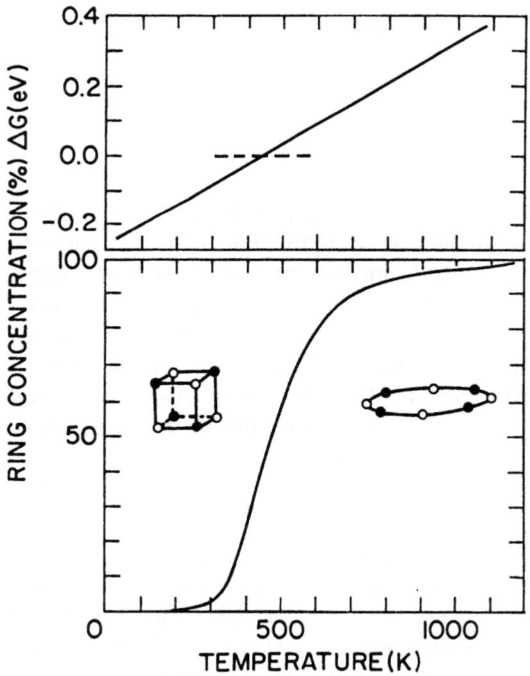

Fig. 2. Free energy difference between the cubic and the ring-shaped cluster of $(NaC\ell)_4$ (top), and the temperature dependence of the relative concentration of the ring-shaped structure (bottom). Data from references 14 and 15.

TABLE I: CLASSIFICATION OF BINDING IN CLUSTERS

TYPE	"CANONICAL" CASES	CAUSE OF BINDING	AVERAGE BINDING ENERGIES	OCCURRENCE
VAN DER WAALS CLUSTERS	RARE-GASES R_n $(N_2)_n$, $(CO_2)_n$ $(SF_6)_n$	DISPERSIVE PLUS WEAK ELECTRO-STATIC	WEAK BINDING $D \leq 0.3$ eV	AGGREGATES OF RARE-GASES AND CLOSED-SHELL MOLECULES
MOLECULAR CLUSTERS	ORGANICS $(M)_n$ $(I_2)_n$	DISPERSIVE ELECTROSTATIC (WEAK VALENCE)	MODERATE BINDING $D \sim 0.3$-1 eV	AGGREGATES OF ORGANIC MOLECULES AND SOME CLOSED-SHELL MOLECULES
HYDROGEN-BONDED CLUSTERS	$(HF)_n$, $(H_2O)_n$	H-BONDING ELECTROSTATIC CHARGE TRANSFER	MODERATE BINDING $D \sim 0.3$-0.5 eV	CLOSED-SHELL MOLECULES CONTAINING H AND ELECTRO-NEGATIVE ELEMENTS
IONIC CLUSTERS	$(NaC\ell)_n$, $(CaF_2)_n$	IONIC BONDS CHARGE-CHARGE INTERACTIONS SHELL EFFECTS (IP + EA)	STRONG BINDING $D \sim 2$-4 eV	METALS FROM LEFT SIDE OF PERIODIC TABLE PLUS RIGHT SIDE ELECTRO-NEGATIVE ELEMENTS
VALENCE CLUSTERS	C_n, S_8, As_4	"CONVENTIONAL" CHEMICAL BONDS	STRONG BINDING $D \sim 1$-4 eV	ISOTERIC MOLECULES AND RADICALS
METALLIC CLUSTERS	Na_n, $A\ell_n$, Cu_n, W_n	"METALLIC" BOND LOW KE OF ELECTRONS	MODERATE TO STRONG BINDING $D \sim 0.5$-3 eV	WIDE OCCURRENCE TO THE LEFT OF B, Si, Ge, Sb, Po IN ROWS 2-6 OF PERIODIC TABLE

has to be incorporated [14], i.e.,

$$G = E - T S$$

$$E = E_0 + \sum_i \hbar\omega_i/2 \tag{I.1}$$

$$S = kT \sum_i \ell n\,[1-\exp(-\hbar\omega_i/kT)]\,,$$

where E_0 is the potential energy, E is the internal energy, and $\hbar\omega_i$ are the characteristic frequencies. The entropy contribution favours the ring structure for $(NaC\ell)_4$ at high temperatures. At 400 K the relative concentration is 50%, the ring structure becoming dominant at temperatures exceeding 900 K. At a finite temperature $(NaC\ell)_4$ clusters may exhibit isomerization phenomena (section II). In this review we shall not consider the rich field of metallic clusters, limiting ourselves to the diverse area of molecular clusters.

It is useful to provide a classification of clusters according to size. The smallest constituent is the dimer (n=2), which acts as an essential and basic ingredient in nucleation, but which should rather be considered separately as a "molecular" entity. Then, in the order of increasing size, we have somewhat arbitrarily for nonmetallic systems: small clusters (n=3-100) and large clusters (n>100), whose characteristics will be briefly considered.

Small clusters involve van der Waals molecules, which are held together by 'weak interactions' in rare gas, molecular and hydrogen bonded systems, as well as ionic and covalent structures, where the chemical bonding corresponds to 'strong interactions'. The structures of small clusters at low temperatures (Fig. 3) are distinct from the solid-state structures [16-19]. For weak interactions, e.g., van der Waals clusters, the solid state structure does not set in this domain, while for strongly interacting ionic clusters the solid state structure sets in for moderately small cluster size (Fig. 3). Apart from chemically neat clusters, the exploration of the properties of heteroclusters (section III) are of considerable interest, as an added guest molecule may act as a microscopic probe for the nuclear motion and the electronic excitations of the cluster. In this context, large van der Waals molecules composed of a single large aromatic molecule bound to a rare-gas atom [20,21,22] (Fig. 4), whose excited-state energetics (Fig. 5) has been thoroughly explored, are of interest as they provide basic information on solvent perturbations as explored from the microscopic point of view. Furthermore, they yield insight into the energetics and dynamics of vibrational motion of rare-gases adsorbed on microsurfaces of graphite, providing a link between cluster and surface phenomena [21,22].

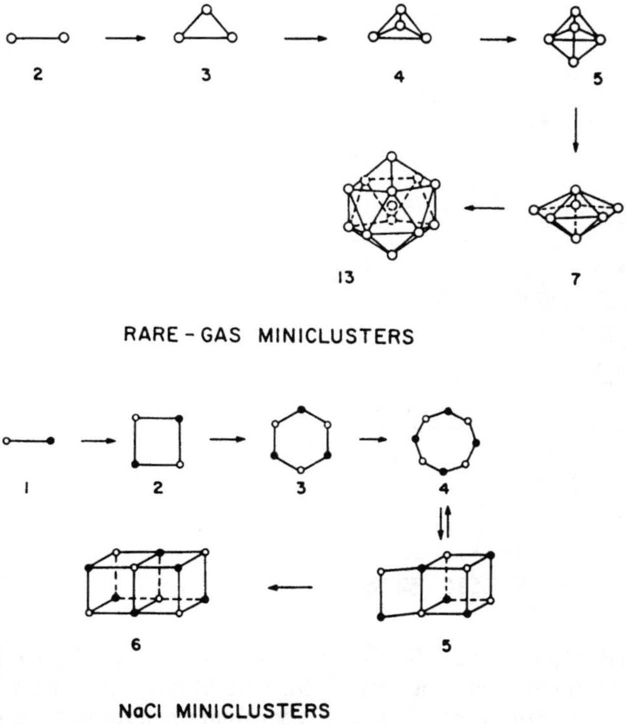

Fig. 3. Structures of small rare-gas and ionic clusters (references 17-19).

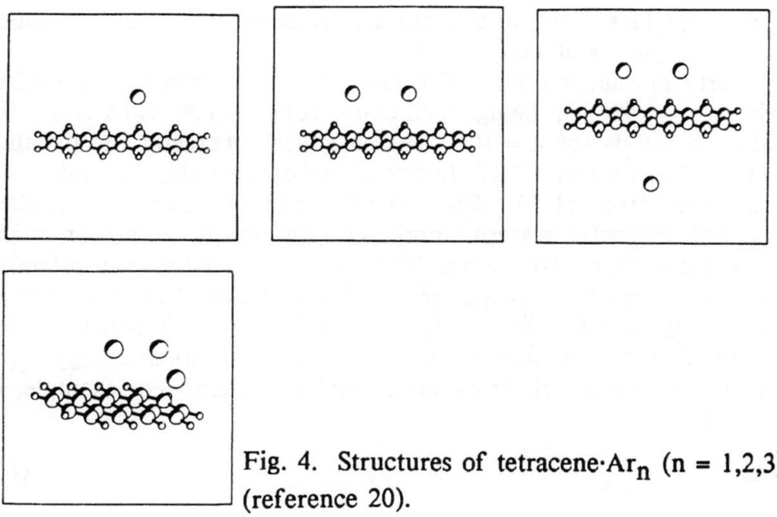

Fig. 4. Structures of tetracene·Ar_n (n = 1,2,3) complexes (reference 20).

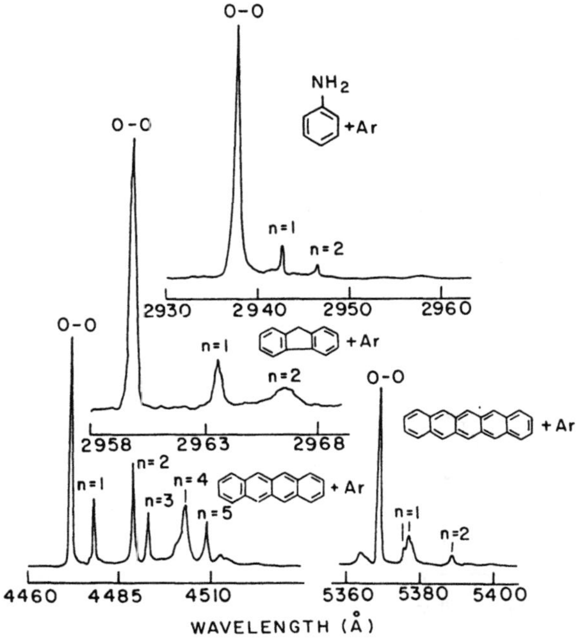

Fig. 5. Laser induced fluorescence (LIF) excitation spectra of $S_0 \rightarrow S_1$ electronic origin (0-0) of several bare molecules M, and $M{\cdot}Ar_n$ complexes in supersonic expansions of Ar (reference 21). The coordination numbers, n, of the $M{\cdot}Ar_n$ complexes are marked.

Let us now consider some unique characteristics of large clusters, which differ from the properties of bulk material.

(A) Large surface/volume ratio. The fraction of the number of surface atoms n_S is $n_S/n = 4/n^{1/3}$, being $n_S/n \simeq 0.4$ for $n = 10^3$, $n_S/n \simeq 0.2$ for $n = 10^4$ and $n_S/n \simeq 0.04$ for $n = 10^6$. The structural, energetic and dynamic properties of the "microsurfaces" of clusters are of considerable interest.

(B) Quantum size effects [1-6]. The dependence of the specific properties, e.g., electric and magnetic susceptibilities, of clusters on their geometrical dimensions orignate from two interrelated causes. Firstly, the effective wavelength of excitations is comparable with the cluster size, resulting in boundary scattering effects. Secondly, the sample size, d, is small, so that the energy gap, Δ, for excitations is comparable to the thermal energy k_BT, to the magnetic interaction μH, or to the spectral resolution. For example, in the case of phonons,

$$\Delta \sim hv/d \quad , \tag{I.2}$$

where $v \sim 10^5$ cm sec^{-1} is the velocity of sound. Thus, quantum size effects will be severe when $\Delta \sim k_BT$ or $d \sim 100$ Å/T(K), so that at T = 10 K we estimate d ~ 10 Å and n ~ 100. On the other hand, for electronic states in the tight binding scheme, one has

$$\Delta \sim zJ/n \tag{I.3}$$

where J is the transfer integral, while z is a numerical constant which takes the value $z = \pi$ for 1 dimension and, in general, we can set z to be approximately equal to the average number of nearest neighbors. Taking J = 1 eV and z = 6, the electronic gap is comparable to the thermal energy when $n \sim 6 \cdot 10^4/T(K)$, so that $n \sim 6000$ at T = 10 K. The different values of n emerging from the estimates of Δ for phonons and for electronic excitations clearly demonstrate that quantum size effects are not universal and depend on the specific physical property.

(C) Thermodynamic size effects [1-6]. An important class of size effects involves the dependence of some intensive thermodynamic properties, e.g., surface tension, vapour pressure and the characteristics of phase transitions on the sample size and shape. The interrelationship between thermodynamic and dynamic size effects will now be considered.

II. ISOMERIZATION AND MELTING OF CLUSTERS

Disorder can be incorporated into clusters by either introducing local defects, e.g., vacancies or dislocations, or by considering the effects of thermal excitations of the nuclear vibrations. An intriguing phenomenon in the category of thermodynamics size effects involves the melting of clusters [23-38,15]. It was predicted in 1909 [23] that the melting temperature of droplets decreases with decreasing size. This prediction has been borne out experimentally [24] for metal clusters (Fig. 6). The applicability of simple

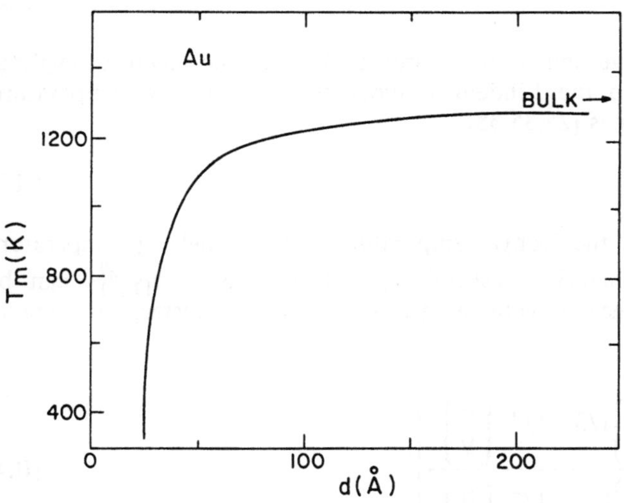

Fig. 6. Size dependence of melting temperature of gold clusters supported on a substrate (reference 24).

thermodynamic arguments [23] is questionable [32], as phase transitions in a finite system are expected to be smeared out by fluctuations. In fact, for finite systems, which are large enough so that the total entropy change per particle accompanying a first-order transition is of the same order of magnitude as in the finite system, taking the mean square fluctuation in temperature $\langle(\Delta T)^2\rangle$ as the intrinsic temperature uncertainty in the system, i.e., $\langle(\Delta T)^2\rangle = (\Delta T_{tr})^2$, one obtains [33,34]

$$\frac{\Delta T_{tr}}{T_{tr}} = \frac{1}{N\sigma} \tag{II.1}$$

where ΔT_{tr} is the temperature spread of the first-order transition, i.e., the width of the finite peak in the curve of specific heat vs. temperature around the peak at the "transition temperature", T_{tr}. σ is the latent entropy per particle, measured in units of Boltzmann's constant, i.e., $\sigma = L/T_{tr}$, where L is the latent heat per particle.

In addition, the construction of theoretical models of melting of small clusters is further complicated by the intrinsic inhomogeneity of the system (exterior, surface, versus interior atoms), and by the spatial and system size dependencies of the spectrum of motional degrees of freedom [25,37]. An alternative approach to cluster melting pertains to the Lindeman hypothesis [24,25], which rests on the notion of the onset of large-amplitude nuclear motion. A quantitative description of size effects on the mean-square atomic vibrational amplitude is given in terms of the ratio of the mean square atomic vibrational amplitude to the squared nearest neighbor separation [25,35,38]

$$\phi^2 \propto T/(c\,\theta_D)^2 \tag{II.2}$$

where θ_D is the Debye temperature and c denotes the nearest neighbor distance. According to the Lindeman hypothesis the melting temperature T_m of the finite cluster is [25,35,38]

$$T_m = T_m^0\,(\theta_D/\theta_D^0)^2 \tag{II.3}$$

where θ_D^0 and T_m^0 are the Debye temperature and the melting temperature of the infinite bulk material, respectively. The ratio of θ_D/θ_D^0 can be inferred from the surface to volume ratio (S/V) of the cluster, resulting in [38]

$$\left(\frac{T_m}{T_m^0}\right) = \left[\frac{1+\frac{1}{8}\left(\frac{\pi}{6}\right)^{1/3} v^{1/3}\left(\frac{S}{V}\right)}{1+\frac{1}{4}\left(\frac{\pi}{6}\right)^{1/3} v^{1/3}\left(\frac{S}{V}\right)}\right]^2 \tag{II.4}$$

where v is the volume per atom. This equation assumes the approximate form

$$\left(\frac{T_m}{T_m^0}\right) = \left(\frac{1+\beta a/R}{1+2\beta a/R}\right)^2 = \left(\frac{1+\beta N^{-1/3}}{1+2\beta N^{-1/3}}\right)^2 \tag{II.5}$$

where $\beta = 0.487$, a is the atomic radius, R is the cluster diameter and N is the number of atoms in the cluster. This result constitutes a 'law of corresponding states' for the size dependent melting temperature. This result clearly applies only to large clusters, for which the concept of a Debye temperature is applicable.

Of considerable interest is the dynamic description of melting [15,27,31,32]. Small van der Waals clusters at finite temperatures were documented by molecular dynamics computer simulations [15,28-31] to undergo a "transition" from a rigid low-temperature form to a nonrigid high-temperature form [27,39]. An attempt was made [27] to attribute these two forms of a finite cluster to two phases of an infinite bulk material on the basis of several of their physical properties, i.e., a break in the caloric equation of state (Fig. 7), the specific heat, the mean coordination number, the radial distribution function, the bond length fluctuation, the power spectrum, the density of vibrational states and the diffusion coefficient. All these attributes, with the exception of the last three, pertain not only to melting but also to the well characterized but little explored phenomenon of cluster isomerization. Indeed, the melting of small clusters may be synonymous with fast isomerization between several distinct nuclear configurations.

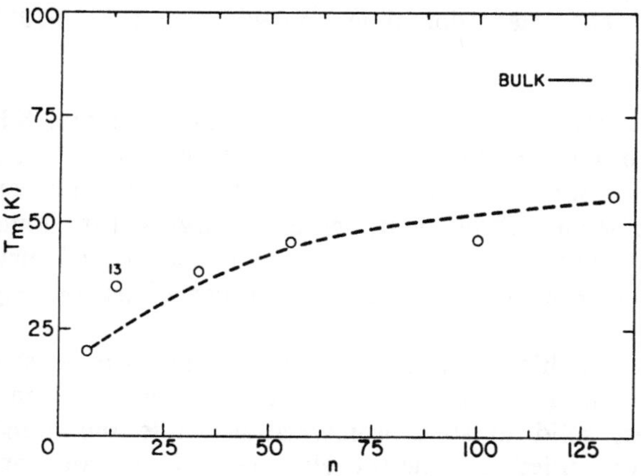

Fig. 7. Melting temperatures, T_m, of Ar_n (n = 7-133) microclusters and small clusters from molecular dynamics computer simulations. T_m is obtained from the break in the caloric curve (references 28-31).

Extensive computer simulation studies were performed for rare-gas clusters [15,27-31]. For these clusters the interaction potentials are "soft" resulting in a large number of structural isomers. For example, the Ar_{13} cluster is characterized by 988 isomeric structures, the lowest energy one corresponding to the icosahedron [40]. In alkali halide clusters the strong interaction potentials result in a more distinct character and a smaller number of isomers. We have undertaken [15,32] molecular dynamics (MD) simulations to study the energetics, structure and dynamics of alkali-halide clusters of variable sizes $(NaC\ell)_n$, for n = 4,16 and 108, over a wide temperature range [15,32]. Our simulations show that the very nature of the phase transformation and the underlying physical processes involved, depend on the size of the system. We conclude that at the lower end of the size spectrum, the phase space of the system is characterized by a small number of stable configurations (solid isomers) between which the system transforms in a diffusionless manner, with temperature dependent rates and branching rations. As the system size is increased the number of accessible conformers increases leading to a hierarchial kinetics of isomerization events which exhibits itself as a broadening of the transition region. For these small clusters coexistence is between solid isomers rather than inter-phase (solid-liquid) coexistence. The latter develops for clusters of sufficient size, characterized by a dense spectrum of accessible states, separated by thermally surmountable barriers [40]. Under these circumstances conventional melting is observed as a sharp transition, the separation of time-scales for inter-well and intra-well dynamics ceases [27,31] and true solid-liquid coexistence is found.

We have performed constant energy MD simulations for $(NaC\ell)_n$ clusters, with interionic potentials consisting of a point Coulomb term and a Born-Mayer repulsion [14,41]. At each temperature the system was equilibrated for a prolonged period (typically 10^{-10} sec) after which data was collected (typically for ~ 10^{-9} sec). The integration time step was chosen as $3 \cdot 10^{-15}$ sec to assure energy conservation.

In Fig. 8, the caloric curves (total energy, U, versus kinetic energy, KE) for $(NaC\ell)_n$ with n = 4, 16 and 108 are shown. The kinetic temperature is defined as $kT = 2U/3(N-2)$, since the clusters are setup initially with zero total angular and linear momenta. Throughout, energy is expressed in atomic units 1a.u. = 2 Rydbergs. For the n = 4 cluster the break in the curve corresponds to isomerization between cube and ring conformations (see Fig. 9

The development of a hierarchy of structural transformations is demonstrated for n = 16, in Fig. 8b. The lower temperature break corresponds to diffusionless solid to solid isomerization while the higher temperature break is accompanied by the development of disorder and diffusion characteristics (see Figs. 11, 12). The caloric curve shown in Fig. 8c for n = 108 cluster is markedly different from those of the smaller systems.

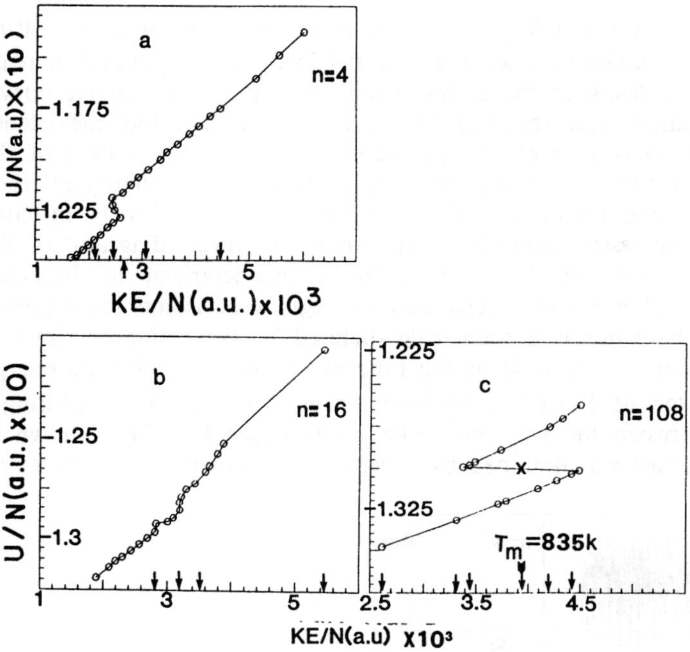

Fig. 8. Caloric curves for $(NaC\ell)_n$ with n = 4,16,108. Arrows denote temperatures in increasing order: (a) 589, 698, 740, 874, 1249K; (b) 636, 726, 806, 1684K; (c) 555, 705, 730, 890, 938K. Note change in character of transition upon increase in size (reference 15).

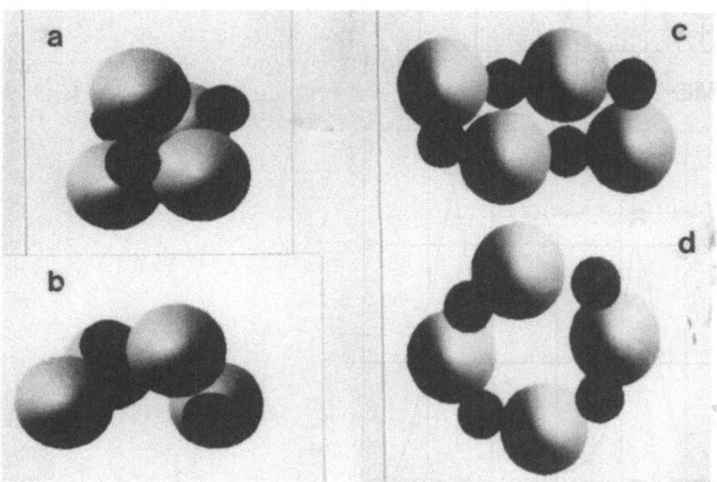

Fig. 9. Cluster geometries for $(NaC\ell)_4$. Large and small balls, $C\ell^-$ and Na^+ ions, respectively. The temperatures are: (a) 589K, (b) 605K, (c) 698K and (d) 698K (reference 15).

We observe in it a "van der Waals loop phenomenon" with the true melting point (solid-liquid coexistence, see Fig. 8c) at 835K. While gradual increase in the total energy leads to the lower (solid) branch of the caloric curve, terminating ultimately and resulting in the liquid branch, the coexistence point (marked by x) is achieved by providing the system, via scaling of the momenta, when on the solid branch at 835K, with sufficient energy to reach the horizontal line (melting of part of the system was observed to be initiated at the surface, progressing inwards). The structural "phase diagram" of the smaller clusters (n = 4 and 16) is seen to be characterized by branches corresponding to stable isomers, separated by regions of isomer-coexistence. Associated with the latter is a time scale dictated by the residence times at the different isomers. A sample of the time evolution of the kinetic energy of the n = 4 cluster, at T = 698 (coexistence region in Fig. 8a), exhibiting a transformation between the ring and cube isomers (see Fig. 9) is shown in Fig. 10a. This record was obtained by short time averaging of KE over sets

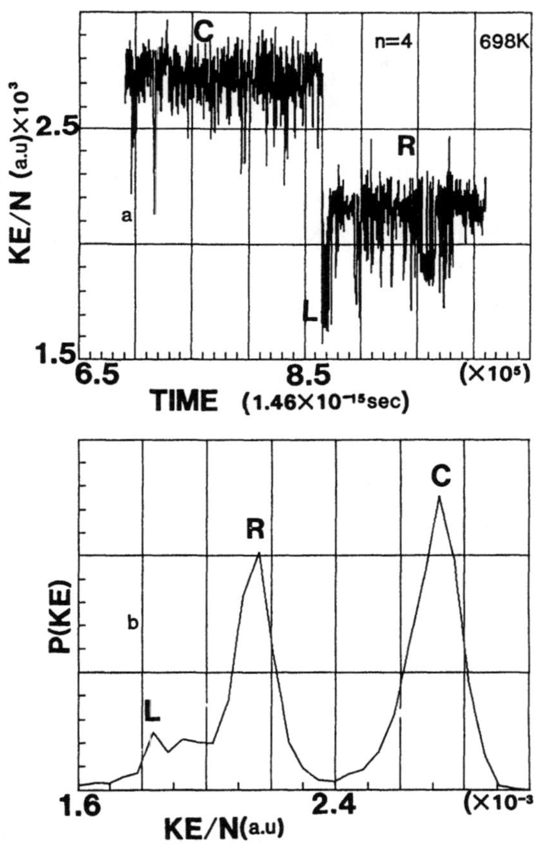

Fig. 10. (a) Sample of KE vs. time for $(NaC\ell)_4$ showing isomerization from cube (C) to ring (R), through a ladder (L) transition state, configurations. (b) Short-time average, KE distribution, P(KE) (reference 15).

of 100 time-steps. The distribution of short-time averages of the kinetic energy (for that point on the caloric curve), shown in Fig. 10b, is trimodal, with each of the peaks corresponding to a distinct structure as noted (transition state (L), cube and ring). Note that the choice of the averaging time span is of importance in obtaining the resolution which affords such structural identification. Away from the coexistence region the distribution becomes, monotonously, unimodal. Further structural characterization of the systems is provided by the function n(r), related to the pair correlation function, and defined as the average number of atoms whose distance from an atom chosen as origin is between r and r+Δr. Additional information is provided by the coordination number function $C(R) = \int_0^R n(r')dr'$. These are shown in Fig. 11 for the n = 4, 16 and 108 clusters, respectively. It is evident that the sharp isomerization for the smallest cluster is accompanied by a sudden decrease from 3 to 2 in the first neighbor coordination while for the intermediate size cluster the decrease in coordination (from ~ 4 at the low temperature) across the transition region is gradual. Finally, for the large (n = 108) cluster the melting is characterized by a sharp drop in the first neighbor coordination number (from ~ 4.5 to ~ 3). The structural transformations are also clearly seen by following the evolution of features in n(r) as a function of temperature. In particular, except for the n = 4 cluster, upon increase in KE the first peak grows broader and decreases in magnitude and further coordination shells lose their structure (more pronounced for the n = 108 cluster). Note also that this characteristic trend, which corresponds to liquefaction, does not occur for the intermediate cluster (n = 16) till past the first, isomerization, break in the caloric curve.

A probe of the dynamical nature of the transition is provided by the density of states (power spectrum) calculated as the Fourier transform of the velocity autocorrelation function [42]. These are given in Fig. 12 for n = 4, 16 and 108, respectively. First, we notice the trend of stabilization at higher temperature of the structural form (solid isomers for the smaller cluster and liquid for the large one) which possesses a higher density of low frequency modes. This can be understood in terms of the vibrational contribution to the free energy [14], which the system attempts to minimize (section I). Secondly, for the n = 4 cluster we do not find any indication of soft mode or onset of diffusion (see value near ω=0 in D(ω) plots). For the intermediate size cluster, however, we do observe what may be construed as soft mode development and onset of diffusion at temperatures above the first, isomerization break in the caloric curve (see Fig. 8b), thus corroborating our distinction between isomerization and melting for intermediate size clusters. True diffusion character is observed for the n = 108 cluster both at coexistence and on the liquid branch.

Experimental exploration of cluster isomerization and melting can rely on structural determination [19]. More interesting is the interrogation of vibrational [43,44] or electronic excitations [45,46] of a probe molecule in a cluster. Inhomogeneous line broadening due to the appearance of isomers, or

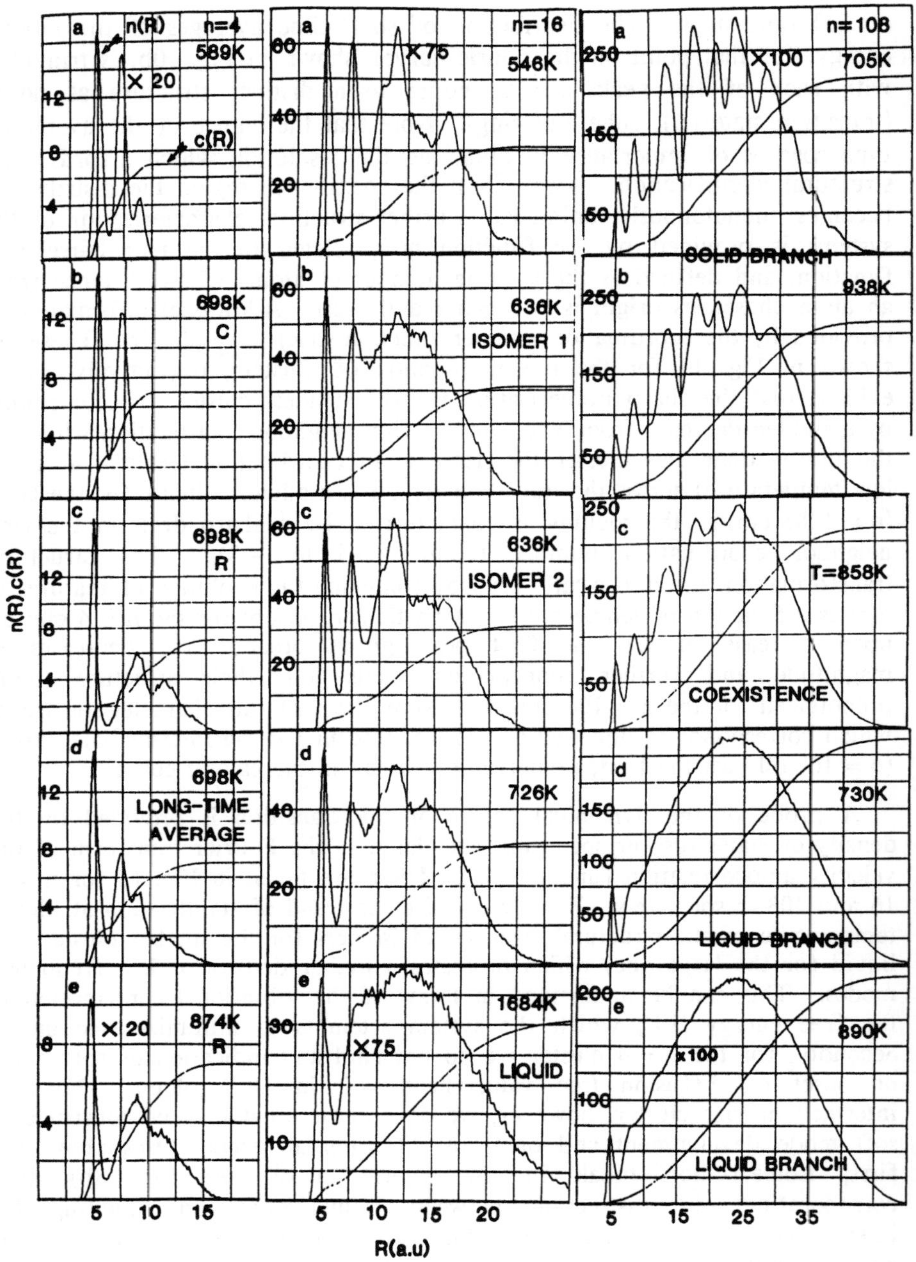

Fig. 11. Particle pair distribution, n(R), and coordination number, C(R), functions for n = 4, 16 and 108 at the indicated temperatures. Note dependencies on size and on temperature (reference 15).

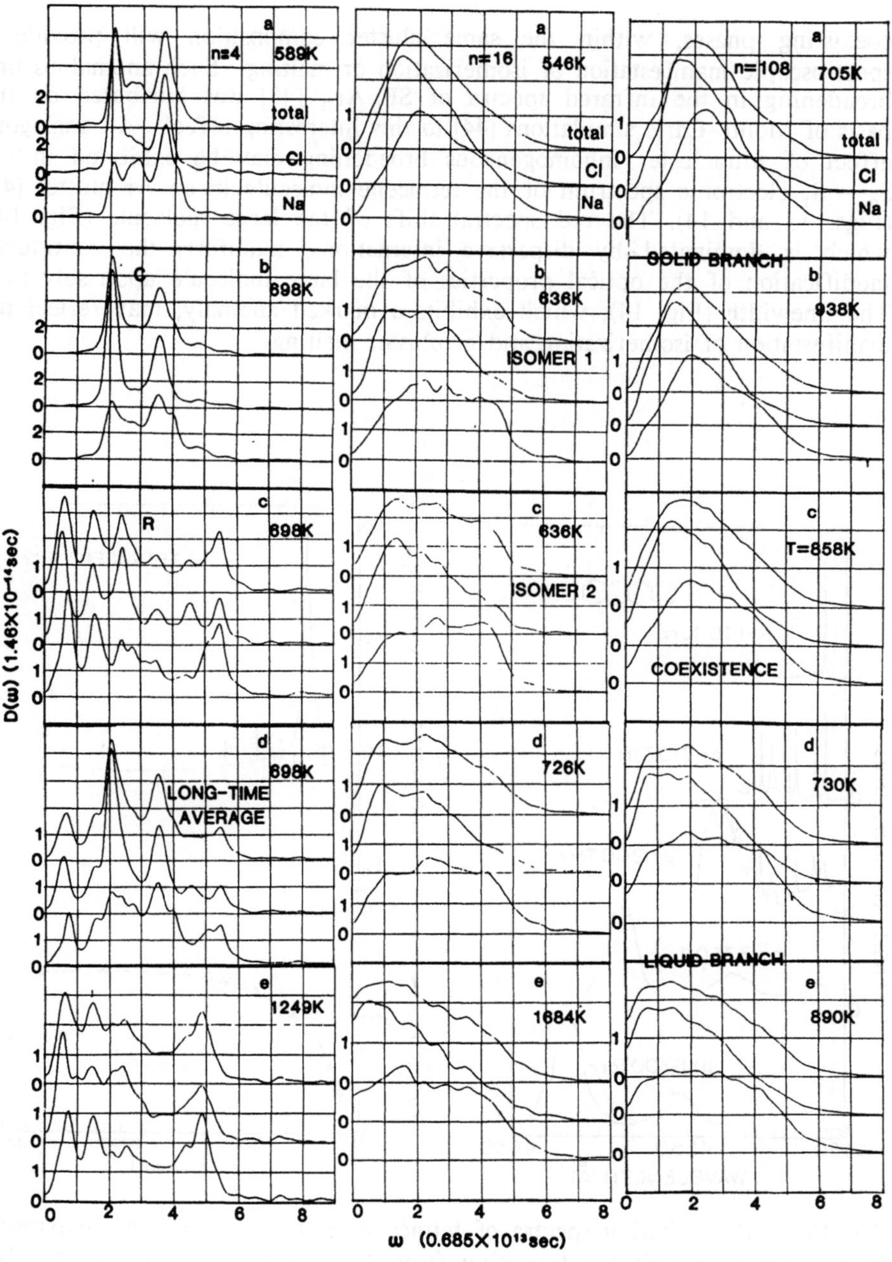

Fig. 12. Densities of states for n = 4, 26 and 108 at the denoted temperatures. Note shift to lower frequencies with increase in temperature, and onset of diffusion for n=16 and 108 at higher temperatures (ref. 15).

coexisting phases, within the same cluster composition will provide a spectroscopic manifestation of isomerization or melting. Such anomalous line broadening in the infrared spectra of $SF_6 \cdot Ar_n$ [43] was attributed on the basis of Monte-Carlo simulations [44] to this interesting effect. An analogous effect of intracluster inhomogeneous broadening may be exhibited in the $S_0 \rightarrow S_1$ electronic spectrum of the tetracene molecule in an Ar cluster [45] (Figs. 13 and 14). The red spectral shift of the large molecule (Fig. 14), which is dominated by dispersive interactions, manifests the continuous modification of the optical properties of the large molecule upon solvation. The linewidth (Fig. 14), which exhibits a marked anomaly, may reflect the manifestation of isomerization and/or cluster melting.

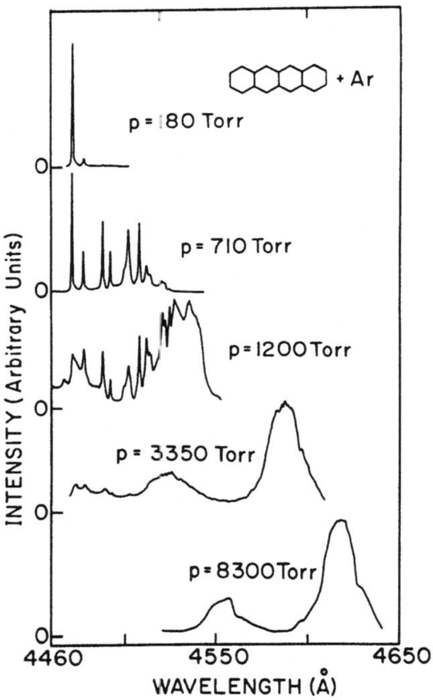

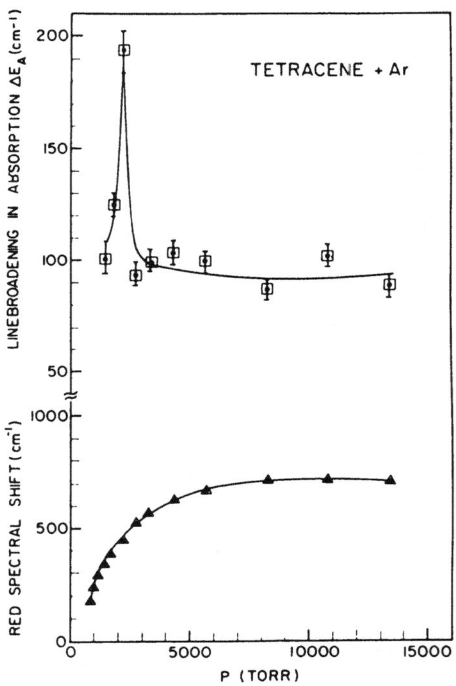

Fig. 13. LIF excitation spectra of tetracene seeded in supersonic expansion of Ar, which exhibit the transition from the isolated molecule (at p = 180 torr) to the molecule solvated in an Ar cluster (p = 8300 torr). Data from reference 45.

Fig. 14. Spectral shifts relative to the 0-0 transition of the bare molecule, and the widths (FWHM) of the 0-0 transition of tetracene in Ar clusters (reference 45).

III. HETEROCLUSTERS

Up to this point we have alluded to some features of chemically neat clusters. The exploration of the properties of doped clusters is of considerable interest, as an added guest molecule may act as a microscopic probe for the molecular structure, the nuclear motion and the electronic excitation of the cluster. Studies of doped clusters provide insight into three classes of phenomena:

(1) The configurational changes induced by doping.
(2) The cluster-guest molecule interactions modify the energies of vibrational and electronic excitations and the intramolecular dynamics (i.e., intramolecular vibrational energy redistribution and interstate electronic relaxation) of the guest molecule.
(3) Clustering around the guest molecule opens up new decay channels for the disposal of vibrational and electronic energy of the guest, e.g., vibrational predissociation in small clusters and vibrational relaxation in large clusters (see section VI).

Some classes of interesting heteroclusters are assembled in Table II. Heteroclusters can be classified according to their charge into neutral, positively charged and negatively charged aggregates. The neutral heteroclusters involve doped vdW clusters [47-50] held together by dispersive and weak electrostatic interactions, as well as doped molecular [52] and hydrogen-bonded clusters [51], which involve an organic or inorganic molecule solvated by nonpolar (e.g., $CC\ell_4$) or by polar (e.g., H_2O) ligands. These interactions are reminiscent of those prevailing in nonionic solutions. Positive heteroclusters span a broad spectrum of systems. The weakest electrostatic interactions prevail in positive vdW heteroclusters [53], involving a positive large ion in a rare-gas cluster, which is held together by charge-induced dipole and dispersive interactions. Similar interactions prevail in He_n^+ clusters [54], where the charge-induced dipole interaction results in electrostriction. Ionic complexes [51,56], e.g., $H^+(H_2O)_n$ and $Na^+(SO_2)_n$ are bound by stronger electrostatic charge-dipole interactions, as those prevailing in ionic solutions. Chemical bonding is exhibited in positive-ion rare-gas clusters of R_n^+, where the positive R_2^+ molecular diatomic ion is formed within the cluster [57,58]. Another extreme case of valence interaction in clusters involves ionic clusters, e.g., the H_n^+ aggregates [59,60], which presumably consist of an H_3^+ ion bound to an ensemble of H_2 molecules. Negative clusters fall into two broad categories. (i) The negative ion clusters involve localization of the excess electron within a single molecule [61,62], resulting in the formation of a "conventional" negative ion, and (ii) the excess electron cluster which involves electron localization within the cluster (section V). We shall now proceed to discuss some features of charged clusters, focusing attention on the formation mechanism of positively charged clusters, and on the nature of extended excess electron states in negatively charged clusters.

TABLE II. Some Heteroclusters

TYPE	EXAMPLES	CAUSE OF BINDING
Doped van der Waals Clusters	$(I_2)R_n$ (n=1-6) $(SF_6)R_n$ (n=1-50) Tetracene·R_n (n=1-200)	Dispersive and Weak Electrostatic
Doped Molecular and H-Bonded Clusters	$(HNO_3)_x(H_2O)_y$ (x=1-6, y=2-20) Tetracene·$(CCl_4)_n$ (n=1-3)	Electrostatic and H-Bonding Dispersive and Weak Electrostatic
Doped Positive van der Waals Clusters	$(C_6F_6^+)\cdot Ar_n$	Dispersive plus Polarization
Positive Ions in van der Waals Clusters	$(He_n)^+$ (n≃68 ^{4}He; n≃85 ^{3}He) $(Ar_n)^+$	Dispersive and Polarization Hole Trapping
Ionic Complexes	$H^+(H_2O)_n$ $Na^+(H_2O)_n$ $Na^+(SO_2)_n$	Charge-Dipole
Simple Ionic Clusters	$(H_n)^+$ (n= 3- 15) (n=500-1000)	Valence and Polarization
Negative Ion Clusters	$(CO_2)_n^-$	Solvation of Negative Molecular Ion
Solvated Electron Clusters	$(H_2O)_n^-$ $(NH_3)_n^-$	Electron Localization in an Extended State

IV. CLUSTER IONIZATION

Selective ionization processes via one-photon or multiphoton ionization of vdW and molecular clusters provide direct information on the energetics of the formation of positively charged clusters. The energetics of ionization can be utilized to elucidate the nature of intramolecular and intermolecular nuclear configurational changes accompanying the formation of positively charged microclusters. In this context, the following mechanisms are relevant.

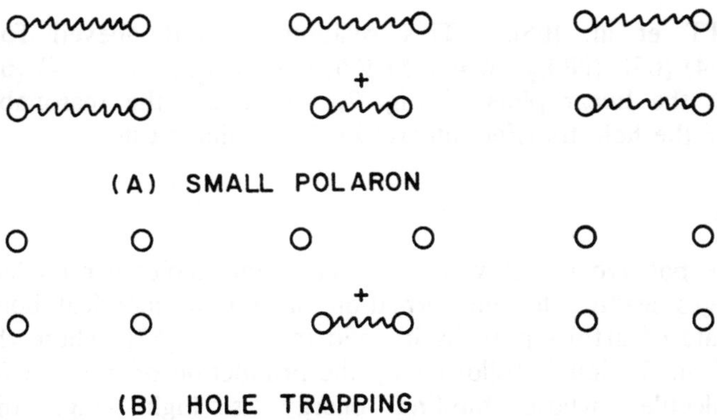

Fig. 15. (a) Small polaron in a cluster of diatomic molecules and (b) hole trapping in a rare-gas cluster.

IV.A Small Polaron Effects

Ionization of a molecule in a cluster may be accompanied by a change in the intramolecular nuclear equilibrium configuration [63]. When such a molecule is ionized within the cluster (Fig. 15), the transfer integral J, which is responsible for the intermolecular hole transfer, corresponds to an electronic contribution, J_e, reduced by the nuclear Franck-Condon factor F, so that $J = J_eF$. An immediate energetic implication of these "dressed" intermolecular interactions, J, involves the lowering of the ionization potential I(n) of the n-th cluster relative to the monomer ionization potential I(1). For a one-dimensional system a simple molecular-orbital calculation results in the relation [64],

$$I(n) = I(1)-2|J| \cos\left(\frac{\pi}{n+1}\right) . \tag{IV.1}$$

Invoking the approximate relation [65] $\cos\left(\frac{\pi}{n+1}\right) = 1 - \frac{1}{n}$ for n = 2-10, results in

$$I(n) = [I(1) - 2|J|] + \left(\frac{2|J|}{n}\right) . \tag{IV.2}$$

For a multidimensional system, we can replace the numerical factor of 2 in Eq. (IV.2) by the number of nearest neighbors, z, so that Eq. (IV.2) is generalized to read:

$$I(n) = [I(1) - z|J|] + \left(\frac{z|J|}{n}\right) . \tag{IV.3}$$

Equation (IV.3) results in the general linear dependence of I(n) versus n^{-1}

proposed by TROTT et al. [65]. This relation is well obeyed for $((CH_3)_2CO)_n$ (n = 1-4) [65], $(NO)_n$ (n = 1-5) [66] and $(CS_2)_n$ (n = 1-4) [67] microclusters. From the linear plots of Fig. 7, we deduce the reasonable value $z|J| \simeq 1$ eV for the hole transfer integral in these microclusters.

IV.B Hole Trapping

The formation of the positive ion may result in large intermolecular nuclear configurational changes leading to the formation of a new chemical bond (Fig. 15). Such a state of affairs prevails in solid rare-gases [57], where the vertical formation of an R^+ ion is followed by the production of a stable R_2^+ diatomic ionic molecule, whose binding energy is high (Fig. 16). Accordingly, the adiabatic ionization potential of the vdW molecule R_2, whose binding energy is ~ 0.01 eV, is lowered by $D_+ \simeq 1$eV relative to the ionization potential of Ar. These local binding effects are reflected in the reduction of the ionization potentials of Ar_n (n = 2,3) vdW molecules [55]. For the larger Ar_n (n = 4-6) clusters [55], additional small polarization effects contribute to the cluster stabilization energy.

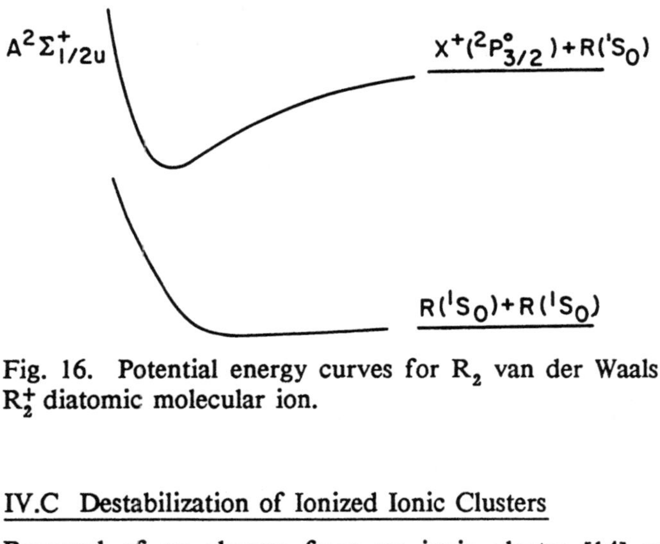

Fig. 16. Potential energy curves for R_2 van der Waals molecule and for the R_2^+ diatomic molecular ion.

IV.C Destabilization of Ionized Ionic Clusters

Removal of an elecron from an ionic cluster [14] results in a substantial decrease in the electrostatic Madelung energy. The ionization potential accompanied by the removal of the charge, Z_j, from the anion located at $\vec{r}_j$ is [14]

$$I = EA + \sum_i \frac{Z_j Z_i}{|\vec{r}_i - \vec{r}_j|} , \qquad \text{(IV.4)}$$

where EA is the electron affinity of the anion. Accordingly, the vertical ionization potential of ionic clusters, which is substantially larger than the electron affinity, increases with increasing n. MARTIN [14] has provided numerical calculations of vertical ionization energies. Furthermore, vertical ionization of ionic clusters results in large configurational changes with the cluster relaxing to a new equilibrium configuration [14].

IV.D Polarization Interactions in Ionic van der Waals (vdW) Systems

The formation of a doped ionic vdW microcluster is accompanied by minor intramolecular and intermolecular configurational changes. The extra stabilization energy of such an ionic microcluster, relative to the corresponding neutral microcluster, originates from charge-induced-dipole polarization interaction. These weak interactions are reflected in the reduction of the ionization potential of the benzene·Ar complex by $\Delta I = -2.5 \cdot 10^{-2}$ eV relative to that of the bare molecule [68], which is in rough agreement with the result $\Delta I = -5 \cdot 10^{-2}$ eV of simple model calculations [69] of these polarization interactions. However, more elaborate data for the ionization potentials of p-xylene·Ar_n (n = 1-6) complexes reveal significant deviations from the simple electrostatic model [70] and a more elaborate treatment of this problem is required.

V. ELECTRON ATTACHMENT TO MOLECULAR CLUSTERS

V.A Prologue

We shall be concerned with the structural and dynamic implications of electron attachment to non-metallic molecular clusters [71-86]. A classification of these systems, according to the strength of the binding in the cluster and the nature of the excess electron state, is provided in Table III. From the point of view of general methodology, electron localization in clusters is interesting because of several reasons. First, quantum phenomena are expected to be pronounced in systems where the electron wavelength is comparable to the cluster size. Second, the excess electron can serve as a probe for the nuclear motion and dynamics of the cluster. Third, the ubiquity of the modes of electron attachment, involving localized and quasifree (extended) states on one hand and bulk and surface states on the other hand, are novel and interesting.

The following physical and chemical phenomena associated with the attachment of an excess electron to clusters are of considerable interest:
(1) The parentage problem. In many interesting cases the excess electron is attached to a cluster, whose individual atomic or molecular constituents do not form a stable negative ion.
(2) Localized and extended states of the excess electron. The spatial extent of the electron charge distribution, i.e., the localization length, relative to the cluster size, provides a distinction between localized and extended states in a finite system.

TABLE III. Electron Attachment to Clusters

TYPE OF CLUSTER	NATURE OF EXCESS ELECTRON STATE	CANONICAL EXAMPLE	CAUSE OF ELECTRO BINDING
Rare Gas $(He)_n$	Surface State localized	$(He)_n^-$	Repulsive + weak polarization
Heavy Rare Gas	Bulk State extended	$(Xe)_n^-$	Dominated by attractive interactions
Van der Waals	Stable negative molecular ion	$(SF_6)_n^-$	Large electron affinity of a single molecule
Van der Waals	Solvated negative ion	$(CO_2)_n^-$	Positive electron affinity of a single molecule. Stabilization of a negative ion by solvation
H-bonded	Diffuse Localized state	$(H_20)_2^-$	Weak charge-dipole
H-bonded	Localized State Surface States Bulk States (large n)	$(H_20)_n^-$ $(n \geq 11)$ $(NH_3)_n^-$ $(n \geq 35)$	Electron-polar molecule attraction in conjunction with large cluster reorganization
Ionic	Electron localization in anion vacancies; surface states	$(Na_n\ C\ell_m)^{m-n+1}$	Electrostatic interactions

(3) Bulk and surface states of the excess electron. The relative energetic stability of these two distinct types of states in different clusters is intriguing.

(4) Cluster reorganization. The energetically stable state of the electron attached to the cluster may involve a nuclear reorganization, where electron attachment is accompanied by a large cluster reorganization energy.

(5) Cluster isomerization induced by electron attachment. Electron

attachment to a cluster may result in a configurational change resulting in a close-energy isomer.

We have recently utilized [83-86] the quantum path-integral molecular dynamics method [87-93] to explore the compositional, structural and size dependence of electron localization in ionic and hydrogen bonded clusters, which provide novel information regarding two central phenomena. First, the electron localization modes in large finite systems are elucidated, resulting in novel phenomena in the formation of energetically stable surface states of excess electrons in clusters. Secondly, electron attachment provides a novel mechanism for the induction of configurational rearrangement of clusters, which is investigated in the general context of the interesting problem of isomerization and melting of small, finite, aggregates, which was discussed in section II.

V.B The Quantum Path Integral Molecular Dynamics Method

The structure and energetics of excess electron states in clusters has been explored using the quantum path-integral molecular dynamics (QUPID) method [87-93]. This approach, which rests on a discrete version of Feynman's path-integral method [87], provides a powerful method for the study of these systems [83-86]. Consider an electron interacting with a set of N particle cluster via a potential V_e which is specified by the Hamiltonian

$$H = -\frac{\hbar^2}{2m}\nabla^2 + V_e(r) \ . \qquad \text{(V.1)}$$

The partition function for such a system is

$$Z = \mathrm{Tr}[\exp(-\beta H)] \, , \qquad \text{(V.2)}$$

where $\beta = 1/k_B T$ is the inverse temperature, and the energy of the system is given by

$$E = -\frac{\partial}{\partial\beta} \ell n Z \ . \qquad \text{(V.3)}$$

An isomorphism between the quantum problem to an appropriate classical problem is established by (a) the factorization of the partition function into P contributions, and (b) by invoking a small β/P expansion for the matrix elements. This procedure allows for the expression of the partition function,

$$Z \simeq \left(\frac{Pm}{2\pi\hbar^2\beta}\right)^{3P/2} \int d\vec{r}_1 \ldots d\vec{r}_P \ \exp[-\beta V_{eff}(\vec{r}_1 \ldots \vec{r}_P)] \ , \qquad \text{(V.4)}$$

in terms of the effective potential

$$V_{eff} = \sum_{i=1}^{P} \left[\frac{Pm}{2\hbar^2\beta^2} (\vec{r}_1 - \vec{r}_{i+1})^2 + \frac{1}{P} V_e(\vec{r}_i) \right] , \tag{V.5}$$

which consists of a superposition of a harmonic potential and the cluster potential.

Equations (V.4) and (V.5) establish an approximate isomorphism between the quantum problem characterized by the Hamiltonian H, Eq. (V.1), and the classical problem defined by the effective potential, Eq. (V.5). In the isomorphic classical problem, the electron is mapped onto a closed chain (necklace) of P pseudo-classical particles (beads). Each point of the necklace exerts two types of interactions (Fig. 17), that is (i) nearest neighbour interactions between the beads in the chain, which are characterized by a harmonic potential with a force constant $Pm/\hbar^2\beta^2$, and (ii) interactions with the cluster through the scaled potential $V_e(\vec{r})/P$. The average energy of the system can then be evaluated from the relation

$$E = \frac{3N}{2\beta} + \langle V \rangle + K_e + \frac{1}{P} < \sum_{i=1}^{P} V_e(\vec{r}_i) > . \tag{V.6}$$

The first two terms in Eq. (V.6) correspond to the kinetic and potential energies of the classical N particle cluster. The electron kinetic energy is [92]

$$K_e = \frac{3}{2\beta} + \frac{1}{2P} \sum_{i=1}^{P} \langle \frac{\partial V_e(\vec{r}_i)}{\partial \vec{r}_i} \cdot (\vec{r}_i - \vec{r}_P) \rangle , \tag{V.7}$$

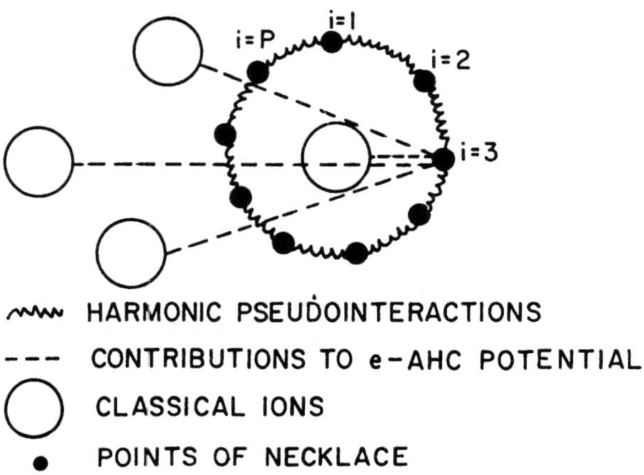

Fig. 17. A schematic representation of the electron necklace.

consisting of a free-particle term, $3/2\beta$, and the contribution K_{int} from the interaction with the ions. The indicated statistical averages $\langle\ \rangle$ are over the Boltzmann distribution as defined in Eq. (V.4). This formalism is converted into a numerical algorithm by noting the equivalence of the sampling described above to that over phase space trajectories generated via molecular dynamics by the classical Hamiltonian,

$$H = \sum_{i=1}^{P} \frac{m^* \dot{\vec{r}}_i^{\,2}}{2} + \sum_{I=1}^{N} \frac{M_I \dot{\vec{R}}^2{}_I}{2} + \sum_{i=1}^{P} \left[\frac{Pm}{2\hbar^2\beta^2} (\vec{r}_i - \vec{r}_{i+1})^2 + \frac{1}{P} V_e(\vec{r}_i) \right] + V\,, \tag{V.8}$$

where the masses m^* and M_I can be chosen arbitrarily , and taken as $m^* =$ 1 a.m.u. and M_I to be the ionic masses. This choice is made such that the internal frequencies of the quantum necklace $\omega \sim [mP/m^*\beta^2\hbar^2]^{1/2}$ will match the frequencies of the nuclear motion.

A measure of the quantum character of the system and the degree of localization of the quantum particle is provided by the complex time correlation function [91]

$$R^2(t-t') = < |\vec{r}(t) - \vec{r}(t')|^2 > \,, \tag{V.9}$$

where $|\vec{r}(t) - \vec{r}(t')|$ is the distance between two points (beads) on the electron path separated by a "time" $(t-t')\epsilon(0,\beta\hbar)$. The value of R at $(t-t') = \beta\hbar/2$, $R(\beta\hbar/2)$, characterizes the breadth of the bead distribution and yields the correlation length. For a free particle, where the bead distribution obeys Gaussian statistics, $R_f = \sqrt{3}\lambda_T$ where λ_T is the classical thermal wavelength. For a localized state of the quantum particle, $R(t-t')$ is dominated by the ground state . As a function of $(t-t')$, $R(t-t')$ starts from zero at $(t-t') = 0$ and rapidly achieves a constant value with $R < R_f$. The rise time, τ, is a rough estimate of the excitation energy $\Delta E = E_1 - E_0$, given by $\Delta E = \hbar/\tau$. Another convenient measure of the quantum character of the system may be inferred from the relative contribution of K_{int} to the kinetic energy K_e, Eq. (V.7). For a classical particle whose thermal wavelength, $\lambda_T = (\beta\hbar^2/m)^{1/2}$, approaches zero the Gaussian factor in Eqs. (V.5) and (V.4) reduces to a delta function and the necklace collapses to a single point ($K_{int} = 0$). Generally, the degree of quantum character depends upon the interplay between the spatial extent of the particle and a characteristic length associated with the rate of change of the potential. A measure of the spatial extent of the localized electron is given by the gyration radius of the electron necklace

$$R_g^2 = \frac{1}{2P^2} < \sum_i \sum_j (\vec{r}_i - \vec{r}_j)^2 > \,. \tag{V.10}$$

V.C Electron Localization in Alkali Halide Clusters

The compositional, structural and size dependence of various electron localization modes in ionic alkali halide clusters (AHC) have been explored using QUPID simulations. The AHCs were chosen for three reasons. First, the nature of interionic interactions in these clusters is well understood [41]. Second, there exists an abundance of model calculations on both neutral and charged AHCs. Third, quite extensive information is available on electron-alkali cation (M^+) and electron-halide anion (X^-) interaction [94].

At this stage it will be convenient to provide an explicit description of the e-AHC system. The AHC, which is treated as a classical system, consists of N ions (N_1 cations and N_2 anions) with masses M_1 and M_2, respectively. The interionic potential within the AHC is

$$V_{AHC} = \sum_{I} \sum_{\neq J} \phi_{IJ} (R_{IJ}) , \tag{V.11}$$

with the interionic potential $\phi_{IJ} (R_{IJ})$ being given by the Born-Meyer potential [41]. The e-AHC potential is

$$V_e(\vec{r}) = \sum_{I} \phi_{eI} (\vec{r} - \vec{R}) , \tag{V.12}$$

consisting of a sum of electron-ion potentials, which are described by a purely repulsive pseudo-potential

$$\phi_{eX^-}(r) = e^2/r \tag{V.13}$$

for electron-anion interaction, and by the local pseudo-potential

$$\begin{aligned} \phi_{eM^+}(r) &= -e^2/R_c \qquad r \leq R_c \\ \phi_{eM^+}(r) &= -e^2/r \qquad r > R_c \end{aligned} \tag{V.14}$$

for the electron-cation interactions.

Numerical simulations were performed for an electron interacting with sodium chloride clusters [85]. The temperature of the system was T = 300K. The electron-cation pseudo-potential parameters were varied by changing the cut-off radius R_c in Eq. (V.14). A typical value chosen for Na^+ was R_c = 3.22 a_0 . Weaker electron-cation interactions, which are appropriate for heavier cations, were mimicked by increasing the value of R_c.

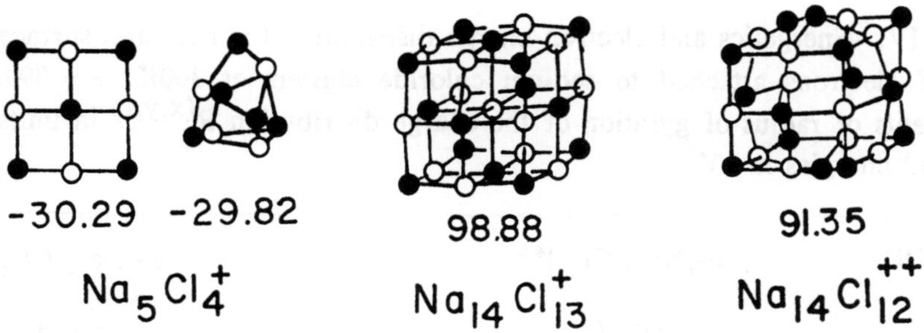

Fig. 18. Configurations of some small and large charged NaCℓ clusters to which an excess electron was attached (reference 14). The numbers refer to the ground state energies in eV units

Structural [14] and classical molecular dynamics [15] calculations have established that when the size of the ionic cluster increases ($N \geq 20$), that NaCℓ crystallographic arrangement is energetically preferred even if the number of positive and negative ions is not equal. We have chosen [85] to study first the interaction of an electron with the positive clusters $[Na_{14}\,C\ell_{13}]^{+}$ and $[Na_{14}\,C\ell_{12}]^{++}$, which exhibit the crystallographic structure without and with a vacancy, respectively (Fig. 18), and which are characterized by a pronounced energetic stability. In Figs. 19 and 20 we present our results for the equilibrium charge distribution of the electron necklace edge points, and for the nuclear configuration of the clusters. The energetics of these systems is summarized in Table IV, where the adiabatic electron binding energy, EABE, of the cluster

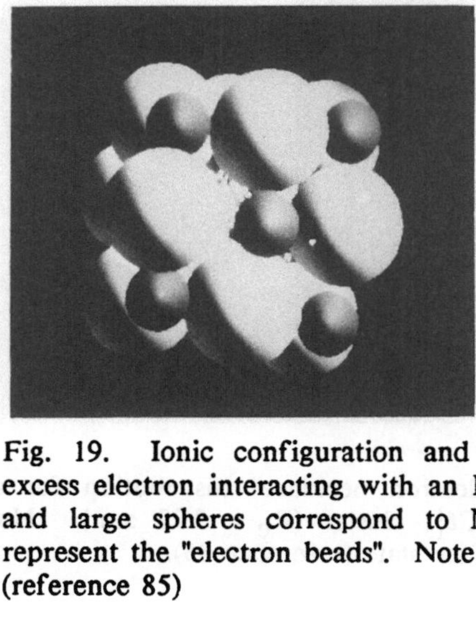

Fig. 19. Ionic configuration and "electron necklace" distribution for an excess electron interacting with an $Na_{14}C\ell_{12}^{++}$ cluster (R_C = 3.22 a.u.). Small and large spheres correspond to Na^{+} and $C\ell^{-}$ ions, respectively. Dots represent the "electron beads". Note electron localization in an anion vacancy (reference 85)

TABLE IV. Energetics and electron charge distribution for bulk and surface states of electrons attached to sodium chloride clusters at 300K, P = 399. Components of radius of gyration of the charge distribution $R_g^{(x,y,z)}$ in units of a_0, and energies in eV.

CLUSTER	$e\text{-}[Na_{14}\,C\ell_{12}]^{++}$	$e\text{-}[Na_{14}\,C\ell_{13}]^{+}$
EVBE	-6.77	-4.34
E_C	2.17	2.87
EABE	-4.6	-1.47
$\left(R_g^x\right)^2$	4.6	5.4
$\left(R_g^y\right)^2$	6.8	7.0
$\left(R_g^z\right)^2$	6.3	9.8

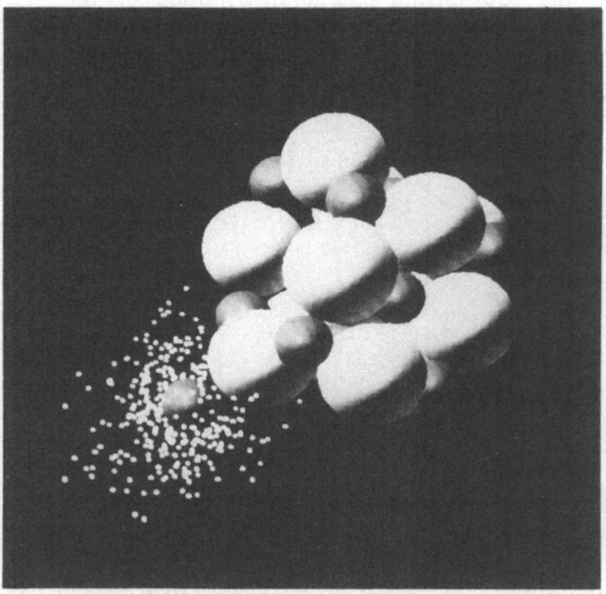

Fig. 20. Ionic configuration and "electron necklace" distribution for an excess electron interacting with $Na_{14}C\ell_{13}^{+}$ cluster (R_C = 3.22 a.u.). Note electron localization in a (compact) surface state (reference 85).

$$EABE = EVBE + E_c \qquad (V.15)$$

is obtained by summing the vertical electron binding energy

$$EVBE = 3/2\beta + K_{int} + P^{-1} \sum_{i=1}^{P} \langle V_e(\vec{r}_i) \rangle , \qquad (V.16)$$

and the cluster reorganization energy

$$E_c = \langle V_{AHC} \rangle - \langle V_{AHC} \rangle_0 , \qquad (V.17)$$

where $\langle V_{AHC} \rangle_0$ is the potential energy of the "bare" AHC in the absence of the electron. As is apparent from Fig. 19, the $[Na_{14}\ C\ell_{12}]^{++}$ cluster stabilizes an internally localized excess electron state. The features of the internal electron localization are:

(1) It occurs in a moderately large cluster.

(2) An anion vacancy is required.

(3) The localized electron state is surrounded by Na^+ ions in an octahedral configuration and by twelve $C\ell^-$ ions.

(4) The ionic configuration is somewhat distorted.

(5) The gain in EVBE exceeds the loss in E_c, i.e., $(EVBE + E_c) < 0$ which favours localization.

(6) The total energy of $e\text{-}[Na_{14}\ C\ell_{12}]^{++}$ is close to that of the $[Na_{14}\ C\ell_{13}]^+$, whereupon the electron (internal) binding energy in the cluster is similar to that of a negative ion.

(7) These results establish the dominance of short-range attractive interactions for internal electron localization.

(8) The spatial extent of the charge distribution is characterized by the radius of gyration (Eq. (V.10)) whose components (Table IV) assume the values 5-10 a_0. The thermal wavelength of a free electron at T = 300 K is $R_T = 56\ a_0$. Accordingly, for the internal excess electron state $R_g << R_T$, demonstrating enhanced localization.

A drastically different localization mode is obtained in the $[Na_{14}\ C\ell_{13}]^+$ system (Fig. 20), where a novel surface state is exhibited. We refer [85] to this state as a cluster surface localized state. Such surface states were considered for macroscopic alkali halide crystals by Tamm [95] about fifty years ago, but were never experimentally documented. The characteristics of external localization of an electron on an Na $C\ell$ cluster are:

(a) It is exhibited in a vacancy-free moderately large AHC.

(b) In such an AHC internal localization is prohibited by a large value of E_c.

(c) The excess electron in $[Na_{14}\ C\ell_{13}]^+$ localizes around Na^+ surface ions leaving a neutral $Na_{13}\ C\ell_{13}$ cluster interacting with a partially neutralized Na atom.

(d) In this system a surface localized electron state is exhibited.

(e) For clusters containing heavier M^+ ions, the surface state will become more extended.

(f) The observation of the localization of an excess electron around a single surface Na^+ ion in the centre of the face of $[Na_{14}\,C\ell_{13}]^+$, rather than a delocalized surface state, manifests symmetry breaking effects induced by vibrational excitations at finite temperature .

In smaller clusters novel effects of dissociative electron attachment and cluster isomerization induced by electron localization were manifested . We have studied the smallest single ionized cluster exhibiting high stability, i.e., $[Na_5\,C\ell_4]^+$, for which the lowest energy configuration is planar with four Na^+ ions at the corners and one at the centre of an approximate square (Fig. 18). This cluster possesses an isomeric structure (less stable by 0.014 a.u.) in which the ions are arranged in a distorted pyramidal configuration [14]. Adding an electron to the ground state planar state, the configuration of $[Na_5\,C\ell_4]^+$ (with the Na^+ pseudo-potential being characterized by $R_C = 3.22$ a_0) transforms the system into a neutral $[Na_4\,C\ell_4]$ cluster with a planar ring structure and a dissociated neutral Na atom (Fig. 21). To demonstrate that this process is driven by the localization of an electron around a single Na^+ ion, we have decreased the cation-electron binding strength, taking R_C = 4.36 a_0 [85]. In this case, extreme localization is not sufficiently counterbalanced by the electron binding. Instead, the planar structure transforms to the isomeric pyramidal configuration with the electron localized as a diffuse cloud about the tip of the pyramid (Fig. 22). Electron localization accompanied by structural isomerization will constitute a prevalent phenomenon for positively charged AHC with smaller electron-cation binding energy, i.e., the heavier alkali metals. In view of the intimate interrelationship between structural isomerization and melting of clusters, it will be interesting to explore the melting of such finite systems induced by electron localization in small neutral ionic clusters.

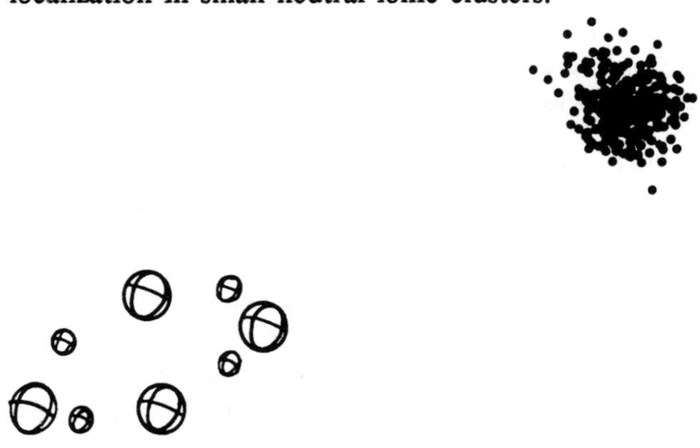

Fig. 21. Ionic configuration and "electron necklace" distribution for an excess electron interacting with $Na_5C\ell_4^+$ cluster and resulting in dissociation detachment (reference 85).

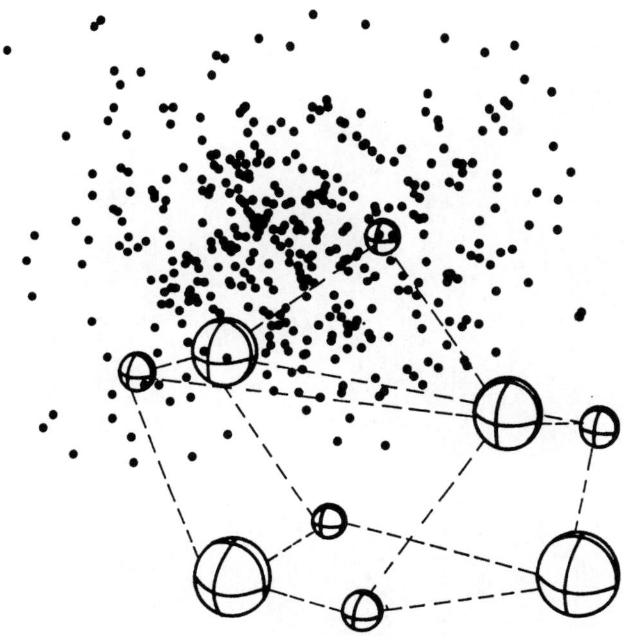

Fig. 22. Ionic configuration and "electron necklace" distribution for an excess electron interacting with a model $Na_5C\ell_4^+$ cluster with R_C = 4.36 a.u. Electron attachment results in cluster isomerization (reference 85).

V.D Cluster Isomerization Induced by Electron Attachment

Cluster structural rearrangement induced by the attachment of an excess electron is prevalent in small ionic clusters. We have found [86] that electron localization induces configurational transformations in small neutral ionic clusters at relatively low temperatures, and that the equilibrium configurations of the cluster with the attached electron do not necessarily relate to those of the parent neutral cluster.

Our QUPID [86] studies of electron interaction with $Na_4\,C\ell_4$ clusters in the temperature range 50-1000K employed the interaction potentials described in Section V.C, modified to include electron-anion polarizability interaction. Equilibrium configurations at various temperatures are shown in Fig. 23. At low temperature ($T \leq 500K$) the electron distribution is extended with an equal probability charge distribution around all the four Na^+ ions, and the ionic configuration is similar to that of the low temperature neutral cluster. At the intermediate temperature domain, $500K \leq T \leq 750K$, a configurational change of the negative cluster is exhibited (see Fig. 23b). This is a distorted planar configuration which is similar to the stable structure of the related classical $Na_4\,C\ell_5^-$ cluster, which we found in simulations performed in the temperature range $50K < T < 1100K$. At the high temperature domain, $750K < T < 1200K$, a coexistence

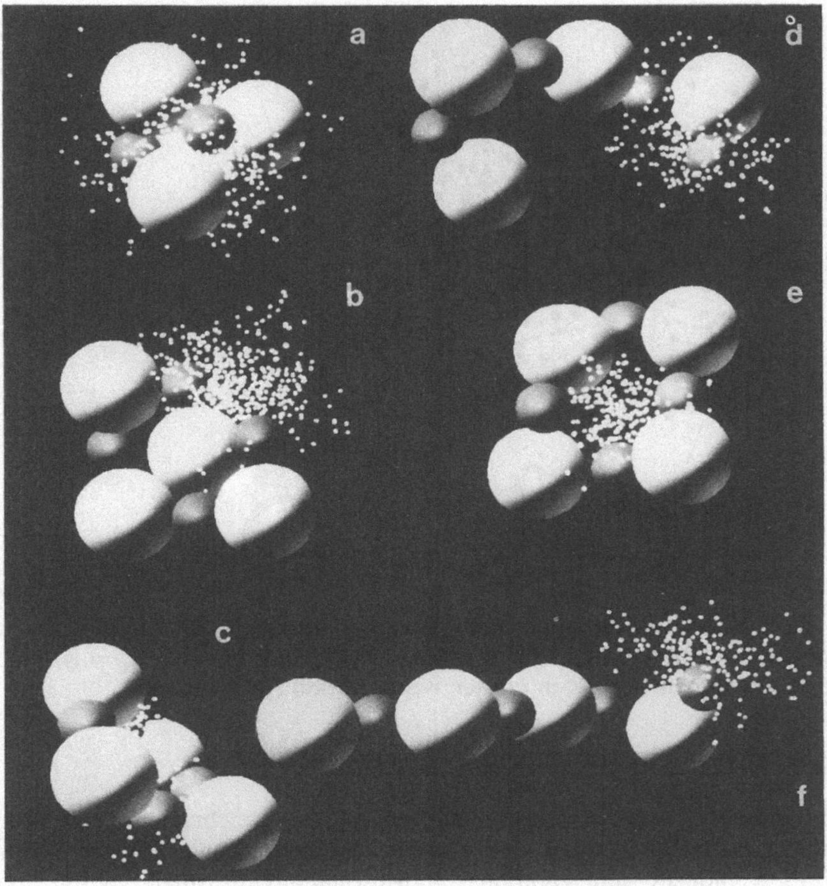

Fig. 23. Cluster configurations and "electron necklace" distribution for an excess electron with a $Na_4C\ell_4$ cluster. (a) 50K, (b) 575K, (c) and (d) 750K, (e) and (f) 1000K. (reference 86).

of several isomers is found. At ~ 750K a boat-like configuration (Fig. 23c) dominates (while a bent chain is also encountered) (Fig. 23d) and at T ~ 1000K an elongated chain is found in coexistence with a planar ring (23e and 23f). We observe a tendency toward spatial localization of the electron upon increase in T. Furthermore, the presence of the excess electron induces two types of configurational modifications.

(i) The localized excess electron can play the role of a negative ion, with appreciable kinetic energy, which is overwhelmed by the potential energy of the electron in the field of the ions. Consequently, new nuclear configurations of the negative cluster, which have no counterpart in the neutral cluster, will be exhibited.

(ii) The partial neutralization of a single positively charged ion by the excess electron results in the appearance of the high-temperature configurations of the neutral parent cluster at substantially lower temperatures for the negatively charged cluster.

The first effect is observed in the intermediate temperature range, while both effects are exhibited in the high-temperature domain. These configurational modifications are either quantitatively or qualitatively different from those in the neutral cluster.

V.E Surface and Internal States of an Excess Electron in Water Clusters

The existence of the solvated electron was experimentally demonstrated in 1863 for liquid ammonia [96] and in 1962 for water [97]. Electron localization in these and other polar fluids is nonreactive, involving cooperative solvation, which originates from the combination of long-range [98] and short-range [99] attractive interactions and is accompanied by a large local solvent reorganization [98,99]. During the last decade the problem of the electron localization in a finite cluster of polar molecules [100,101] constituted an experimental and conceptual challenge for two reasons. Firstly, such systems allow for the elucidation of microscopic solvation effects on an excess electron in a well defined finite solvent environment. Second, this problem falls into the general category of size effects on chemical processes pertaining to the central issue: what is the minimal (water or ammonia) cluster size which sustains a bound state of an excess electron?

The advent of supersonic and cluster beams allowed for the experimental interrogation of size effects on electron attachment to water and ammonia clusters [76-80]. The following information is available:
(1) The water dimer constitutes the smallest water cluster which attaches an electron, resulting in a weakly bound $(H_2O)_2^-$ state, with a binding energy of ~ 17 meV .
(2) Strongly bound $(H_2O)_n^-$ clusters are observed for $n \geq 11$. This critical cluster size does not exhibit a deuterium isotope effect.
(3) The stable $(H_2O)_n^-$ $(n \geq 11)$ clusters are characterized by a large vertical electron binding energy EVBE = -0.75eV for n = 11 to EVBE = -1.12eV for n = 19.
(4) Nonreactive electron localization in water clusters was experimentally documented to originate either from electron binding during the cluster nucleation process or by electron attachment to preexisting clusters.
(5) A striking difference is exhibited between the lowest coordination number for the formation of $(H_2O)_n^-$ $(n \geq 11)$ and of $(NH_3)_n$ $(n \geq 35)$.

The experimental observation of stable $(H_2O)_n^-$ $(n \geq 11)$ clusters poses a challenging theoretical problem, as the origin of the energetic stability of

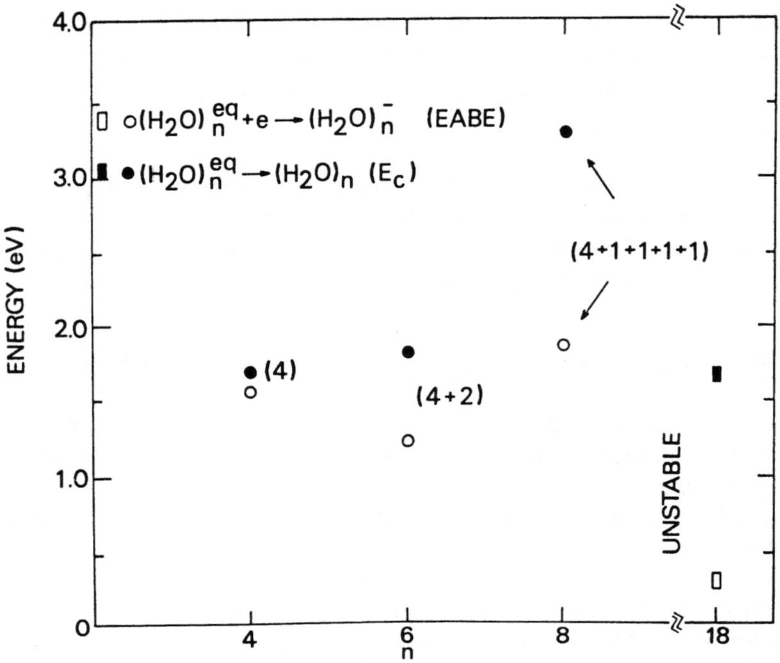

Fig. 24. Adiabatic electron binding energy and cluster reorganization energy. O and ● data from quantum chemistry calculations (references 81,82). □ and ■ data from QUPID simulations.

these excess electron clusters is not clear. Quantum mechanical calculations of electron localization in $(H_2O)_6^-$ and $(H_2O)_8^-$ have established the occurrance of large cluster reorganization energies [81,82] and positive adiabatic electron binding energies. In Fig. 24 we have assembled the EABE data, which correspond to the energetic change in the process

$$(H_2O)_n^{(eq)} + e \rightarrow (H_2O)_n^- , \qquad (V.18)$$

together with the cluster reorganization energies obtained from quantum mechanical calculations [81,82], as well as from QUPID simulations, for an interior of $(H_2O)_{18}^-$ cluster. These data reveal that:

(1) The cluster reorganization energy is sizeable so that $E_C > |EVBE|$.

(2) For the $(H_2O)_n^-$ (n = 6,8,18) with an internal excess electron state, EABE > 0 and reaction (V.18) is strongly endothermic, precluding the existence of such excess electron clusters.

These conclusions are in contrast with experiment. These theoretical studies followed faithfully the conventional wisdom in the field of solvated electron theory, invoking the implicit assumption that the excess electron state in $(H_2O)_n^-$ constitutes an interior localization mode. On the basis of QUPID simulations it was established that electron localization in small [84] (n = 2) and medium sized [83] (n ≃ 8-10) water clusters does not involve a precursor of the hydrated electron, but rather constitutes a novel excess electron surface state on polar clusters.

A key issue in modelling the system by QUPID simulations rests on the choice of interaction potentials. Fortunately, for neutral small water clusters, interaction potential functions which provide a satisfactory description for a range of properties are available. LANDMAN, BARNETT and JORTNER [83,107] used the RWK2-M model for the intra- and inter-molecular interactions [102]. For the electron-water interaction they have constructed a pseudo-potential (Fig. 25) in the spirit of the density functional theory, which consists of Coulomb, polarization, exclusion, and exchange contributions [83,107],

$$V(\vec{r}_e,\vec{R}_O,\vec{R}_1,\vec{R}_2) = V_{coul} + V_p + V_e + V_x \quad . \tag{V.19}$$

The position of the oxygen and hydrogen nuclei of the water molecule are given by $(\vec{R}_O,\vec{R}_1,\vec{R}_2)$ and $\vec{r}_e$ is the position of the electron. The Coulomb interaction is

$$V_{coul}(\vec{r}_e,\vec{R}_O,\vec{R}_1,\vec{R}_2) = -\sum_{j=1}^{3} q_j e/\max(|\vec{r}_e-\vec{R}_j|,R_{cc}) \quad , \tag{V.20}$$

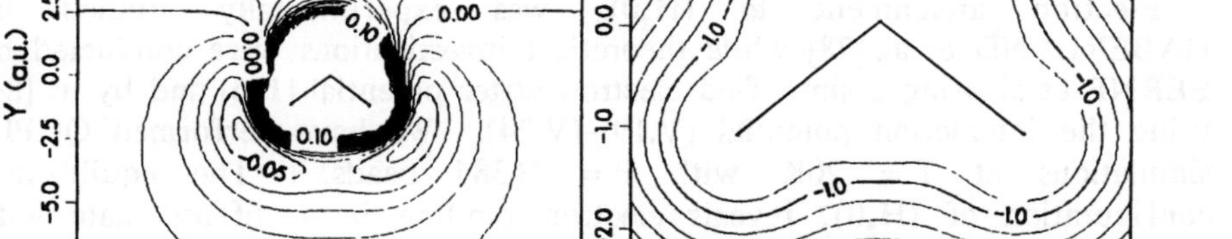

Fig. 25. Contours of electron-water interactions, left, and of the electronic density ($Log_{10}\rho(r)$), right, in the plane containing the nuclei. The oxygen is located at the origin (reference 83).

where $\vec{R}_3 = \vec{R}_O + (\vec{R}_1+\vec{R}_2-2\vec{R}_O)\delta$ is the position of the negative point charge of the RWK2-M model and $R_{CC} = 0.5a_0$. The values $q_1 = q_2 = 0.6e$, $q_3 = 01.2e$, $\delta = 0.22183756a_0$ were chosen to give a good representation of the dipole and quadrupole moments of H_2O. The polarization interaction is given by

$$V_P(\vec{r},\vec{R}_O) = -0.5\alpha e^2 / \left(\left| \vec{r}_e - \vec{R}_O \right|^2 + R_P^2 \right)^2 , \tag{V.21}$$

where $\alpha = 9.7446$ a.u. is the spherical polarizability of the water molecule. The form of V_P and the value of $R_P = 1.6a_0$ were chosen to fit the adiabatic polarization potential as calculated by Douglass et al. [103]. The exclusion, V_e, and exchange, V_X, contributions both require the electron density $\rho(\vec{r},\vec{R}_O,\vec{R}_1,\vec{R}_2)$, of the water molecule [104] which, in the regions of importance, is adequately approximated by the simple expression (see Fig. 25),

$$\rho(\vec{r};\vec{R}_O,\vec{R}_1,\vec{R}_2) = 8\, a_0^{-3} e^{-3|\vec{r}-\vec{R}_O|/a_0} + a_0^{-3} \sum_{j=1}^{2} e^{-3|\vec{r}-\vec{R}_j|/a_0} . \tag{V.22}$$

The repulsion, due to the exclusion principle, is modeled as a "local kinetic energy" term [105],

$$V_e(\vec{r}_e,\vec{R}_O,\vec{R}_1,\vec{R}_2) = 0.5\, e^2\, a_0 (3\pi^2 \rho)^{2/3} . \tag{V.23}$$

The exchange interaction is modeled via the local exchange approximation,

$$V_X(\vec{r}_e,\vec{R}_O,\vec{R}_1,\vec{R}_2) = -\alpha_X e^2\, a_0\, (3\pi^2 \rho)^{1/3}/\pi \tag{V.24}$$

The parameter α_X was taken to be $\alpha_X = 0.3$ in order to obtain good agreement between our simulation results and the SCF results of Rao and Kestner [81] for $(H_2O)_8^-$ at a fixed configuration of the water molecules.

Electron attachment to $(H_2O)_2$ was experimentally studied by HABERLAND et al. [78] while theoretical investigations were conducted by BERNE et al. using a simplified electron water potential [106] and by us [84] using the interaction potential (V.19)-(V.24). We have performed QUPID simulations at $T = 20K$ with $P = 16384$ beads. The equilibrium configuration of $(H_2O)_2^-$ reveals electron binding in a diffuse state with EVBE $\simeq$ -3 meV, EABE $\simeq$ -3 meV and $E_C \simeq 0$. The diffuse charge distribution is characterized by $R_g = 36\ a_0$. During the initial equilibration runs, the system of the water dimer and necklace spent a long time in an apparent metastable state in which the bead distribution was more compact (R_g = 16-30 a.u.) and during which the dimer configuration fluctuated between that of the neutral dimer and a distorted configuration. The dipole moment during this period fluctuated between 1.2 and 1.7 a.u., the vertical binding energy being about -13 meV to -26 meV. It is interesting to note that the results obtained from this "metastable state" are consistent with an estimate of the binding energy (17 meV) obtained from field detachment [78]. Our QUPID simulations [84] predict the existence of a metastable configuration of $(H_2O)_2^-$ with EVBE comparable to the experimental result. From the point of view of general methodology the diffuse electron state in equilibrium and metastable $(H_2O)_2^-$ constitutes a "molecular surface" state of the excess electron.

We have embarked upon an investigation of the energetics and geometry of medium sized $(H_2O)_n^-$ (n = 8-32) clusters [83]. In correspondence with the alternative experimental preparation methods [76-80] we invoked two initial conditions: (i) First condensing the water molecules around a classical negatively charged particle with a radius of $5a_0$, and subsequently replacing the classical particle with the electron necklace. (ii) Placing a compact distribution of beads next to an equilibrated neutral cluster. For the smaller clusters $n \leq 12$ a surface state develops rapidly, regardless of the initial setup of the calculation, while for n = 18 and 32 (i) and (ii) yield an "internal" and "surface" state, respectively.

In Table V we summarize the energetic data for the electron vertical binding energy, EVBE = $PE(e\text{-}(H_2O)_n) + K_{int}$ where PE is the averaged interaction potential energy between the electron and the water molecules, the cluster reorganization energy, E_C, and the electron adiabatic binding energy, EABE = EVBE + E_C. The neutral cluster reference states were obtained by simulated annealing. The lower energy configuration for each cluster size was then used to calculate E_C (the difference between the

TABLE V. Energetics and excess electron charge distribution for $(H_2O)_n^-$ clusters. All calculations, except where indicated, were at T = 79°K, using P = 4096 beads for the electron necklace. Energies in eV, radius of gyration, R_g, of the electron charge distribution in a_0 units.

Cluster	$(H_2O)_8^-$ Diffuse	$(H_2O)_{12}^-$ Surface	$(H_2O)_{18}^-$ Surface	$(H_2O)_{18}^-$ Internal	$(H_2O)_{32}^-$ Surface	$(H_2O)_{32}^-$ Internal	$(H_2O)_{32}^-$ Internal*	$(H_2O)_{64}^-$ Internal*	$(H_2O)_{128}^-$ Internal*
EVBE	-0.190	-0.97	-1.31	-1.96	-1.15	-2.40	-2.47	-3.20	-3.55
E_c	0.136	0.871	1.333	2.204	1.05	2.38	2.43	2.70	2.56
EABE	-0.054	-0.136	0.023	0.245	-0.10	-0.023	-0.037	-0.48	-0.99
R_g	10.6	6.1	5.5	4.1		4.0	3.8	3.8	
$\frac{R(\beta\hbar/2)}{R_f}$	0.28	0.15	0.14	0.11		0.10	0.2	0.19	

* Calculated at 300°K with P = 1024

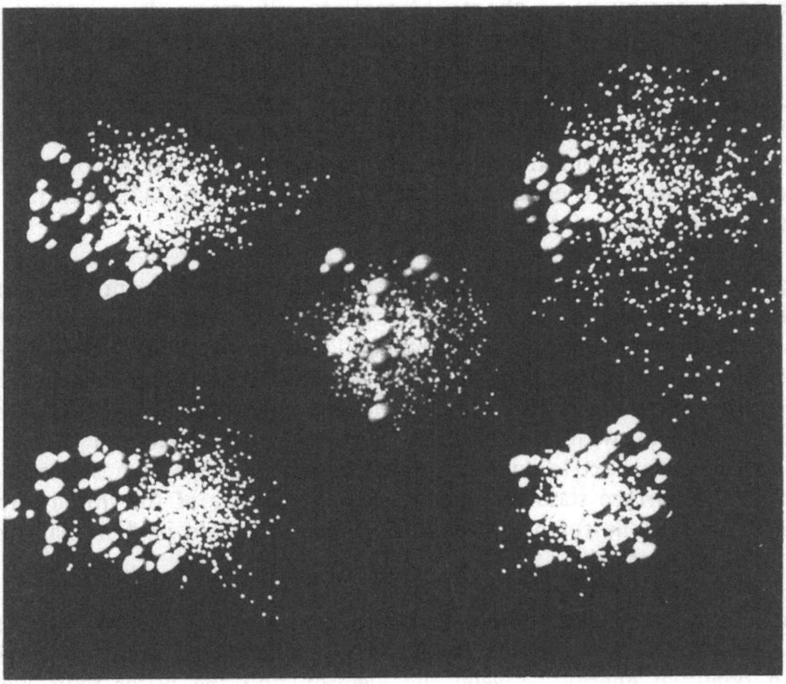

Fig. 26. Cluster configurations of $(H_2O)_n^-$, from QUPID simulations. Balls large and small, correspond to oxygen and hydrogen, respectively. The dots represent the electron (bead) distribution. Shown at the center is $(H_2O)_8^-$, for a static molecular configuration as in reference 81. From top right and going counterclockwise: (i) diffuse surface $(H_2O)_8^-$; (ii) surface state of $(H_2O)_{12}^-$; (iii) surface state of $(H_2O)_{18}^-$ and (iv) internal state of $(H_2O)_{18}^-$ (reference 83)

molecular potential energies of the negatively charged and neutral clusters). The energetic stability of the negatively charged cluster, with respect to the equilibrium neutral cluster plus free electron, is inferred from the magnitude and sign EABE (negative values corresponding to a stable bound state). The bead distributions for the excess electron (Fig. 26) are characterized by the gyration radius R_g and the complex time correlation function $R(t-t')$, Eqs. (V.10) and (V.9), respectively.

From these results we assert that there is a remarkable quantitative difference between internal and surface states of the excess electron in medium sized water clusters. The value of E_c is considerably lower for a surface state than for an internal state, insuring relative energetic stability of the former (Table V). As is apparent for $(H_20)_{18}^-$, |EVBE| is considerably higher (and outside the range of the experimental values) for the interior state, however, the high value of E_C results in EABE = 0.245 eV, precluding a stable internally localized state. On the other hand, for the electron surface state of $(H_20)^-_{18}$ the value of EABE is close to zero favoring this mode of localization. For $(H_20)^-_{12}$, only a surface state is found. Finally, for $(H_20)^-_8$, a very small electron binding energy is found and the state is characterized by a diffuse charge distribution (Fig. 26).

From these results we conclude that:
(1) The electron localization mode in medium sized $(H_20)^-_n$ clusters ($8 < n < 32$) involves the formation of surface state. The surface localization mode is favored due to the dominating contribution of nearest water-electron interactions and optimization of cluster rearrangement energy.
(2) The onset of electron localization in a tightly bound state in $(H_20)^-_n$ clusters is exhibited for $n > 8$, in accord with experiment [76-80] ($n > 11$).
(3) The vertical electron binding energy for the cluster-electron-surface-state in $(H_20)^-_{12}$ (EVBE = -0.97eV) and $(H_20)^-_{18}$ (EVBE = -1.31eV) are in adequate agreement with the experimental photoelectron spectroscopic data [80] (EVBE = -0.83eV for n = 12 and EVBE = -1.1eV for n = 18).

We have demonstrated the prevalence of the surface localization mode in medium sized $(H_20)^-_n$ clusters, which bears an analogy to vacancy-free alkali halide clusters (Section V.C). The surface state of medium-sized $(H_20)^-_n$ (n = 8-32) is qualitatively different from the localization mode of the hydrated electron in the bulk and does not constitute the precursor of the celebrated hydrated electron.

With increasing the size of the water cluster beyond n = 18 the internal electron localization mode becomes energetically favored. For n = 32 the values of EABE for the surface and internal excess electron states are identical within experimental uncertainty (Table V). For $(H_20)^-_n$ (n = 32,64 and 128) at 300K the internal, localized excess electron state is energetically stable (Table V). The energetics data (Table V) for the internal state exhibit a gradual increase in the magnitudes of the vertical and adiabatic binding energies with increasing n. From the empirical plot (Fig. 27) of EABE vs 1/n for internally localized states in large $(H_20)^-_n$ clusters (which includes also the unstable n = 18 internal state), we note that the convergence to the bulk limit is slow with the cluster size, demonstrating the importance of long-range interactions. It is significant that the empirical EABE vs n^{-1} plot of Figure 27 at $n = \infty$ extrapolates to the heat of solution of the hydrated electron in the bulk $\Delta H(eq) = -1.7eV$ [108]. The internal electron localization modes in large water clusters constitute the precursor of the hydrated electron.

The "transition" from surface to internal excess electron localization mode in water clusters is manifested by the size dependence of the adiabatic binding energies. A courageous extrapolation of the values of EABE for

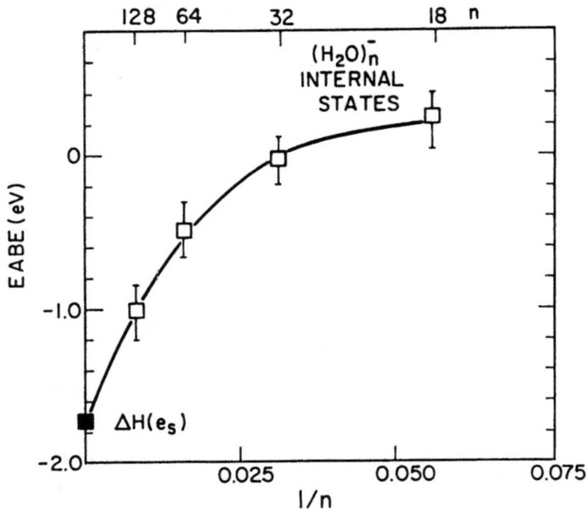

Fig. 27. The adiabatic electron binding energies for internal excess electron states in large $(H_2O)_n^-$ clusters from QUPID simulations. The result for n = 18 corresponds to an unstable state (see text). The black square at n = ∞ corresponds to the experimental heat of solution of an electron in water (reference 107).

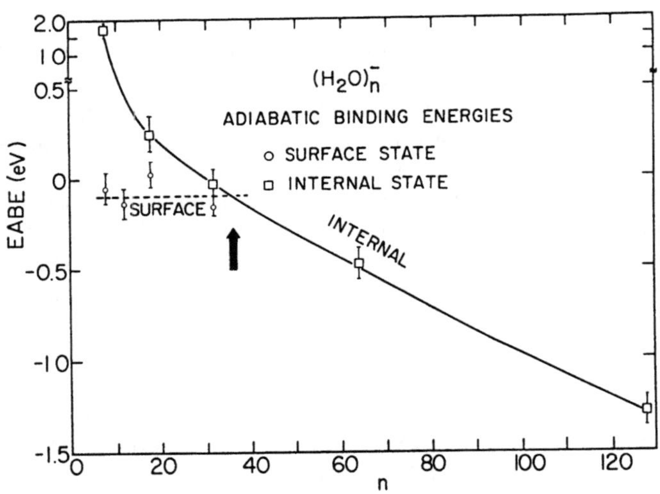

Fig. 28. The size dependence of adiabatic binding energies of excess electron in water clusters. Energetically stable states are exhibited for medium-sized clusters (8 < n < 32) while internal states are favored for n > 32. The crossover from surface to internal excess electron state is marked by an arrow (reference 107).

surface (small n) and internal (large n) excess electron states (Fig. 28) indicates that the transition from surface to bulk electron state is exhibited in the range n = 30-36. This transition provides the signature of the onset of a significant contribution from long-range, large-polaron attractive interactions for localized electron stabilization in large water clusters. Consequently, we infer that long-range attractive interactions play an important role in electron localization in the bulk. In this context we conjecture that the striking difference [76] between the lowest coordination number for electron localization in water ($n \geq 11$) and that in ammonia ($n \geq 35$) clusters, may originate from a weaker electron-molecule interaction in ammonia, which renders the surface state unstable in $(NH_3)_n$ for small n. Electron localization in $(NH_3)_n$ may require the buildup of long-range attractive interactions thus resulting in large coordination numbers.

These studies of size effects on electron localization in and on water clusters bridge the gaps between molecular, surface and condensed matter phenomena.

VI. CLUSTER DYNAMICS

Intramolecular dynamics in isolated molecules falls into the category of "few-body" relaxation in a small molecule, or the relaxation of "few-levels", which are photoselected and subsequently interrogated, in a dense vibrational-rotational manifold of a large molecule [109]. A gradual transition from intramolecular dynamics of finite molecular systems to dynamics in infinite systems, i.e., condensed phases, emerges from the exploration of cluster dynamics. The processes of energy acquisition, storage and disposal in clusters are of particular interest for the elucidation of dynamic size effects in large finite systems. In this context, and of considerable importance, is the face of vibrational excitation of a molecule in the cluster, or of the cluster intermolecular mode(s), which can be accomplished by collisional excitation, photoselective vibrational excitation, electronic excitation followed by intramolecular radiationaless transition [110] or by exciton trapping [111,112,113]. In charged clusters, vibrational excitation can originate from ionization followed by hole trapping in rare-gas clusters [114-117] as well as in alkali-halide clusters and from electron attachment to alkali-halide clusters (Chapter V). Vibrational energy relaxation in van der Waals molecules and larger clusters falls into two categories:

(i) Reactive vibrational predissociation [118-129], which results in vibrational-translational, vibrational-rotational and vibrational-vibrational energy exchange.

(ii) Nonreactive vibrational energy redistribution, resulting in energy exchange between an intramolecular vibration and low frequency intermolecular modes of the cluster [112,113,122].

Extensive experimental and theoretical studies [118-129] of reactive vibrational predissociation of small van der Waals molecules, e.g., HeI_2 or $(N_2O)_2$ focused attention on the reactive channel. A central conclusion emerging from these studies pertains to the failure of the statistical theories of unimolecular reactions to describe vibrational predissociation of small van der Waals molecules. The lack of vibrational energy randomization in small van der Waals molecules provides a touchstone for mode-selective reactive dynamics. Sequential vibrational energy redistribution followed by presumably non-statistical dissociation seems to be ubiquitous in van der Waals molecules built from polyatomic molecules and in clusters of diatomics [123-129], e.g., $(C_2H_4)_2$, $(NH_3)_2$, $(\text{benzene})_2$ and $(HF)_n$ (n = 2-6). An apparent dichotomy [129] between the dephasing lifetimes of some van der Waals molecules (10^{-12} sec), which originates from line broadening and their vibrational predissociation lifetimes (> 10^{-9} sec) originates from the dominant contribution of the nonreactive vibrational relaxation to dephasing [129].

Information regarding vibrational energy degradation in larger clusters stems from molecular dynamics (MD) computer simulations of the dissociation of ground state Ar_n (n=1-6) clusters [122], which can be accounted for in terms of the statistical theory of unimolecular reactions, and which implies vibrational energy randomization in these clusters [122]. Another significant MD study of vibrational energy degradation in clusters involves the dynamic implications of hole trapping, i.e., the formation of R_2^+ dimer ions in rare gas clusters [115,116]. These MD studies [115,116] demonstrated the occurrence of complete vibrational relaxation of the dimer ion in Xe_{13}^+ clusters on a time scale of 1 nsec, with ~ 3 dissociation events of Xe atoms being exhibited on a time scale of 150 psec past ionization. These studies, which are of considerable interest for the elucidation of the issue of the effects of fragmentation on the "magic numbers" of rare-gas clusters [114], did not address the applicability of statistical unimolecular theories; however, the observation of a retardation in the fragmentation with increasing cluster size is in accord with such approaches.

The nonreactive and reactive processes induced by the degradation of electronic-vibrational energy into vibrational energy in clusters are of considerable interest. In this context, we have undertaken the study of the dynamical consequences of the trapping of electronic excitations in rare-gas clusters (RGCs) [112,113]. Extensive information is currently available regarding exciton trapping in neat solid and liquid rare-gases [111] and on the trapping of electronic-vibrational excitations in impurity pairs in solid and liquid rare-gas alloys. Exciton or excitation trapping in neat heavy rare-gas liquids and solids, or in impurity dimers of heavy rare gases, i.e., Ar, Kr and Xe, exhibits two-centre localization, resulting in the formation of an electronically excited diatomic rare-gas excimer molecule [111]. The localization of atomic impurity excitation in dilute solid and liquid rare-gas alloys was also explored in considerable detail, being accompanied by large local configurational changes [111]. The nature of electronic excitations in pure and mixed RGCs can be inferred from the information currently available regarding the lowest electronic excitations in pure or doped solid and liquid rare-gases. The lowest electronic excitations in pure solid and liquid rare-gases can be described either in terms of intermediate Wannier states with large central cell corrections, or by strongly perturbed Frenkel excitons. Similarly, the lowest electronic excited states of an impurity atom in liquid and solid rare-gas alloys are amenable to description either in terms of Wannier impurity states with large central cell correction, or a strongly perturbed atomic excitation. Adopting the latter, tight binding description, the lowest electronic excitations R^*, of an atom (R = Ar,Kr,Xe) in neat and mixed RGCs can be described in terms of the atomic $^1S_0 \rightarrow {}^3P_1$ and $^1S_0 \rightarrow {}^1P_1$ excitations, which are modified by large nonorthogonality corrections [111]. The energetic separation between these two electronic excitations corresponds to the spin-orbit splitting. This tight binding description of the atomic electronic excitations in RGCs rests on a decidedly molecular picture. An analogous molecular point of view will be adopted for exciton trapping in a neat RGC or excitation trapping in an impurity pair in a mixed RGC, which results in excimer formation. The excimer diatomic molecule R_2^* (R = Ar,Kr,Xe) constitutes a localized state which is characterized by a large binding energy, and which is initially produced in a highly vibrationally excited state. Energy exchange between electronically excited states of impurity atoms or impurity excimers and the cluster atoms in RGCs falls into two classes:

(1) Short range repulsive interactions between a spatially expanded Rydberg type excitation of R^* and R_2^* and the cluster atoms result in vibrational energy flow into the cluster.

(2) Vibrational relaxation of the excimer results in exchange of vibrational energy between the R_2^* internal mode and the cluster intermolecular modes.

We explored the dynamic implications of energy exchange between an electronically excited rare-gas impurity atom (R^*) or excimer (R_2^*) and host RGCs [112,113]. We have investigated the mechanisms of cluster vibrational excitation induced by vibrational energy flow from R^*, due to short range repulsive interactions, as well as the consequences of excimer formation in RGCs. We have applied [112,113] the classical molecular dynamics (MD) method to simulate electronically excited states of mixed $XeAr_{n-1}$ and Xe_2Ar_{n-2} clusters (with n = 13 and 55). Our work constitutes an application of the MD technique for the dynamical evolution of electronically-vibrationally excited states of clusters.

For our mixed rare-gas clusters the Ar-Ar, Xe-Xe and Xe-Ar pair-interactions in the ground electronic state were fitted, on the basis of experimental data, to a Lennard-Jones (LJ) potential [130-133]. We have used for the elementary units of energy and length, the values of the LJ parameters for the Ar-Ar interaction in the ground electronic state, namely ϵ/k_B = 120°K and σ = 3.40Å For the Ar mass, we take m = 39.95 a.m.u.

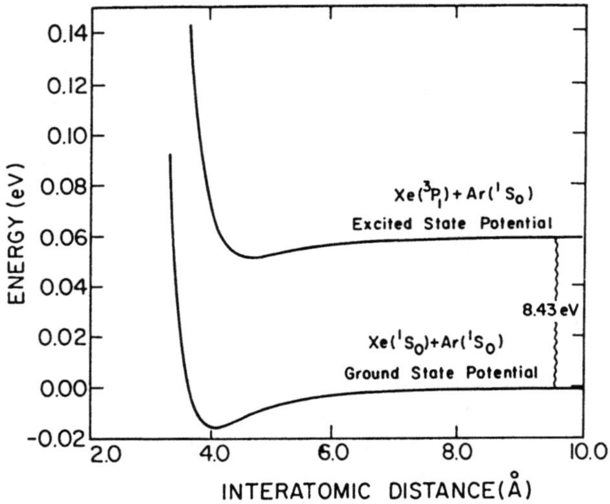

Fig. 29. The potential energy curves for the Xe-Ar potential in the ground- and the lowest $^3P_1$ excited-electronic states (potential parameters from refs. 131,132).

The resulting time unit is 1 t.u. = $(m\sigma^2/\epsilon)^{1/2}$ = 2.14263 · 10^{-12} sec. For the interaction potential between an electronically excited Xe atom (Xe^*) in its Xe ($^3P_1$) excited state and a ground state Ar atom, we have used the Lennard-Jones potential parameters extracted from spectroscopic measurements of the lowest resonance line of atomic Xe, Xe($^1S_0$) → Xe($^3P_1$) in dense supercritical and subcritical fluid Argon [131,132] (Fig. 29).

The electronically excited state of the Xe_2^* in the $^3\Sigma_u$ state, Fig. 30, was described by a potential curve of a diatomic molecule which was fitted to a Morse potential. From the spectroscopic date [134] for the $^3\Sigma_u$ state of Xe_2^* we have obtained the following Morse parameters for this state: dissociation energy D_e = 11609°K (95.94 ϵ_{Ar}/k_B), B = 4.228 and equilibrium distance R_e = 2.999Å The potential parameters for the $^1\Sigma_u$ state (Fig. 30) are similar to those of the $^3\Sigma_u$ state [134]. The interaction between the Ar atoms is not influenced by the electronic excitation of the Xe atom.

Following equilibration of the ground state system, electronic excitation was achieved (at the time t = 0) by the instantaneous switching-on of the excimer potential between the pair of Xe atoms and of the potentials between the excited Xe (Xe^*) or the excimer (Xe_2^*) and the ground-state atoms comprising the rest of the cluster. We used a fifth-order predictor-corrector method. The integration time step in the ground state was 1.6 · 10^{-14} sec and in the excited state 1.6 · 10^{-16} sec.

We have performed extensive studies of $Xe_m^*Ar_{n-m}$ with n = 13, 55 and m = 1,2, with the Xe atoms located at either the center or on the outer shell. In the case of Xe^*Ar_{12}, with the Xe^* centrally located, the excited atom escaped from the cluster on a time scale of 50 ± 20 psec (Fig. 31).In view of

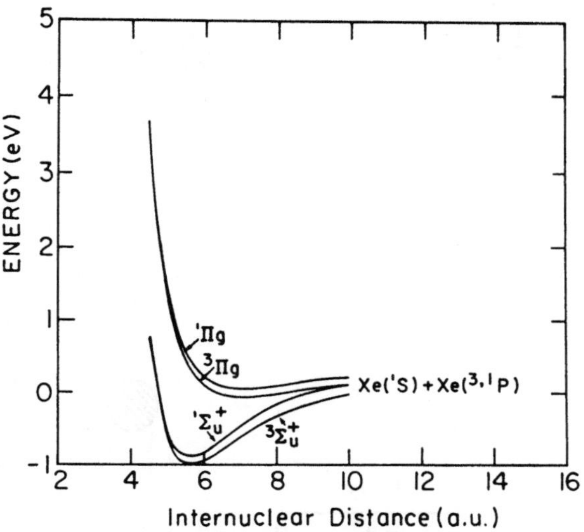

Fig. 30. The four lowest Xe_2^* potential energy curves without spin-orbit coupling. Notice that the lowest $^3\Sigma_u$ and $^1\Sigma_u$ curves lie close to each other and are very similar (from reference 134).

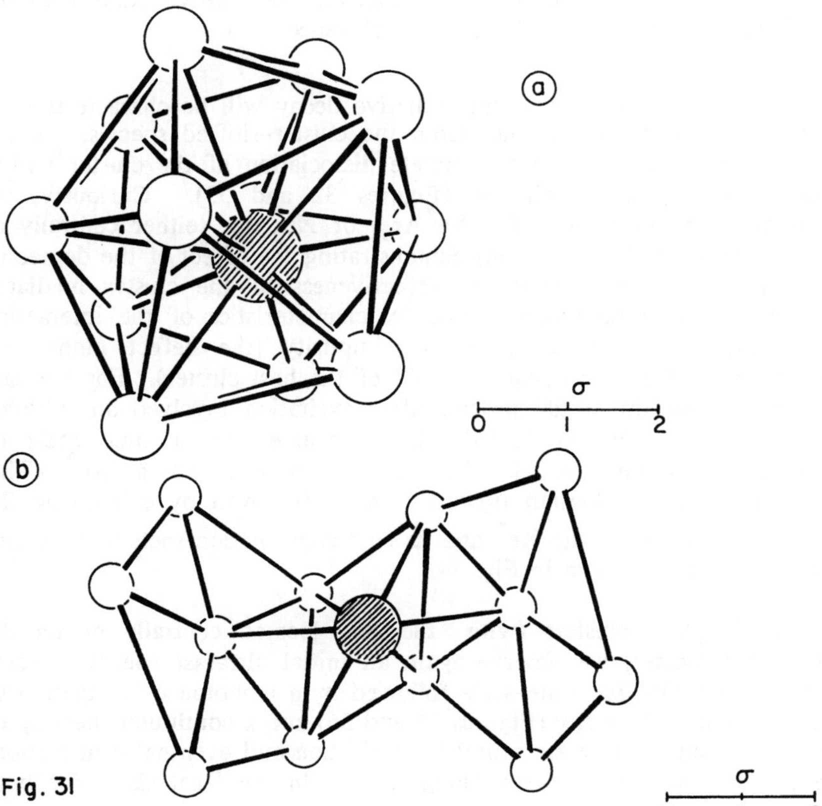

Fig. 31

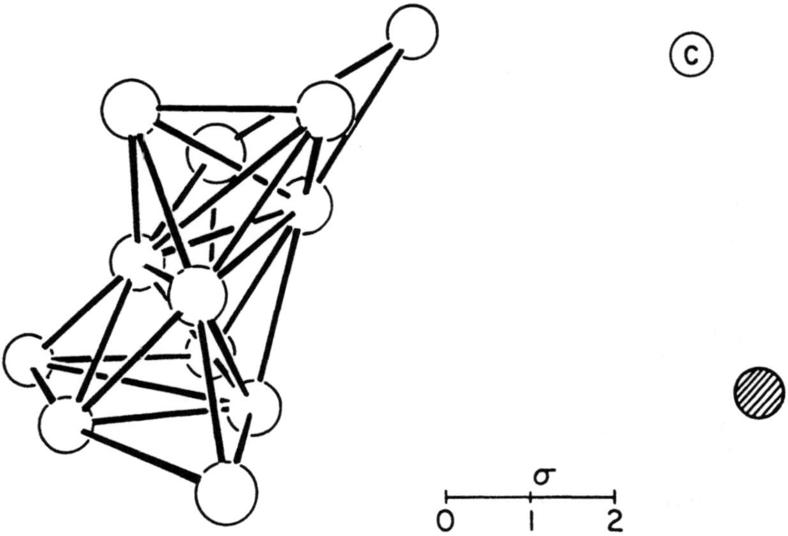

Fig. 31. (a) The ground state $XeAr_{12}$ configuration at t = 24 K. Empty and hashed balls correspond to Ar and Xe, respectively. (b) The configuration of the cluster after 1.1 psec past the excitation. (c) The configuration after the escape of the Xe^* atom after 55±25 psec (reference 113).

the fast escape, we predict that the radiative decay will be characteristic of the free excited atom rather than from the cluster-shifted species. In the case of $Xe_2^*Ar_{11}$ we observed a complete dissociation of the cluster within 200 psec following the excitation (Figures 32 and 33). Curiously, we observed no such phenomena for Xe^*Ar_{54} or $Xe_2^*Ar_{53}$ (either centrally or peripherally locating the excitation) demonstrating the effect of the degree of aggregation. In these systems the effectiveness of the cluster mediated vibrational relaxation dominates, aided by characteristics of the interaction potentials and the heavy mass of the impurity (the defect modes are embedded within the vibrational manifold of the host cluster). For the case of the Xe^* system the initial process after excitation involved an ultrafast energy transfer to the Ar "bath" which occurred on a time scale of (250 ± 100) fsec, accompanied by "bubble" formation around the Xe^*. This bubble formation, or dilation process, can be followed by calculating the average distance between the Xe^* and its first coordination shell of Ar atoms as a function of time shown in Fig. 34.

For the $Xe_2^*Ar_{53}$ clusters (with the Xe_2^* located centrally or on the surface of the cluster) we observe again an initial ultrafast energy transfer occurring on the 350 fsec time-scale followed by a monotonic relaxation of the excess vibrational energy (Figures 35 and 36) and a continuous heating of the Ar-bath, resulting in a structural isomerization and eventual evaporations of one to four Ar atoms at very long times. In the final state, for both

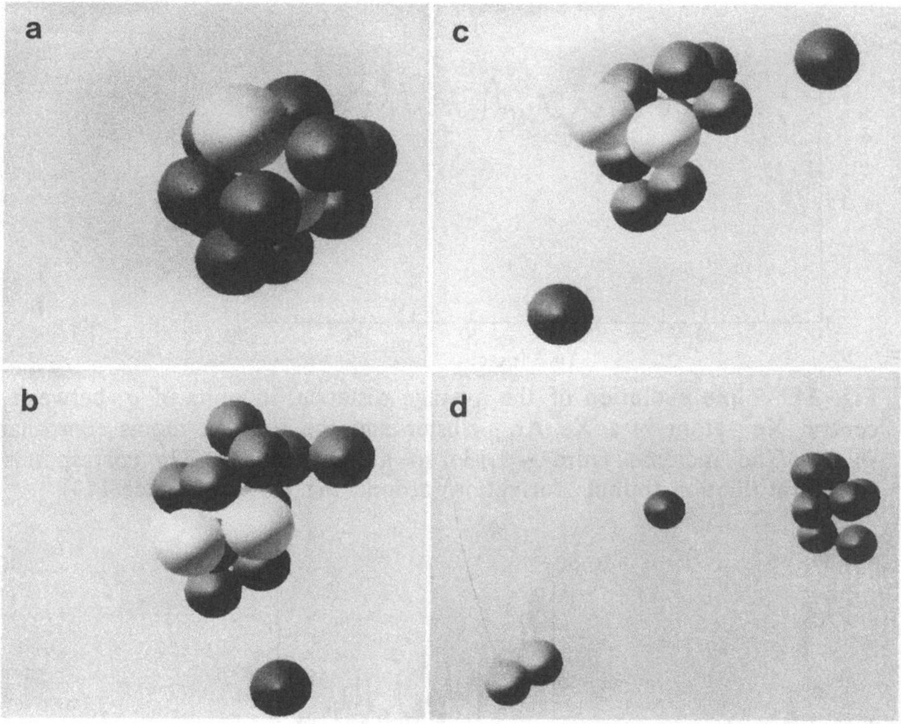

Fig. 32. Steps in the fragmentation of a $Xe_2^*Ar_{11}$ cluster (a) $t = 0^+$, (b) t = 1.81 psec, (c) t = 3.32 psec, (d) t = 56.7 psec (reference 113).

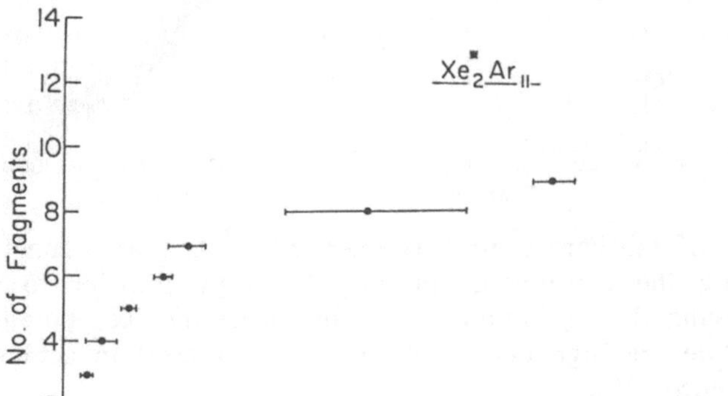

Fig. 33. Number of fragments as a function of time (in psec) for $Xe_2^*Ar_{11}$, averaged over initial equilibrium ground state cluster configurations. The horizontal bars denote the statistical uncertainty of the fragmentation times. Note the change in rate with time (reference 113).

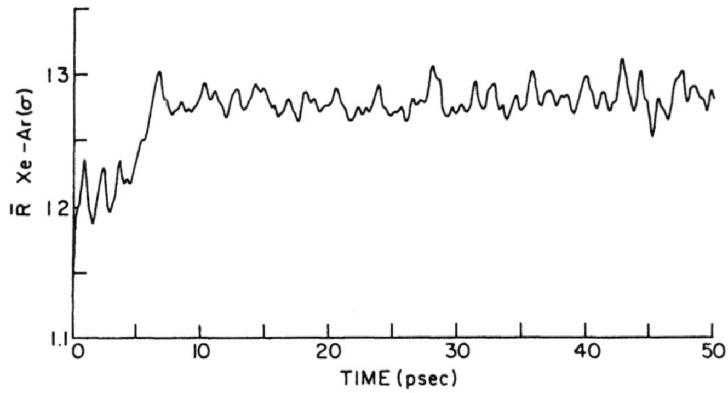

Fig. 34. Time evolution of the average distance, in units of σ, between the central Xe^* atom in a Xe^*Ar_{54} cluster and the first Ar atoms coordination shell. The increase from $\sim 1.13\sigma$ to a value of $\sim 1.27\sigma$ corresponds to structural dilation (bubble formation) around the Xe^* (reference 113).

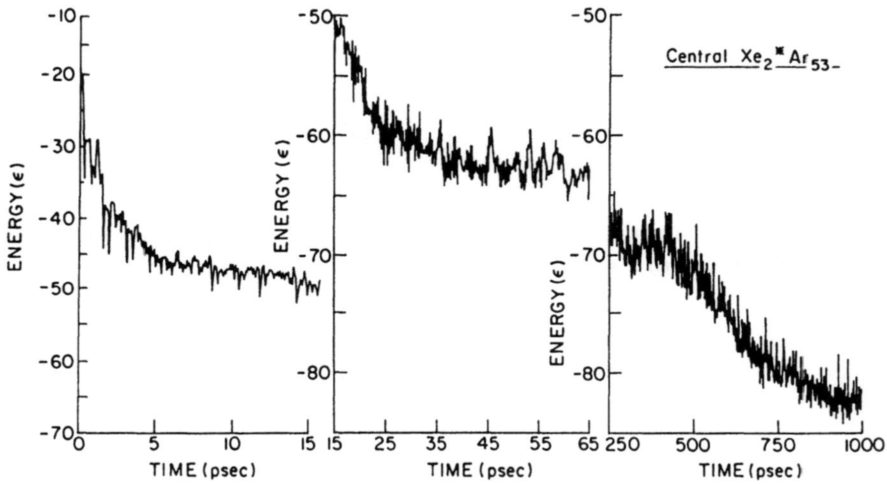

Fig. 35. Total energy of the bare central excimer in $Xe_2^*Ar_{53}$ as a function of time, demonstrating the continuous, prolonged energy transfer to the bath-subsystem, following the excitation. The bottom of the Xe_2^* potential well is at $\sim -96\epsilon$. Note the high degree of vibrational relaxation achieved after long times (reference 113).

initial locations of the excimer, the Xe_2^* is found on the surface of the cluster (Fig. 37). A quantitative information concerning mass transfer of the excimer to the cluster surface was inferred from the time dependence of the distance between the centre of mass of the excimer and the centre of mass of the Ar atoms in the cluster (Fig. 38).

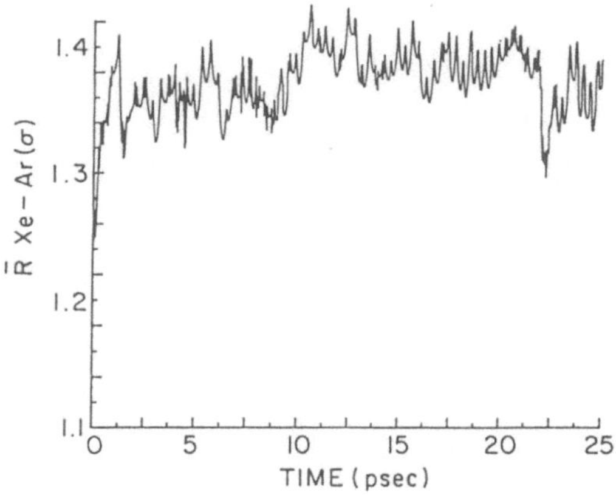

Fig. 36. The average distance, in units of σ, between one of the Xe_2^* excimer atoms and the first Ar coordination shell in $Xe_2^*Ar_{53}$ for a central excimer (reference 113).

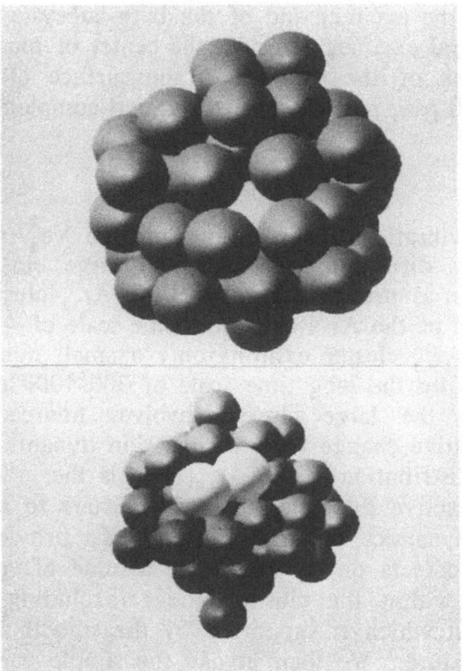

Fig. 37. Cluster configurations for $Xe_2^*Ar_{53}$, for an initially centrally located Xe_2^* at $t = 0^+$ (immediately past the excitation), on top, and after 1 nsec, bottom. The light atoms are the Xe_2^* and the dark ones are the Argon atoms (reference 113).

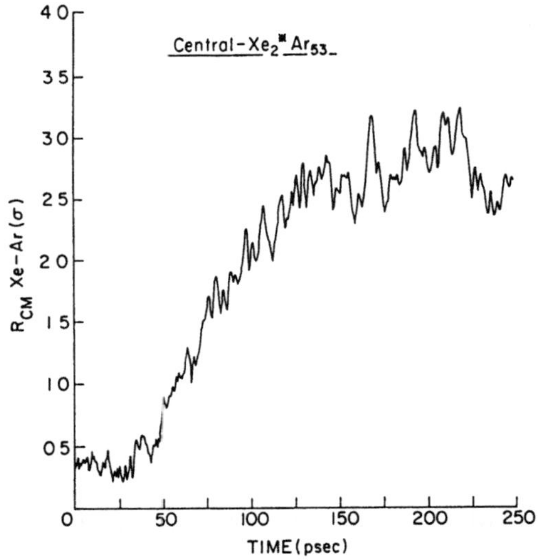

Fig. 38. Excimer migration. The time evolution of the distance between the corresponding centers of mass of the excimer and of the bath-subsystem in central $Xe_2^*Ar_{53}$. Initially, the central excimer was near the center of mass of the bath-subsystem. The migration of the excimer to the surface of the cluster begins on a time scale of 60 psec past the excitation, and completes in ~ 200 psec (reference 113).

The dynamic consequences of vibrational energy flow into the $Xe_2^*Ar_{n-2}$ (n = 13,55) cluster are drastically different for small and large clusters. Reactive vibrational predissociation is prevalent for small $Xe_2^*Ar_{11}$ clusters, resulting in a complete dissociation of the Ar atoms on a time scale of ~ 120 psec. In contrast, the large $Xe_2^*Ar_{53}$ cluster exhibits only a small number (~ 3 events) of Ar fragmentations on the long time scale of 300-1000 psec. The major dynamic process in the large cluster involves nonreactive vibrational relaxation. The qualitative change from dissociation dynamics in $Xe_2^*Ar_{11}$ to vibrational energy redistribution in $Xe_2^*Ar_{53}$ reflects the gradual transition from molecular type reactive dynamics in small clusters to non-reactive condensed matter type dynamics in large clusters. To provide a heuristic description of the size effects on dynamics we assume effective vibrational energy redistribution within the cluster modes (excluding the excimer), prior to any dissociation, which is supported by the smooth time dependence of the cluster temperature. We then invoke the simple version of the statistical unimolecular theory for the time, τ_D, for the first cluster dissociation event ,

$$\frac{1}{\tau_D} = A \left(\frac{E_v - E_t}{E_v}\right)^{s-1} , \qquad (VI.1)$$

where A is the frequency factor, E_t is threshold energy, E_v the excess vibrational energy and s the number of vibrational degrees of freedom. For large s, which is of interest to us, Eq. (VI.1) results in a simple rate equation,

$$\frac{1}{\tau_D} = A \exp\{-E_t/kT_{eff}] \quad , \qquad \text{(VI.2)}$$

with the effective temperature ,

$$kT_{eff} = \frac{E_v}{s-1} \quad . \qquad \text{(VI.3)}$$

The first dissociation time depends, of course, on the cluster size and its excess vibrational energy. The MD data reveal that the first dissociation event of $Xe_2^*Ar_{11}$ occurs at τ_D = 2.5±0.5 psec when the cluster excess vibrational energy is $E_v \sim 20\epsilon$, while for $Xe_2^*Ar_{53}$ we obtained τ_D = 400±100 psec when $E_v \simeq 60\epsilon$. The ratio of the τ_D values can be accounted for in terms of Eq. (VI.2) with the parameters $E_t = 5\epsilon$, which reproduces faithfully the binding energy of a surface Ar atom, while the absolute value of τ_D result in the reasonable value of $A = 2 \cdot 10^{15}$ sec^{-1}. Accordingly, the reactive dissociation process will be practically switched-off when $\tau_D \geq \tau_{rad}$, where τ_{rad} = 10 nsec is a characteristic radiative excimer decay lifetime, in its $^3\Sigma_u$ state [111]. This state of affairs will prevail for $E_v = 50\epsilon$ when $s \geq 170$, i.e., $n \geq 60$, which provides the signature for the transitionfrom molecular to condensed matter vibrational dynamics in clusters.

These MD studies [113] unravel some novel dynamic implications of energy exchange in electronically-vibrationally excited rare-gas clusters:
(1) Vibrational energy flow from electronically excited R^* and electronically-vibrationally excited R_2^* into the cluster, which originates from short-range repuslive interactions (for both R^* and R_2^*) and from vibrational energy degradation (of R_2^*).
(2) Reactive and nonreactive processes in clusters. The elucidation of the nature of the relaxation process and its dependence on the electronic-excitation of R^* or R_2^*, on the excess vibrational energy of R_2^*, on the cluster structure, composition and size is of considerable interest.
(3) Size effects on intramolecular dynamics in clusters. The reactive vibrational predissociation process manifests the dynamic consequences of vibrational energy flow in small clusters, while vibrational relaxation of an excimer in condensed phases can be viewed as nonreactive vibrational energy redistribution. Vibrational energy flow in clusters is expected to undergo a gradual transition from reactive molecular-type behavior in small clusters to nonreactive solid-state type behavior in large clusters, manifesting the transition from molecular to condensed matter relaxation in large finite systems.

(4) Relaxation of surface and bulk states. The vibrational energy flow from an excimer into a cluster is expected to be qualitatively different when the excimer is in the interior of the cluster or when it is located on the cluster surface. Thus, the exploration of clusters provides a new insight into condensed phase dynamics, making contact with surface phenomena. This distinction between bulk and surface dynamics pertains both to reactive processes in small clusters and to nonreactive vibrational ralaxation in large clusters.
(5) Configurational cluster relaxation around electronic excitations. The short range repulsive interaction between the Rydberg type excitation of R^* or R_2^* and the neighboring cluster atoms, resulting in cluster dilation around the electronically excited state. Such a local dilation will result in the formation of a "microscopic cavity" or a "bubble" surrounding the electronic excitations in the cluster. The local structural dilation of the cluster around the excited electronic state is responsible for vibrational energy flow into the cluster. Such configurational relaxation of the cluster bears a close analogy to medium dilation around electronically excited states of impurity atoms in rare-gas solids [111], which cause large Stokes shifts and are responsible for excessive homogeneous line broadening in absorption. Local dilation effects, which are expected in clusters, are also analogous to "bubble" formation around electronically excited states in liquid Helium [111].
(6) Mass transport of electronic excitations in clusters. When nonreactive vibrational energy flow from an electronic excitation of an impurity to the cluster prevails, local heating of the cluster will result in excessive cluster configurational changes. The most extreme configurational change involves cluster melting (Chapter II). Under these circumstances excessive diffusion of the electronically excited impurity may take place. This mass transport process may be of considerable interest for the interrogation of excited-state chemical processes in large finite systems.

The progress in our understanding of the evolution of quantum, thermodynamic, dynamic and chemical size effects in clusters is expected to provide a central contribution to the merging between the microscopic and macroscopic points of view in chemical dynamics.

Acknowledgments. This research was supported by the U.S.-Israel Binational Science Foundation and by the U.S. DOE Grant No. FG05-86ER45234.

REFERENCES

1. Proceedings of International Meeting on Small Particles and Inorganic Clusters. J. Phys. (Paris) Colloque C2 38 (1977).
2. Proceedings of the Second International Meeting on Small Particles and Inorganic Clusters, Lausanne, 1980, Surface Sci. 106, 1-608 (1981).
3. Proceedings of Bunsengesellschaft Discussion Meeting on Experiments on Clusters, Koningstein, 1983, Ber. Bunsenges. Phys. Chem. 88 (1984).

4. Proceedings of the Third International Meeting on Small Particles and Inorganic Clusters, Berlin, 1985, Surface Sci. 165, 1-1072 (1985).
5. J. Friedel: J. Phys. (Paris) C2 38, 1 (1977).
6. J. Jortner: Ber. der Bunsenges. Phys. Chem. 88, 188 (1984).
7. R.L. Whetten, D.M. Cox, D.J. Trevor, A. Kaldor: Phys. Rev. Lett. 65, 1494 (1985).
8. M.E. Geusic, M.D. Morse, R.D. Smalley: J. Chem. Phys. 82, 590 (1985).
9. E.W. Becker, K. Bier, W. Henkens: Z. Physik. 146, 333 (1956).
10. O.F. Hagena, W. Obert: J. Chem. Phys. 56, 1793 (1972).
11. B. Donn, J. Hecht, R. Khanna, J. Nuth, D. Stranz, A.B. Anderson: Surface Sci. 106, 576 (1981).
12. Polycyclic Aromatic Hydrocarbons and Astrophysics, ed. by A. Leger, L. d'Hendecourt, N. Boccara, NATO ASI series, Vol. 191 (D. Reidel Company, Dordrecht 1987).
13. G.H. Herbig: Astrophys. J. 196, 129 (1975).
14. T.P. Martin: Phys. Reports 95, 169 (1983).
15. J. Luo, U. Landman, J. Jortner: In Physics and Chemistry of Small Clusters, ed. by P. Jena (Plenum, New York 1987).
16. H.P. Blates: J. Phys. (Paric) C2 38, 153 (1977).
17. J. Barker: J. Phys. (Paris), C2 38, 37 (1977).
18. M.R. Hoare: Adv. Chem. Phys. 40, 49 (1979).
19. J. Farges, M.F. de Feraudy, B. Raoult, G. Torchet: J. Chem. Phys. 78, 5067 (1983).
20. M.J. Onderchen, Z. Berkovitch-Yellin, J. Jortner: J. Am. Chem. Soc. 103, 6586 (1981).
21. U. Even, A. Amirav, S. Leutwyler, M.J. Onderchen, Z. Berkovitch-Yellin, J. Jortner: J. Farad. Discuss. Chem. Soc. 73, 153 (1982).
22. S. Leutwyler, J. Jortner: J. Phys. Chem. in press, (1987).
23. P. Pawlow: Z. Phys. Chem. 66, 545 (1909).
24. D.A. Buffat, J.P. Borel: Phys. Rev. A13, 2289 (1976).
25. P.R. Couchaman, C.L. Ryan: Phil. Mag. A37, 369 (1978).
26. K. Hoshino, S. Shimamura: Phil. Mag. A40, 137 (1979).
27. G. Natanson, F. Amar, R.S. Berry: J. Chem. Phys. 78, 399 (1983).
28. C.L. Briant, J.J. Burton: J. Chem. Phys. 63, 2045 (1975).
29. J.B. Kaelberer, R.D. Etters: J. Chem. Phys. 66, 3233 (1977).
30. R.D. Etters, J.B. Kaelberer: J. Chem. Phys. 66, 5112 (1977).
31. J. Jellinek, T.L. Beck, R.S. Berry: J. Chem. Phys. 84, 2783 (1986).
32. U. Landman, R.N. Barnett, C.L. Cleveland, J. Luo, D. Scharf, J. Jortner: In The Few Body Problem, ed. by D. Micha, AIP Proceedings(New York 1987).
33. Y. Imry: Phys. Rev. B21, 2042 (1980).
34. Y. Imry, D.J. Bergman: Phys. Rev. A3, 1416 (1971).
35. R.W. Hockney, J.W. Eastwood: In Computer Simulation Using Particles (McGraw-Hill, New York 1981) pp. 488-498.
36. P.R. Couchman, W.A. Jessor: Nature 269, 481 (1977).
37. R. Balian, C. Bloch: Ann. Phys. N.Y. 60, 401 (1970).

38. P.R. Couchman: Phil. Mag. 40, 637 (1979).
39. This is reminiscent of the picture advanced by F.M. Stillinger, T.A. Weber for a liquid, see: Kinam A5, 159 (1981); Phys. Rev. A25, 978 (1972).
40. M.R. Hoare, J.A. McInnes: Adv. in Phys. 32, 791 (1983).
41. F.G. Fumi, M.P. Tosi: J. Phys. Chem. Solids 25, 31, 45 (1964).
42. D.A. McQuarrie: In Statistical Mechanics (Harper and Row, New York 1976), Chap. 21,22.
43. T.E. Gough, D.G. Knight, G. Scoles: Chem. Phys. Lett. 97, 155 (1983).
44. D. Eichenauer, R.J. Le Roy: Phys. Rev. Letters 57, 2920 (1986).
45. A. Amirav, U. Even, J. Jortner: J. Phys. Chem. 86, 3345 (1982).
46. S. Leutwyler: In Large Finite Systems, ed. by J. Jortner, A. Pullman, B. Pullman (D. Reidel Publishing Co., Dordrecht 1987).
47. D.H. Levy: Adv. Chem. Phys. 47, 323 (1982).
48. S.S. Kim, G.D. Stein: J. App. Phys. 51, 6419 (1980).
49. A. Amirav, U. Even, J. Jortner: J. Chem. Phys. 75, 2489 (1981).
50. A. Amirav, U. Even, J. Jortner: J. Phys. Chem. 85, 309 (1982).
51. A.W. Castleman, B.D. Kay, V. Hermann, P.M. Holland, T.D. Märk: Surface Sci. 106, 179 (1981).
52. U. Even, J. Jortner: J. Chem. Phys. 78, 3445 (1983).
53. M. Heaven, T.A. Miller, V.E. Bondybey: J. Chem. Phys. 76, 3831 (1982).
54. J. Gspann, H. Vollmar: J. Low Temperature Phys. 45, 343 (1981).
55. P.M. Dehmer, S.T. Pratt: J. Chem. Phys. 76, 843 (1982).
56. J.Q. Searay, J.B. Fenn: J. Chem. Phys. 61, 5282 (1974).
57. K. Schwentner, E.E. Koch, J. Jortner: In Photophysics and Photochemistry in the Vacuum Ultraviolet, ed. by S.P. McGlynn, G.L. Findley, R.H. Huebner (D. Reidel Publishing Co., Dordrecht 1985), p. 515.
58. H. Haberland: Surface Sci. 156, 305 (1985).
59. Y. Chanut, J. Martin, R. Salin: Surface Sci. 106, 563 (1981).
60. P.R. W. Henkes, U. Pfeiffer: Ber. Bunsenges. Phys. Chem. 88, 258 (1984).
61. C.E. Klotz, R.E. Compton: J. Chem. Phys. 67, 1779 (1977).
62. C.E. Klotz, R.E. Compton: J. Chem. Phys. 69, 1636 (1978).
63. T. Holstein: Ann. Phys. N.Y. 8, 343 (1959).
64. W. Kantzmann: In Quantum Chemistry (Academic Press, New York 1957).
65. W.M. Trott, N. Blais, E.A. Walters: J. Chem. Phys. 69, 3150 (1978).
66. S.H. Linn, Y. Ono, C.Y. Ng: J. Chem. Phys. 74, 3342 (1981).
67. Y. Ono, S.H. Linn, H.F. Prest, M.E. Gress, C.Y. Ng: J. Chem. Phys. 73, 2523 (1980).
68. K.H. Fung, W.E. Henke, H.L. Selzle, E.W. Schlag: J. Phys. Chem. 85, 3560 (1981).
69. U. Even, S. Leutwyler, Z. Berkovitch-Yellin, J. Jortner: J. Chem. Phys. 78, 309 (1983).

70. P.D. Dao, S.Morgan, A.W. Castleman: Chem. Phys. Lett. 113, 219 (1985).
71. R.N. Compton, P.W. Reinhardt, C.D. Cooper: J. Chem. Phys. 63, 3821 (1975).
72. A.R. Rossi, K.D. Jordan: J. Chem. Phys. 70, 4422 (1979).
73. C.E. Klotz, R.E. Compton: J. Chem. Phys. 67, 1779 (1977).
74. C.E. Klotz, R.E. Compton: J. Chem. Phys. 69, 1636 (1978).
75. M. Armbuster, H. Haberland, H.G. Schindler: Phys. Rev. Lett. 47, 323 (1981).
76. H. Haberland, C. Ludewigt, H.G. Schindler, D.R. Worsnop: Surface Sci. 156, 157 (1985).
77. H. Haberland, H. Langosch, H.G. Schindler, D.R. Worsnop: Ber. Bunsenges. Phys. Chem. 88, 270 (1984).
78. H. Haberland, H.G. Schindler, D.R. Worsnop: J. Chem. Phys. 81, 3742 (1984).
79. J.V. Coe, D.R. Worsnop, K.H. Bowen: J. Chem. Phys. (1987).
80. M. Knapp, O. Echt, D. Kreisle, E. Recknagel: J. Chem. Phys. 85, 636 (1986); J. Phys. Chem. (preprint, 1986).
81. B.K. Rao, N.R. Kestner: J. Chem. Phys. 80, 1587 (1984).
82. N.R. Kestner, J. Jortner: J. Phys. Chem. 88, 3818 (1984).
83. R.N. Barnett, U. Landman, C.L. Cleveland, J. Jortner: Phys. Rev. Lett. (in press).
84. U. Landman, R.N. Barnett, C.L. Cleveland, D. Scharf, J. Jortner: J. Phys. Chem. in press (1987).
85. U. Landman, D. Scharf, J. Jortner: Phys. Rev. Lett. 54, 1860 (1985).
86. D. Scharf, U. Landman, J. Jortner: J. Chem. Phys. 87, 2716 (1987).
87. R.P. Feynman, A.R. Hibbs: In Quantum Mechanics and Path Integrals (McGraw Hill, New York 1965).
88. D. Chandler, P.G. Wolynes: J. Chem. Phys. 79, 4078 (1984).
89. D. Chandler: J. Phys. Chem. 88, 3400 (1984).
90. M. Parrinello, A. Rahman: J. Chem. Phys. 80, 860 (1984).
91. A.L. Nicholas, D. Shandler, V. Singh, D.M. Richardson: J. Chem. Phys. 81, 5109 (1984).
92. M.F. Herman, E.J. Bruskin, B.J. Berne: J. Chem. Phys. 76, 5150 (1982).
93. D. Callaway, A. Rahman: Phys. Rev. Lett. 49, 613 (1982).
94. R.W. Shaw: Phys. Rev. 174, 769 (1968).
95. N.F. Mott, R.W. Gurney: In Electronic Processes in Ionic Crystals (Oxford University Press, Oxford 1946).
96. W. Weyl: Ann. Phys. 197, 601 (1863).
97. E.J. Hart, J.W. Boag: J. Am. Chem. Soc. 84, 4090 (1962).
98. J. Jortner: J. Chem. Phys. 30, 839 (1959).
99. D.A. Copeland, N.R. Kestner, J. Jortner: J. Chem. Phys. 53, 1189 (1970).
100. A. Gaathon, G. Czapski, J. Jortner: J. Chem. Phys. 58, 2648 (1973).
101. Early attempts to observe $(H_2O)_n^-$ clusters in supersonic beams include

those by D.R. Herschbach of Harvard University (private communication), and of R. Compton of Oak Ridge National Laboratory (private communication).
102. J.R. Reimers, R.O. Watts, M.L. Klein: Chem. Phys. 64, 95 (1982); J.R. Reimers, R.D. Watts, ibid. 85, 83 (1984). This paper as well as the description of the potential in the first one (Eq. 13 and Table 1) contain several ambiguities and typographical errors. When corrected their results are reproduced.
103. C.H. Douglass, Jr., D.A. Weil, P.A. Charlier, R.A. Eades, D.G. Truhlar, D.A. Dixon: In Chemical Application of Atomic and Molecular Electrostatic Potentials, ed. by P. Politzer, D.G. Truhlar (Plenum, New York 1981), p. 173.
104. C.W. Kerr, M. Karplus: In Water, ed. by F. Franks (Plenum, New York 1972), p. 21; M.W. Ribarsky, W.D. Luedtke, U. Landman: Phys. Rev. B32, 1430 (1985).
105. J.N. Bardsley: Case Studies in Atomic Physics 4, 299 (1974); G.G. Kleiman, U. Landman: Phys. Rev. B8. 5485 (1973).
106. A. Wallquist, D. Thirumalai, B.J. Berne: J. Chem. Phys. 85, 1583 (1986).
107. U. Landman, R.N. Barnett, J. Jortner: J. Chem. Phys. (submitted).
108. G. Lepoutre, J. Jortner: J. Phys. Chem. 76, 683 (1972).
109. J. Jortner, R.D. Levine: In Photoselective Chemistry, ed. by J. Jortner, A.D. Levine, S.A. Rice (Wiley, New York 1981), p. 1.
110. J. Jortner, R.D. Levine, S.A. Rice: Advances in Chemical Physics (in press, 1987).
111. N. Schwentner, E.E. Koch, J. Jortner: In Electronic Excitations in Condensed Rare Gases (Springer Verlag, Berlin 1985).
112. D. Scharf, J. Jortner, U. Landman: Chem. Phys. Lett. 126, 495 (1986).
113. D. Scharf, J. Jortner, U. Landman: J. Chem. Phys. (submitted).
114. H. Haberland: Surface Sci. 156, 305 (1985).
115. J.J. Sâenz, J.M. Soler, N. Garcia: Surface Sci. 156, 121 (1985).
116. J.J. Sâenz, J.M. Soler, N. Garcia, O. Echt: Chem. Phys. Letters 109, 71 (1984).
117. E.E. Polymeropoulos, J. Brickmann: Surface Sci. 156, 563 (1985).
118. D.H. Levy: Adv. Chem. Phys. 47, 323 (Wiley Interscience, New York 1981).
119. A. Beswick, J. Jortner: Adv. Chem. Phys. 47, 363 (1981).
120. K.C. Janda: Adv. Chem. Phys. 60, 201 (1985).
121. A. Amirav, U. Even, J. Jortner: Chem. Phys. Lett. 72, 16 (1980).
122. J.W. Brody, J.D. Doll: J. Chem. Phys. 73, 2767 (1980).
123. M.F. Vernon, J.M. Lisy, A.S. Kwok, D.J. Krajnovich, A. Tramer, Y.R. Shen, Y.T. Lee: J. Phys. Chem. 85, 3327 (1981).
124. T.E. Gough, R.E. Miller, G. Scoles: J. Chem. Phys. 69, 1588 (1978).
125. M.P. Casassa, D.S. Bomse, J.L. Beauchamp, K.C. Janda: J. Chem. Phys. 72, 6805 (1980).
126. M.A. Hoffbauer, K. Liu, C.F. Giese, W.R. Gentry: J. Chem. Phys. 78, 5567 (1983).

127. M.J. Howard, S. Burdenski, C.F. Giese, W.R. Gentry: J. Chem. Phys. 80, 4137 (1984).
128. A. Mitchell, J.J. McAuliffe, C.F. Giese, W.R. Gentry: J. Chem. Phys. 83, 4271 (1985).
129. R.D. Johnson, S. Burdenski, M.A. Hoffbauer, C.F. Giese, W.R. Gentry: J. Chem. Phys. 84, 2624 (1986).
130. W.D. Kristensen, E.J. Jensen, M.J. Cotterill: J. Chem. Phys. 60, 4161 (1974).
131. I. Messing, B. Raz, J. Jortner: J. Chem. Phys. 66, 4577 (1977).
132. I. Messing, B. Raz, J. Jortner: J. Chem. Phys. 66, 2239 (1977).
133. A.E. Sherwood, J.M. Prasunitz J. Chem. Phys. 41, 429 (1964).
134. W.C. Ermler, Y.S. Lee, K.S. Pitzer, N.W. Winter: J. Chem. Phys. 69, 976 (1978).

Electronic, Structural and Thermal Properties of Microclusters from the Car-Parrinello Method

W. Andreoni

IBM Zürich Research Division, Zürich Research Laboratory,
Säumerstraße 4, CH-8803 Rüschlikon, Switzerland

A brief sketch is given of the Car-Parrinello method (Unified Molecular Dynamics — Density Functional Theory) and its special advantages in the application to the study of microclusters. A brief summary of results recently obtained on small clusters of silicon and sodium is also presented: These concern electronic and structural properties at T = 0 as well as at finite temperature. It is also possible to distinguish different regimes in the temporal evolution of the atomic displacement, which reveals 'fluid-like' and 'solid-like' behavior at different temperatures.

The intention of this lecture is

i) to briefly describe a new theoretical method introduced recently by CAR and PARRINELLO [1] under the name of Unified Approach for Molecular Dynamics and Density Functional Theory, with the aim of clarifying its special advantages in the study of the physics of microclusters compared with other approaches, and

ii) to present some results of current calculations [2-4] on electronic, structural and thermal properties of small aggregates of silicon and sodium.

1. OUTLINE OF THE CAR-PARRINELLO METHOD

For a complete description of the method and its applications, we refer the reader to the original papers [1,2,5]. Here, we shall only recall the main steps in the development of the theoretical framework, and emphasize special features.

In [1], CAR and PARRINELLO introduce a classical Lagrangean for the mixed quantum-classical system of electrons and ions, the former being described by one-electron orbital amplitudes $\{\psi_i\}$ and the latter by the coordinates $\{\vec{R}_I\}$:

$$L = \sum_i 1/2\mu \int_\Omega d^3r\,|\dot{\psi}_i|^2 + \sum_I 1/2\, M_I \dot{\vec{R}}_I^2 + \sum_\nu 1/2\mu_\nu \dot{\alpha}_\nu^2 - E\left(\{\psi_i\}, \{\vec{R}_I\}, \{\alpha_\nu\}\right), \tag{1}$$

and specify the classical potential energy to be the density functional of Density Functional Theory (DFT) [6]:

$$E\left(\{\psi_i\}, \left\{\vec{R}_I\right\}, \{\alpha_\nu\}\right)$$

$$= \sum_i \int_\Omega d^3 r\, \psi_i^* \left(-\hbar^2/2m\nabla^2 \right)\psi_i + U\left(\rho_e(\vec{r}), \left\{\vec{R}_I\right\}, \{\alpha_\nu\}\right), \tag{2}$$

with

$$\rho_e(\vec{r}) = \sum_i |\psi_i|^2; \quad N = \int_\Omega d^3 r\, \rho_e(\vec{r}), \tag{3}$$

and the variables $\{\alpha_\nu\}$ describing possible external constraints to which the system may be subject [7]. The electron wave functions satisfy the orthonormality conditions

$$\int_\Omega d^3 r\, \psi_i^*(\vec{r}, t)\psi_j(\vec{r}, t) = \delta_{ij}. \tag{4}$$

M_I is the (physical) mass of the ion I, while μ and μ_ν are fictitious masses associated with the electronic variables and the α_ν, respectively.

From (1), a set of coupled equations of motion is derived:

$$\mu \ddot{\psi}_i(\vec{r}, t) = -\delta E/\delta\psi_i^*(\vec{r}, t) + \sum_K \Lambda_{ik}\psi_k(\vec{r}, t) \tag{5a}$$

$$M_I \ddot{\vec{R}}_I = -\nabla_{R_I} E \tag{5b}$$

$$\mu_\nu \ddot{\alpha}_\nu = -\left(\partial E/\partial\alpha_\nu\right), \tag{5c}$$

where Λ is the matrix of the Lagrange multipliers introduced to satisfy the orthonormality conditions (4).

We are interested in solving for the equilibrium states of the physical system at various temperatures (and in particular at T = 0). This implies that the trajectories generated by (5) are kept on the Born-Oppenheimer surface: This is accomplished by using a relatively small value for the fictitious mass μ associated with the electronic variables.

As usual in Molecular Dynamics (MD) simulations, statistical mechanical properties can be calculated as temporal averages over these trajectories, taking care that the time interval over which the averages are made are larger than the characteristic relaxation times of the system. In particular, the temporal average of the classical kinetic energy $\langle K \rangle = \langle L + E \rangle$ is used to define the temperature T of the system. T can be varied by rescaling all the velocities.

By varying T in an appropriate manner, one can subject the system to specific thermal treatments, such as quenching and annealing. In particular, Dynamical Simulated Annealing (DSA) can be realized in the search for the ground state (equilibrium state of lowest energy at T = 0). This entails starting with the system at some point on the Born-Oppenheimer surface, then heating it up in such a way that it loses memory of the initial configuration.

Finally, it is cooled down very slowly, to let it explore a large region of the configurational space and thus avoid, as much as possible, becoming trapped in local minima.

The connection between the Car-Parrinello method and traditional approaches to the physics of microclusters, such as electronic-structure calculations using DFT and molecular dynamics using empirical potentials, can be seen very easily. In fact, at equilibrium ($\ddot{\psi} = 0$), the Newton equations for the one-electron orbitals in (5a) are equivalent to the Kohn-Sham equations up to a unitary transformation, and the eigenvalues of the matrix Λ reduce to the one-electron energies of the occupied states ε_i. The link with classical MD can easily be seen by recalling that here the potential energy surface Φ, which in standard simulations is derived from some simple and empirical prescription for the interatomic potentials, is calculated explicitly as

$$\Phi\left(\left\{\vec{R}_I\right\}\right) = \underset{\{\psi_i\}}{\mathrm{MIN}} \; E\left(\{\psi_i\}, \left\{\vec{R}_I\right\}\right), \tag{6}$$

where MIN means global minimum with respect to all electronic configurations.

Finally, the advantages presented by the use of the Car-Parrinello method emerge very clearly:

i) If one is concerned only with the calculation of the electronic structure at a given ionic configuration, the procedure is equivalent to the solution of Kohn-Sham equations through diagonalization, but computationally more efficient.

ii) For the determination of the equilibrium structure of the cluster (global minimum of the potential energy surface), the DSA is obviously more powerful than any other method traditionally adopted in *ab initio* approaches, e.g., steepest descent around an ionic configuration of given symmetry.

iii) One can also investigate thermal and dynamical properties of the aggregate with the same accuracy that one could so far achieve only in the study of static properties at T = 0. This constitutes an obvious advantage with respect to classical molecular dynamics, especially for the simulation of materials with either covalent or metallic bonding, where interatomic interactions depend crucially on the local environment. Also, it is evident that the possibility of following the variation of the electronic structure with temperature is an extra gift of the method.

In the calculations, one uses the usual approximations:

i) The Hohenberg-Kohn functional in (2) is expressed in the local density approximation (LDA).

ii) The core electrons are considered to be frozen, and their effect is resumed in a nonlocal norm-conserving pseudopotential.

2. SOME RECENT RESULTS

Here, we shall give some technical details of the calculations, and describe some of the salient results we have recently obtained on silicon and sodium clusters. For a more complete presentation, we refer the reader to [2] and [3] and especially to forthcoming publications [4].

2.1 Technical Remarks

For the nonlocal norm-conserving pseudopotential, we chose a suitable prescription proposed by VON BARTH and CAR [8], and for the exchange-correlation density functional the results of the Monte Carlo simulation of CEPERLEY and ALDER [9] for the electron liquid, as interpolated by PERDEW and ZUNGER [10]. Periodic boundary conditions were used. In particular, a supercell with face-centered-cubic-type periodicity and an edge of 35 a.u. for silicon and 44 a.u. for sodium was chosen. The ψ_i were expanded in plane waves with an energy cutoff of about 6 Ry in the two cases. The differential Eqs. (5) were solved with the Verlet algorithm [11], and by using $\Delta t = 1.2 \cdot 10^{-16}$ sec as integration time step. Adiabatic motion was controlled by using a value $\mu = 300$ a.u. for the fictitious mass of the electrons. We note that Δt is reduced by one order of magnitude with respect to classical molecular dynamics simulations. No external constraints α were imposed on the cluster. The cooling rates we typically adopted in a DSA run were of 10^{14} K/sec.

2.2 Equilibrium Structures at T = 0

We first applied the DSA to very small silicon clusters, with 3, 4 and 5 atoms, and found that our results [2] for the equilibrium geometries were in agreement with those obtained in previous LDA [12] and configuration interaction [13] calculations. Then, we made a thorough investigation for Si_{10}, for which only limited *ab initio* calculations existed. In addition to the DSA, we used less expensive techniques, such as rapid quenching and combined electronic and ionic steepest descent, to explore the potential energy surface of the cluster in some detail, and gain confidence in the results of the annealing procedure. We found several minima, corresponding to different structures with cohesive energies ranging over 0.1 eV/atom [4]. This is a rather interesting result in view of the special stability of Si_{10} suggested by several experiments [14]. The configuration we found with lowest energy is a slightly distorted tetracapped triangular prism which is shown in Fig. 1(a). The structure closest in energy is the tetracapped octahedron already proposed by RAGAVACHARI [13], which is higher by less than 0.01 eV [Fig. 1(b)]. We have noticed that all lowest-energy geometries are rather compact, and tend to have the majority of bond angles around 60°. The structure with lowest energy, however, presents a significant peak of the bond angle distribution function at 109° ('bimodal' angular distribution). Also, the two structures competing for the ground state, in Fig. 1, have the atoms confined to a narrow shell with respect to the center of mass.

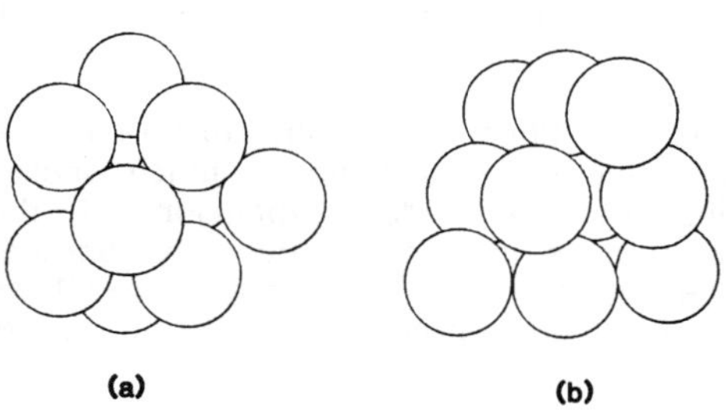

Fig. 1. Structures of lowest energy for Si_{10}: (a) result of simulated annealing, and (b) tetracapped octahedron

2.3 Finite Temperature Properties

The physical quantities we calculated were obtained as temporal averages over time intervals of typically 1 psec, after equilibration runs of 0.2 psec. In particular, we have studied the variation with temperature of both radial and angular distribution functions g(r) and g(ϑ) of the ions, as well as of the spherical average $\bar{\rho}_e$ of the electron density $\rho_e(\vec{r})$.

In Fig. 2, the spherically averaged electronic and ionic distributions $r^2\bar{\rho}_e(r)$ and $r^2\bar{\rho}_I(r)$ of Si_{10} are plotted as a function of the distance from the center of mass, at T = 2000 K. Apart from broadening, both are still rather similar to the low-temperature analogs. In particular, the ionic density, although broadened, keeps the shell structure of the lowest-energy configurations. $\bar{\rho}_e(r)$ strongly peaks where the ions are located, and spills outside the ionic shells, mostly owing to the contribution of the highest-energy orbitals.

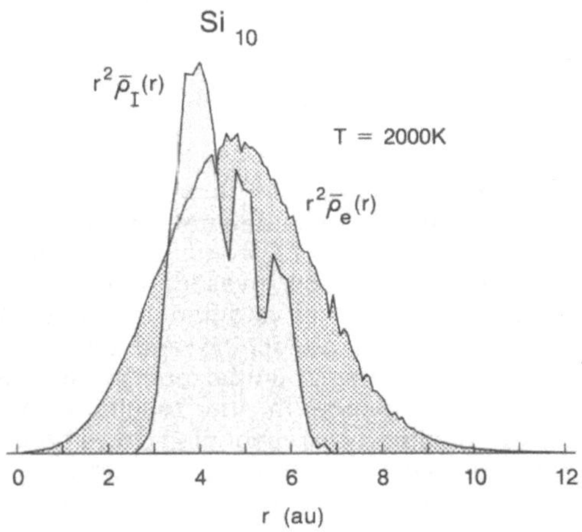

Fig. 2. Spherical average of the electronic ($r^2\rho_e$) and ionic ($r^2\rho_I$) distributions for Si_{10} at T = 2000 K. r = 0 is the center of mass of the cluster. The units of $\bar{\rho}_e$ and $\bar{\rho}_I$ are arbitrary and different from one another

The effect of the temperature increase from 1600 K to 3300 K in Si_{10} appears to be reflected in a decrease of the probability of finding tetrahedral bond angles and in a slight narrowing of the nearest-neighbor distance. This suggests a tendency to more compact atomic configurations, in analogy to what occurs at the solid-liquid transition of the homogeneous phase. From the comparison of $\bar{\rho}_e(r)$ and $\bar{\rho}_I(r)$, a strong ionic-electronic correlation is still recognizable, although weakened.

In contrast, by examining the same quantities in sodium clusters of the same size, we have found that the ionic-electronic correlation is much weaker at any temperature $T \gtrsim 200$ K, and that a trivial expansion of the nearest-neighbor distance follows the increase in temperature. The angular distribution function is essentially concentrated around 60° at any temperature, in contrast to the bimodal distribution of silicon aggregates mentioned above. This is illustrated in Fig. 3.

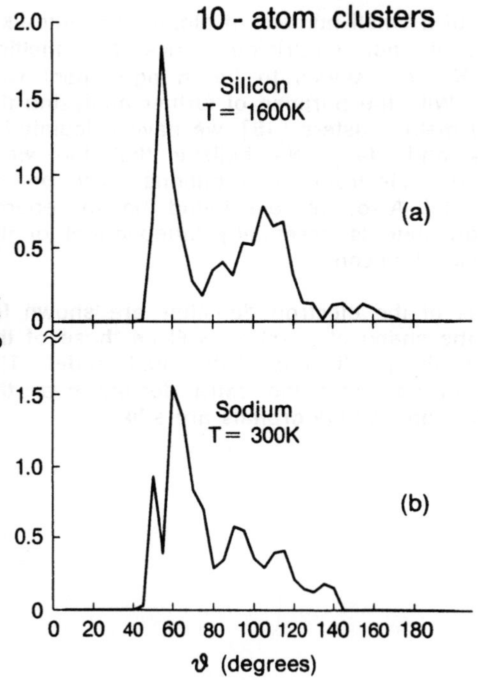

Fig. 3. Angular distribution function for (a) Si_{10} at T = 2000 K, and (b) Na_{10} at 300 K. From [3]. © 1987 Royal Swedish Academy of Sciences

We have also looked at the variation of the atomic Mean Square Displacement (MSD) as a function of time. In Na_{10}, this seems to have diffusive behavior at any temperature at least above 200 K, while for Si_{10} we have been able to distinguish two distinct regimes (Fig. 4): one (T = 3300 K) with diffusive behavior and the other, at lower temperatures, with only sporadic jump-like movements of the ions. Further analysis of the MSD is discussed in [3] and [4].

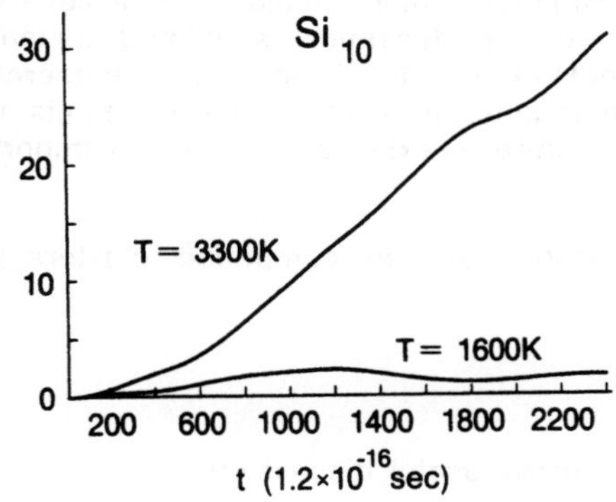

Fig. 4. Mean square displacement of Si_{10} as a function of time for two different temperatures. From [3]. © 1987 Royal Swedish Academy of Sciences

The fact that sodium aggregates of a small number of atoms are 'fluid-like' at and below room temperature is not surprising, since the melting temperature of the crystal is 371 K. This seems to be in agreement with suggestions from experiments [15]. With the purpose of further analyzing the validity of the shell model for alkali metal clusters [16], we have calculated a few electronic properties for Na_8 and Na_{10}. We believe that the weak correlation found between ionic and electronic distributions may be an argument in support of the shell model. Also, we have found that the energy of the highest occupied one-electron state is essentially independent of the ionic positions, in contrast to the case of silicon.

In Fig. 5, the spherical averages of the electron densities are shown for Na_8 and Na_{10}. In the case of Na_8, the shape of $\rho_e(r)$ as well as those of the single orbitals are in agreement with the predictions of the shell model. The difference between Na_8 and Na_{10}, in particular the extra density near the origin, is simply due to the extra electrons whose orbitals are s-like.

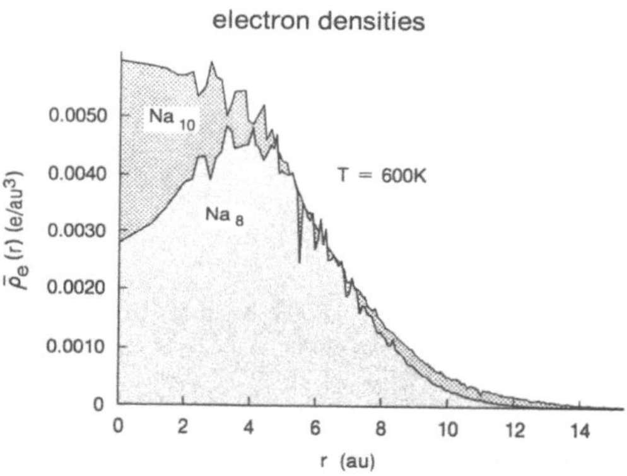

Fig. 5. Radial average of the electronic density for Na_8 and Na_{10} at T = 600 K

3. CONCLUSIONS

In this lecture, we have simply given a taste of what can be calculated with the Car-Parrinello method in the study of microclusters. Some interesting results are coming out of these investigations, in particular, the discovery of new relevant structures with the aid of the dynamical simulated annealing strategy, and the possibility of looking in detail at finite temperature properties. It would be especially interesting to contrast these results with experiments where the effect of temperature is expected to play an important role.

We are now extending these investigations to compound clusters and larger sizes.

ACKNOWLEDGMENT

It is a pleasure to thank P. Ballone for many useful discussions.

REFERENCES

1. R. Car, M. Parrinello: Phys. Rev. Lett. 55, 2471 (1985)
2. R. Car, M. Parrinello, W. Andreoni: In Proc. 1st NEC Symposium on Fundamental Approaches to New Material Phases (Microclusters), Tokyo, Japan, ed. by S. Sugano (Springer, Heidelberg 1987), p. 134
3. W. Andreoni, P. Ballone: Physica Scripta (to be published)
4. P. Ballone, W. Andreoni, R. Car, M. Parrinello: To be published
5. R. Car, M. Parrinello: In Proc. XVIII Int. Conf. on the Physics of Semiconductors, Stockholm, Sweden, 1986, Vol. 2 (World Sci. Pub. Co., Singapore 1987) p. 1165
6. P. Hohenberg, W. Kohn: Phys. Rev. B 136, 864 (1964); W. Kohn, L.J. Sham: Phys. Rev. A 140, 1133 (1965); see also W. Kohn, P. Vashishta: In Theory of the Inhomogeneous Electron Gas, ed by S. Lunqvist and N.H. March (Plenum, New York 1983), p. 79
7. See, for instance: H.C. Andersen: J Chem. Phys. 72, 2384 (1980); M. Parrinello, A. Rahman: Phys Rev. Lett. 45, 1196 (1980)
8. U. von Barth, R. Car: Unpublished; P. Giannozzi Private communication
9. D.M. Ceperley, B.J. Alder: Phys. Rev Lett. 45, 566 (1980)
10. J.P. Perdew, A. Zunger: Phys Rev. B 23, 5048 (1981)
11. L. Verlet, Phys. Rev. 159, 98 (1967)
12. D. Tománek, M.A. Schlüter. Phys. Rev. Lett. 56, 1955 (1986)
13. K.J. Ragavachari: J. Chem. Phys. 84, 5672 (1986)
14. L A. Bloomfield, R.R. Freeman, W.L. Brown. Phys. Rev. Lett. 54, 2246 (1985); W. Begemann, K H. Meiwes-Broer, H O Lutz: Phys. Rev. Lett. 56, 2248 (1986); S.C. O'Brien, Y. Liu, Q. Zhang, J.R. Heath, F.K. Tittel, R.F. Cur, R.E. Smalley: J. Chem. Phys. 84, 4074 (1986); T.P. Martin, H. Schaber: J. Chem. Phys. 83, 855 (1985)
15. M.M. Kappes, M. Schar, P. Radi, E. Schumacher: J. Chem. Phys. 84, 1863 (1986); M.M. Kappes, M. Schar, E. Schumacher: J. Phys. Chem. 91, 658 (1987)
16. W.D. Knight, K. Clemenger, W.A. de Heer, W.A. Saunders, M.Y. Chou, M.L. Cohen: Phys. Rev. Lett. 52, 2141 (1984); W.D. Knight, W.A. de Heer, K. Clemenger, W.A. Saunders: Solid State Commun. 53, 445 (1985)

Quantum Chemical Investigation of Alkali Metal Clusters and Its Use for Interpretation and Predictions of Experiments in Cluster Research

J. Koutecký[1], *V. Bonačić-Koutecký*[1], *I. Boustani*[1], *and P. Fantucci*[2]

[1]Institut für Physikalische und Theoretische Chemie,
Freie Universität Berlin, Takustr. 3, D-1000 Berlin 33, Germany
[2]Dipartimento di Chimica Inorganica e Metallorganica, CNR,
Università di Milano, Via Venezian 21, I-20133 Milano, Italy

1. INTRODUCTION

The analysis of the quantum chemical results for alkali metal clusters (Li, Na) as well as the careful comparison of experimer data with theoretical predictions is particularly illuminating for the general understanding of the cluster structure. Evident] the valence electrons predominantly determine the leading featu of alkali metal clusters. This essential fact simplifies signifi cantly the treatment which can be applied in theoretical studie Furthermore, the predictions based on the theoretical investigat of Li and Na clusters can be,in principle, used for the interpre- tation of the experimental results for K and Rb clusters, and wi some caution, for Ib clusters. Since all other factors seem to b of secondary importance, simple generally valid rules determinir the electronic and geometrical structure of clusters can be dete from the adequate analysis of the theoretical results for the sp cial simple case of Li clusters. The rules governing the electrc and geometric structure of clusters can be summarized as follows /1-3/:

1) Basic features of molecular orbitals (MO's) in one-electron a proximation as well as properties of natural orbitals (eigenfunc tions of the density operator) in many electron theories are of a minant importance. Mainly the nodal properties, degeneracies (or near degeneracies) and the occupancies of these one-electron fun tions can determine cluster geometry. The partial occupation of which is characteristic for the highly symmetrical (and conseque ly very compact) geometries of clusters with small number of val ce electrons can lead to deformations towards less compact geome tries. For this reason small alkali metal clusters with the nuc- learity less or equal to six prefer planar arrangement.

The overall character of the eigenvalue spectrum of considere one-electron functions is quite independent of the detailed form of the potential effectively acting upon the electrons in the system.

2) Topological properties are also very important because the "bonds" in the metal clusters are delocalized (cf. e.g. /4,5/). It should be emphasized that the "chemical bonds" at least in alkali metal clusters are not delocalized two-center bonds, and therefore the full analogy could not be drawn with delocalized bonds in the π-electron systems.

3) Compactness of cluster geometry is another favorable factor for the cluster stability. For this reason, e.g. clusters exhibiting five-fold symmetry /6,7/ can be very stable.

4) The use of the polarization functions in the AO-basis set is essential for the proper description of the interactions in clusters /8,1/. This fact is connected with the special character of the electron delocalization in clusters and probably with the outset of the metallic state.

5) Cluster geometries of cationic and anionic clusters attempt to minimize Coulomb repulsion of the charge density. This factor is mainly important for anionic clusters /9/.

2. GENERAL FEATURES OF LI CLUSTERS

The results obtained from the ab initio CI quantum chemical treatments for Li clusters are shown in fig. 1 /3,9/. The characteristic favorable geometries of neutral, cationic and anionic Li clusters can be well understood applying the rules mentioned in introduction The quantum chemical methods adopted (AO basis set, HF, MRD-CI, energy minimization procedure etc.) are described and discussed in another contribution in this book /11/.

The stabilities of Li_n, Li_n^+ and Li_n^- clusters, characterized by corresponding binding energies per atom E_b/n, E_b^+/n and E_b^-/n, defined as:

$$E_b/n = (nE_1-E_n)/n \, , \tag{1}$$

$$E_b^+/n = [(n-1)\ E_1 + E_1^+ - E_n^+]/n \, , \tag{2}$$

$$E_b^-/n = [(n-1)\ E_1 + E_1^- - E_n^-]/n \, , \tag{3}$$

respectively, are generally increasing functions of the nuclearity. E_n, E_n^+ and E_n^- label the total energy of the neutral, cationic and anionic cluster of nuclearity n, respectively. The increase of binding energy per atom is especially pronounced for $n \leq 8$. For these small clusters the average number of nearest neighbours rapidly increases although reaching only the value 4.5 for Li_8. This value should be compared with the values of 12 and 8 in the fcc and bcc crystal lattice, respectively. The further increase of the average coordination number with growing nuclearity is less pronounced and, moreover, the definition of nearest neighbours is not any more unique

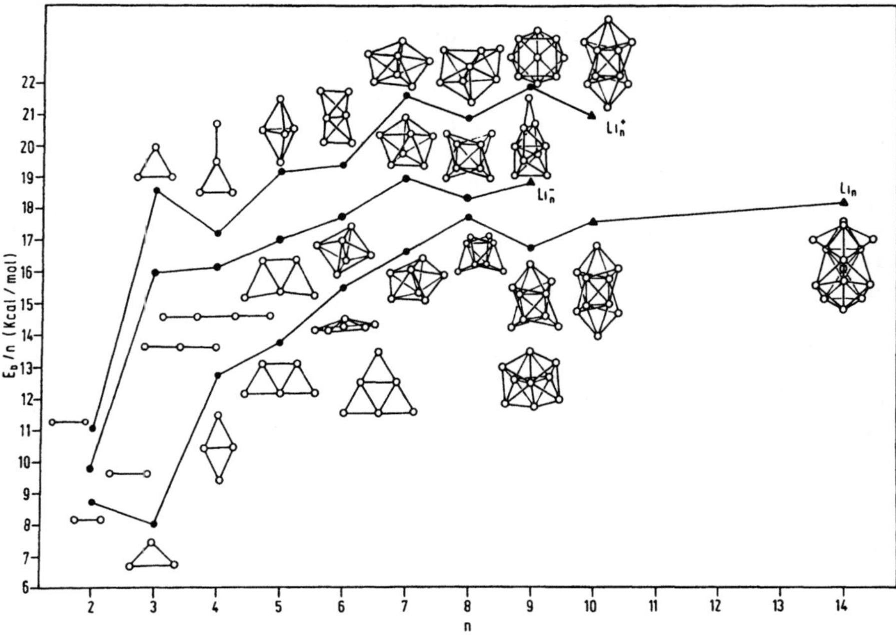

Fig. 1. Atomization energies per atom (binding energy per atom E_b/n) of Li_n, Li_n^+ and Li_n^- clusters as a function of the number of atoms in cluster (n). MRD-CI extrapolated energies with Davidson corrections for cluster geometries found with the SCF energy optimization are plotted. Topologies of the SCF energy optimized clusters are shown. The results for which the normal frequencies have not been determined are labeled by ▲

This is possible to see on the example of the nearly spherical structure of Li_{20}, which is only a tentative geometry for this relatively large cluster obtained from the HF energy optimization starting from the fixed geometry of Li_8-T_d subsystem and adding twelve Li atoms on the outer sphere (compare Li_{20}-T_d (ss) in fig. 4). This Li_{20}-cluster /2,10/ is very compact but some of the individual interatomic distances differ substantially from each other and consequently the coordination cannot be uniquely defined.

Moreover, minima on the E_b/n, E_b^+/n or E_b^-/n curves for Li_3, Li_9, Li_4^+, Li_8^+ and Li_8^- and maxima for Li_8, Li_3^+, Li_7^+, Li_9^+ as well as for Li_7^- can be easily seen in fig. 1. The clusters Li_3^+, Li_9^+, Li_7^- and Li_8 have "closed-shell" electronic structures. Generally, E_b/n, E_b^+/n and E_b^-/n as functions of n_e, which labels the number of va-

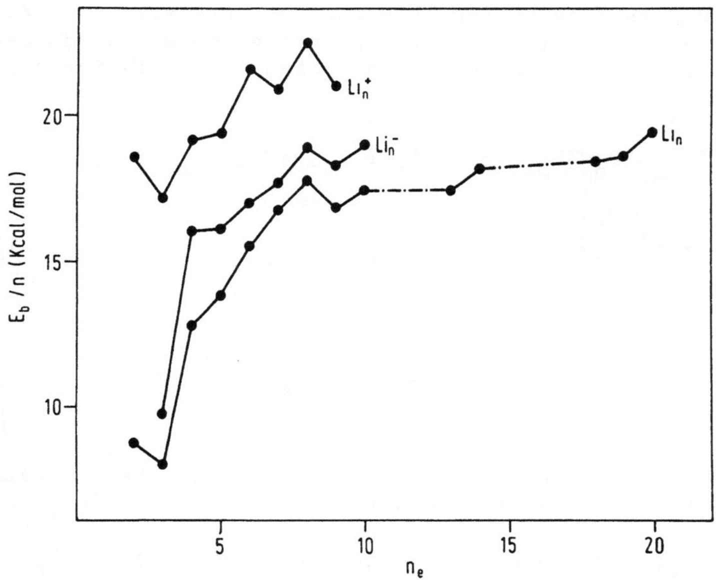

Fig. 2. E_b/n, E_b^+/n and E_b^-/n of Li clusters as functions of the number of valence electrons n_e

lence electrons, have steeper slopes in the intervals (n_e=2k-1, n_e=2k) than in intervals (n_e=2k, n_e=2k+1), where k is an integer. In other words, the atomization energies per atom increase more in the intervals with the clusters having odd number of valence electrons on their left and the clusters with even number of valence electrons on their right. Consequently, if the atomization energies per atom (E_b/n, E_b^+/n, E_b^-/n) are plotted as functions of the valence electrons present in clusters, the similarities of all three functions are clearly recognizable (fig. 2). This feature is not as trivial as it seems at first sight, since the similarity between the curves representing atomization energies per atom for Li_n^+ and Li_n^- clusters as functions of n_e is present only after inclusion of correlation effects /9/. In the one-electron approximation the curve E_b^+/n (n) differs substantially from the E_b^-/n (n) curve.

The functions E_b/n (cf. Eq. 1-3), which are good measures of the deviations from the linear increase of the binding energy per atom with increasing number of the cluster atoms, are quantities derived from the primarily calculated total energy E_n of cluster and give more insight into the cluster stability. The total energy E_n increases with the cluster size, of course, but it is not suitable for direct drawing of conclusions about physical properties of clusters. A detailed study of differences in slopes of the E_b/n

and E_n functions is more appropriate for interpretation of experimental data and their predictions. For example, the differences between the slopes in the neighbouring intervals (second differences):

$$\Delta^2 E_n = E_{n+1} + E_{n-1} - 2E_n = (E_{n+1} - E_n) - (E_n - E_{n-1}) = \Delta E_{n+1,n} - \Delta E_{n,n-1} \quad (4)$$

$$\Delta^2 E_n^+ = E_{n+1}^+ + E_{n-1}^+ - 2E_n^+ = \Delta E_{n+1,n}^+ - \Delta E_{n,n-1}^+ , \quad (5)$$

and

$$\Delta^2 E_n^- = E_{n+1}^- + E_{n-1}^- - 2E_n^- = \Delta E_{n+1,n}^- - \Delta E_{n,n-1}^- , \quad (6)$$

can be considered as a measure of the stability of the neutral, cationic and anionic clusters, respectively (fig. 3). $\Delta E_{k+1,k}$, $\Delta E_{k+1,k}^+$ and $\Delta E_{k+1,k}^-$ represents the slope of the curve E_n, E_n^+ and E_n^-, respectively, in the interval (n=k, n=k+1).

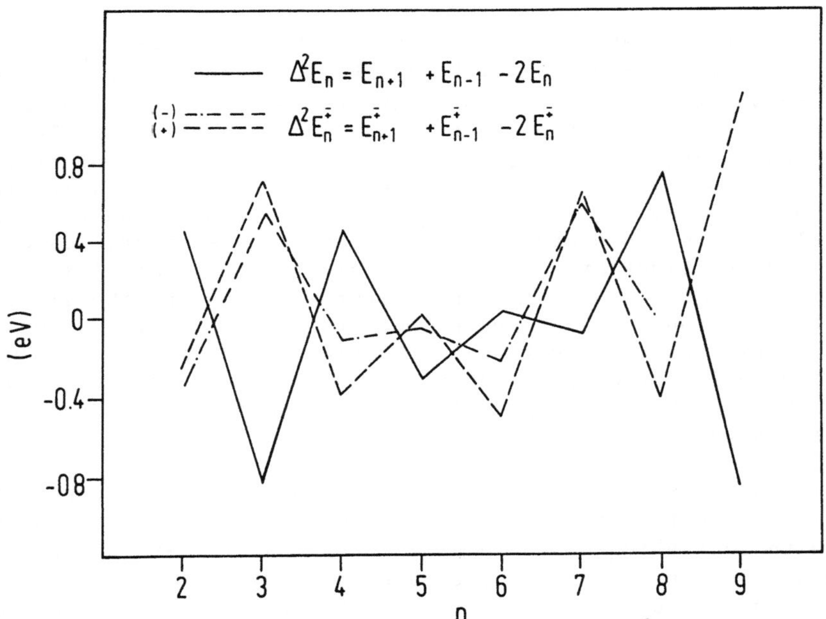

Fig. 3. Second differences $\Delta^2 E_n$, $\Delta^2 E_n^+$ and $\Delta^2 E_n^-$ for Li_n, Li_n^+ and Li_n^- clusters as functions of the number of atoms n

3. RELATION BETWEEN GEOMETRIC AND ELECTRONIC STRUCTURE OF ALKALI CLUSTERS

It exists an apparent and unnecessary controversy in the cluster theory /12-13/. The problem can be formulated as follows: Is it necessary to take into account the detailed geometric form of clusters explicitly or is it sufficient to study the mere electronic properties assuming only a general cluster shape (e.g. spherical /12,14/ or spheroidal /15/) without localization of the atomic nuclei (jellium model)? The simple jellium model in which the spherical cluster form is assumed has shown very interesting and important aspects of the cluster theory. Mainly the existence of "magic numbers" can be understood as a consequence of the completed electron shells. Naturally, the "super atom" model, which exhibits some striking similarities to models of atom and nucleus, enjoys a deserved popularity among the researchers in cluster science. The shell model reflects some basic, generally valid properties of the electronic structure of an arbitrary (three-dimensional) cluster system. Mainly, the consequences of the nodal properties and degeneracies of the molecular orbitals are reflected in this model because they are necessarily reproduced by an arbitrary suitable cluster model. The Knight's shell model /12/ has, moreover, a big advantage in its simplicity.

On the other hand, the shape of a cluster in this model is a priori given and also the deformation of the spheric form to spheroidal shape in the more sophisticated theory /15/ underestimates the importance of the position of nuclei for the general properties of the electronic cloud present in metal clusters. The jellium model is mainly appropriate for larger clusters and the extension of its applicability to very small clusters seems to be in some respects artificial. The splitting of the originally degenerated MO levels for the highly symmetric cluster geometries i.e., the planar and linear forms of small alkali metal clusters, can, of course, be mimicked by a spheroidal cluster form.

As has been mentioned in the introduction, the general features of this splitting are not dependent on the detailed form of the potential well in which the electrons can move in the framework of the chosen cluster model. Especially, the potential resulting from the explicit consideration of the Coulombic potentials due to the atomic nuclei and the potential in the appropriate jellium or plasma model can yield very similar MO spectra. In particular, in the "lower part" of the spectrum only the symmetry and the general dimensions of the potential well are of importance. More difficult and more interesting are near degeneracies due to the special character of the potentials or approximate cluster symmetries. This is probably the case of Li_{20} with nearly degenerated six eigenvalues (five of them originate from "1d" and one from "2s" orbital). In contrast, it is easy, for example, to split the three-fold degenerate "p-type" MO level simply by assuming that one of the three characteristic dimensior of the potential well is longer or shorter than the other two. In the first case the nondegenerate level will be necessary lowe than the two-fold degenerate levels and vice versa. Nevertheless, it seems to be less artificial to consider generally valid effect (e.g. Jahn-Teller and pseudo-Jahn-Teller effects) to explain the mutual interdependence of the electronic and geometric structure of small clusters.

The role of occupation of the molecular orbital levels is clea ly demonstrated on the examples of Li_4, Li_8 and Li_{20} clusters (fig. 4). The very compact tetrahedral and square forms of Li_4 have only partly occupied degenerate MO energy levels corres-

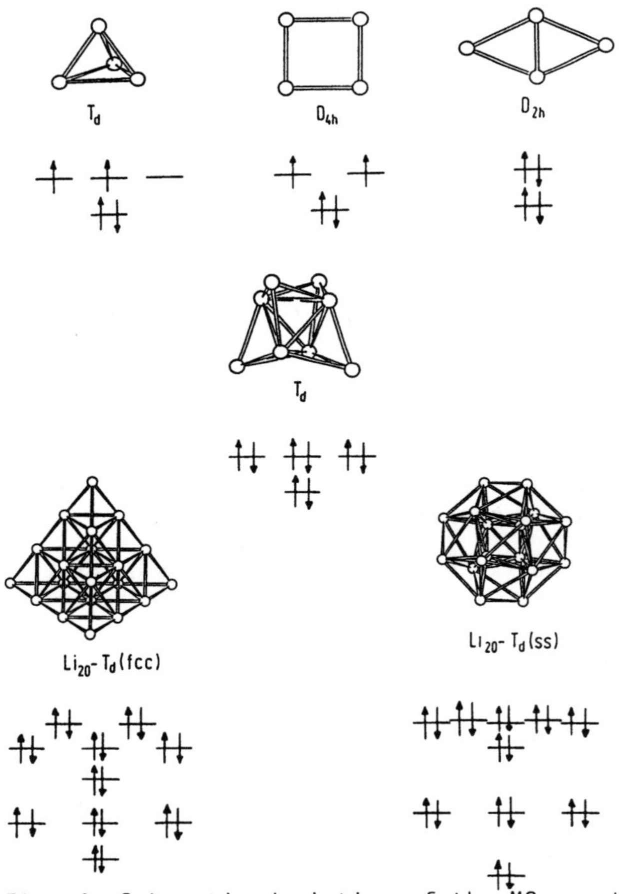

Fig. 4. Schematic depiction of the MO spectra of some forms of Li_4, Li_8 and Li_{20}

ponding to molecular orbitals with a single nodal plane. Consequently, the pseudo Jahn-Teller effect /6/ gives rise to energetically favorable deformation towards a less compact and less symmetric rhombic form /13,18,19/. On the other hand, the very stable T_d form of the approximately spherical Li_{20} cluster with E_b/n = 19.4 kcal/mol has nearly degenerate six MO levels /10/.

The interplay of electronic and geometric features gives rise to very symmetrical and stable cluster isomers of Li_{20} (and Li_8 as well). It is necessary to emphasize that only inclusion of electron correlation effects through the CI procedure stabilizes the Li_{20}-T_d (ss) isomer more than the Li_{20}-T_d (fcc) which is a section of the fcc lattice /10/.

In general, the electron shell model is an extremely useful approach which has played a pioneering role in cluster science, but it is difficult to use this approach for explanation of more delicate cluster features.

The advantage of simple models is that they can explain in a clear, transparent and easily understandable manner some relevant and important facts which follow from the basically realistic features abstracted from the more complicated nature of physical phenomena. A necessary cost for the simplicity of the model is just the useful simplification from which some one-sided features follow. This is also the case of the very useful and illuminating shell model of clusters. If the geometric and other conditions (e.g. full occupation of degenerated MO's) are satisfied, the "super atom" cluster model works excellently.

4. THE USE OF RESULTS OF QUANTUM CHEMICAL INVESTIGATION FOR INTERPRETATION AND PREDICTION OF EXPERIMENTS IN CLUSTER SCIENCE

The verification of the results obtained from the usual quantum mechanical methods by their comparison with experimental data requires a proper interpretation of experiments. Moreover, the proper choice of theoretical quantities used for such a comparison is necessary and equally indispensable for the proper interpretation of the experimental results and their relevance to the investigated cluster properties. It is important to realize that the quantum chemical studies based, in principle, on the time-independent Schrödinger equation in the Born-Oppenheimer approximation can be compared with cluster experiments only if some simple features determine the usually quite complicated experimental process. Mainly, the experimentally measured processes should be roughly determined by the time-independent properties of clusters. It means that dynamics in the proper sense does not essentially change the validity of these theoretical predictions.

The "magic numbers" /12,18/ of the cluster abundances detected by the appropriate devices are in the literature very often related to the cluster stability. The processes of preparatior and detection of clusters are very complicated, and they include many individual steps which depend strongly on the very details of the experimental equipment used and the experimental conditic employed. The process is, of course, basically different if the originally produced clusters are neutral or charged.The neutral clusters are usually ionized before detection and the energy of ionizing species is very important for the possibility of subsequent dissociation and fragmentation processes /20,21/. Among other experimental conditions merely the pressure of carrier gas and the shape of the nozzle in a set-up using the adiabatic expansion should be mentioned.

The cationic alkali metal clusters and alkali metal clusters which origin from the fragmentation of neutral clusters due to photoionization with high energy photons exhibit prevalently large abundances for odd nuclearities equal 3,9,19,21 /20,21,22 The high stability of corresponding closed shell clusters with 2,8,18 and 20 electrons can be explained with the jellium shell model. If the energy of the ionizing species is not sufficient to cause fragmentation, the "magic numbers" 2,8,20 etc. are foun The large peaks for clusters with four alkali metal atoms have been very often found, but they are usually not mentioned. The spherical jellium model does not predict the magic number four.

A striking feature of the second differences $\Delta^2 E_n$ for the neutral Li clusters is the presence of an expressed maximum for n=8 followed by a large drop of $\Delta^2 E_n$ for Li_9 as well as a well developed maximum for n=4, again followed by a drop for n=5. On the other hand, $\Delta^2 E_n^+$ for the cationic Li clusters exhibits maxim for n=3 and 9 as well as a maximum for n=7 (cf. fig. 3).

The definitions of $\Delta^2 E_n$ and $\Delta^2 E_n^+$ allow also for an interpretation of these quantities as differences of the dissociation energy of the cluster with (n+1) atoms and the dissociation energy of the cluster with n-atoms:

$$\Delta^2 E_n = (E_{n-1} + E_1 - E_n) - (E_n + E_1 - E_{n+1}) , \qquad (7)$$

$$\Delta^2 E_n^+ = (E_{n-1}^+ + E_1^+ - E_2^+) - (E_n^+ + E_1^+ - E_{n+1}^+) . \qquad (8)$$

It can be assumed that a large dissociation energy of the cluster with the nuclearity n and small dissociation energy of the (n+1) cluster indicates the occurence of the large abundance of the cluster with the nuclearity n. It goes without saying that this measure of the cluster stability is very vague and does not take dynamic processes into account at all. The large abundances found for Na, K and Rb clusters correspond to the nuclearity for which the maxima on the $\Delta^2 E_n$ and $\Delta^2 E_n^+$ curves of fig. 3 exist and therefore these maxima can be considered as manifestation of special stability of neutral alkali metal clusters with n=4 and 8 and of cationic alkali metal clusters with n=3,7 and 9.

Let us emphasize once more that the experimental confirmation of theoretical predictions should be taken with caution because of the complicated interpretation of the experiments and to some extend arbitrary definition of the compared theoretical quantities. The results for Li_n and Li_n^+ clusters can be compared with experiments for Na, K and Rb clusters since the valence s electrons are responsible for the similar electronic structure of these clusters. This is also confirmed by the far-reaching similarities between the results obtained from the ab initio all electron CI as well as effective core potential-CI investigation of Na clusters /23,24/ and the ab initio-CI calculations for Li clusters /3/.

The choice of fragmentation channels of photoionized alkali metal clusters /20,21/ is a second property for which a theoretical criterion can be found. If one assumes that the relative values of the fragmentation energies of a cationic cluster with n-atoms determine the fragmentation channel, then the quantity $\delta_{n;k,l}$ defined as,

$$\delta_{n;l,k} = (E^+_{n-k} + E_k - E^+_n) - (E^+_{n-l} + E_l - E^+_n) =$$

$$(E^+_{n-k} - E^+_{n-l}) + (E_k - E_l) \quad , \tag{9}$$

can be considered as the appropriate criterion for the choice between the fragmentation channel $A^+_n \rightarrow A^+_{n-k} + A_k$ and $A^+_n \rightarrow A^+_{n-l} + A_l$. In the case that k=1 and l=2, then

$$\delta_{n;2,1} = (E^+_{n-1} - E^+_{n-2}) - (E_2 - E_1) = \Delta E^+_{n-1,n-2} - \Delta E_{2,1} \; . \tag{10}$$

Consequently, the criterion $\delta_{n;2,1}$ for the dissociation channels,

$$A^+_n \rightarrow A^+_{n-1} + A_1 \text{ and } A^+_n \rightarrow A^+_{n-2} + A_2 \quad , \tag{11}$$

compares the slope of the E^+_n curve in the interval (n-2, n-1) with the slope of the E_n curve in the interval (1,2). If $\delta_{n;2,1} > 0$, it means $\Delta E^+_{n-1,n-2} > \Delta E_{2,1}$ the dissociation channel $A^+_n \rightarrow A^+_{n-2} + A_2$ is favored.

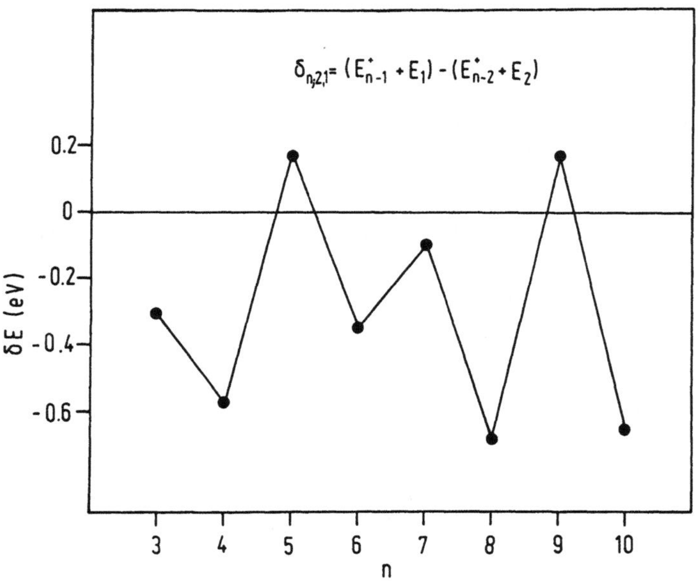

Fig. 5. The quantity $\delta_{n;2,1}$ as function of the number of atoms in Li clusters

The plot of the theoretically obtained quantity $\delta_{n;2,1}$ for Li clusters is shown in fig. 5. The $\delta_{n;2,1}$ values for n=5 and 9 are positive and for n=7, $\delta_{7;2,1}$ is nearly equal zero. The experimental results of Brechignac et al. show also the preference of channel $A^{+}_{n} \rightarrow A^{+}_{n-1} + A_1$ for clusters with even number of Na or K atoms (4,6,8) but the preference for the channel $A^{+}_{n} \rightarrow A^{+}_{n-2} + A_2$ for Na^{+}_{5}, Na^{+}_{7}, K^{+}_{5}, K^{+}_{11} /20,21/. This agreement is supported by calculations of Na^{+} and K^{+} clusters with other quantum mechanica methods /21/.

The appearance ionization potentials of clusters offers anothe possibility for the comparison of experimental and theoretical results. The dependence of vertical and adiabatic ionization potentials on the cluster nuclearity n (fig. 6) for Li clusters, calculated with the MRD-CI procedure, shows two main features: An overall decreasing tendency and some characteristic superimposed oscillations. The minima for n=3,7 and 9 are quite developed. The minimum for n=5 is less pronounced. The rough similarity of the curve of fig. 6 to the dependence of $\Delta^2 E_n$ on n (fig. 3) documents the profound connection between ionization po tential and the electronic structure. The drastic change of character of the geometry of the cluster in the interval between n=5 and 7 from two to three dimensional species leads to a somehow irregular behaviour of $\Delta^2 E_n$ as well as of ionization potenti Notice, the predicted values of the vertical ionization potentia for the Li tetramer, pentamer, hexamer and octamer are very comparable. The overall decrease is remarkable only for adiabatic p

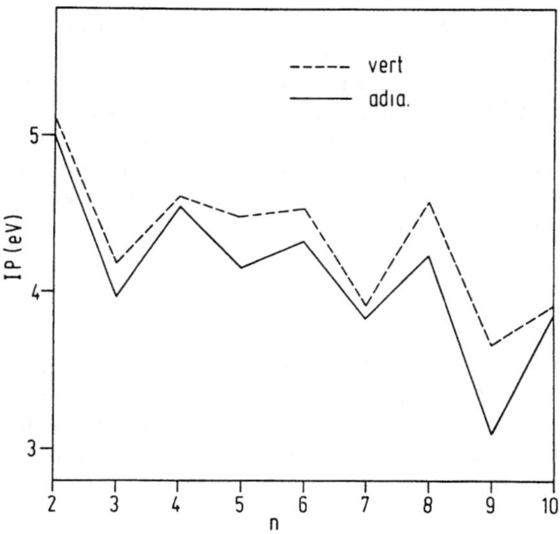

Fig. 6. Vertical and adiabatic ionization potentials of Li clusters

tentials. This circumstance can be of some interest for the proper interpretation of the appearance potentials of clusters. The preliminary MRD-CI calculations of IP's for Na clusters yield very similar results, as can be expected /23/.

The calculated normal frequencies can be useful for the interpretation of the results of the infrared and Raman spectroscopy of clusters. According to Moskovits /25/ the vibration spectra of alkali metal tetramers can be well interpreted if the rhombic shape of tetramers is assumed. The ESR measurements indicate the form of a pentagonal bipyramid (D_{5h}) for the Li heptamer /26,27/. The detailed analysis of the experiments on alkali metal trimers is also in agreement with the general theoretical prediction of the C_{2v} shape of Na_3 /28/. Further experiments for determination of cluster form are very desirable and nowadays feasible. The proper use of the theory can be very helpful to answer some fundamental questions concerning the general nature of small clusters.

5. CONCLUSIONS

The consideration of the mutual dependence of the electronic and geometric structure of clusters yields an interesting insight into the specific nature of clusters.

The most important factors determining the cluster properties are of such a general and in principle simple nature that various quantum chemical methods can yield good description of some properties of alkali metal clusters (cf. e.g. 4, 11, 24, 29).

The calculated slopes of the energy dependence of neutral and charged alkali metal clusters on nuclearity can be used to estimate the relative stability of neutral and cationic clusters and their tendency to fragmentize. Appropriate quantum chemical investigations are capable to yield theoretical predictions of som other essential properties like ionization potentials, normal vibrational frequencies and further characteristics important for various spectroscopies of clusters.

ACKNOWLEDGEMENT:

This work has been partly supported by the Sfb 6 "Structure and Dynamics of Interfaces" and Sfb 337 "Energy- and Charge-Transfer in Molecular Aggregates" and by the CNR "Italian National Resea Council". Thanks are extended to W. Pewestorf and G. Pacchioni, who actively participated in the cluster research.

REFERENCES

1. J. Koutecký and P. Fantucci, Chem. Rev. 86, 539 (1986)
2. J. Koutecký, V. Bonačić-Koutecký, I. Boustani, P. Fantucci and W. Pewestorf: In "Large Finite System". The 20th Jerusalem Symposium on Quantum Chemistry and Biochemistry, D. Reide Publ., Dordrecht, 1987.
3. I. Boustani, W. Pewestorf, P. Fantucci, V. Bonačić-Koutecký and J. Koutecký, Phys. Rev. B35, 9437 (1987).
4. a) Y. Wang, T. F. George, D. M. Lindsay and A. C. Beri, J. Ch Phys. 86, 3493 (1987)
   b) D. M. Lindsay, Y. Wang and T. F. George, J. Chem. Phys. 86, 3500 (1987)
5. J. Koutecký, D. Plavšić and D. Döhnert, Croat. Chim. Acta, 62, 461 (1983)
6. P. Fantucci, J. Koutecký and G. Pacchioni, J. Chem. Phys., 80, 325 (1984)
7. G. Pacchioni and J. Koutecký, J. Chem. Phys., 81, 3588 (1984)
8. J. Koutecký and P. Fantucci, Z. Phys. D: Atoms, Molecules and Clusters, 3, 147 (1986)
9. I. Boustani and J. Koutecký, to be published
10. I. Boustani and J. Koutecký, to be published
11. P. Fantucci and J. Koutecký, this book
12. W. D. Knight, W. A. De Heer, W. A. Saunders, K. Clemenger, M. Y. Chou and M. L. Cohen, Chem. Phys. Lett., 52, 2141 (1984)
13. M. M. Kappes, P. Radi, M. Schar and E. Schumacher, Chem. Phys Lett., 119, 11 (1985)
14. W. Eckart, Phys. Rev. B29, 1558 (1984)
15. K. Clemenger, Phys. Rev. B32, 1359 (1985)
16. H.-O. Beckmann, J. Koutecký and V. Bonačić-Koutecký, J. Chem. Phys. 73, 5182 (1980)

17. J. Koutecký, G. Pacchioni, G.-H. Jeung and E. Ch. Hass, Surf. Sci. 156, 650 (1985)
18. O. Echt, K. Sattler and E. Recknagel, Phys. Rev. Lett., 47, 1124 (1981)
19. J. Koutecký and G. Pacchioni, Ber. Bunsenges. Phys. Chem., 88, 233 (1984)
20. C. Bréchignac and Ph. Cahudzac, Z. Phys. D: Atoms, Molecules and Clusters, 3, 121 (1986)
21. C. Bréchignac, Ph. Cahudzac, J.-Ph. Roux, D. Pavoloni and F. Spiegelmann, J. Chem. Phys., in press
22. N. D. Bhaskar, R. P. Früholz, C. M. Klimacek and R. A. Cook, Phys. Rev. B, in press
23. V. Bonačić-Koutecký, P. Fantucci and J. Koutecký, to be publish
24. J. L. Martins, J. Buttet and R. Car, Phys. Rev. B31, 1804 (1985)
25. M. Moskovits, private communications.
26. D. A. Garland and D. M. Lindsay, J. Chem. Phys. 78, 2813 (1983)
27. G. A. Thompson, E. Tischler and D. M. Lindsay, J. Chem. Phys. 78, 5946 (1983)
28. a) A. Herrmann, S. Leutwyler, E. Schumacher and L. Wöste, Chem. Phys. Lett. 52, 418 (1977);
b) W. H. Gerber and E. Schumacher, J. Chem. Phys. 69, 1692 (1978);
c) M. Broyer, G. Delacretaz, P. Labastie, J. P. Wolf and L. Wöste, private communication.
29. M. J. McAdon and W. A. Goddard, Phys. Rev. Lett. 55, 2563 (1985)

Part IV

Fragmentation

Fragmentation, Reaction and Photodetachment of Mass Separated Cluster Ions

W. Begemann, S. Dreihöfer, G. Ganteför, H.R. Siekmann, K.H. Meiwes-Broer, and H.O. Lutz

Fakultät für Physik, Universität Bielefeld,
D-4800 Bielefeld 1, Fed Rep. of Germany

Mass separated metal and semiconductor cluster ion beams have been produced by sputtering as well as laser evaporation. During and after the production process, sputtered clusters fragment at a high rate. Cooling of laser evaporated clusters in a He carrier gas results in stable configurations. The spontaneous fragmentation, fragmentation in collisions, chemical reactions with clusters, and photoelectron spectroscopy of cluster ions is the main topic of the present investigations.

INTRODUCTION

The interaction of free size-selected clusters with laser fields or collision partners allows insight into fundamental properties like electronic and geometrical structure, reactivity etc. There have been a number of studies involving neutral clusters in which it is difficult to unambiguously assign the collision partners. Working with cluster ions reduces this ambiguity because a defined cluster size can be selected; in addition, the charged products do not have to be ionized prior to product mass analysis.

The paper presents investigations of a variety of sputtered as well as laser-evaporated cluster ions (Al_n, Si_n, Cu_n, W_n, Cu_nF_m, Al_nF_m). After the study of sputtered clusters and their unimolecular decomposition in section I, section II describes the "reflectron" mass spectrometer as is used for fragment analysis. The results of collision-induced fragmentation of negatively and positively charged sputtered clusters, including first results of charge exchange, are given in the third chapter. In the last part the production of cold metal and compound cluster ions is described; in this section we also present the spectroscopy of photodetached electrons of negatively charged metal clusters.

I. SPUTTERING OF CLUSTER IONS AND UNIMOLECULAR DECOMPOSITION

Sputtering of solid targets by high energy ion bombardment produces neutral as well as charged clusters. Several attempts have been made to describe the production process theoretically; two different growth models are discussed:

a) The whole cluster emerges from the surface; in this case the constituents must stem from neighbouring lattice places. /1/

b) Sputtered atoms from one collision cascade recombine above the surface in the vacuum. /2/

In both cases the relative velocity of the combining atoms or molecules must be low so that the total energy of the nascent cluster is lower than its dissociation energy. None of the models has been uniquely favoured by experimental results so far; measurements of kinetic energy distributions of sputtered metal clusters show a decrease in width and mean energy with increasing number of atoms n /3/. From this it has been concluded that the low energy part of the collision cascade is mainly responsible for cluster emission.

Experimental work on sputtering suffers from the fact that the produced species is internally excited and shows a strong tendency to fragment. In order to shed more light on this spontaneous behaviour we first investigated the unimolecular decomposition of sputtered metal and silicon cluster ions. The sputtered ion beam apparatus is described in detail elsewhere /4/. In short, the experiments are performed in a five-stage differentially pumped UHV machine (see Fig. 1) with pressures of $<10^{-8}$ mbar (base pressure) and $<10^{-7}$ mbar (during sputtering) inside the target chamber. Under these conditions the unimolecular decay is not disturbed by collisional relaxation or target contamination. Double differential pumping between target and scattering chamber turns out to be a prerequisite for chemical reaction experiments on sputtered clusters: A contamination of the sputter target by e.g. a partial pressure of only approximatively 1×10^{-7} mbar oxygen changes the cluster ion production conditions severely so that the sputtered material contains a considerable amount of metal oxides. These oxides might partly be transmitted by the mass separator; in this case a clear distinction between gas-phase reaction products and sputtered oxides would be impossible.

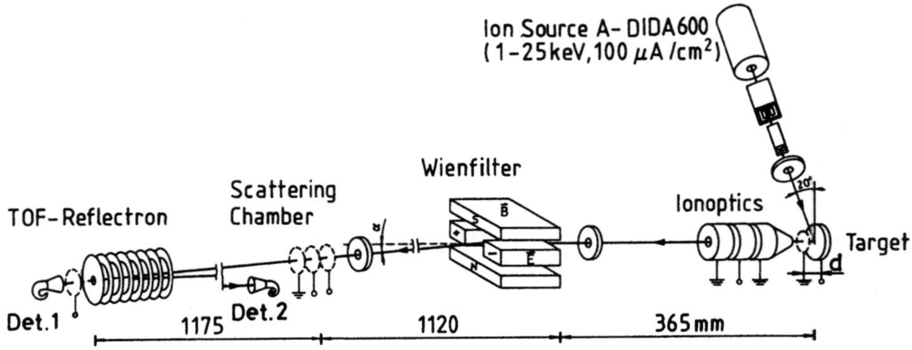

Fig. 1 . Ion beam apparatus for the investigation of sputtered cluster ions

Neutral and charged clusters are produced by about 20 keV Xe^+ ion bombardment of thick targets. The cluster ions are then accelerated from the floated target (V_{acc} = 500 ... 4000 V, typically 1800 V) and focussed through a Wien filter into the scattering chamber.

A time-of-flight "reflectron" serves mainly for the analysis of collision-induced fragments or chemical reaction products and is described in chapter III. For the investigation of the unimolecular decomposition detector 1 (see Fig.1) is utilized with a repeller grid energy analyzer mounted in front. This setup defines a time window in which the decay is measured: After acceleration over a preselected distance d (1.7 or 18.5 mm) the cluster ions undergo velocity selection while passing the Wien filter (COLUTRON 600 B). Only those clusters (of specific mass) are selected which do not fragment for at least a time t_0 after they were produced at or immediately in front of the target, t_0 being the time needed by the cluster beam to reach its final energy; it varies, e.g., (for V_{acc} = 1.3 kV) between 35 ns (Al^+, d = 1.7 mm) and 2.7 μs (Cu_{21}^+, d = 18.5 mm). All particles which fragment before acceleration has been completed have different velocities and are deflected out of the beam by the Wien filter. Fragments being produced after acceleration keep the parent's velocity and will be transmitted. Therefore, the ion beam just in front of the detector contains the parent ions and their decay products which arise in a time window which is opened at t_0 and closed when the particles reach the detector. The repeller-grid in front of the multiplier separates fragment ions from the cluster ion beam. Intensity steps appear at voltages sufficient for repelling the fragments and reveal uniquely the unimolecular decay into defined fragments for given time windows.

Wien Filter Mass Spectra and Decay Rates

The ion beam composition is studied either by chopping the cluster beam and performing a time-of-flight analysis in the reflectron, or by scanning the deflection voltage of the Wien filter and monitoring the transmitted ion current.

The second method yields spectra with lower resolution but higher intensity. Fig. 2 shows as examples three Wien filter spectra of negatively charged aluminium, silicon and copper clusters.

Mass selected cluster ion beams can thus be used for further investigation. The mass range available in this experiment depends on the maximum size of clusters produced by sputtering and on the Wien filter resolution (which could be increased by inserting an energy filter between target and Wien filter).

For unimolecular decay studies the ion beam current for the mass of interest is monitored by detector 1 (Fig.1); scanning

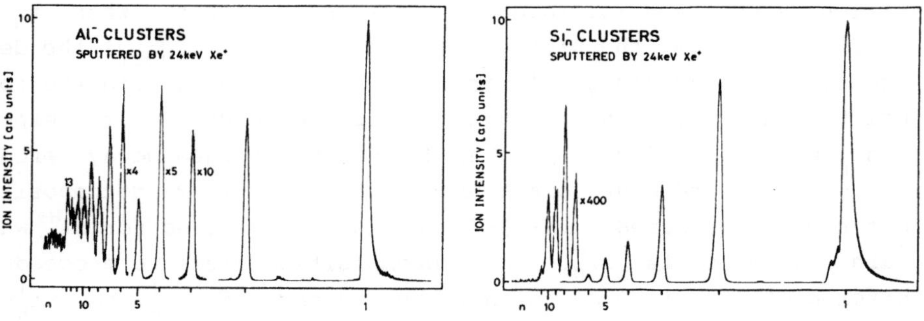

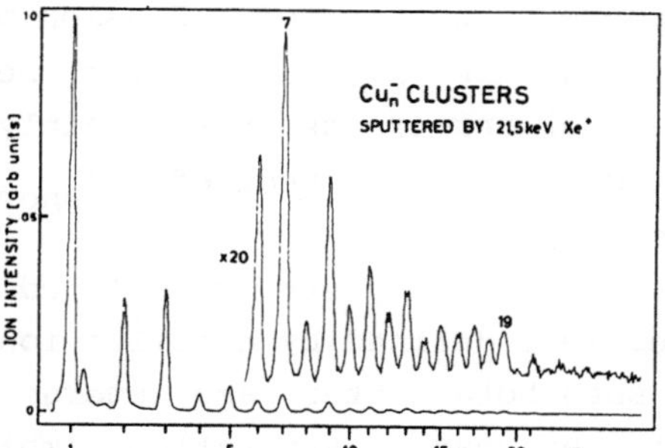

Fig. 2.
Wien filter spectra of sputtered Al_n^-, Si_n^-, and Cu_n^-

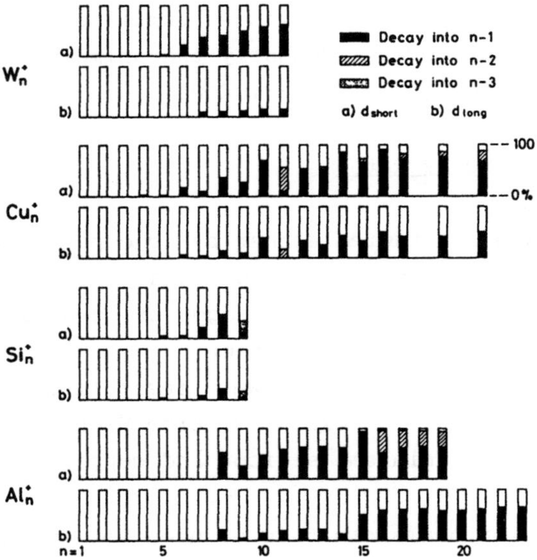

Fig. 3. Decay (in %) of sputtered Al_n^+, Si_n^+, Cu_n^+ and W_n^+ into positively charged fragments for the long (d_{long} = 18. mm) and short (d_{short} = 1. mm) acceleration distances. Statistical error ± 3.5%

the repeller grid voltage results in intensity steps and gives the decay rates which are shown in Fig. 3 for Al_n^+, Si_n^+, Cu_n^+ and W_n^+.

In all cases small cluster ions with n ≤ 3 are practically stable within the experiment's time windows Δt. Strong decomposition is observed for larger n, increasing with n. When t_0 is shifted towards the production event ($d_{long} \rightarrow d_{short}$) the decay probability also rises. A variation of Δt by enlarging the total drift length from 1.5 to 2.5 m affects the outcome of the experiment only insignificantly. The dominating charged decay product is the (n-1)-fragment (black bars in Fig. 3), (n-2)-docomposition occurs in some cases. (n-3)-fragments are observed for Si_9^+ whose main fragment is Si_6^+. This decomposition seems to proceed via fission of Si_9^+ since no significant decay of Si_8^+ to Si_6^+ is observed. For metal clusters there is no hint of a fission process: After some initial cooling time t_0, sputtered metal cluster ions apparently loose further excess energy mainly by evaporation of single neutral atoms. If t_0 is reduced (d_{short}), two-atom decomposition becomes more prominent.

The decay probabilities for Cu_n^+ have been calculated by KLOTS /5/: An "evaporative ensemble", defined by a collection of excited cluster ions, starts to shed both matter and energy by evaporation of neutral atoms. Thermodynamic evaporation combined with classical unimolecular decay theory yields decay rates which are in good agreement with the mean n-dependence of the measured Cu_n^+ decomposition for $n \geq 7$. These findings again support the assumption that excited copper cluster ions evaporate single neutral atoms, cool down and gain stability.

A number of consequences arise from the results given above. For a thorough discussion we refer to earlier publications /4,6/; only a few aspects will be given here:

a) Mass spectra of sputtered cluster ions are qualitatively mirror images of the decay probabilities. The intensity distributions turn out to be a result of decomposition cascades in which clusters with high stability survive and show up more abundantly. Therefore, the shape of a mass spectrum is a function of the time between cluster production and detection. Similar observations have been made by ENS et al. on sputtered alkali halide cluster ions /7/.

b) The investigation of the sputtering process will lead to unambiguous results only when the time dependence of metastable decomposition is properly taken into account. In other words: The clusters whose kinetic energy or internal energies are investigated some ns or μs after their production might not be those which have been born on the target surface.

c) Mass spectra after ionization (by electrons or light) of cold neutral metal clusters often show fine structure which is similar to those of sputtered cluster ions. The spectra of cold metal cluster ions emerging directly from a source without post-ionization (c.f. chapter IV), however, display no fine structure. We conclude that even "careful" ionization may lead to pronounced fragmentation resulting in the observed fine structure.

d) The metastable decompositions of metal and silicon clusters cease after several μs. It is possible to investigate stable sputtered cluster ions when they have had a chance to cool down by evaporation.

e) The unimolecular decay by evaporation of neutral atoms is a quite general behaviour also of non-metal cluster ions: Measurements of MARK et al. /8/ and CASTLEMAN et al. /9/ show a preferred (n-1) decomposition for rare gas and for hydrogen-bonded systems.

II. A TIME-OF-FLIGHT REFLECTRON FOR MASS ANALYSIS OF FRAGMENTS AND REACTION PRODUCTS

A time-of-flight (TOF) reflectron is a mass spectrometer with bent drift line (see Fig. 1). This instrument was developed by MAMYRIN et al. /10/ and has mainly been used because of its high mass resolving power in laser ionization experiments of molecules /11/. Now the TOF reflectron becomes of growing importance for the spectroscopy and fragment analysis of clusters /4,12-14/.

In the experiment described in chapter I we have the special situation that a continous beam of charged clusters has to be analyzed, possibly together with charged fragments produced by collisions, light or chemical reaction. For time-of-flight analysis, such a continuous cluster beam must be pulsed, which leads to ion bunches with considerable spread in translational energy. A TOF reflectron is able to compensate for this energy width; we have studied this compensation both theoretically and experimentally. Mass spectra with improved resolution are presented and discussed. In addition, the usefulness of the reflectron for fragment analysis is tested by a computer simulation of decomposing cluster ions as well as by a collision induced fragmentation experiment.

Mass Spectroscopy in the TOF Reflectron

Ion pulses of defined mass pass the field-free drift region and are rapidly decelerated in the first retarding potential U_1, see Fig. 4.

They then enter the reflecting field (U_2-U_1) with low velocity, where their differences in flight time are corrected. The

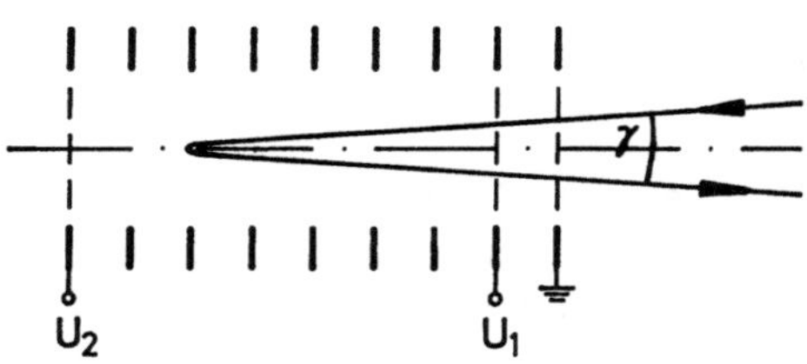

Fig. 4 . Retarding (U_1) and reflecting (U_2) voltages in a TOF reflectron

reflected ion beam (inclined at an angle of $\gamma=3.5^0$ to the incident beam) is detected by a tandem channel plate.

For the calculation of the total flight time (from pulse formation to detection) the flight times in each stage (two-stage acceleration, field-free drift, retardation, reflection, reacceleration and second field-free drift) are added and calculated as function of the voltage ratio $U_1/(U_2-U_1)$, the reflecting voltage U_2 and the ion energy E_{ion}. The results for an optimized ratio $U_1/(U_2-U_1)=2$ are depicted as contour lines in Fig. 5.

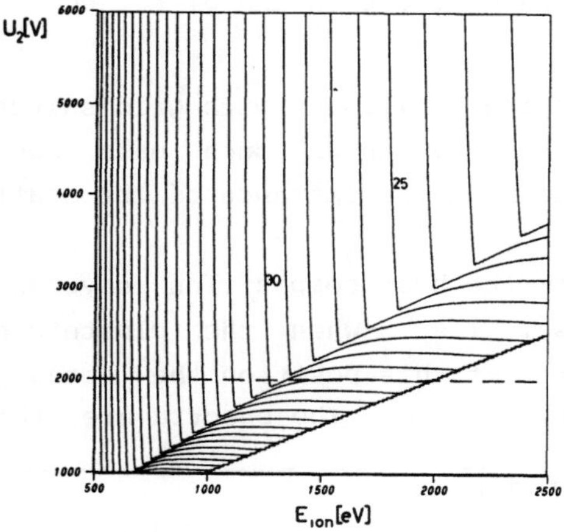

Fig. 5.
Calculated contour map of the flight time (in μs) from start to detection in the TOF reflectron for Cu^+, as function of the reflection voltage U_2 and the ion energy E_{ion}. The spacing between the contour lines is 1 μs

Compensation of energy spread in the ion bunches occurs when the contour lines are parallel to the x-axis. As an example for U_2=2000 V (a typical value in these experiments, dashed line in Fig. 5), ions with E_{ion} ranging from 1400 to 1900 eV need the same flight time of 31 ± 0.1 μs. Consequently, ion bunches with an energy width of up to 500 eV should be focussed to a 0.1 μs bunch.

As experimental verification, an Al^+ ion bunch with an artificially large energy width of 400 eV is produced. Fig. 6 shows the resulting TOF spectra in forward direction (detector 1) and reflected (detector 2).

The broad peak (Fig.6a) is focussed by the TOF reflectron into a line with 0.3 μs width for optimized reflection voltages (Fig.6b). In practical use, the energy spread is much smaller; for the analysis of the sputtered cluster ion beam composition a number of different pulse-forming methods (including single or

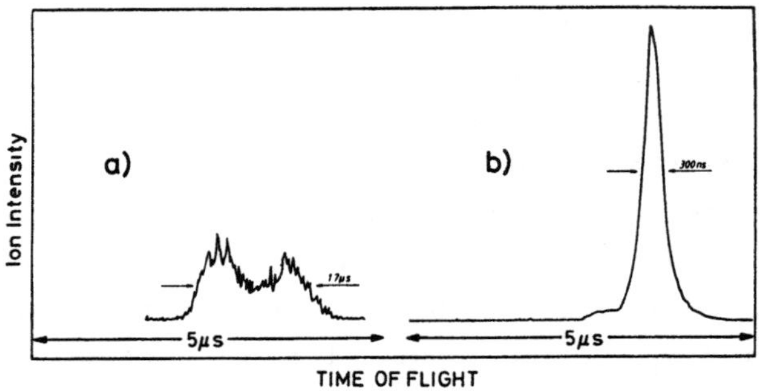

Fig. 6. TOF spectra of Al^+ bunches having an energy width of 400 eV, recorded in a) forward direction and b) reflected

double grid switching, chopping combined with grid switching etc.) were used. As examples Fig. 7 shows such mass spectra obtained by chopping the ion beam at the entrance of the scattering chamber.

Note that the spectra show the beam composition several µs after the sputtering process, i.e. when the clusters have reached the scattering chamber. At this position the metastable decomposition has practically ceased. The spectra are, there-

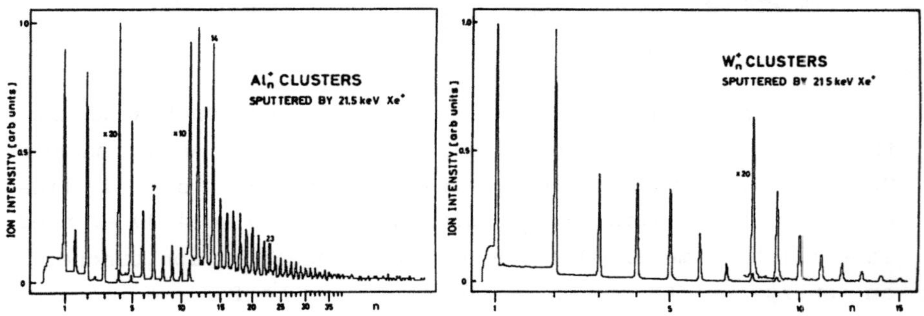

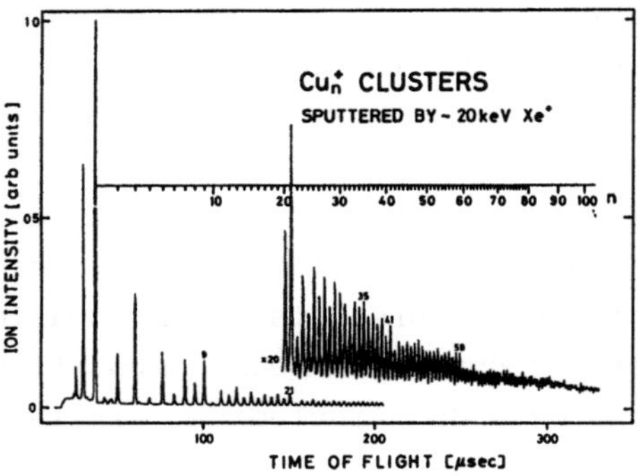

Fig. 7. Mass spectra of sputtered Al_n^+, W_n^+ and Cu_n^+ clusters, recorded in the TOF reflectron

fore, the result of a number of decay processes and reveal the thermodynamic stability of the species as has been discussed in the previous section. This is especially obvious in the Al_n^+ and Cu_n^+ spectra where pronounced fine structures occur: Aluminium shows enhanced intensities for n=7, 14 and 23; the copper spectrum is governed by a pronounced even-odd intensity alternation which can be attributed to the increased stability of clusters possessing spin-paired binding electrons. In addition pronounced lines arise at n=9, 21, 35, 41, 59 and are partly followed by strong intensity drops.

Most of these intense lines correspond to the "shell closings" as has been discussed by HERMANN et al. /15/, ECKARDT /16/, KNIGHT et al. /17/ and CHOU et al. /18/. This model regards a metal cluster as a conducting drop in which the positive charge of the n atom cores is homogeneously distributed. The valence electrons are then filled into this potential well so that a neutral "jellium" is formed. As a consequence of the spherical symmetry the electronic eigenstates have the same degeneracies as those of a hydrogen atom. The spatially extended positive charge distribution, however, causes a different ordering pattern. The quantum mechanics of this structureless problem has been worked out in different approximations /16,18/. The resulting closed-shell clusters (containing 2, 8, 18, 20, 34, 40, 58 ... electrons) have enhanced relative binding energies. Although these calculations mainly concern alkali metal clusters, nearly the whole rich structure in the spectra of sputtered positively and negatively charged copper clusters can be explained (except the even-odd-alternation). Even for Al_n^+ a part of intensity abnormalities match with the shell model. Sputtered cluster ion mass spectra taken by other authors with different methods also show pronounced fine structures for a number of metals /19,20/. A strong correlation between the shapes of the spectra and shell closings is observed.

Fragment Analysis in the TOF reflectron

The TOF reflectron can be operated in several modes: For the investigation of cluster collisions at variable ion energies, the beam is decelerated into a cell, crosses the gas target and is reaccelerated for TOF analysis. In this mode the instrument works as a mass spectrometer. In another mode cluster ion pulses hit without deceleration onto the gas target so that the parent ions as well as their fragments have nearly the same velocity. According to the fragment/parent mass ratio the fragments have lower energy, but reach the reflector at the same time as the parents. Since the ions' penetration depth into the reflecting field is a function of their energy, their time of arrival at the detector is a function of their mass.

For a thorough understanding of the fragment mass spectra computer simulations of the decay of copper cluster ions were performed. Fig. 8 shows as an example such a spectrum for Cu_7^+.

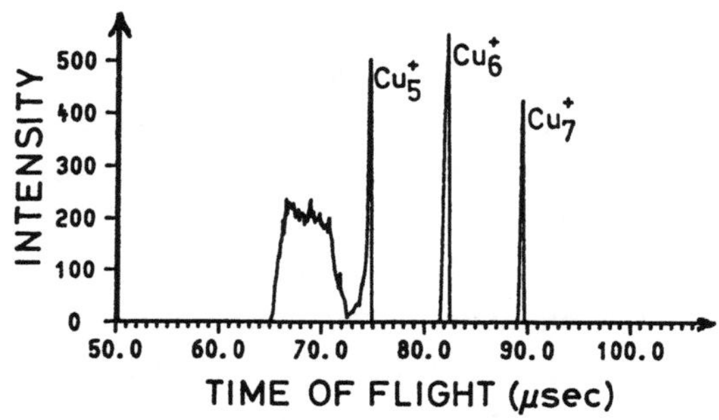

Fig. 8. Simulated fragment spectrum of the decay of $Cu_7{}^+$ in the field-free drift tube of a TOF reflectron

It is supposed in the calculations that all fragments $Cu^+ \ldots Cu_6^+$, as well as the parent ions, appear with equal strength. The parent ion energy (1800 eV) has an assumed width of 10 eV. Note that the parent mass has reduced intensity since the reflecting voltage is chosen to be U_2=1800 V, thus preventing part of the parent ions from being reflected. Only the fragment masses Cu_5^+ and Cu_6^+ are resolved as their energy is sufficient to overcome the retarding potential U_1, see Fig. 4. The smaller fragments ($n<5$) are reflected at U_1 and appear as a broad unresolved line between 65 μs and 72μs. Lowering U_1 and U_2 (for high resolution the voltage ratio has to be kept constant, i.e. $U_1/(U_2-U_1)=2$) the contributions of the smaller fragments will be resolved whereas the large masses vanish since their energy is now higher than eU_2.

In an experimental verification, 1.8 keV Cu_9^+ cluster pulses hit onto a $5x10^{-5}$ mbar argon gas. The resulting fragment spectra

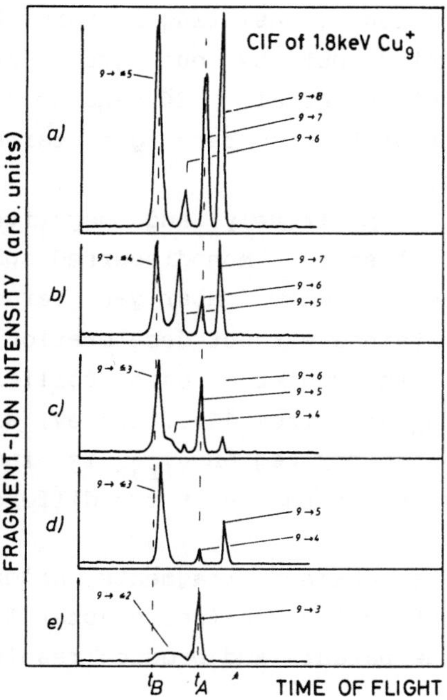

Fig. 9.
TOF reflectron fragment spectra of 1.8 keV Cu_9^+ collisions in $5x10^{-5}$ mbar argon. Successively lowering the reflecting voltages (a...e) makes different fragment lines appear at position t_A

are given in Fig. 9 as a function of the reflecting voltage U_2 (a...e). Parts a) and b) show spectra in which large fragments are resolved, whereas all contributions from the small fragments form a broad line near t_B. For an intensity analysis the reflecting voltages are tuned so that the fragment line of interest is shifted to a defined position t_A in the TOF spectra.

The lines corresponding to smaller fragments are successively extracted out of the unresolved peak near t_B and integrated when they reach t_A. With this method "collision induced fragmentation" cross sections of mass selected metal and silicon ions can be determined.

III. COLLISION INDUCED FRAGMENTATION AND NEUTRALIZATION OF METAL AND SILICON CLUSTER IONS

Collisions of (hydrogen) cluster ions in a low pressure gas target have first been investigated by VAN LUMIG et al. /21/. They called the resulting break up of the clusters "fragmentation". Fragmentation of metal dimers was investigated by ERVIN et al. /22/ and yielded bond energies. Collision induced fragmentation (CIF) of metal cluster ions (consisting of more thar two atoms) was first carried out by our group /4/. Recently, JARROLD et al. /23/, HANLEY et al. /24/ and we /25/ studied the fragmentation behaviour of Al_n^+ in rare gas targets using different collision energies.

In this paper we discuss the CIF of positively and negatively charged metal and semiconductor clusters. Monodispersed sputtered cluster ions impinge onto the $1...10 \times 10^{-5}$ mbar gas target. CIF cross sections (σ_{frag}) are obtained without deceleration of the 1.8 keV cluster ion beam, i.e. the center-of-mass collision energies E_{CM} vary e.g. for $Cu_n \rightarrow Ar$ from 117...431 eV, with n=9...2. It turns out that in this energy region σ_{frag} is independent of E_{CM}. For product analysis alternatively two different methods are applied:

a) Surviving parent ions as well as charged fragments undergo mass analysis in a time-of-flight reflectron (see chapter II); the fragments are separated from the parents and produce resolved lines in the spectrum.

b) A repeller grid energy analyzer separates CIF products from surviving parent clusters. This setup is also utilized to study the charge exchange of negative clusters in a gas target.

Experimental raw data are relative fragment intensities resulting from defined parent cluster ions as function of gas

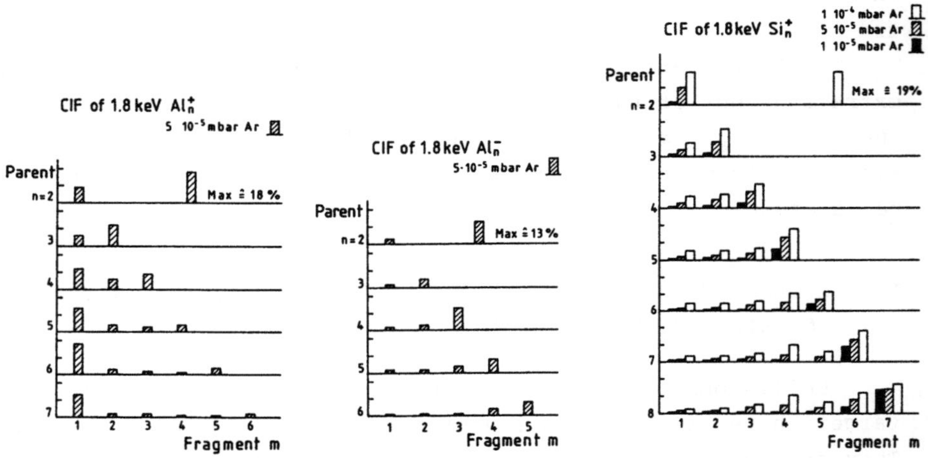

Fig. 10. Relative fragment intensities of cluster ion collisions in argon gas targets. The values are normalized to the intensity of all transmitted ions

pressure. Fig. 10 shows such fragment patterns for Al_n^+, Al_n^- and Si_n^+ collisions in argon.

A dominant charged fragment of Al_n^+ is Al^+, i.e. the clusters lose one constituent, the charge remaining on the small fragment. Similar observations have been made by other authors where E_{CM} was reduced to $\lesssim$ 5 eV /23,24/. CIF of <u>negatively</u> charged aluminium clusters proceeds also mainly by the loss of one constituent, but now the charge remains on the large fragment. For $Si_n^+ \rightarrow Ar$ also mainly (n-1,2) fragments dominate. The same holds true in CIF of all other positively charged metal cluster ions we studied so far, like Cu_n^+, W_n^+ and Mo_n^+ /4/. The fine structure of the CIF patterns, of course, is different for these systems. The extraordinary behaviour of Al_n^+ might be due to the fact that - in contrast to other metal clusters - the ionization potential of Al_n <u>increases</u> with n as has been calculated by UPTON /26/.

For the determination of CIF cross sections the pressure dependence has to be taken into account (see Fig. 10): In the case of, e.g., Si_4^+, the patterns for the different pressures remain roughly unchanged. The Si_8^+ pattern, however, shows a strong dependence on the gas pressure, thus indicating an influence of secondary collisions. Therefore, we calculate first the apparent cross sections σ_{app} by means of the fragment intensity sums which result from defined parent ions n, for a set of different pressures:

$$\sigma_{app} = - \frac{1}{Nl} * \ln \frac{I_p}{I_p + I_{\Sigma frag}} \quad .$$

N gives the gas density, l the collision chamber length, I_p and $I_{\Sigma frag}$ are the intensities of the transmitted parent ions and the fragments, respectively. With this method the influence of the cluster beam absorption (scattering, neutralization), which turns out to be a strong function of n, is minimized: The signal is normalized to $I_p + I_{\Sigma frag}$, i.e. to the sum of all <u>transmitted</u> ions.

For larger clusters, σ_{app} varies dramatically with n. A linear extrapolation of σ_{app} to zero pressure gives the total CIF cross sections $\sigma_{\Sigma mn}$ as shown in Fig. 11 a,b for silicon and copper.

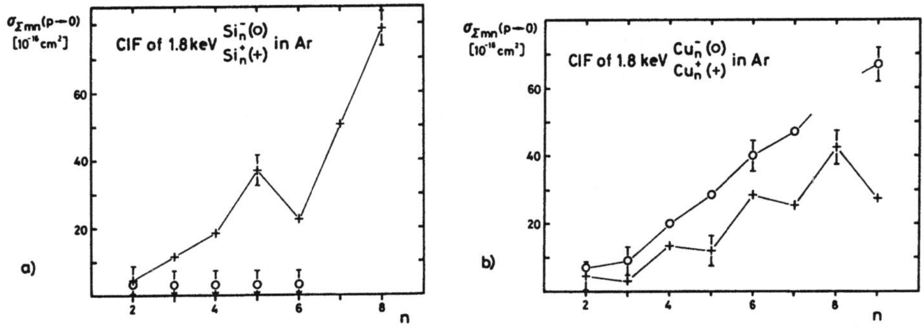

Fig. 11. Total CIF cross sections $\sigma_{\Sigma mn}$, i.e. the cross sectior for fragmentation of an n-cluster into all m fragments.
a) Silicon in argon. The values for Si_n^- are less thar $5x10^{-16}$ cm^2, for all n.
b) Copper in argon. The role of the charge is obvious

$\sigma_{\Sigma mn}$ shows for Si_n^+ a nearly exponential rise with n. Superimposed is a pronounced minimum for Si_6^+, being in coincidence with the enhanced binding energy as has been calculatec previously /27/. The increase of $\sigma_{\Sigma mn}$ with n is much steeper than for all metals which have been investigated so far. As ar example Fig.11b gives the results for cationic and anionic copper clusters: Cu_n^+ shows a marked even-odd variation of $\sigma_{\Sigma mn}$; this behaviour is a consequence of the enhanced binding energies of spin-paired metal clusters. The oscillations are nearly washec out for Cu_n^-. In a separate experiment the transmission of copper cluster ions through an argon gas target has been measured to be practically independent of the cluster charge; thus the geometrical cross sections are not strongly dependent on the sign of the charge. Therefore, we conclude that the differences in $\sigma_{\Sigma mn}$ are mainly a result of the different binding energies.

For <u>negatively</u> charged silicon clusters $\sigma_{\Sigma mn}$ is lower thar $5x10^{-16}$ cm^2 for all n, thus being in complete contrast to all metals studied so far. The absorption, however, turns out to be significantly larger than that of positively charged silicon clusters. This is a first hint of an effective neutralization of Si_n^- in Ar. In the next step a repelling voltage is applied in front of the detector 1 in order to directly extract the contribution of neutral particles in the beam. As an example Fig.1:

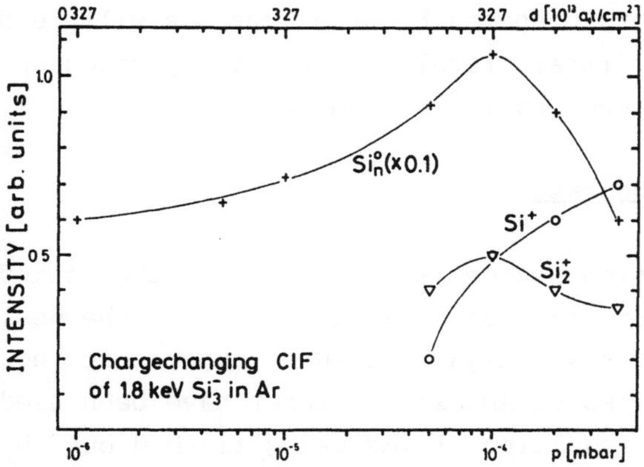

Fig. 12. Neutralisation and re-ionization of Si_3^- in Ar as function of gas pressure p, or thickness d. Upper curve: Intensity of neutral particles. Lower curves: Positive ion signal after collisions of negatively charged Si_3 with the Ar target

gives the neutral particle intensity after collisions of 1.8 keV Si_3^- in Ar.
The neutral particle intensity rises smoothly with p to a maximum at 1×10^{-4} mbar, followed by a fast drop. At the same time the neutrals are ionized to Si_2^+ in secondary collisions. Further pressure increase leads to fragmentation of Si_2^+ and ends up in the production of Si^+. For all n studied so far the largest ionized clusters are the (n-1)-mers. With this crude method (neutralization and subsequent ionization in the argon gas target) it is not uniquely proved that the charge exchange of Si_n^- leads to monodispersed neutral beams. But from the comparison of the CIF and transmission behaviour we find that the cross section for neutralization exceeds $\sigma_{\Sigma mn}$ by about a factor of five to ten. This makes Si_n^- a promising candidate for the production of mass selected neutral clusters.

IV. COLD METAL AND COMPOUND CLUSTER IONS: PRODUCTION AND PHOTOELECTRON SPECTROSCOPY

The experiments described above have been performed on sputtered, i.e. excited cluster ions. To shed more light on the role of the cluster's temperature we also produce cluster ions in a pulsed laser vaporization source. Differences between such cold clusters and clusters prepared by other methods will be dis- cussed. In addition, first results on the spectroscopy o photoelectrons of these species are presented.

Cold Cluster Ion Beam Apparatus

The laser vaporization source is similar to the standar technique of laser vaporization of a target rod of the desire metal mounted in a pulsed, high pressure supersonic nozzl /28,29/. Different lines of an excimer laser have been used t vaporize the metal, the resulting plasma being flushed out by a pulse of ≈1 atm. helium or neon through a channel of variabl length into high vacuum. We do not use a post-ionizer but stud the charged clusters as they emerge from the source. In order t avoid any disturbance of the supersonic expansion, the are between nozzle and skimmer is carefully shielded from electri fields.

The ion beam composition is analyzed in a time-of-fligh mass spectrometer which works collinearly to the source direc- tion as shown in Fig. 13. Appropriate switching of th accelerator entrance grid allows short ion pulses to drift int the starting area and to be accelerated in two stages for optimu time resolution. After having passed the ion optics the cluster are detected in a channeltron.

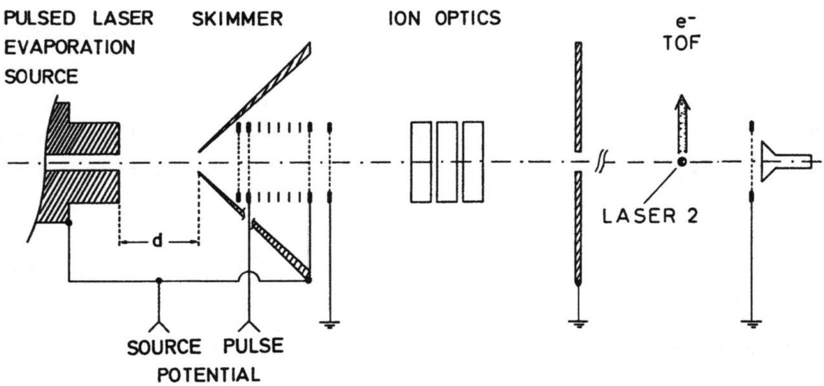

Fig. 13. Experimental setup for the investigation of cluster ion emerging from a laser evaporation source

For the spectroscopy of photoelectrons a second (excimer) laser is used. Its delay is tuned in order to be in coincidence with the arrival of a defined ion bunch in the starting area of a TOF electron spectrometer. Without further acceleration the electrons start in a strongly divergent magnetic field and are adiabatically guided into a weak field drift tube /30/.

Spectra of Cold Cluster Ions

Fig. 14 shows as examples the mass spectra of Cu_n^+ and Al_n^- emerging from the source without post-ionization.

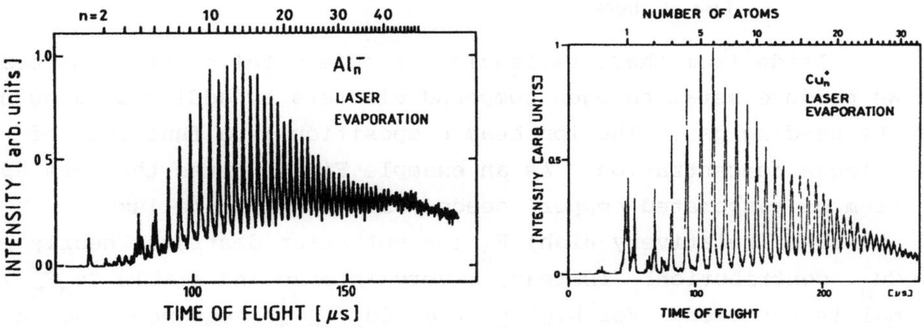

Fig. 14. Cluster spectra of cold Cu_n^+ and Al_n^- obtained directly from the source

The envelope of the intensity distributions can be changed by adjusting the pulse timing, but the fine structures are not affected. For Cu_n^+, e.g., such a spectrum is in contrast to those of hot copper cluster ions which have been heated either by ionization, or during the production by sputtering /25/: Hot Cu_n^+ show a marked even-odd intensity distribution as well as several intensity steps in the spectrum which we have attributed to the unimolecular decomposition of excited cluster ions. Such a structure is completely missing in Fig. 14, thus proving that the temperature of Cu_n^+, obtained directly from the laser evaporation source, is at least below the boiling point.

With this experimental setup and an appropriate line shape analysis, it is possible to determine the translational temperature T_{trans}: e.g., for Cu^+ a value of $T_{trans} = 6\pm2$ K is obtained. Note that copper ions produced without seeding gas have a mean energy of 7 eV.

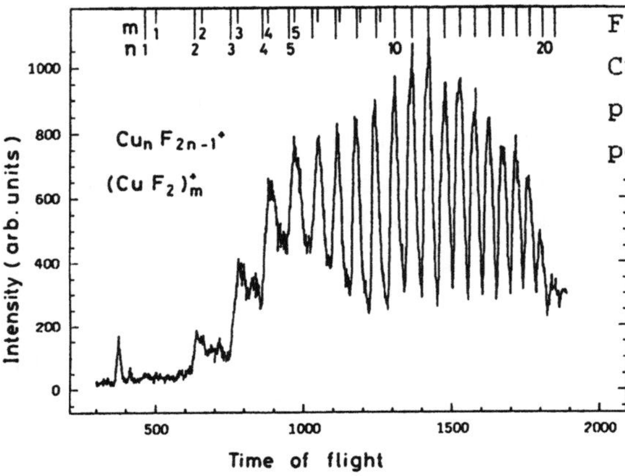

Fig.15. $Cu_nF_m^+$ compound cluster ior produced by flushing the cop per plasma in a F_2/He mixture

Aside from characterization of pure metal cluster ion beams we produce metal-halogen compound clusters by adding a halogen t the seeding gas. The ion beam composition is a function of th halogen concentration. As an example Fig.15 gives the mass spec trum of evaporated copper, seeded with 5% F_2 in helium.

The (relatively high) F_2 concentration destroys nearly an Cu_n^+ contribution. Instead, a very intense and stable $Cu_nF_m^+$ sig nal is observed. For high n the $Cu_nF_{2n-1}^+$ sequences dominate $(CuF)_n^+$ appears with minor abundance only for low n. In contras to spectra of (ionized) copper halide clusters $Cu_nX_m^+$, X=Cl, Br I, studied by MARTIN et al. /31/, low n clusters have wea intensity (which is no artifact of the mass spectrometer). I addition, also in contrast to the behaviour of other copper hal ides, copper binds two fluorine atoms; in the strongest sequenc one of the F atoms is missing to give the positive charge.

At the moment it cannot be decided whether the compoun cluster ions stem from laser-evaporated CuF_2 species or are pro ducts of chemical reactions of copper clusters in a F_2-ric environment. The residence time of the ion pulse in the source however, is for Cu_n^+ about twice as long as for $Cu_nF_m^+$. So th compound clusters seem not to be produced as "rigid plasma" which might be a hint to CuF_2 being evaporated from the surfac and growing in the extender or/and in the supersonic expansion.

Photoelectron Spectroscopy of Al_n^-

Most studies of the electronic properties of larger cluster were limited to a determination of ionization potentials. Eve such experiments might not uniquely reveal physical properties of clusters of defined size as the neutral beams are not mass selected; the nascent ions can decompose on their way to the detector and eventually prevent the ionization threshold from being assigned to the correct mass. Therefore, we decided to investigate the energy of photodetached electrons of negatively charged clusters <u>with defined mass.</u> Fragmentation after detachment does not influence the outcome of the experiment as the electron's energy is an image of the ion's ground state before light interaction.

Only recently the groups of LINEBERGER /32/ and SMALLEY /33/ investigated the photoelectrons of Cu_n^-. In these experiments the anions have been produced continously in a cold cathode discharge /32/ or after post-ionization of clusters in a pulsed laser vaporization source /33/. The resulting electron affinities show a pronounced even-odd variation with decreasing amplitude for increasing n; strong minima are observed for n=2,4,8 /32/. A qualitatively similar n-dependence is seen for the intensity of <u>sputtered</u> Cu_n^-, see Fig.2, thus showing directly the correlation between electron binding energy and the survival of excited negative cluster ions.

We investigate the energies of photoelectrons which are detached from <u>cold</u> metal cluster anions. Different lines of an excimer

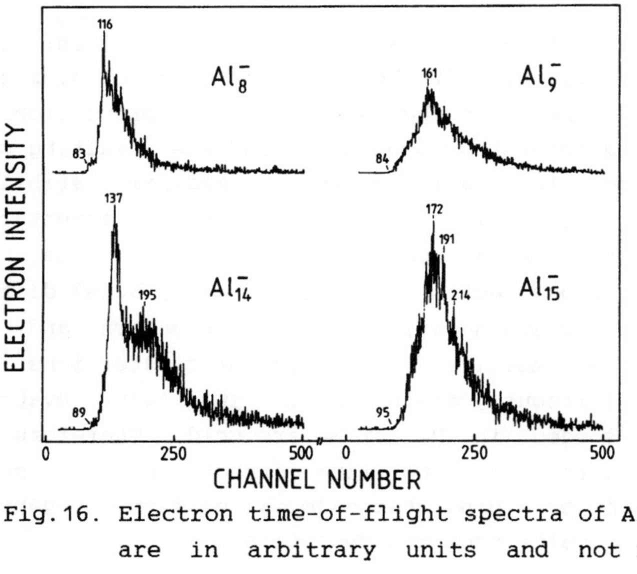

Fig.16. Electron time-of-flight spectra of Al_n^-. The intensities are in arbitrary units and not related to detachment cross sections

laser serve as light source. As an example Fig.16 compares som electron TOF spectra from Al_n^-.

The data are obtained at excitation energies of 3.5 eV (Xe excimer laser) and give four typical spectra for some selecte aluminium cluster ions: The electron appearance thresholds ar all in the same spectral region which corresponds to upper limit of adiabatic electron affinities of 1.5 ... 2.0 eV. The shape of the photoelectron bands vary irregularly with n. It seem that the electronic band development of aluminum clusters doe not proceed continuously but rather stepwise.

A thorough discussion of our photoelectron results would g too far for the scope of this paper and will be published else where /34/. Instead, we report on two general observations fo Al_n^- photoelectrons: First, the electron thresholds show an over all increase with n, for $2 \leq n \leq 20$, ranging from 1.4 to 2.5 eV The latter value is still far away from the work function of alu minium (4.2 eV).
Second, some bands show narrow lines. The origin of suc structures (isomers, excited states of neutral clusters, ...) ha to be clarified by additional theoretical and experimenta effort.

V. CONCLUSION

Mass identified cluster ions provide an excellent tes ground for the investigation of the size-dependence of geometri and electronic structures. In the case of cluster production b sputtering, the unimolecular decomposition and the resulting mas spectra give important information on the thermodynamic stabilit and thus the binding energy. Collision induced fragmentatio (CIF) of mass selected cluster ions confirms those results; i addition, the CIF cross sections reveal a (partly huge) differ ence between positively and negatively charged clusters and, o the other hand, between metal and semiconductor cluster ions.

Ions produced (without post-ionization) in a laser evapora tion source and seeded in He or Ne are cold. Therefore th resulting mass spectra show no fine structures or "magi numbers". This might be a hint that a deviation from homogeneou mass distributions results from decompositions.

Addition of a halogen to the seeding gas produces very intense metal fluoride cluster ion beams. Such compound clusters are cold and thus preferred for chemical reaction experiments.

The powerful tool of photoelectron spectroscopy has been applied to cold metal cluster anions. For aluminum we observe huge variations in the photoelectron band shapes, the thresholds show an overall rise with n.

ACKNOWLEDGEMENT

We thank Y.Y. Liu and R. Hector for helpful discussions and are obliged to W.C. Lineberger for communicating his results prior to publication. Financial support by the Deutsche Forschungsgemeinschaft is gratefully acknowledged.

REFERENCES

/ 1/ G. Staudenmaier, Rad. Eff. 18 (1973) 181

/ 2/ G.P. Können, A. Tip, A.E. de Vries, Rad. Eff. 21 (1974) 269 and 26 (1975) 23
W. Gerhard, H. Oechsner, Z. Phys. B 22 (1975) 41

/ 3/ R.F.K. Herzog, W.P. Poschenrieder, F.G. Satkiewicz, Rad. Eff. 18 (1973) 199
K. Wittmaack, Phys. Lett. 69A (1979) 322
For a review on cluster emission in sputtering see:
W.O. Hofer, Proc. NATO Advanced Study Institute on Erosion and Growth of Solids Stimulated by Atom and Ion Beams, eds. G. Kiriakidis et al. (M. Nijhoff, 1986) p.1

/ 4/ W. Begemann, S. Dreihöfer, K.H. Meiwes-Broer, H.O. Lutz, Z. Phys. D 3 (1986) 183

/ 5/ C.E. Klots, Z. Phys. D5 (1987) 83

/ 6/ W. Begemann, K.H. Meiwes-Broer, H.O. Lutz, Phys. Rev. Lett. 56 (1986) 2248

/ 7/ W. Ens, R. Beavis, and K.G. Standing, Phys. Rev. Lett. 50 (1983) 27

/ 8/ T.D. Märk, P. Scheier, K. Leiter, W. Ritter, K. Stephan, and A. Stamatovic, Int. J. Mass Spectr. Ion. Proc. 74 (1986) 281

/ 9/ A.W. Castleman, Jr., and R.G. Keesee, Chem. Rev. 86 (1986) 589

/10/ B.A. Mamyrin, V.I. Karataev, D.V. Schmikk, V.A. Zagulin Sov. Phys.-JETP 37 (1973) 45

/11/ U. Boesel, H.J. Neusser, R. Weinkauf, E.W. Schlag, J. Phys Chem. 86 (1982) 4857

/12/ A.W. Castleman, Jr., O. Echt, S. Morgan, P.D. Das R.J. Stanley, Ber. Bunsenges. Phys. Chem. 89 (1985) 281

/13/ M.A. Johnson, M.L. Alexander, and W.C. Lineberger, Chem Phys. Lett. 112 (1984) 285

/14/ Th. Leisner, O. Echt, D. Kreisle, M. Knapp, an E. Recknagel, to be published

/15/ A. Hermann, E. Schumacher, L. Wöste, J. Chem. Phys. 6 (1978) 2327

/16/ W. Eckardt, Phys. Rev. B 29 (1984) 1558

/17/ W. Knight, K. Clemenger, W. de Heer, W. Saunders, M. Chou and M. Cohen, Phys. Rev. Lett. 52 (1984) 2141
W. Saunders, private communication

/18/ M.Y. Chou, M.L. Cohen, Phys. Lett. 113A (1986) 420
M.Y. Chou, A. Cleland, and M.L. Cohen, Solid State Comm. 5 (1984) 645

/19/ L. Wöste, P. Fayet, W. Saunders, private communications

/20/ I. Katakuse, T. Ichihara, Y. Fujita, T. Matsuo, S. Sakurai and H. Matsuda, Int. J. Mass Spectr. Ion Proc. 69 (1986 109

/21/ A. Van Lumig and J. Reuss, Int. J. Mass Spectr. Ion Phys 27 (1978) 197

/22/ K. Ervin, S.K. Loh, N. Aristov, and P.B. Armentrout, J Phys. Chem. 87 (1983) 3503

/23/ M.F. Jarrold, J.E. Bower, J. Chem. Phys. 86 (1987) 3876

/24/ L. Hanley, S.A. Ruatta, S.L. Anderson, to be published

/25/ W. Begemann, S. Dreihöfer, K.H. Meiwes-Broer, H.O. Lutz i "The Physics and Chemistry of Small Clusters", Eds P. Jena, S. Khanna and B. Rao, NATO ASI Series, Plenum Pres (1987) 269

/26/ T.H. Upton, Phys. Rev. Lett. 56 (1986) 2168

/27/ K. Raghavachari, V. Logovinsky, Phys. Rev. Lett. 2 (1987) 2853

/28/ D.E. Powers, S.G. Hansen, M.E. Geusic, D.L. Michalopoulos and R.E. Smalley, J. Chem. Phys. 78 (1983) 2666

/29/ L.A. Bloomfield, M.E. Geusic, R.R. Freeman, and W.L. Brown Chem. Phys. Lett. 121 (1985) 33

/30/ P. Kruit, F.H. Read, J. Phys. E. Scientific Inst. 16(1983) 313

/31/ T.P. Martin and A. Kakizaki, J. Chem. Phys. 80 (1984) 3956

/32/ D.G. Leopold, J. Ho, and W.C. Lineberger, J. Chem. Phys. 86 (1987) 1715

/33/ O. Cheshnovsky, P.J. Brucat, S. Yang, C.L. Pettiette, M.J. Craycraft, and R.E. Smalley, in "The Physics and Chemistry of Small Clusters", Eds. P. Jena, S. Khanna and B. Rao, NATO ASI Series, Plenum Press (1987)

/34/ G. Ganteför, K.H. Meiwes-Broer, H.O. Lutz, to be published (1987)

Dissociation of Photoionized Mass Selected Alkali Clusters

C. Brechignac, P. Cahuzac, and J.P. Roux

Laboratoire Aimé Cotton, Bât. 505, Campus d'Orsay, F-91405 Orsay Cedex, France

1. INTRODUCTION

With one valence electron per atom, alkali clusters are the simplest metal clusters. They can be used as a prototype to follow the nature of the bonds, the electron delocalization, the binding energy and the stability as a function of cluster size. These properties are important to understand the cohesion of the bulk.

From a theoretical point of view, alkali clusters have been extensively studied by different methods, namely Hückel calculations [1], local spin density calculations [2-3], Jellium calculations [4] which can be used for large clusters. More sophisticated ab initio calculations have been focused on smaller systems [5-8]. These calculations deal with geometrical structures, stabilities and binding energies.

From an experimental point of view, owing to a low melting point and a photoionization threshold in the near U.V. range, alkali clusters are easily formed using molecular beam technics and can be ionized using conventional lasers in order to be mass-selected [9-12]. Fragmentation experiments [13-20] are appropriate to give information on cluster stabilities and on their binding energies. The dissociation channels and the associated dissociation lifetime give insight on the fragmentation mechanisms and the relaxation of energy within a finite system.

So, alkali clusters are good examples for a confrontation between theory and experiment. In a first set of experiments the study of the unimolecular dissociation of mass-selected photoionized alkali clusters is presented. The well defined, relatively low ionizing photon energy prepares the cluster ions in their electronic ground states, but with sufficient internal energy to dissociate. Such a dissociation is found to occur within a time scale 10^{-6}, 10^{-5} s. The observed fragmentation channels are likely to be related to the lowest dissociation channels corresponding to fragments in their ground states. For homogeneous cluster ions a quantitative comparison with ab initio calculations [21] shows that adiabatic dissociation channels are governed by energetics. Moreover for heterogeneous cluster ions adiabatic dissociation channels reveal to what extent the charge distribution plays a role in fragmentation processes. In a second set of experiment, ionized alkali clusters are photoexcited high above the dissociation limits of their ground states. A fast dissociation occurs with a lifetime less than 10^{-9} s, and is found to evaporate neutral atoms. The number of ejected atoms is proportional to the photon energy, leading to a mean value for the relative binding energy per atom in agreement with available calculations [1]. However, the number of ejected atoms as a function of cluster size for both homogeneous and heterogeneous clusters and the dissociation patterns for heterogeneous clusters cannot be explained by a statistical distribution of the deposited energy.

2. EXPERIMENT

Experimental set up is shown in figure 1. The alkali clusters are produced by an adiabatic expansion of metal vapor through a nozzle of 0.1 mm diameter. The stagnation pressure of the metal vapor in the oven is kept at a few hundred torrs. Different alkalies are used : sodium, potassium, mixture of sodium-potassium or potassium-cesium. A pulsed U.V. laser beam crosses the cluster beam at right angles providing the ionization of neutral species. Different wavelengths are used in order to provide a one photon ionization. The photon energy lies between the ionization and the dissociation thresholds of each cluster. Photoionization takes place in a low field acceleration region; the extracted cluster ions are accelerated by a three grids system up to 2 or 3 kV and propagate along the axis of a tandem time-of-flight [21]. A couple of deflection plates is set up on the way down to the time-of-flight and operates as a gate to select one mass to proceed further.

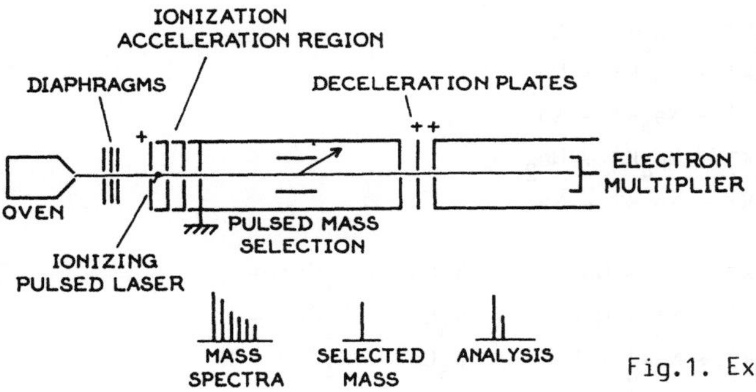

Fig.1. Experimental set up

In a first step the ion package is slowed by deceleration plates, so that parents and fragments resulting from decomposition during the first field free drift tube are mass separated by dispersion in a second time-of-flight and detected by an electron multiplier. This procedure allows us to investigate the unimolecular dissociation which follows the photoionization.

In a second step the chosen ion package is photoexcited, within the decelerating-accelerating grid system, by a second pulsed laser beam which crosses the cluster ion beam at right angles. For this purpose we used either the flux of N_2 laser (6 ns pulse length) providing up to 2 mJ per pulse at 3.68 eV or the flux of an N_2 pumped dye laser oscillating at 2.7 eV. Parent ions and their corresponding photofragment ions are mass-discriminated by the second time-of-flight and detected by the electron multiplier. In fact, a chosen ion package is constituted with primary parents and their fragments provided from unimolecular dissociation in the first field free drift tube. After the deceleration region, parents and fragments have slightly different velocities. A correct timing of the excitation laser is chosen in order to favor the photodissociation of parents and minimize the photodissociation of unimolecular fragments.

3. UNIMOLECULAR DISSOCIATION OF ALKALI CLUSTER IONS

This part deals with unimolecular dissociation of mass selected photoionized alkali clusters Na_n^+, K_n^+, $Na_{n-1}K^+$, NaK_{n-1}^+, $K_{n-1}Cs^+$, with $n \leq 15$. In all cases the ionizing energy is chosen to lie below (within .2 eV) the dissociation limit of the ground state of each ion. So, for all these polyatomic ions part, if not all, of the excess of ionizing energy is redistributed between the numerous vibrational modes of the electronic ground state of the cluster ion and is added to the internal (thermal) energy. The equipartition of this energy leads to an energy per mode which is lower than the binding energy of the cluster, but the total energy lies above its first dissociation limit. So, there is a statistical probability of localizing sufficient vibrational energy in a dissociation coordinate leading to dissociation. We have shown that such a dissociation occurs in a time range of a few tenths of μs [21]. So there is a high probability that the cluster ion will decompose in the drift tube of the first time-of-flight. We found that in our experimental conditions the dimers and trimers are stable but a significant decomposition is observed for $n \geq 4$. The dissociation rate is smaller than 10 % for cluster ions with $4 \leq n \leq 9$ and about 25 % for larger clusters with $n \geq 10$. The predominantly observed dissociation channels are indicated below :

$Na_4^+ \rightarrow Na_3^+ + Na$	$Na_3K^+ \rightarrow Na_2K^+ + Na$	
$Na_5^+ \rightarrow Na_3^+ + Na_2$	$Na_4K^+ \rightarrow Na_2K^+ + Na_2$	
$Na_6^+ \rightarrow Na_5^+ + Na$	$Na_5K^+ \rightarrow Na_4K^+ + Na$	
$Na_7^+ \rightarrow Na_5^+ + Na_2$	$Na_6K^+ \rightarrow Na_4K^+ + Na_2$	
$Na_8^+ \rightarrow Na_7^+ + Na$	$Na_7K^+ \rightarrow Na_6K^+ + Na$	
$K_4^+ \rightarrow K_3^+ + K$	$Na\ K_3^+ \rightarrow K_3^+ + Na$	$K_3Cs^+ \rightarrow K_2Cs^+ + K$
$K_5^+ \rightarrow K_3^+ + K_2$	$Na\ K_4^+ \rightarrow K_3^+ + NaK$	$K_4Cs^+ \rightarrow K_2Cs^+ + K_2$
$K_6^+ \rightarrow K_5^+ + K$	$Na\ K_5^+ \rightarrow NaK_4^+ + K$	$K_5Cs^+ \rightarrow K_4Cs^+ + K$
$K_7^+ \rightarrow K_6^+ + K$	$Na\ K_6^+ \rightarrow NaK_5^+ + K$	$K_6Cs^+ \rightarrow K_5Cs^+ + K$
$K_8^+ \rightarrow K_7^+ + K$	$Na\ K_7^+ \rightarrow NaK_6^+ + K$	$K_7Cs^+ \rightarrow K_6Cs^+ + K$
		$K_6Cs_2^+ \rightarrow K_5Cs_2^+ + K$
$K_9^+ \rightarrow K_8^+ + K$	$Na\ K_8^+ \rightarrow NaK_7^+ + K$	
$K_{10}^+ \rightarrow K_9^+ + K$	$Na\ K_9^+ \rightarrow NaK_8^+ + K$	$K_9Cs^+ \rightarrow K_8Cs^+ + K$
$K_{11}^+ \rightarrow K_9^+ + K_2$	$Na\ K_{10}^+ \rightarrow NaK_8^+ + K_2$	
$K_n^+ \rightarrow K_{n-1}^+ + K\ (n \leq 15)$		

The main result is that for all clusters there is a predominant dissociation channel associated to the ejection of either a neutral monomer or a neutral dimer which generally favors the most stable cluster ions having even number or/and closed shell electrons. However, in several cases a second dissociation channel is observed with much smaller intensity. ($\leq$ 10 % of the first one). We systematically studied the second dissociation channel for K_nCs^+ and showed that it leads to the evaporation of Cs which would correspond to cluster having a higher internal energy. In order to put into evidence to what extent the energetics govern the fragmentation channels, we compared the experimental results to the adiabatic dissociation channels obtained from different available energy calculations.

3.1. Homogeneous clusters

Up to now the energy calculations have been done mostly for homogeneous clusters.

Consider the two dissociations channels :

$$A_n^+ \rightarrow A_{n-1}^+ + A \qquad (1)$$

$$A_n^+ \rightarrow A_{n-2}^+ + A_2 \qquad (2)$$

The predominant adiabatic dissociation channel corresponds to the lowest value for the sum of the total energies of the fragments. So the ejection of a dimer leads to $E(A_{n-2}^+) + E(A_2) < E(A_{n-1}^+) + E(A)$ (3), where $E(A_n^+)$ is the total energy of the cluster ion A_n^+ corresponding to the energy channel for the process $A_n^+ \rightarrow nA^+$. $E(A)$ and $E(A_2)$ are the total energies for the monomer and the dimer respectively. Considering now the energy difference $\delta = E(A_{n-1}^+) + E(A) - E(A_{n-2}^+) - E(A_2)$. This equates to $\delta = \Delta E_{n-1,1}^+ - BE(A_2)$, where $\Delta E_{n-1,1}^+$ is the energy change for the dissociation $A_{n-1}^+ \rightarrow A_{n-2}^+ + A$, and $BE(A_2)$ is the binding energy change for the dissociation $A_2 \rightarrow A + A$. From inequality (3) $\delta > 0$ corresponds to an ejection of a dimer whereas $\delta < 0$ corresponds to the ejection of a monomer.

The values of δ obtained from different energy calculations for Na_n^+ and K_n^+ [1,3,21] are plotted in figure 2. They predict, on the whole, the experimentally observed predominant dissociation channels, and show that the unimolecular dissociation channels are governed by energetics. In more detail, for even numbered cluster ions the evaporation of a monomer is found to prevail, in agreement with the experimental observations. For odd numbered cluster ions the evaporation of a dimer dominates for A_5^+ and A_{11}^+ since the binding energies of A_4^+ and A_{10}^+ are smaller than the binding energy of A_2, but for A_7^+ and A_9^+ the theoretical values of δ are close to zero, which means that the two channels (1) and (2) are very close in energy. In this last case the theory is not accurate enough to predict the experimental results. However for K_9^+ we have seen the adiabatic dissociation $K_9^+ \rightarrow K_7^+ + K_2$ with 1 % of dissociation rate.

3.2. Heterogeneous clusters

The unimolecular fragmentation channels which have been observed experimentally for heterogeneous cluster ions are similar to the ones observed for homogeneous clusters. This indicates that the unimolecular dissociation of mixed clusters is also closely related to their total energies. However, except for the alkali trimers, and for the tetramers Na_3K^+ and NaK_3^+, no energy calculation has been done for mixed clusters so far [8].

If we consider a heterogeneous cluster ion $A_pB_{n-p}^+$, constituted with kinds of atoms A and B, which evaporates a neutral monomer, there are two possible adiabatic dissociation channels :

$$A_pB_{n-p}^+ \rightarrow A_pB_{n-p-1}^+ + B$$

$$\rightarrow A_{p-1}B_{n-p}^+ + A \quad .$$

The predominant one corresponds to the lower energy cost and the evaporation of A as a predominant channel leads to :

$$E(A_{p-1}B_{n-p}^+) + E(A) < E(A_pB_{n-p-1}^+) + E(B) \qquad (2) \quad ,$$

whereas the evaporation of B leads to :

$$E(A_pB_{n-p-1}^+) + E(B) < E(A_{p-1}B_{n-p}^+) + E(A) \qquad (3) \quad .$$

$E(A_xB_y^+)$ is the total valence energy of the cluster $A_xB_y^+$,

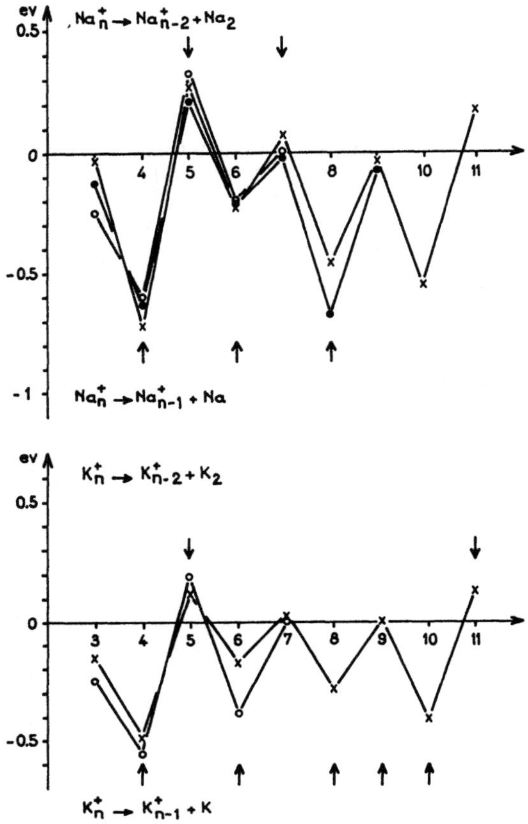

Fig.2. Calculated energy differences δ between the two adiabatic dissociation channels:
$A_n^+ \rightarrow A_{n-2}^+ + A_2$ and $A_n^+ \rightarrow A_{n-1}^+ + A$ plotted as a function of n for Na and K.
x values taken from ref. [1].
• values taken from ref. [3].
o values taken from ref. [21].
The vertical arrows represent the experimental dissociation channels

$E(A) = -IP(A)$ and $E(B) = -IP(B)$, which are the total valence energies of A and B.
Assuming that $IP(B) > IP(A)$; (2) and (3) are equivalent to :
$E(A_pB_{n-p-1}^+) - E(A_{p-1}B_{n-p}^+) > IP(B) - IP(A)$ (2') for the evaporation of A
$E(A_pB_{n-p-1}^+) - E(A_{p-1}B_{n-p}^+) < IP(B) - IP(A)$ (3') for the evaporation of B.
This indicates that the evaporation of either A or B from $A_pB_{n-p}^+$ depends on the energy difference of the two successive energy levels $E(A_pB_{n-p-1}^+)$ and $E(A_{p-1}B_{n-p}^+)$ as compared to $IP(B) - IP(A)$. Figure 3 shows a schematic diagram of the energies.

We have shown [22] that if the charge is equally distributed among all the atoms the energy levels are equidistant and their distance is larger than $IP(B) - IP(A)$, which leads to an evaporation of A for all the heterogeneous clusters. On the other hand if we suppose, in an extreme

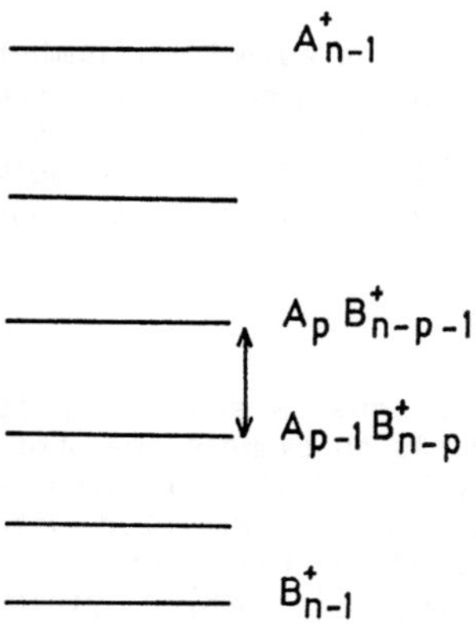

Fig.3. Schematic diagram of total energies of binary alkali cluster ions constituted with n atoms

case, that the charge is located among the A atoms (which are the most electropositive atoms) the energy levels of mixed clusters are shifted towards the B energy level. This leads to an ejection of A for $A_{n-1}B^+$ and an ejection of B for AB^+_{n-1} in agreement with the experimental results. In fact the total localization of the charge on the most electropositive atoms is unrealistic but a partial localization would be sufficient to draw the same conclusions. This has been recently confirmed by ab initio calculations for the trimers Na_2K^+ and NaK_2^+ [23] which show a partial localization of the charge on potassium atoms.

4. PHOTO-INDUCED DISSOCIATION OF ALKALI CLUSTER IONS

The photo-induced dissociation of clusters differs from their unimolecular dissociation since the photon energy is larger than the binding energies of clusters, so that the clusters are excited in highly excited electronic states. For such states nothing is known. However, a question arises : are the photo-induced dissociation channels identical to the adiabatic unimolecular dissociation ones ? This would be the case if the photon energy was redistributed among the vibrational modes of the cluster. In order to get insight into the photo-induced dissociation mechanism, mass selected potassium cluster ions K_n^+ (with $n \leq 41$) are photoexcited with a photon energy either $h\nu_1 = 2.7$ eV or $h\nu_2 = 3.68$ eV. In these cases the energy per mode is less than the atomization energy per atom for all clusters with $n \geq 6$ and the photofragmentation patterns are obtained for several laser flux densities. Moreover, the photo-induced fragmentation of heterogeneous clusters, in which either one Na or Cs atom is substituted with a potassium atom, allows deeper understanding of the statistical aspect of the photofragmentation mechanism. The case of n = 7 is discussed below.

4.1. Homogeneous K_n^+ clusters

At low flux density the observed photofragment distributions are spread over less than three consecutive cluster ion masses. The "mean" of the distribution depends upon the photon energy. The corresponding number of lost mass units smoothly increases with parent cluster size and comes up to a constant value which is proportional to the photon energy. This behavior, which is similar to the one obtained by Lineberger et al. [17] for $(CO_2)_n^+$, shows that the photodissociation of K_n^+ proceeds via a sequential loss of neutral potassium atoms : $K_n^+ + h\nu \rightarrow K_{n-p}^+ + pK$. From the asymptotic number of the potassium atoms lost, we can deduce the mean binding energy per atom which is found to be 0.63 eV for $8 \leq n \leq 15$ and 0.69 eV for $16 \leq n \leq 40$. This is in agreement with the expected smooth evolution toward the bulk value (0.94 eV per atom) and with the recent atomization energy of Lindsay et al. [1] who found 0.60 eV for $8 \leq n \leq 14$.

4.2. Heterogeneous NaK_{n-1}^+ and $K_{n-1}Cs^+$ clusters

The photodissociation of such heterogeneous cluster ions is studied up to n = 10. They all present the same kind of features. As an example the photo induced dissociation of n = 7 with a photon energy $h\nu$ = 3.68 eV is presented in figure 4. The photofragmentation patterns are constituted with monomer, dimer and trimer ions, with higher intensities for the dimers. This similarity between the fragmentation patterns for NaK_6^+, K_7^+ and K_6Cs^+ yields that the photodissociation of heterogeneous alkali clusters proceeds likely via sequential ejection of neutral atoms as has been shown for homogeneous clusters, and they might present close atomization energies. However, the main difference in these three photofragmentation spectra is that heterogeneous fragments are more abundant in K_6Cs^+ dissociation than in

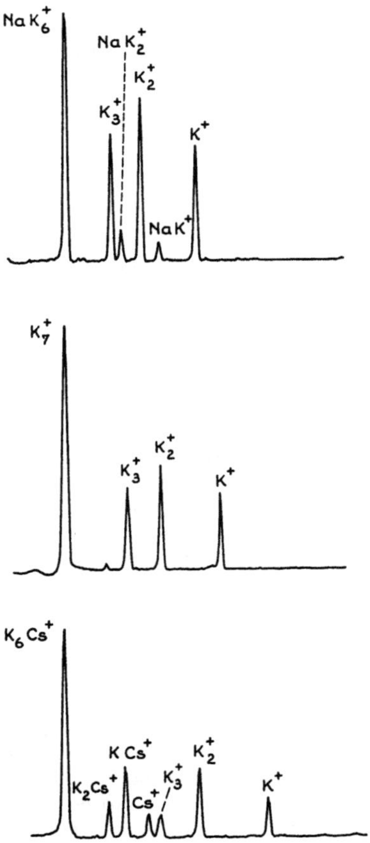

Fig.4. Photofragmentation spectra of NaK_6^+, K_7^+ and K_6Cs^+ with a photon energy $h\nu$ = 3.68 eV

NaK_6^+ dissociation. This indicates that it is easier to photoeject sodium atom from NaK_6^+ than cesium atom from K_6Cs^+.

Let us consider now the ejection of p monomers from XK_{n-1}^+.

$$XK_{n-1}^+ + h\nu \rightarrow XK_{n-1-p}^+ + pK$$
$$\rightarrow K_{n-p}^+ + (p-1)K + X.$$

The statistical probability to eject X within p ejections is p/n, leading an intensity ratio $\frac{K_{n-p}^+}{XK_{n-1-p}^+} = \frac{p}{n-p}$. The corresponding experimental ratios for X = Na and X = Cs with n = 7 are reported below.

		$\frac{K^+}{X^+}$	$\frac{K_2^+}{XK^+}$	$\frac{K_3^+}{XK_2^+}$
Calculated	$\frac{p}{n-p}$	6	2.5	1.33
Experimental	X = Na	>50	9	4
	X = Cs	1.7	1	0.6

This shows that the evaporation rate is not under statistical control.

5. CONCLUSION

These results show the different behavior of the two dissociation processes studied.

For homogeneous clusters, unimolecular dissociation leads to two dominant dissociation channels either the loss of a neutral monomer or a neutral dimer, which are not present in photo-induced dissociation experiments. In this last case the fragment masses have a constant displacement from the parents for masses with $n \geq 7$.

However, the difference is more pronounced for heterogeneous clusters in which the substituted foreign atom plays a role as a probe for dissociation processes. This is typically the case for NaK_n^+ which evaporates a potassium atom in unimolecular dissociation but ejects preferentially a sodium atom in photo-induced dissociation.

Since the unimolecular dissociation is shown to be an adiabatic process governed by energetics, the photo-induced dissociation does not lead to the same conclusion. This would like to show that the redistribution of the photon energy after the excitation is not likely valid in photodissociation.

REFERENCES

[1] D.M. Lindsay, Y. Wang, T. George : J. Chem. Phys. (1987) in press.

[2] J. Flad, H. Stoll, H. Preuss : J. Chem. Phys. 71 3042 (1982).

[3] J.L. Martins, J. Buttet, R. Car : Phys. Rev. B. 31 1804 (1985).

[4] Y. Ishii, S. Ohnishi, S. Sugano : Phys. Rev. B.: 33 5271 (1986).

[5] J. Koutecký, P. Fantucci : Chem. Rev. B. 32 2058 (1985).

[6] I. Boustani, W. Pewestorff, P. Fantucci, V. Bonačić -Koutecký and J. Koutecký : Phys. Rev. (1987) in press.

[7] B.K. Rao, P. Jena : Phys. Rev. B. 32 2058 (1985).

[8] D. Pavolini and F. Spiegelmann : J. Chem. Phys. (1987) in press.

[9] M.M. Kappes, M. Shär, P. Radi, E. Schumacher : J. Chem. Phys. 84 1863 (1986) and references therein.

[10] K.I. Peterson, P.D. Dao, R.W. Farley, R.W. Castleman : J. Chem. Phys. 80 1780 (1984).

[11] W.D. Knight, K. Klemenger, W.A. de Heer, W.A. Saunders, M.Y. Chou and M.L. Cohen : Phys. Rev. Lett. 52 2141 (1984).

[12] C. Bréchignac, Ph. Cahuzac and J.Ph. Roux : Chem. Phys. Lett. 127 445 (1986).

[13] A.J. Stace and A.K. Shukla : Chem. Phys. Lett. 85 157 (1982).

[14] O. Echt, D. Kreisle, M. Knapp, E. Recknagel : Chem. Phys. Lett. 108 401 (1984).

[15] O. Echt, P.D. Dao, S. Morgan and A.W. Castleman : J. Chem. Phys. 82 4076 (1985).

[16] W. Begemann, K.H. Meiwes-Broer and H.O. Lutz : Phys. Rev. Lett. 56 2248 (1986).

[17] M.L. Alexander, M.A. Johnson and W.C. Lineberger : J. Chem. Phys. 82 5288 (1985).

[18] L.A. Bloomfield, R.R. Freeman, W.L. Brown : Phys. Rev. Lett. 54 2246 (1985).

[19] P.J. Brucat, L.S. Zheng, C.L. Pettiete, S. Yang and R.E. Smalley : J. Chem. Phys. 84 3078 (1986).

[20] M.E. Geusic, M.F. Jarrold, T.J. Mc Ilrath, R.R. Freeman, W.L. Brown : J. Chem. Phys. 86 3862 (1987).

[21] C. Bréchignac, Ph. Cahuzac, J.Ph. Roux, D. Pavolini and F. Speigelmann : J. Chem. Phys. to be published.

[22] C. Bréchignac, Ph. Cahuzac and J.Ph. Roux : J. Chem. Phys. to be published.

[23] D. Pavolini, F. Speigelmann : J. Chem. Phys. (1987) in press.

Van der Waals Clusters: Metastable Decay, Coulomb Explosion, and Electron Attachment

O. Echt

Department of Physics, University of Konstanz, P.O. Box 5560,
D-7750 Konstanz, Fed. Rep. of Germany

1. ABSTRACT

The size distributions of van der Waals clusters in general, and of rare gas clusters in particular, often display local intensity anomalies. Mass spectra of clusters of argon, krypton, and xenon, e.g., all show a pronounced intensity drop beyond cluster size n = 309. We discuss the origin of these magic numbers and their possible relation to the stability and the geometrical structure of these clusters. The processes occurring in a rare gas cluster after ionization are discussed in detail; experimental techniques for the study of delayed dissociation are discussed together with some prototypical results. An extension of these methods has made it possible to study the channels for delayed, thermally activated fission of multiple charged van der Waals clusters. Our present knowledge about these metastable species is summarized briefly. Triply charged clusters of CO_2, e.g., preferentially decay into a large doubly charged plus a very small singly charged fragment. Finally we discuss the information that can be extracted from electron attachment to van der Waals clusters, if the ion yield is determined as a function of the electron energy. The resonances of those ions that are known to be formed by attachment to monomers are increasingly redshifted with increasing cluster size due to the electronic solvation shift. Non-reactive attachment which leaves intact the intramolecular bonds is also possible, because new, competing decay channels open up in clusters. Furthermore, new, strong resonances for attachment of slow electrons are found for several cluster species.

2. RARE GAS CLUSTERS: MAGIC NUMBERS, FRAGMENTATION AND METASTABLE DECAY

Clusters of rare gas atoms have been among the most thoroughly studied aggregates during the past two decades. The pairwise additivity of the interaction potential between atoms (or molecules) in van der Waals-bonded systems, together with the spherical symmetry of the potential for rare gases, make these species an ideal play-ground for theoretical studies. Models of hard spheres may be used to determine the energy of different isomeric structures for a cluster of n atoms. Obviously, the energetically favourable structure should feature a minimum volume and surface of the aggregate, but these quantities are ill-defined for small clusters. A simple approach is to search for those structures that maximize the number of "bonds" or nearest-neighbor pairs.

With this technique, Mackay /1/, following a suggestion by Buckminster Fuller, has pointed out that very dense clusters which have the symmetry of the icosahedral point group (I_h) may be built from n = 13, 55, 147, 309, 561,... spheres. In the smallest of these aggregates, 12 atoms (spheres) are on the surface, forming the corners of 20 equilateral triangles, and one additional atom is located at the center. Each surface atom is surrounded by other surface atoms plus the central one. Therefore, the total number of bon is 12*5/2 + 12 = 42. The corresponding number for closed-packed structures (face-centered cubic or hexagonal) is only 36, because each surface atom is surrounded by only 4 other surface atoms plus the central one. Larger cluste with the same icosahedral symmetry can be built from this seed structure by successively adding dense packed layers of atoms to the 20 facets with a fcc stacking sequence (ABCAB...). The number of atoms in the i-th layer (i = 1, 3,...) is $10\ i^2 + 2$ /1/. Models of these icosahedral aggregates with 3 and 4 complete shells are displayed above a mass spectrum of krypton clusters in Fig. 2.

Thus, on the basis of our simple bond-counting technique, the minimum-energy configuration of a rare gas cluster having n = 13, 55, 147,... atoms should be a perfect icosahedron. Why, then, is this peculiar packing of atom which exhibits long-range order but no translational symmetry, not present i the solid bulk phase of the rare gases (which crystallize in fcc, except for He which is hcp)? The icosahedral structure inevitably introduces a strain, because the radial distance between an atom of the first shell and the centr atom is shorter by 5 % than the distance between neighboring atoms <u>in</u> the first shell. This small distortion hardly changes the nearest-neighbor inter action for any physically reasonable type of potential. The binding energy o a dimer, e.g., interacting via a Lennard-Jones potential, would be reduced b only 6% if the interatomic distance were expanded by 5% with respect to its equilibrium distance r_e. This is to be contrasted to the binding energy of t atoms at a distance 1.41 r_e (representing a second nearest neighbor pair in fcc or hcp structures), which is 76% lower than the binding energy at r_e.

The difference in bond lengths between inter-shell atom pairs and intra-shell atom pairs will, however, continue to exist for larger icosahedral clusters. Eventually, the distance between atoms in the i-th shell will be much larger than that between, say, the central atom and an atom in the firs shell; and a crystalline structure will be lower in energy than the icosahed ral one.

The cluster size where the break-even in energy will occur is still unknown. Obviously, one would have to calculate the total energy of the cluste in different isomeric structures, taking into account the interaction of eacl atom with <u>all</u> the other atoms in the corresponding cluster. This can be done very rapidly in a computer. The problem, however, is to find favourable isom ric structures or, more precisely, to find all local minima in the 3n-6 dimensional energy surface. For many purposes, it might be sufficient to determine the global, i.e. the deepest minimum, but there is no fool-proof procedure for this other than to systematically search for all local minima. Hoare and McInnes /2/ have endeavored to determine all local minima for clusters of size $n \leq 13$. The number of metastable isomers increases dramatically with n: They found only 2 isomers for the hexamer, but 988 isomers for the 13-mer, if a Lennard-Jones potential was applied! This does not necessarily imply that large, physical clusters at finite temperature will hardly ever be found in their most stable structure: Hoare and McInnes have shown that the global minimum for the 13-mer, which is the icosahedron, is significantly deeper (by about three times the energy unit of the Lennard-Jones potential) than the next deepest local minimum. Furthermore, most of the metastable isomers appear to be artifacts of the particular choice of the interaction potential. Only 36 isomers of the 13-mer were recovered if a Morse potential was used instead!

Nonetheless, exhaustive searches for global minima in the energy surface have been restricted to clusters of size $n \lesssim 25$. An exhaustive review has been presented by Hoare /3/. In some cases, additional isomers have been discovered in later work /4/, but only for those cluster sizes (n = 17 and 24) where several isomers of nearly identical energy are possible candidates for the ground state structure. There is no doubt that the icosahedral structures form the global minimum for n = 13 and 55 /3,5/ and even for n = 147 /6/. A 13-atom- or 55-atom-cuboctahedron (fcc-structure) would,in fact,spontaneously collapse into the icosahedral structure /7/. Northby has recently attempted to determine all ground state structures and energies for $n \leq 147$ by random distribution and subsequent relaxation of atoms added to a rigid icosahedral core of 13 and 55 atoms, respectively /8/. Allpress and Sanders have calculated the energies of some large, selected (complete-shell) clusters and conclude that clusters of icosahedral symmetry compare favourably with fcc-clusters for sizes up to at least 2000 atoms, but the clusters were not fully relaxed in this study /9/. Stein and coworkers /10/ have extrapolated the calculated energies of radially relaxed, complete-shell clusters to large cluster sizes; they find that the fcc structure becomes more favourable than the icosahedral one beyond n = 3030. The number would decrease if the zero-point motion were taken into account.

How much have experiments contributed to our knowledge about rare gas clusters? Farges and coworkers, in a series of beautiful experiments, have crossed collimated beams of neutral argon clusters, formed in a supersonic beam, by an energetic electron beam /11/. The resulting diffraction pattern, arising from a wide distribution of cluster sizes, is compared with patterns calculated for cluster models of a given size and structure. The main conclusions from their work are as follows: The structure of Ar_n, $20 \lesssim n \lesssim 50$, is "polyicosahedral", i.e. it is composed of 13-atom icosahedra either joined one to the other or interpenetrating each other. Larger clusters have a "multilayer icosahedral" structure, i.e. the atoms which are in excess of a complete-shell icosahedron are distributed on energetically favourable surface sites. The topology of these sites depends on the number of excess atoms, i.e. on whether the outermost layer is nearly empty or nearly full (also see /8/). For very strong expansions, the structure of the clusters finally converts to fcc; the structural transition seems to occur at n ~ 750.

In contrast, Stein and coworkers /10/ conclude from their electron diffraction data that Ar clusters undergo a smooth transition from icosahedral to fcc over the range of 1500 to 3500 atoms. It is also interesting to note that the (vibrational) temperature of large Ar clusters is, according to their analysis, only 10 K. This is 3 times lower than the temperature determined by Farges et al. /11/ with the same technique!

The only other experimental source of information on rare gas clusters is mass spectrometry: Clusters with a wide size distribution are generated in a supersonic beam. The collimated neutral beam is then subjected to electron impact ionization; the resulting cations are accelerated and analyzed in a time-of-flight mass spectrometer (Fig. 1) or in a quadrupole or magnetic sector field mass spectrometer. Spectra of Ar clusters had been known for a long time /12, 13 and references therein/. These spectra, however, showed no features in the size distribution that might have been caused by the particular stability of certain cluster sizes such as n = 13 or 55. The only persistent feature was a local minimum in the intensity distribution for n = 20. Therefore, it was surprising that a large number of local intensity anomalies showed up in mass spectra of xenon clusters /14/. Clusters with n = 13, 19, 23, 25, 55, 71, 81, 87 and 147 atoms were at least a factor of 2 more abundant than the next larger cluster; they were termed "magic". These magic clusters do not always form local maxima in the mass spectra, because, depending on expansion conditions, the overall distribution may be a rapidly decreasing function with increasing n.

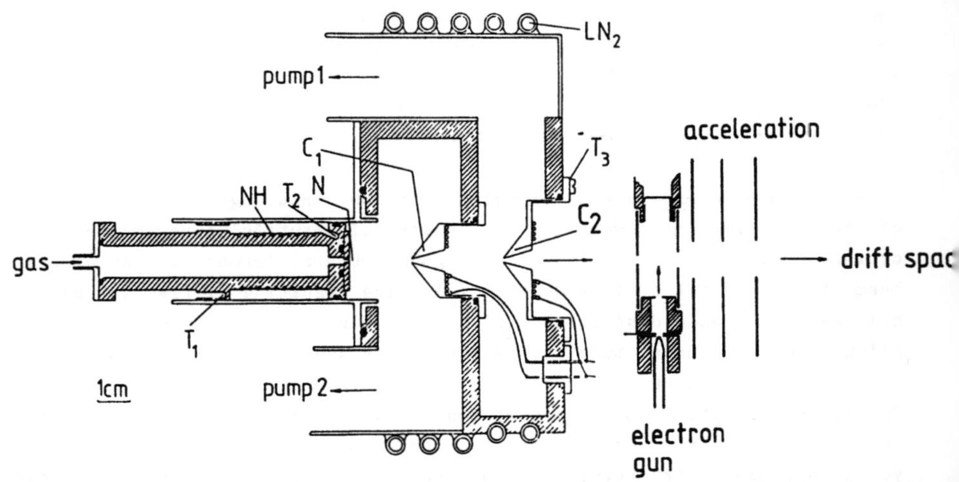

Figure 1. Source for van-der-Waals clusters, generated by adiabatic expansion of gases or vapors, and ionizer for the time-of-flight mass spectrometer

Some researchers have erroneously concluded that these "magic numbers" depend on expansion conditions because, for a mild expansion, they no longer form local maxima in the rapidly decreasing size distribution. The magic numbers do, however, always cause strong local deviations from the smoothly varying size distribution, hence they are well-defined. If evaluated properly, these local deviations in the distribution of Ar and Xe clusters are found to be independent of stagnation temperature T_o, stagnation pressure p_o, and electron energy. Even different kinds of ionization (Penning ionization or one-photon ionization in case of Ar_n /15/, multiphoton ionization in case of Xe_n /16/) were found to leave the local deviations from the overall size distribution quantitatively unchanged.

What is the origin of these magic numbers? Echt et al. noted the agreement between observed magic numbers and those cluster sizes which were calculated, or, by extrapolation, expected to be particularly stable <u>in comparison to the next larger cluster</u> /14/, i.e. for which the heat of evaporation at 0 K displays abrupt drops with increasing n. For example, an atom added to any surface site of a closed-shell icosahedron (n = 13, 55, 147,...) will form only three additional "bonds". In contrast, removal of an atom from a closed-shell icosahedron results in the loss of six bonds. Similar discontinuities in the heat of evaporation are calculated to occur if subshells are closed at n = 19 (double icosahedron) or n = 23 (triple icosahedron) /3/, 17/; the corresponding partial closures in the next shell would occur at n = 71 and 81. All these sizes were indeed observed to be much more intense than the next larger Xe cluster /14/. Recent experiments in our laboratory have revealed that dramatic intensity drops also occur beyond n = 309 (icosahedron with 4 complete shells) for Ar, Kr, Xe, and beyond n = 561 (5 complete icosahedral shells) for Ar /18/. A mass spectrum of krypton clusters, shown in Fig. 2, demonstrates this point /18/.

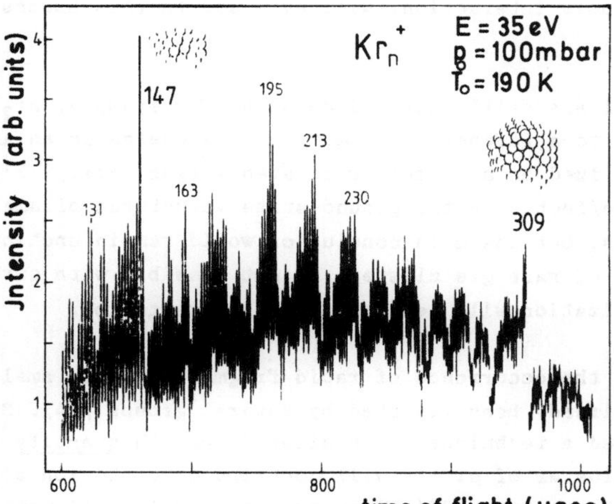

Figure 2. Mass spectrum of large krypton clusters. The particularly large intensity drops beyond n = 147 and 309 suggest a complete-shell icosahedral structure for these cluster ions /18/

Echt et al. originally attributed the abundance anomalies to thermal evaporations from neutral clusters; these decays might enrich the more stable sizes once the clusters have left the nucleation zone. Initially the clusters are hot, because no carrier gas, which might remove the heat of condensation, is used in these experiments. The kinetic energy of the rare gas clusters is $5*kT_o/2$ (neglecting velocity slip and the conversion of heat of condensation into translational energy). Thus, if rare gas clusters (Ne, Ar, Kr or Xe) are formed in a neat expansion through a nozzle cooled to approximately the boiling point of the corresponding gas, and if they are subjected to analysis by ionization or electron diffraction at a distance of 10 cm from the nozzle throat, about 0.4 ms will have elapsed between cluster formation and analysis. This is ample time for clusters to cool by monomer evaporation and, possibly, to enrich stable cluster sizes. Klots has recently elaborated on the thermodynamic properties of such an "evaporative ensemble" of clusters /19/.

An obvious problem with this interpretation, which assumed that mass spectra mirror the abundance of neutral clusters before ionization, is the absence of the expected magic numbers in mass spectra of small ($n \lesssim 80$) Ar or Kr clusters. Some systematic differences between the various rare gas clusters might arise from three-body contributions to the interaction potential /20/, but the fragmentation following ionization is the now widely accepted mechanism which controls the mass spectra of rare gas clusters and, probably, of other van der Waals clusters as well. Haberland /21/ has pointed out that a charged rare gas dimer in its electronic ground state is i) very tightly bound (1.0 to 1.2 eV for Xe through Ne /22/) and ii) that its equilibrium distance is shorter by 30 % to 50% (Xe through Ne) than in the neutral dimer. Hence, vertical ionization of a dimer into the ionic ground state would necessarily produce a highly excited dimer ion. The vibrational energy (~1 eV) of this dimer, if formed in a cluster after charge localization, would greatly exceed the heat of vaporization of the most weakly bound atoms in a cluster of size n $\geq$ 3. The relaxation of the neutral atoms towards the charge, due to the attractive charge-induced dipole interaction, will even add to this excess energy.

Some details of this model are still under discussion. For example, Ar_2^+ appears to bind "chemically" to a further Ar atom, i.e. the charge in an argon cluster may actually be localized on a trimer or an even larger entity /23/. This would possibly have an effect upon the ground state structures of small Ar_n^+ and their total energies, but the main conclusion would remain unchanged: Adiabatic, "soft" ionization of rare gas clusters is not possible with current techniques; and vertical ionization will lead to rapid fragmentation.

Experimental evidence for the occurrence of rapid fragmentation in small van der Waals clusters had already been reported by several groups /24/. Buck and coworkers finally employed a technique which allowed them to quantify the probability that a neutral cluster of size x will contribute to the ion signal at size n ($n \leq x$) in the mass spectrum /25/. They crossed a well-collimated beam of neutral van der Waals clusters with a nearly monoenergetic collimated He beam. By measuring the angular and velocity distribution of the scattered clusters, the different neutral cluster species could be separated from each other, and their intensity could be determined independently from the subsequent mass spectrometric detection process. They found that argon clusters of size $3 \leq x \leq 6$ predominantly feed the dimer ion channel /25/.

How rapidly do the ionized clusters decompose? Haberland estimated that energy randomization from the initially excited dimer ion into the van der Waals modes is completed within 10^{-8} s /21/. (Prompt dissociation, within < 1 ps, will, of course, occur if the ionizing electron or photon excites a repulsive electronic state.) A detailed molecular dynamics /26/ simulation of the energy randomization and monomer evaporation revealed that the time scale is ~0.1 ns; essentially all of the excess energy has been removed by evaporations within 1 ns, but the cluster ion will still contain enough energy to evaporate another atom at much later times (see Fig. 3).

Size analysis in a mass spectrometer typically requires 1 to 100 μs, depending on the type of spectrometer and on the cluster size. One can identify unimolecular decompositions of cluster ions if they occur on this time scale. Even the time scale of 0.1 μs is accessible, as demonstrated recently for small xenon clusters ionized by multiphoton ionization /16/. The most general

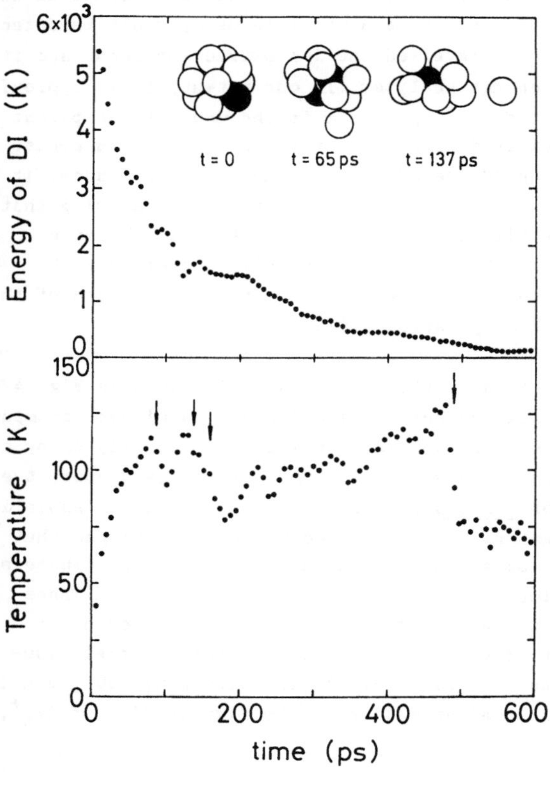

Figure 3. Top: Calculated decrease of the vibrational energy of Xe_2^+, formed in a Xe_{13} cluster at time zero by vertical ionization. Bottom: The cluster temperature (neutral atoms only) rapidly increases because of energy transfer from the excited dimer ion; occasionally it decreases if a monomer is evaporated (indicated by arrows, also see "snapshots" of Xe_n^+). Results are from a molecular dynamics simulation by Soler et al. /26/

technique consists of using two mass spectrometers in tandem. For example, t] magnetic field in a double focussing mass spectrometer may be tuned to selec a monodisperse cluster ion beam of well defined kinetic energy; unimolecular decompositions between the magnet exit slit and the entrance slit of the ele tric sector field are detected if the potential of the electrostatic analyze is scanned /27/. Likewise, a triple quadrupole mass spectrometer may be used where the second quadrupole is simply used as an ion guide, and to provide a means of shifting the effective time window for the decay /28/. A time-of flight mass spectrometer offers several means to detect delayed dissociation Once the ions have acquired their full kinetic energy $q\ U_o$ upon passing through the ion extraction lens, their "parent" mass m_p may be determined fr their time-of-flight through the field-free drift tube. If they decay into a charged daughter of mass m_d, its velocity will be unchanged because of momentum conservation, but its kinetic energy will be only $q\ U_o\ m_d/m_p$. (We have neglected the additional kinetic energy imparted to the two fragments in the center-of-mass reference frame. This energy has been found to be ~5 meV for products from Ar_n^+ /29/, clearly negligible in comparison to the kinetic energy of the parent ion in the lab frame, which is ~2 keV in our studies).

The daughter ions may be excluded from the mass spectrum by applying a repulsive potential slightly less than U_o to a fine grid in front of the i detector (i.e. at the end of the drift tube). Using this technique, Kreisle al. /30/ have determined the probability that a parent ion Xe_n^+ decays in th drift tube. The effective time window, where a possible decay would be detected, opens when the parent ion has traversed the ion extraction lens, and it closes when the ion passes the energy analyzer. In our set-up, it is approximately equivalent to $0.2\sqrt{m} \leq t \leq 2\sqrt{m}$, where m is the mass of the parent ion in amu, and t (in μs) is measured with respect to the ionization event. order to correct for the size dependence of the duration of this window, it seems more appropriate to display dissociation rates by simply assuming that the decay is exponential in the time window of interest.(Actually, ample experimental evidence exists that the decay rate actually decreases with increasing time /16, 27/. This to be expected, because the temperature of th cluster ions decreases with each evaporation /19/).

The decay rates for Xe_n^+ obtained by this method are displayed in Fig. 4 /30/. It is obvious that each magic number mentioned above is either accompanied by a local minimum in the evaporation rate (n = 19, 23, 25, 71) or by a maximum for the next larger cluster (n = 13, 55). These findings support the conclusion that the stability of charged clusters controls the local anomali in mass spectra of rare gas clusters. It is also worth being mentioned that most of these magic numbers are observed in spectra of Ar_n^+ if the clusters are directly formed by nucleation at Ar^+ in the supersonic jet /31/. These spectra actually reveal a rich pattern in the size distribution, even for small clusters, where Ar_n^+, formed by ionization of preformed neutral clusters, is nearly void of magic numbers /12, 13/. In the size range $80 \lesssim n \lesssim 1$ both kinds of techniques reveal local anomalies in the distribution of Ar_n^+,

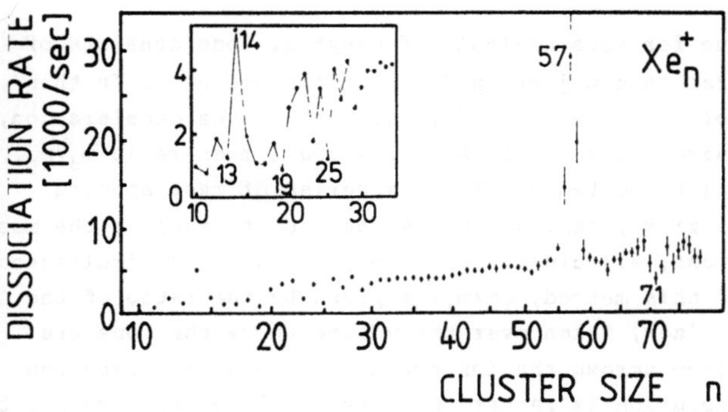

Figure 4. Effective rates for dissociation of Xe_n^+ in the field-free region of a time-of-flight mass spectrometer /30/

but the magic numbers are not always the same. Many of the intensity extrema in the Ar_n^+ distribution, obtained by ionization of neutral clusters, occur at a size n + 1, if n denotes the position of the corresponding extremum in the distribution of Ar_n^+, borne in the charged state, or in the distribution of Xe_n^+ borne neutral. Märk and Scheier /27/ have recently recorded mass spectra of neon clusters, borne neutral; their most pronounced magic numbers (at n = 14 and 56) are also 1 unit above those expected for sphere packings. Finally, however, all these magic numbers converge to a common set: Mass spectra of Xe_n^+, Kr_n^+ (Fig. 2) and Ar_n^+ (borne charged or neutral) all show a pronounced intensity drop beyond n = 309, where closing of the fourth icosahedral shell would occur. Even nitrogen clusters, $(N_2)_n^+$, show the same effect /18/! The deviation (or absence) of magic numbers in clusters of Ne_n^+, Ar_n^+, Kr_n^+, $n \lesssim 150$, borne neutral, from those to be expected from soft-sphere-packing studies /3, 8/ is usually attributed to the distortion introduced into the cluster by the dimer ion /26, 32/. The shrinkage of the dimer upon ionization is least for xenon, hence spectra of Xe_n^+ are probably the best indicator of how the heat of vaporization and, indirectly, how the structure of rare gas clusters changes with increasing size.

We have mentioned above that ionized rare gas clusters, as well as other van der Waals clusters, have a large probability to lose a monomer in the mass spectrometer. If one performs a mass analysis of the daughter ions, one does indeed find that cluster ions may also lose two or even more monomers. In a time-of-flight mass spectrometer this mass analysis can be accomplished by two different methods. One is to subdivide the field-free drift tube into two segments kept at different electric potentials. If, for example, an ion decays in the first segment and the potential in the second segment is higher, such that the ions are decelerated, fragment ions from the same precursor will disperse in time, the lightest fragment ions being decelerated most; the neutral fragments will not be affected at all. Standing and coworkers used this method in order to unravel the origin of magic numbers in sputtered CsI cluster ions /33/. A variant of this method, exploiting deceleration of ions in an electrostatic mirror, was used by Castleman and coworkers: they observed loss of up to 5 NH_3 molecules from $(NH_3)_nH^+$, formed by multiphoton ionization /34/.

An alternative technique for mass analysis of daughter ions consists of using a very short deceleration gap just in front of the detector. In this case, the time-of-flight of the ions is hardly affected by the deceleration, but ions of low kinetic energy, i.e. those having a small mass ratio m_d/m_p, are prevented from reaching the detector. Thus, a series of mass spectra recorded with different barrier potentials U_b may be used to analyse the mass of all possible fragment ions; all cluster sizes may be analysed simultaneously /35/. The resolution of this method, $\Delta m/m$, is given by the ratio of the potential spread in the ionizer, taken over the volume where the ions are formed, to the potential drop across the ion extraction lens. For electron impact ionization, the resolution is rather poor, typically a few percent, but for ionization with a well-focussed laser beam much better resolution can be achieved.

A serious short-coming of this energy-barrier technique is that the dominant contribution to the ion intensity always comes from the stable parent ions, which possess the highest kinetic energy. A more appealing technique for time-of-flight mass spectrometry has been introduced recently by Leisner et al. /18/. They simply detect those ions which are reflected off the first barrier potential U_{b1}. If another, slightly lower barrier potential U_{b2} is applied to a grid in front of the detector, one has, in essence, combined a TOF-mass spectrometer with an electrostatic analyzer: Only daughter ions of energy $q\,U_{b2} \leq q\,U_d \leq q\,U_{b1}$ will be detected, and their time-of-arrival will be still (nearly) identical to that of their precursor ions, which are detected if $U_{b2} < U_0 < U_{b1}$. Fig. 5 (top) illustrates this point: The top spectrum has been recorded with $U_{b2} = 0$, $U_o < U_{b1}$. Hence, charged fragments from $(CO_2)_n^+$ contribute to the parent ion mass peaks, as they would do in a conventional mass spectrum. In the bottom spectrum, we have chosen U_{b1} slightly below 0.75 U_o, and U_{b2} slightly above. Hence, only those daughter ions are detected that carry a mass (i.e. kinetic energy) equal to 0.75 of their

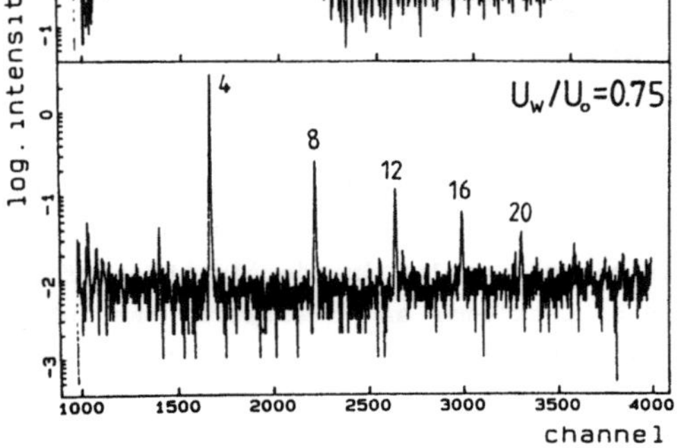

Figure 5. Top: Conventional time-of-flight mass spectrum of $(CO_2)_n^+$ (with some doubly charged clusters showing up beyond $(CO_2)_{22}^+$). Bottom: Energy-filtered spectrum of daughter ions, carrying 75% of the parent mass. Note the logarithmic intensity scale /18/

precursor. The largest cluster that significantly contributes to this metastable spectrum is a daughter from the 24-mer, which has lost 6 monomers in the drift tube.

From these and similar measurements of, say, NH_3 clusters /34/, Xe clusters /30/, Ar and Ne clusters /27/, it appears that the main decay channel of singly charged van der Waals clusters, formed by ionization of pre-existing neutral clusters, is loss of one monomer. Metastable loss of two monomers may also occur, especially if the resulting daughter ion is "magic" /27, 30/, but the main contribution to this and higher channels (i.e. loss of several monomers) arises from collisions with background molecules. The decay pattern of nitrogen clusters, however, is completely different: Michl and coworkers observed that $(N_2)_n^+$, formed by sputtering solid nitrogen, have <u>three</u> major decay channels: i) loss of one N_2, ii) loss of several N_2 molecules, the exact number (2 to 5) depending on parent size (varying from 4 to 27), and iii) total disintegration into N_2^+ or $(N_2)_2^+$ /28/. The sources of energy for these three dissociation channels are likely to be: i) vibrational energy, randomly distributed over the van der Waals modes (normal metastable decay), ii) relaxation of a quantum of vibrational energy from a N_2 molecule, iii) release of an energy quantum stored in an electronically metastable state. Decay channels (i) and (ii) are found to prevail if neutral N_2 clusters are ionized by electron impact /18/. For large clusters, release of <u>several</u> vibrational quanta from N_2 into the lattice occurs "simultaneously" (within the time-of-flight through the drift tube.) Thus, a quasi-periodic pattern appears in the distribution of daughter ions. $(N_2)_{40}^+$, e.g., loses either 1 to 2 or ~6 or ~11 or ~16 or ~20 molecules in the drift tube, depending on how many vibrational quanta have been released /18/.

This effect shows up so nicely in N_2 clusters because the vibrational energy quantum of N_2, 0.29 eV, obviously greatly exceeds the energy required for desorption of N_2 (which is 0.0735 eV for bulk nitrogen). The periodic pattern in the metastable spectrum also reveals that once the initial excess energy has been released into the lattice, it will be converted into evaporations within a time much shorter than the transit time of the ion through the spectrometer. A similar conclusion has been drawn from photofragmentation experiments on charged CO_2 clusters /36/.

3. DISSOCIATION OF MULTIPLY CHARGED CLUSTERS

"In consequence of electrical repulsion, a charged spherical mass of liquid, unacted upon by other forces, is in a condition of unstable equilibrium... When (the charge) is great, the spherical form is unstable... Under these circumstances the liquid is thrown out in fine jets, whose fineness, however, has a limit". These words were written by Lord Rayleigh in his treatise "On the Equilibrium of Liquid Conducting Masses charged with Electricity" /37/; they provide an excellent description of what is currently termed "Coulomb

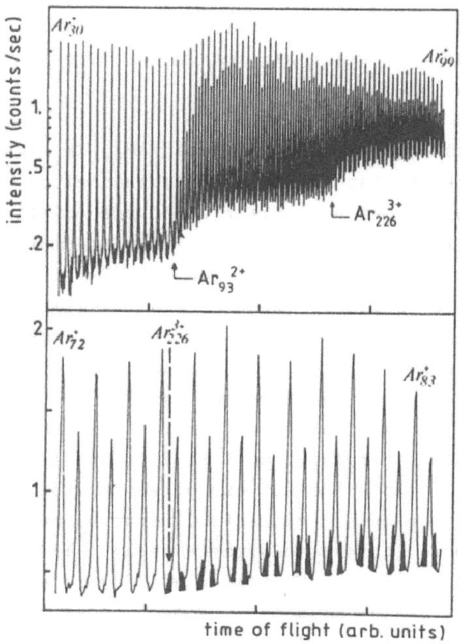

Figure 6. High-resolution spectrum of argon clusters, revealing the critical sizes for doubly and triply charged clusters. A time-of-flight mass spectrometer, equipped with an reflectron, was used for this study /18/

explosion" of microclusters. It is intuitively clear that a small, doubly positively charged cluster will disintegrate into singly charged fragments if the cohesive forces cannot compensate the Coulomb repulsion between the two holes. Thus, one may expect that doubly or, more generally, z-fold charged clusters will undergo rapid fission if their diameter is less than a critical size $n_c(z)$. This is indeed borne out in the experiment: Multiply charged small cluster ions A_n^{z+} are conspicuously absent in mass spectra of van der Waals clusters, but they are observed beyond a certain size limit /38/. Fig. 6, e.g., displays a time-of-flight mass spectrum of argon clusters, born neutral and ionized by electron impact at 400 eV /18/. The smallest doubly charged cluster observed is Ar_{93}^{2+}; triply charged clusters are identified for $n \geq 226$ (see bottom spectrum, which displays part of the top spectrum). These critical sizes are remarkably well defined; Märk and coworkers, e.g., using a double focussing mass spectrometer, have reported $n_c(2) = 91$ a: $n_c(3) = 226$ for argon /27, 39/. An exhaustive review of this subject has been published recently /41/.

An illuminating "microscopic" view of the dissociation dynamics of doubly charged clusters has been obtained by Berne and coworkers /40/ in a molecular dynamics study of Xe_n^{2+}. Fig. 7 shows their results: Neutral Xe_n is first relaxed in this study and then, at time zero, two of the atoms are suddenly charged by vertical ionization. The separation of the charges rapidly increases with time because the Coulomb repulsion distorts the cluster (see "snapshots"). Xe_{51}^{2+} eventually dissociates, but Xe_{55}^{2+} does not on the time scale of this "experiment" (1 reduced time unit = 2.76 ps).

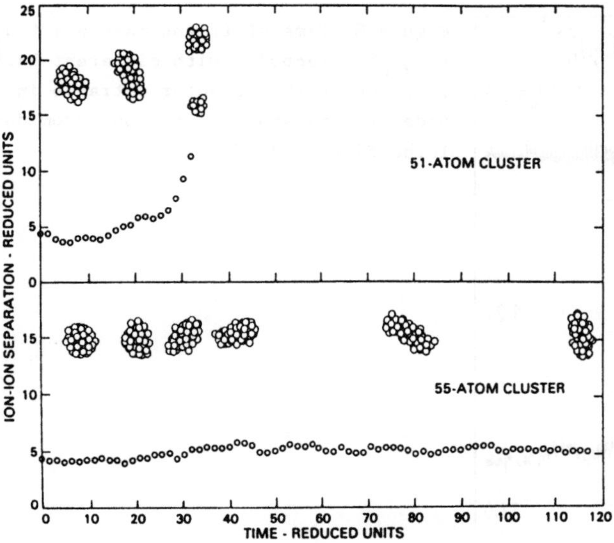

Figure 7. Molecular-dynamics simulation of the fission of doubly charged xenon clusters, assumed to be formed from relaxed Xe_n by vertical ionization at time zero. The time scale extends over 330 ps. Taken from Gay and Berne /40/

Until recently, however, the experiments were restricted to a determination of "critical sizes". Some systematic behaviour has emerged from these studies: The critical size of doubly charged van der Waals-bonded or hydrogen-bonded clusters, e.g., increases linearly with the quantity $(d_c\ T_b)^{-1}$, where d_c is the diameter of the cluster (calculated from n_c, assuming bulk density and a spherical shape), and T_b is the standard boiling temperature of the bulk /41, 42, 27/. Furthermore, the critical size of triply charged clusters, known for 17 different materials, is 2.3 $\pm$ 0.2 times larger than the corresponding $n_c(2)$ /41, 42, 27/. This value agrees well with a theoretical value obtained by Bennemann and coworkers /43/, based on the assumption that multiple charged van der Waals clusters break "symmetrically", i.e. that any charged fission fragment has (approximately) the same size-to-charge ratio, n_d/z_d, as its parent, n_p/z_d.

Direct experimental tests of this assumption have been performed only recently. These studies have exploited the fact that the smallest triply charged clusters which are observable in a conventional mass spectrum, i.e. which have just survived the time required for mass analysis, have large probability to undergo fission into two charged fragments at slightly later times. This means that the randomized (thermal) energy in the cluster is still large enough to drive the cluster over the fission barrier, in agreement with theoretical considerations /40/.

The experimental studies of this delayed fission have employed either a time-of-flight mass spectrometer equipped with an electrostatic barrier in front of the detector or with an electrostatic mirror /18, 41, 44/, or a double focussing sector field mass spectrometer /27, 45/. The principle of the experiment is as described in chapter 2. One has to be aware, however, that

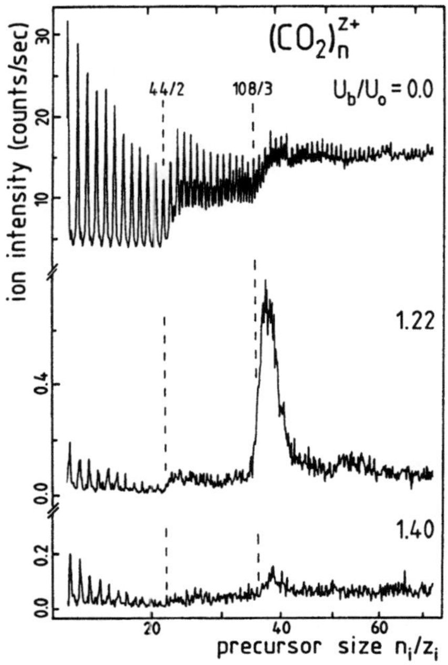

Figure 8. Time-of-flight mass spectra of $(CO_2)_n^{z+}$, recorded with different barrie voltages at the detector entrance in order to analyze product ions from delayed fission /46/

the kinetic energy of a fully accelerated parent ion (size n_p, charge z_p q) : z_p q U_o, and that an electrostatic analyzer actually determines $E_d/q\ z_d$, i.e the ratio of the kinetic energy to the charge of the ion at the instant whe it is analyzed. Hence, in a time-of-flight mass spectrometer with an integra or differential energy analyzer, one obtains, in general, $(n_d/z_d):(n_p/z_p)$. This simplifies to the ratio n_d/n_p, used in chapter 2, if only singly charge ions occur.

As an example, Fig. 8 shows three mass spectra of $(CO_2)_n^{2+}$ (z = 1,2,3), recorded with different potentials applied to the barrier grid in front of t detector. In the upper spectrum, the barrier potential is zero; one observes doubly charged clusters beyond $n_c(2) = 44$ and triply charged clusters beyond $n_c(3) = 108$. The middle spectrum in Fig. 8 reveals the high yield of fragmen ions, resulting from fission of triply charged CO_2 clusters of size n, just above 108. This spectrum has been recorded by applying a repulsive potential $U_b = 1.22\ U_o$ to the barrier grid. Therefore, only large fragment ions of reduced charge $z_d < z_p$ can be detected. The criterion for detectability of fragments is $(n_d/z_d):(n_p/z_p) \geq U_b:U_o$. Thus, from a series of spectra as show in Fig. 8, middle and bottom, the distribution of n_d/n_p is obtained if z_d i determined from high resolution fragment spectra /45/.

We find that triply charged carbon dioxide clusters preferentially decay into doubly charged clusters which carry ~92 % of the initial mass; their size distribution is surprisingly narrow, see Fig. 9. Similar results were obtained for ammonia, whereas ethene clusters /46/ and nitrogen clusters /18/

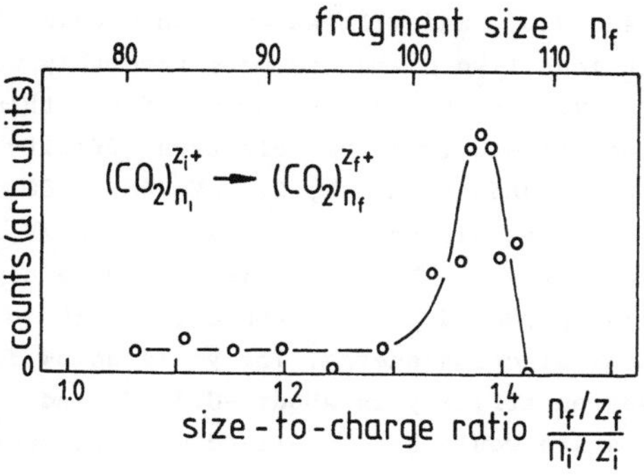

Figure 9. Size distribution of doubly charged fragments, originating from triply charged CO_2 clusters of size $108 \lesssim n_i \lesssim 120$ /44/

tend to fission less asymmetrically /18/. In contrast, doubly charged clusters do not undergo delayed fission. They only decay by evaporation of neutral monomers. The corresponding decrease in size should be accompanied by a reduction in the height of the fission barrier, but the concomitant decrease in vibrational excitation obviously more than compensates for this, the net effect of delayed evaporations being a reduced tendency for fission.

The recently discovered absence of delayed fission in triply charged clusters of Ar and Kr /18, 27/ is possibly related to the fact that these clusters have no means to store vibrational energy in intramolecular modes (other than in Rg_2^+). In cases of singly charged nitrogen clusters (see chapter 2) we noted that the energy in these intramolecular modes did provide enough energy for the instantaneous loss of several N_2 molecules; intra- and intermolecular vibrations were not in thermal equilibrium.

The critical sizes and the preferred fission channels of triply charged clusters can be understood in the framework of a simple continuum approximation, which allows to estimate the fission barrier as a function of the size and charge state of the parent ion and daughter ion, respectively. All large clusters favor asymmetric fission, but small ones do not. This can be understood by realizing that in order to minimize the increase in surface energy upon fission, one of the fragments should be very small. On the other hand, the emission of a charged monomeric fragment would be energetically unfavorable, because the solvation energy for this ion would be lost. These two competing processes favor an extremely asymmetric fission for large clusters, but a symmetric one for clusters of size ~30 or smaller /47/.

4. ELECTRON ATTACHMENT TO VAN DER WAALS CLUSTERS

Electron attachment to small molecules like CO_2 or N_2O does not result in long-lived anions with the corresponding composition, because i) their adiabatic electron affinities are very small (N_2O) or even negative (CO_2), whereas ii) their vertical electron affinities are strongly negative. The transient anions formed by resonant electron capture are, therefore, highly excited (by approximately 3 eV in both CO_2 and N_2O), and will autodetach within ~10^{-14} s. Dissociation into stable product ions like O^- may compete favorably with autodetachment only if the electron exceeds the energy threshold for this process. This condition is fulfilled for N_2O (vertical electron affinity -2.2 eV, the reaction $N_2O + e \rightarrow N_2 + O^-$ is endothermic by 0.2 eV /48/). CO_2 constitutes a different case: Its vertical electron affinity is -3.8 eV, but the heat of the endothermic reaction $CO_2 + e \rightarrow CO + O^-$ is even more negative by 0.2 eV. Thus, the only anion being produced by electron attachment to CO_2 is O^-, with a maximum yield at 4.4 eV electron energy. The yield below 3.8 eV is zero, although the adiabatic electron affinity is about -0.6 eV, and although CO_2^- in its vibronic ground state would be long-lived with respect to autodetachment /48/.

This picture changes drastically if electrons are attached to CO_2 clusters. We observe ions with the composition $(CO_2)_n^-$, with n being as small as 1, because i) rapid intermolecular fragmentation may quench autodetachment, ii) the electron energy need not exceed the vertical electron affinity of the parent cluster because, in contrast to O^- production, no highly endothermic reaction occurs, and iii) the increase in the number of degrees of freedom helps to dissipate the excess energy (difference between the electron energy and the adiabatic electron affinity). Consequently, CO_2^- is observable if electrons are attached to clusters rather than to monomers /49/. Its yield (Fig. 10 top, left spectrum) exhibits a resonance at 3.4 eV (corrected electron energy), well below the vertical electron affinity of the monomer, be-

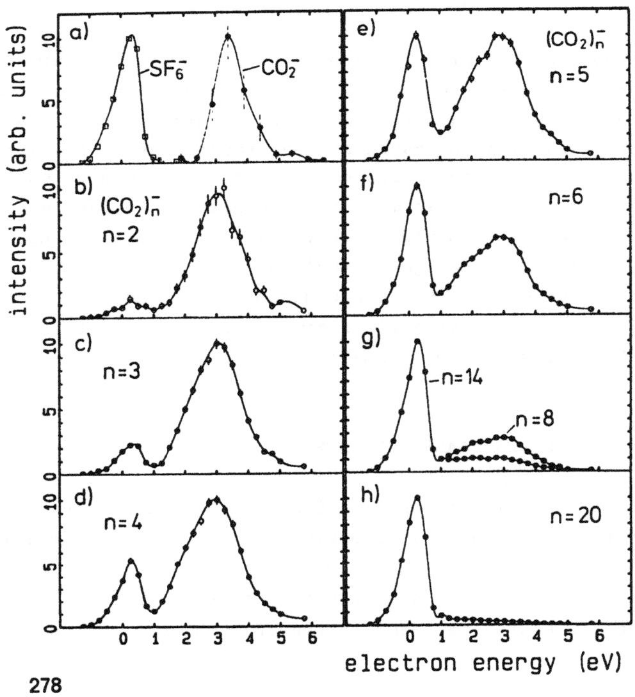

Figure 10. Yield of $(CO_2)_n^-$, with n ranging from 1 to 20, versus electron energy. The SF_6^- yield (top left trace) indicates the position of a "true" 0 eV resonance, with the shape of the resonance being controlled by the energy distribution of the electrons

cause the precursor of CO_2^- is a cluster (whose size, unfortunately, cannot be determined unambiguously).

We are thus witnessing the change of the vertical electron affinity with increasing cluster size or, in other words, the electronic solvation shift of CO_2^- in a cluster: The energy difference between $(CO_2)_n$ (having relaxed inter- and intramolecular coordinates) + e and $(CO_2)_n^-$ (geometry of the neutral precursor) decreases due to the rapid response of the electrons of the individual, linear CO_2 molecules to the incoming excess electron. (The information obtained from photoelectron spectroscopy /50/ is complementary to this quantity, because it reflects the energy difference between the geometrically relaxed cluster anion and a neutral cluster of the same geometry). Accordingly, the resonance in the $(CO_2)_n^-$ yield is increasingly red-shifted with increasing size n. The concurrent increase in the width of the resonance is possibly related to the fact that different precursor sizes contribute to the same cluster ion size.

The most remarkable feature in Fig. 10, however, is the appearance of an additional resonance for electrons of low energy. The existence of this new resonance, not known from experiments on CO_2 monomers, was established first by Mark and coworkers for the trimer and larger cluster anions /51/. Fig. 10 suggests that this resonance already occurs in the dimer ion. Anyhow, this new feature becomes dominant in the yield curves of clusters containing more than ~5 molecules. Absolute cross sections cannot be determined, because the partial densities of the neutral clusters in the beam are, of course, unknown. However, by comparing with positively charged clusters produced by electron impact with identical parameters being used for the cluster source, we find that attachment of low-energy electrons to large CO_2 clusters is at least as efficient as electron impact ionization at 30 eV.

The exact position of the low-energy resonance has to be determined by comparison with the yield curve of other well-known anions. The top left spectrum in Fig. 10 displays the yield curve of SF_6^-, obtained from attachment to SF_6, which is known to have an extremely large cross section for attachment of electrons at energies $\lesssim$ 20 meV. Thus, a comparison of the yield curves in Fig. 10 reveals that the $(CO_2)_n^-$ resonance occurs very close to zero eV, with an experimental uncertainty of 0.1 eV. The observation of $(CO_2)_n^-$, $n \geq 2$, at ~0 eV electron energy implies that the adiabatic electron affinities of neutral CO_2 clusters as small as the dimer is non-negative, although that of the monomer is -0.6 eV /48/. This conclusion is in agreement with a calculation of the energetics of $(CO_2)_2^-$ /52/, but at variance with a more recent theoretical investigation /53/.

What is the nature of this resonance at low energy? The strongly negative vertical electron affinity of CO_2 makes it unlikely that the incoming electron directly couples with an individual molecule in the cluster. Antoniewicz et al. /54/ have pointed out that a polarizable cluster will always provide a long range attractive potential for an electron outside of the cluster. The binding energy for these surface electrons will depend on the size and the dielectric constant of the cluster. Even in the extreme case of helium, a bound surface state for electrons is known to exist at 0.7 meV below the vacuum level, at least for planar He films, corresponding to infinite cluste size /55/. The transfer of the excess energy of the incoming electron to the cluster, which is required for electron attachment, may be accomplished by excitation of vibrational van der Waals modes, in analogy to the electron-ripplon coupling in helium films.

Another interesting system featuring a resonance for attachment of electrons at ~0 eV is water. The behaviour of low-energy excess electrons in liquid water has been subject to numerous studies (see, e.g., references cit in /56/). The adiabatic electron affinity of H_2O is extremely close to zero, whereas the energy of a solvated electron in liquid bulk water is approximat ly 1.5 eV below the vacuum level. This implies that sufficiently large water clusters should strongly bind an excess electron. In spite of this, several attempts to generate $(H_2O)_n^-$ by attachment of electrons to free water cluste failed /57/. Haberland and coworkers finally succeeded in producing $(H_2O)_n^-$, ≥ 11, and $(D_2O)_n^-$, $n \geq 12$, by injecting electrons into the high-density regi of adiabatically expanding water vapor /58/. Their findings suggested that hot, thermally disordered clusters may temporarily capture electrons, with subsequent relaxation, stabilization and cluster growth being guaranteed by frequent collisions, while relatively cold, structurally relaxed clusters would repel electrons (i.e. their vertical electron affinity being strongly negative, considerable molecular reorientation is necessary to accommodat the excess electron). These ideas were supported by theoretical studies of neutral and charged water clusters /59/. Also, they were in line with estimates, according to which the concentration of preexisting deep traps for excess electrons in polar liquids is exceedingly small /60/; these estimates, though, have been questioned recently /61/.

Recent experiments in our laboratory, however, have revealed that $(H_2O)_n^-$, $n \geq 11$, can be formed very efficiently by electron attachment to free cluste in a molecular beam if, and only if, the energy of the electrons is close to zero /56/. A typical mass spectrum is displayed in Fig. 11 (middle) for deuterated water. It is devoid of small water cluster ions; but the intensity of $(D_2O)_n^-$ rises steeply above $n = 11$ and levels off at $n \sim 20$, depending slightly on the expansion conditions. These distributions are essentially identical to those obtained by injection of electrons into a water jet, if neat water is expanded in both experiments.

Thus, the efficient capture of electrons at free water clusters may indeed be related to the above-mentioned surface states /54/, which may at least provide an initial, weakly bound state until accommodation of the electron into a more tightly bound state is made possible by reorientation of the individual molecules. From quantum path integral molecular dynamics simula-

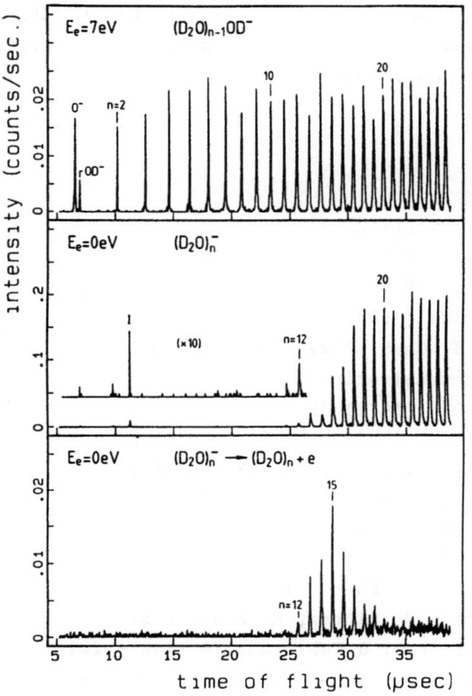

Figure 11. Mass spectra of deuterated water cluster anions, obtained by electron attachment to pre-existing clusters. Top spectrum: Ions $(D_2O)_{n-1}OD^-$, n $\geq$ 1, are formed at 7 eV electron energy. Middle spectrum: At ~0 eV electron energy, we observe $(D_2O)_n^-$, n $\geq$ 11. The bottom spectrum displays the intensity distribution of neutrals, arising from cluster anions in the drift tube by autodetachment (distribution of parent ions as shown in the middle spectrum)

tions it has indeed been concluded that the energetically most favorable structure of $(H_2O)_n^-$ in this size range corresponds to an electron being trapped at the surface of a distorted cluster, rather than in the cluster /62/.

The bottom spectrum in Fig. 11 displays the intensity of neutrals, being produced by unimolecular reactions of $(D_2O)_n^-$ in the drift tube of the time-of-flight mass spectrometer. These neutrals are mainly produced by autodetachment and, to a lesser extent, by "evaporation" of neutral monomers /56/. In general, cluster ions may stabilize against autodetachment by evaporation of neutrals, if the corresponding activation energy is less than the electron affinity. The electron affinity of the water dimer, however, is only about 0.01 eV; it appears to increase rather slowly with cluster size /63/. The heat of sublimation for bulk water, however, is 0.47 eV; the internal temperature of the neutral clusters, before electron attachment, will be correspondingly high. Therefore, small water clusters have no means to stabilize against autodetachment; we may tentatively identify the cluster size $15 \lesssim n \lesssim 20$ as defining the region where the electron affinity of water clusters eventually becomes larger than their heat of sublimation.

ACKNOWLEDGEMENT

The author is grateful to his co-workers T. Leisner, O. Kandler. M. Knapp, D. Kreisle, W. Miehle and E. Recknagel for permission to present previously unpublished data. Close collaboration with T.D. Märk (Innsbruck), and J.J. Saenz and J.M. Soler (Madrid) is also appreciated. This work was financially supported by the Deutsche Forschungsgemeinschaft.

REFERENCES

1. A.L. Mackay, Acta Cristallogr. 15 (1962) 916
2. M.R. Hoare and J.A. McInnes, Adv. Phys. 32 (1983) 791
3. M.R. Hoare, Adv. Chem. Phys. 40 (1979) 49
4. L.T. Wille, Chem. Phys. Lett. 133 (1987) 405
5. G.L. Griffin and R.P. Andres, J. Chem. Phys. 71 (1979) 2522
6. E. Blaisten-Barojas, private communication
7. M.R. Hoare and P. Pal, Nature 236 (1972) 35
8. J.A. Northby, preprint
9. J.G. Allpress and J.V. Sanders, Aust. J. Phys. 23 (1970) 23
10. J.W. Lee and G.D. Stein, J. Phys. Chem. 91 (1987) 2450
11. J. Farges, M.F. de Feraudy, B. Raoult and G. Torchet, J. Chem. Phys. 78 (1983) 5067; ibid 84 (1986) 3491
12. T.A. Milne and F.T. Greene, J. Chem. Phys. 47 (1967) 4095
13. T.D. Märk, P. Scheier, K. Leiter, W. Ritter, K. Stephan and A. Stamatovic, Int. J. Mass Spectrom. & Ion Proc. 74 (1986) 281
14. O. Echt, K. Sattler and E. Recknagel, Phys. Rev. Lett. 47 (1981) 1121; O. Echt, A. Reyes Flotte, M. Knapp, K. Sattler and E. Recknagel, Ber. Bunsenges. Phys. Chem. 86 (1982) 860
15. H.P. Birkhofer, H. Haberland, M. Winterer and D.R. Worsnop, Ber. Bunsenges. Phys. Chem. 88 (1984) 207
16. O. Echt, M.C. Cook and A.W. Castleman, Chem. Phys. Lett. 135 (1987) 229
17. J. Farges, M.F. Feraudy, B. Raoult and G. Torchet, Surf. Sci. 156 (1985) 370
18. T. Leisner, O. Echt, O. Kandler, W. Miehle and E. Recknagel, unpublished results
19. C.E. Klots, Z. Phys. D5 (1987); and references therein
20. E.E. Polymeropoulos and J. Brickmann, Chem. Phys. Lett. 92 (1982) 59; ibid, 96 (1983) 273
21. H. Haberland, Surf. Sci. 156 (1985) 305
22. A.A. Radzig and B.M. Smirnov, "Reference Data on Atoms, Molecules and Ions", Springer Series in Chem. Phys. 31, 1985
23. H.U. Böhmer and S. Peyerimhoff, Z. Phys. D4 (1987) 195
24. K. Stephan and T.D. Märk, Chem. Phys. Lett. 90 (1982) 51; T.E. Gough and R.E. Miller, Chem. Phys. Lett. 87 (1982) 280; J. Geraedts, S. Stolte and J. Reuss, Z. Phys. A 304 (1982) 167
25. U. Buck and H. Meyer, J. Chem. Phys. 84 (1986) 4854
26. J.M. Soler, J.J. Saenz, N. Garcia and O. Echt, Chem. Phys. Lett. 109 (1984) 71

27. T.D. Märk, Int. J. Mass Spectrom & Ion Proc., in press, and references therein
28. T.F. Magnera, D.E. David and J. Michl, Chem. Phys. Lett. 123 (1986) 327
29. A.J. Stace, J. Chem. Phys. 85 (1986) 5774
30. D. Kreisle, O. Echt, M. Knapp and E. Recknagel, Phys.Rev. A33 (1986) 768
31. I.A. Harris, K.A. Norman, R.V. Mulkern and J.A. Northby, Chem. Phys. Lett. 130 (1986) 316; Phys. Rev. Lett. 53 (1984) 2390
32. J.J. Saenz, J.M. Soler and N. Garcia, Chem. Phys. Lett. 114 (1985) 15
33. W. Ens, R. Beavis and K.G. Standing, Phys. Rev. Lett. 50 (1983) 27
34. O. Echt, P.D. Dao, S. Morgan and A.W. Castleman, Jr., J. Chem. Phys. 82 (1985) 4076
35. O. Echt, D. Kreisle, M. Knapp and E. Recknagel, Chem. Phys. Lett. 108 (1984) 401
36. M.L. Alexander, M.A. Johnson and W.C. Lineberger, J. Chem. Phys. 82 (1985) 5288; Phys. Rev. Lett. 57 (1986) 976; P.C. Engelking, J. Chem. Phys. 85 (1986) 3103
37. Lord Rayleigh, Phil. Mag. 14 (1882) 185
38. K. Sattler, J. Mühlbach, O. Echt, P. Pfau and E. Recknagel, Phys. Rev. Lett. 42 (1981) 160; O. Echt, K. Sattler and E. Recknagel, Phys. Lett. 90A (1982) 185
39. P. Scheier and T.D. Märk, Chem. Phys. Lett. 136 (1987) 423
40. J.G. Gay and B.J. Berne, Phys. Rev. Lett. 49 (1982) 194
41. O. Echt in "15th Int. Conf. Physics of Electronic and Atomic Collisions", Brighton, 22 - 28 July 1987, book of invited papers, North Holland, in press
42. O. Echt, in "Physics and Chemistry of Small Clusters", NATO ASI Series (P. Jena, B.K. Rao and S.N. Khanna, eds.), 1987, p. 623
43. D. Tomanek, S. Mukerjee and K.H. Bennemann, Phys. Rev. B28 (1983) 665
44. D. Kreisle, O. Echt, M. Knapp, E. Recknagel, K. Leiter, T.D. Märk, J.J. Saenz and J.M. Soler, Phys. Rev. Lett. 56 (1986) 1551
45. D. Kreisle, K. Leiter, O. Echt and T.D. Märk, Z. Phys. D3 (1986) 319; K. Leiter, D. Kreisle, O. Echt and T.D. Märk, J. Phys. Chem. 91 (1987) 2583
46. D. Kreisle, Ph.D. Thesis, Universität Konstanz, 1986
47. R. Casero, J.J. Saenz and J.M. Soler, unpublished results
48. References to thermodynamic data of CO_2, N_2O and their anions are to be found in ref. 49.
49. M. Knapp, O. Echt, D. Kreisle, T.D. Märk and E. Recknagel, Chem. Phys. Lett. 126 (1986) 225
50. J.V. Coe, J.T. Snodgrass, C.B. Freidhoff, K.M. McHugh and K.H. Bowen, Chem. Phys. Lett. 124 (1986) 274
51. A. Stamatovic, K. Leiter, W. Ritter, K. Stephan and T.D. Märk, J. Chem. Phys. 83 (1985) 2942
52. Y. Yoshioka and K.D. Jordan, J. Am. Chem. Soc. 102 (1980) 2621
53. S.H. Fleischman and K.D. Jordan, J. Phys. Chem. 91 (1987) 1300
54. P.R. Antoniewicz, G.T. Bennett and J.C. Thompson, J. Chem. Phys. 77 (1982) 4573

55. A.J. Dahm and W.F. Vinen, Physics Today, February 1987, p. 43; and references therein
56. M. Knapp, O. Echt, D. Kreisle and E. Recknagel, J. Phys. Chem. 91 (1987) 2601
57. C.E. Klots, Radiat. Phys. Chem. 20 (1982) 51; R.N. Compton, in "Electronic and Atomic Collisions", edited by N. Oda and K. Takayanagi (North-Holland, Amsterdam, 1980), p. 251; D.R. Herschbach, in a private communication to J. Jortner
58. H. Haberland, H. Langosch, H.-G. Schindler and D.R. Worsnop, J. Phys. Chem. 88 (1984) 3903
59. N.R. Kestner and J. Jortner, J. Phys. Chem. 88 (1984) 3818; B.K. Rao and N.R. Kestner, J. Chem. Phys. 80 (1984) 1587
60. M. Tachiya and A. Mozumder, J. Chem. Phys. 61 (1974) 3890
61. J. Schnitker and P.J. Rossky, J. Chem. Phys. 86 (1987) 3471; J. Schnitker, P.J. Rossky and G.A. Kenney-Wallace, J. Chem. Phys. 85 (1986) 2986; M. Hilczer, W.M. Bartczak and M. Sopek, J. Chem. Phys. 85 (1986) 6813
62. R. N. Barnett, U. Landman, C.L. Cleveland and J. Jortner, submitted for publication
63. A. Wallqvist, D. Thirumalai and B.J. Berne, J. Chem. Phys. 83 (1986) 1583; U. Landman, J. Jortner and D. Scharf, presented at the Int.Symp. Phys. Chem. of Small Clusters, Richmond, Oc 30 - Nov 3, 1986

Part V

Chemistry

The Electronic Structure of Clusters

P.S. Bagus

IBM Research, Almaden Research Center, 650 Harry Road, San Jose, CA 95120-6099, USA

1. INTRODUCTION

This paper deals with the determination of the electronic wavefunctions of clusters or, in other words, the theoretical study of cluster electronic structure. In principle, all properties of a system can be obtained from its quantum mechanical wavefunction, often as expectation values of appropriate operators. However, while these properties can be measured, the wavefunction itself cannot be determined from experimental observations. It is difficult and requires many assumptions, which may or may not turn out to be valid, to infer characteristics of the wavefunction from the observed properties. Thus, it is essential to obtain wavefunctions theoretically in order to understand the full and correct physical significance of most observables in terms of the nature of the interactions between the atoms of which a cluster is composed.

The theoretical approach used in this paper is the ab initio molecular orbital, MO, approach. It has had very substantial successes for determining the properties of atoms and molecules /1/. When ab initio MO methods are used for clusters, the clusters are treated simply as large molecules. Although the largest number of the applications of MO theory have been to small systems /1/, e.g. atomic dimers and trimers, the computational power of supercomputers has made rigorous ab initio calculations on large systems possible. For example, LÜTHI and ALMLÖF /2/ have recently reported self-consistent field, SCF, wavefunctions and properties for a Buckminster Fullerene cluster /3/, C_{60}, and for even larger systems. Thus, it is possible to obtain good quality electronic wavefunctions for quite large systems. Further, the availability and power of very fast computers is increasing.

Cluster wavefunctions have been used to describe the bulk and surface regions of crystals as well as to describe the properties of molecular clusters. For descriptions of representative MO cluster model studies of surface properties see Refs. 4 to 9. The MO cluster approach has been used to study both the surfaces of metals /4,5,7,8/ and the surfaces of covalently bonded semiconductors /5,6,9/. Cluster models have been used rather extensively to describe the bulk properties of ionic materials; for oxide insulators see, for example, Refs. 10 and 11. Examples of the use of ab initio MO wavefunctions for molecular clusters will be given in several papers in these proceedings in addition to this one (those by P. Fantucci, by J. Koutecky and P. Fantucci, and by G. Pacchioni).

There is a great deal in common among these three classes of applications. In particular, in all three the concern is to derive or deduce common types of information from the cluster wavefunctions. One wishes to understand the nature of the interaction, the chemical bond, between atoms or units in the cluster. This bonding may be between the component units of a "pure" cluster or it may be between the cluster and an atomic or molecular adsorbate on the surface of a cluster or of a crystal. Following are three examples of the kinds of questions about these interactions that MO cluster wavefunctions may be used to resolve. One, the extent of the hybridization of electrons from their orbital angular momentum in the free atom to other angular momentum. For example, the valence s electrons of alkaline earth atoms are substantially hybridized with p character in clusters /12/. Another kind of hybridization in clusters is the mixing of valence level s and p character into the d shells of transition metals. Two, the nature of the bonding of adsorbed CO to a metal surface or cluster and the respective importance of σ donation and π back-donation /7,13,14/. And three, the characterization of an adsorbate bond as being covalent or ionic /6, 15-17/. Having established the nature of the bond, one wishes to understand the consequences of the bonding for a range of observed properties including, certainly, photoemission spectra /18/, vibrational frequencies /16,19/, and geometrical parameters /6,17/. Examples of the analysis of MO cluster wavefunction to answer questions for each of the areas mentioned above will be given in this paper. Thus, the MO wavefunctions appropriate to clusters, surfaces, and bulk can be used to resolve similar questions; it is also very likely that there will be similar qualitative answers to these questions for all three cases. This is expected to be particularly true for the adsorbate-cluster interaction. The details of the adsorbate-cluster geometry and bond strength may depend strongly on the cluster size and shape, see the paper by D. E. Cox in these proceedings. However, the general character and features of this interaction are almost certain to be quite similar to that for the interaction of the adsorbate with a crystal surface. For this reason, a part of this paper will deal with the properties of selected atomic and molecular adsorbates on surfaces.

Before going on, it is important to stress the philosophy of the approach which my collaborators and I have used in our theoretical studies of clusters. Our primary objective is to use the results obtained from an analysis of ab initio wavefunctions to gain physical insight into the meaning and significance of observed properties. The object of our cluster studies is not to exactly reproduce experimental results. The object is to identify and to characterize the physical mechanisms which lead to the observed results. For several reasons, the ab initio MO cluster method is very well suited for this.

(1) It does not use parameters adjusted to experiment /1,20/. There are many parameters involved in the basis sets used to describe the orbitals and in the pseudo-potentials, called effective-core potentials (ECP's), which are often used to treat the core electrons /21-23/. The basis set parameters are chosen from criteria related to the variational energy of the system and its component units. The ECP parameters are normally chosen to make the properties of valence orbitals of the isolated atom, treated with a pseudopotential, as close as possible to those determined from all-electron atomic wavefunctions. In both cases, these criteria use only comparisons among calculated wavefunctions, primarily atomic wavefunctions. Further, it is extremely difficult, if not impossible, for an investigator to allow his choice of parameters to be influenced by his initial beliefs about the character of the interactions in the system.

(2) Electron correlation effects can be included in a controlled fashion, in particular with multi-configuration self-consistent field, MCSCF, wavefunctions but also with more general configuration interaction, CI, wavefunctions. In principle, exact cluster wavefunctions may be obtained by using sufficiently large CI expansions. Much more important, however, it is possible to identify and treat individually particular electron correlation effects; in this way, the importance of a specific correlation effect can be determined. For the metal-carbonyl bond , a subject to be discussed later in this paper, a major correlation effect is to increase the metal $d\pi$ to CO $2\pi^*$ covalent dative bonding /14,24/. While this correlation affects the precise values of the bond energy and distance, it does not change the general form of the bond from the SCF description /7,25/. Thus, MO theory does not use correlation as a magic black box that provides the correct answer. It is possible to relate electron correlation to specific kinds of physical behavior and mechanisms.

And finally (3), new techniques for the analysis of the electronic wavefunctions make it possible to quantify the importance of various conceptual models used to interpret experimental data. In order to evaluate the contribution of different types of charge rearrangements to the binding energy and other properties of an interaction, the constrained space orbital variation, CSOV, method /13/, has been developed. The CSOV method makes it possible to separate the various kinds of intra-unit and inter-unit charge redistributions involved in a chemical interaction and to determine the importance of each.

The CSOV has been used to examine the nature of metal-CO bonding and has led to a surprising reinterpretation of the importance of σ donation and π back-donation and of the significance of observed vibrational /19/ and photoemission /18/ shifts. An approach, somewhat complementary to the CSOV method, involves the projections of the orbitals of a component onto the wavefunction of the total cluster. The expectation value of the projection operator for a given orbital is a measure of the extent to which this orbital is occupied in the cluster. Projection has shown /16,17/ that CN is ionic, CN^-, when it interacts with a metal atom or surface; on the other hand, the CN interaction with an organic unit is covalent.

The outline of this paper is as follows. In the next section, a summary of some of the key aspects of MO theory is reviewed; particular emphasis is placed on those aspects related to points discussed later in connection with the applications considered. In this section, the CSOV and projection approaches to the analysis of the cluster wavefunctions will be reviewed. In Sec. 3, some features and consequences of cluster bonding will be presented. In particular, the relationship among hybridization, coordination, and cluster stability will be examined. In Sec. 4, the ionization potentials, IP's, of the core and valence level electrons of clusters will be examined. For the core level IP's, it will be shown that two initial state effects, coordination of the ionized atom and 3d → 4sp hybridization, have important consequences for the shifts of the IP's from their values for either isolated atoms or bulk materials /26/. For the valence level IP's, the results provide a test of the energy distributions predicted by the shell model /27/. They strongly suggest that the pattern of cluster sizes with special stability /27/ may be connected with the particular geometric structure of the atoms of the cluster rather than with the energy levels of a simple potential well. In Sec. 5, the interaction and bonding of clusters with molecular adsorbates will be examined and the features of ionic and covalent bonds will be distinguished. The specific examples treated are for adsorbate interactions with surfaces, but the same features identified for this case should also be important for the adsorbate-cluster interaction. Finally, in Sec. 6, some conclusions about the value of theoretical studies of the electronic structure of clusters will be summarized.

2. SUMMARY OF MOLECULAR ORBITAL THEORY

2.1 General Considerations

It is standard practice /1,20/ to use a Born-Oppenheimer Hamiltonian to obtain electronic wavefunctions $\Psi(x_1, ...x_n; X_1, ...X_M)$ where x_i are the space and spin coordinates of the n electrons and X_I are the space coordinates of the M fixed nuclei. The many electron wavefunction Ψ is built from spin orbitals ϕ which are written as products of space and spin; $\phi = \varphi\alpha$ or $\phi = \varphi\beta$, where $\varphi(x,y,z)$ is the spatial part of the orbital. In addition, space and spin symmetry and equivalence restrictions /28/ are usually applied to the orbitals and the orbitals form an orthonormal set. This is true for all the wavefunctions described in this paper. In virtually all cases, the orbital, φ, is expanded in terms of basis functions χ,

$$\varphi_i = \Sigma_p c_{ip} \chi_p,$$

where the c_{ip} are the expansion coefficients. The χ_p are factored into the radial and angular parts,

$$\chi = f(r) S_{lm}(\theta, \phi),$$

where the S_{lm} are real or complex spherical harmonics and the radial function has an exponential character. In this paper, and in most work on systems larger than diatomic molecules /1,20/, the radial function has a Gaussian form,

$$f(r) \propto r^n \mathbf{exp}(-\alpha r^2),$$

centered, normally, on one or another of the atomic nuclei; see Ref. 29 for citations to compilations of basis sets used for molecular calculations. These orbitals are the elements of determinants which are many-electron products and are anti-symmetric for the exchange of the coordinates of any two electrons. A configuration state function, ψ, which is an eigenfunction of spin and transforms according to one of the irreducible representations of the point group symmetry of the nuclear centers /30/, is a linear combination of these determinants. It represents a particular distribution of the electrons into shells belonging to the irreducible representations.

The many-electronic wavefunction, Ψ, is written as a linear combination of the configuration state functions,

$$\Psi = \Sigma C_i \psi_i. \tag{1}$$

The self-consistent field, SCF, wavefunctions is a single configuration, $\Psi = \psi$. the optimum orbitals, φ_i, used to construct ψ are determined by solving equations, the Fock equations, which are given by application of the variational principle to the expectation value of the energy of ψ /31/. These equations include the exact, non-local, exchange operator but do not, by definition /32/, include electron correlation effects. In effect, in the SCF approximation each electron moves in the average field of the other electrons.

Electron correlation is introduced in a controlled but limited fashion with multi-configuration SCF, MCSCF, wavefunctions. An important use of MCSCF wavefunctions is to treat near degeneracy effects. For example, the ns and np levels of alkaline earth atoms are close in energy and are often considered as degenerate /33/. A wavefunction for the ^1S state /33/ which includes the mixing of the ns and np levels, $\Psi = C_1(.....ns^2) + C_2(.....np^2)$, can give a much better representation than one configuration SCF $\Psi = (.....2s^2)$. The complete active space SCF /34/, CASSCF, approach to MCSCF wavefunctions is particularly well suited for treating the near degeneracy problem. The CASSCF wavefunction is a linear combination of the configurations that can be formed by distributing the electrons to be correlated over the correlating orbitals in all symmetry allowed ways. The MCSCF wavefunction for the treatment of near degeneracy effects is normally a linear combination of between 2 and ~100 configuration state functions. In this case, the variational principle is used to obtain optimum mixing coefficients, C_i of Eq. (1) as well as the optimum orbitals used to construct the ψ_i. As with the SCF procedure, the exact, non-local, exchange operator is used in the equations for the optimum orbitals /34/.

Finally, the detailed treatment of electron correlation required to obtain highly accurate theoretical values for binding energies, excitation and ionization energies, and bond lengths is obtained with general CI wavefunctions. In the CI method, the orbitals used to construct the ψ_i in Eq. (1) are taken from SCF or MCSCF wavefunctions and are normally fixed; optimum values of only the linear mixing coefficients, C_i, are obtained. New methods have been developed /20,35/ to solve the variational equations for the C_i and very large expansions of Eq. (1), $\sim 10^6$ terms, can be handled. Although a CI wavefunction is required for exact numerical values of many properties, the same general features of the chemical bond are often found with SCF, MCSCF, and CI wavefunctions; see for example Refs. 7,12,24 and 25. The computation and the analysis of SCF and MCSCF wavefunctions are far easier than for CI wavefunctions. Thus, extensive use of SCF wavefunctions will be made in the examples presented later in this paper. In fact, unless the specific interest centers about bond energies or distances, only SCF wavefunctions will normally be used.

The approximate treatment of the core electrons of the atoms which compose a molecule or cluster substantially reduces the computational resources required to determine the electronic wavefunction. With such approximate treatments, it is possible to treat systems with much larger numbers of atoms than if all the electrons are treated on an equal footing. The most common approximation for the core electrons is a pseudopotential which includes the effect of the core electrons in a one-electron operator; only the valence electrons are explicitly retained in the wavefunction. The essence of these so called effective-core potentials, ECP's, is the assumption that the core electrons can be frozen because they do not participate in the chemical bond between the atoms. For these frozen cores, the Coulomb, exchange, and core-valence orthogonality are approximated by the ECP. Several forms for the ECP have been developed and representative examples of these forms are given in Refs. 21 to 23. In these ECP's /21-23/, the pseudopotential parameters are determined from the properties of the isolated atom. Taking a more rigorous approach, WHITTEN and his collaborators /36/, have explicitly treated the multicenter character of the Coulomb and exchange operators. In other extensions of the ECP formalism, the polarization of the cores away from their spherical free atom form has been considered /37/.

The number of electrons included in the ECP as core electrons is an important parameter. The Cu atom is used as an example to discuss three possible choices of core electrons: (i) 10 Ne core electrons $1s^22s^22p^6$; (ii) 18 Ar core electrons ... $3s^23p^6$; and (iii) 28 Ar core plus the $3d^{10}$ electrons. For these different choices, the number of electrons explicitly included in the wavefunctions for the 29 electron Cu atom are 19 electrons for the Ne core ECP, 11 electrons for the Ar core ECP, and 1 electron for the Ar + $3d^{10}$ core ECP. The computational requirements for different steps in the calculation of an ab initio MO electronic wavefunction scale as powers of the number of basis functions /38/ and this is directly related to the number of electrons. This scaling ranges between N^2 and N^5. Thus, the smaller the number of electrons explicitly included in the wavefunctions, the greater the saving due to the use of the ECP. However, there is a strong possibility of artifacts arising from ECP's which include too many electrons in the core /39/. For transition metal atoms, if the d electrons are included in the ECP, choice (iii) above, their participation in the covalent bonding is not allowed. BAGUS et al. /39/ have shown that this leads to incorrect results for the metal-CO interaction. They have also pointed out that there may be serious errors even if only the Ar core ECP, choice (ii) above, is used. This occurred for the CuO molecule and may be related to the fact that the spatial extent of the 3s and 3p shells of Cu is similar to that of the 3d shell.

BAGUS et al. /39/ have developed a technique for using ECP's which has been shown /7,39/ to closely reproduce the results obtained with all electron wavefunctions where ECP's are not used. Their approach focuses on examining the nature of the bonding with an adsorbate. In particular, their tests have been made for the metal-CO interaction. The atoms involved in this interaction have all their electrons included in the wavefunction. The other atoms in the cluster are considered to be environmental atoms and only those electrons that are in the highest energy orbitals, for Cu only the atomic 4s electron, are included in the wavefunction. While this approach retains most of the advantages of the use of ECP's, it avoids many of the problems related to their use. Recently, this approach has been extended so that the deep core levels of the atoms directly involved in the interaction with the adsorbate are also included in the ECP /15/. The environmental atoms still have only their highest energy electrons explicitly retained in the wavefunction /40-42/.

The brief description above has focussed on the broadest features of the methods of MO theory. The intent has been to give the reader an introduction to the use and limitations of the various types of wavefunctions, SCF, MCSCF, and CI, and an introduction to terminology which is often encountered in the description of MO wavefunctions. Details are given in the references cited; see, in particular, the surveys of Refs. 1(a) and 20. Atomic units where the electron mass and charge and $\hbar$ are set to 1 are standardly used. In these units, distance is in bohrs, 1 bohr = 0.529 Å, and energy is in hartrees, 1 hartree = 27.212 eV. Another common energy relation is 1 eV = 8066 cm^{-1}.

2.2 Constrained Space Orbital Variation, CSOV

The CSOV method makes it possible to determine the energetic and other consequences of different contributions to a chemical interaction /13/. The interaction of interest is taken to be between the A and B sub-units of the total system denoted AB where, for most applications /6,7,13-15,18,19,39/, A has been an adsorbate (atom or molecule) interacting with a cluster B. In the following, the basic ideas of the CSOV approach, as it applies to SCF wavefunctions, are reviewed. Extensions to MCSCF wavefunctions have been given /24/. The CSOV analysis consists of a series of steps where the AB wavefunction is computed with two constraints imposed. In a given CSOV step, denoted V(X;Y), the orbitals arising from the X sub-unit, X = A or B, are varied, while those arising from the other sub-unit are fixed. (More precisely, the density of the sub-unit is fixed.) The second constraint is related to the variational space, Y, of the orbitals X. The variational space is normally taken as either the space arising from only the sub-unit being varied or from the full AB system. The basis set expansion of the orbitals is used to define the spaces. Thus, basis functions centered on the atoms associated with sub-unit A are considered to define the variational space arising from this sub-unit. The series of CSOV steps used to describe ionic or dative covalent bonding are given below. If open shells of A and/or B are involved in the covalent bonding of AB, the additional steps must be used /15,24/.

Step 0. This is the starting point denoted as Frozen Orbital, FO, or (more rigorously) Frozen Density, FD. Here, the fixed orbitals of the sub-units A and B are used to form an anti-symmetric many-electron determinental wavefunction. While the overlap of the A and B orbitals must be taken into account in evaluating the properties of this wavefunction, no changes in the A and B orbitals are allowed and no chemistry is possible. Note that as a numerical convenience /13/, the orbitals of A and B are orthogonalized but there is no physical significance to the orthogonalization at this FO step. With suitable formulas /43/, the energy and other properties of the FO wavefunction could be computed directly with the non-orthogonal orbitals. There are two features that determine the interaction at this step. The purely electrostatic interaction of the A and B charge distributions can contribute to an attractive or a repulsive interaction. The non-bonding overlap and inter-penetration of the orbitals of the sub-units contributes to a repulsive interaction /44/.

Step 1. - V(B;B). In this step, the orbitals of A are kept fixed and the orbitals of B are varied in the space of B. Physically, this describes the intra-unit polarization of the sub-unit B due to the presence of the fixed, frozen, sub-unit A.

Step 2.- V(B;All). The A sub-unit orbitals are still fixed but now the B orbitals are varied in the full space of the total AB system. The difference between the properties of the wavefunctions for steps V(B;All) and V(B;B) are due to donation; that is, to charge transfer and dative covalent bonding between the occupied orbitals of B and the virtual, unoccupied, orbitals of A.

Step 3. - V(A;A). The B occupied orbitals are fixed as they were determined at the preceding step, V(B;All), and the A orbitals are varied in the space of A. This describes the intra-unit polarization of A due to the presence of B.

Step 4. - V(A;All). The B orbitals remain fixed and the A orbitals are varied in the full space of AB. This step measures the importance of donation and dative covalent bonding from sub-unit A to sub-unit B.

This series of steps neglects the interplay between the charge rearrangements which are allowed in the different CSOV steps. If this interplay is important, it can be taken into account by performing an additional cycle of CSOV steps /45/; that is, using the orbitals determined at step 4, V(A;All), to restart step 1. In most cases, this has not been necessary. The measure of the extent to which this set of CSOV steps describes the A-B interaction is obtained by comparing the properties at step 4, V(A;All), with those of the un-constrained SCF calculation. When the differences are small, the CSOV steps have included all major features of the interaction. It is to be stressed that the unique virtue of the CSOV analysis is that with it, it is possible to determine the importance of individual intra-unit polarization and inter-unit donation charge rearrangement. With the un-constrained, full variation SCF wavefunction this decomposition is not possible.

2.3 Orbital Projection

A measure of the extent to which an orbital, φ, is contained in a general, many electron wavefunction, Ψ, is given by the expectation value P_φ of the projection operator, $P_{op}(\varphi)$,

$$P_{op}(\varphi) = \varphi\varphi^\dagger \tag{2}$$

$$P_\varphi = <\Psi \mid P_{op}(\varphi) \mid \Psi>$$

where φ is the spatial part of a spin orbital. The orbital which is projected is selected on the basis of physical interest. In the example of a system with two sub-units, discussed in Sec. 2.1, one might project orbitals taken from the wavefunction of one of the isolated sub-units. Typically, one might want to determine how the occupation of A orbitals had changed from isolated A to adsorbed A on B (the system AB /16,17/). One might also project the orbitals of constituent atoms on to the wavefunction of a cluster of these atoms. Neglecting orbital degeneracy, it is easy to show /17/ that $0 \leq P_\varphi \leq 2$. If $P_\varphi \simeq 2$, the orbital φ is occupied by 2 electrons and is fully contained in Ψ. If $P_\varphi \simeq 0$, φ is not contained in Ψ and has a 0 occupation. For intermediate values of P_φ, φ is partially occupied in Ψ. The partial occupation can arise for several reasons which include the following: (1) The orbital is involved in a covalent bond with orbitals of other sub-units of the system. (2) There has been charge transfer out of or into this orbital from/to other sub-units. And (3), the orbital of the isolated sub-unit has become hybridized with other orbitals of the same sub-unit. For the interpretation of intermediate values of P_φ, it is necessary to take into account the fact that the orbitals arising from different isolated sub-units have non-zero overlap with each other /16,17/.

3. FEATURES OF BONDING IN METAL CLUSTERS

3.1 Hybridization and Binding Energy for Be and Mg Clusters

In this sub-section, properties of Be and Mg dimers, tetramers, and bulk will be compared. The dimer properties are taken from extensive CI results of LIU and McLEAN /46/. The tetramers have a tetrahedral, T_d, geometry and both CI and SCF wavefunctions have been obtained by BAUSCHLICHER et al. /47/. To model the bulk situation, they obtained SCF wavefunctions for X_{13}, X = Be and Mg, clusters. Bulk geometries and bond distances were used for these clusters which have a central atom with the bulk, 12-fold coordination and 12 edge atoms with 5-fold coordination. Results for X_2, X_4, and bulk are given in Table 1. The calculated CI binding energies, D_e, per atom for the dimer and tetramer are compared to the experimental bulk heats of sublimation. The weakly bound Van der Waals dimers are much less stable than the tetramers which, in turn, are less stable than the bulk. The Be systems are more stable than the Mg systems; however, while $D_e(Be_4)$ is 5 times larger than $D_e(Mg_4)$, bulk Be is only twice as stable as bulk Mg.

These stabilities closely follow the extent of the ns → np hybridization given, in Table 1, by the Mulliken gross atomic np populations for the X_4 and X_{13} SCF wavefunctions. Note that the ratio of the Be 2p and Mg 2p populations is very similar to the ratio of the D_e's for both X_4 and bulk. The s → p hybridization is necessary to form directed bonds between the metal atoms and it is not surprising that it is a good guide to the relative stabilities; a larger np population is fully consistent with a stronger bond. It is also reasonable that the extent of the s → p hybridization should be related to the coordination of the atom; i.e. to the number of nearest-neighbors that it can form directed bonds with. (Note that in counting numbers of nearest neighbors, the small differences in bond distance for the hcp crystal structure will be neglected. However, the average next-nearest-neighbor distance, NNN, is considerably larger than the average nearest-neighbor distance, NN; NNN = $\sqrt{2}$ NN.) From Table 1, it can be seen that the np population grows monotonically from the 3-fold coordinated edge atoms in X_4 to the 5-fold coordinated edge atoms of X_{13} to the 12-fold coordinated central atoms of X_{13}. However, in all cases the Be 2p population is larger than the Mg 3p population.

Table 1. Binding energies, D_e, bond distances, r_e , and np populations for dimers, X_2, tetramers, X_4, and bulk or X_{13} clusters for Be and Mg. The ratios of the Be and Mg values are given under R[Be/Mg] The X_2 and X_4 values for D_e and r_e are from accurate CI wavefunctions and the bulk values are from experiment. The X_4 and X_{13} np populations are from SCF wavefunctions for the clusters and the atomic sizes, $<r>_{ns}$, are for atomic SCF orbitals. See Ref. 12 for sources of these values

	Be	Mg	R[Be/Mg]
	D_e per atom (kcal/mol)		
X_2	1.16	0.61	1.9
X_4	14.80	2.98	5.0
Bulk	78	35	2.2
	np population per atom		
X_4	1.06	0.22	4.8
X_{13}(central)	1.44	0.85	1.7
X_{13}(edge)[a]	1.11	0.40	2.8
	bond distance - r(bohr)		
X_4	3.92	6.02	1.5
Bulk[a]	4.26	6.05	1.4
$<r>_{ns}$	2.65	3.25	1.2

[a] Average value

There is an energetic cost as well as a benefit to the s → p hybridization /12,47/. This cost as measured by the atomic $ns^2(^1S) \rightarrow nsnp(^3P)$ excitation energy, is almost the same for Be and Mg, 2.7 eV, and hence it cannot explain the different hybridization. However, note that the bond distances, Table 1, for the tetramer and for bulk are larger for Mg than for Be. This is, in part, because the size of the outermost atomic orbital, as measured by its SCF expectation value of r, is larger for Mg 3s than for Be 2s. But while Mg $<r>_{3s}$ is 20% larger than Be $<r>_{2s}$, the Mg bond distances are larger than the Be bond distances by 40% and 50% for X_4 and bulk respectively. The repulsion due to the non-bonding overlap of core orbitals on one Mg atom with the core and valence orbitals on a neighboring Mg atom may well contribute to this bond distance increase. For a larger bond distance relative to the atomic size, the bonding overlap will be smaller and the bond weaker. A weaker bond is less able to pay the energetic cost of hybridization and the np population will be smaller. It is reasonable to expect a synergistic coupling of bond length and hybridization acting to further reduce the bond strength.

3.2 D Shell Hybridization for Cu Clusters

Here we consider evidence for a reduction in the 3d occupation for Cu_2 and Cu_{24} clusters shown schematically in Figure 1. These clusters were originally chosen to model chemisorption at a bridge site of a Cu(100) surface /8,42/ and have the geometry appropriate to an unreconstructed and unrelaxed surface. The central atoms of the first layer which form the bridge site /8/ are described with a Ne core ECP /42,48/ while the remaining 22 environmental atoms of the Cu_{24} cluster are described with an Ar plus $3d^{10}$ core ECP /7,39,42/. For both Cu_2 and Cu_{24}, a 2 configuration CASSCF /34/ wavefunction is obtained. This is chosen to account for the $4s\sigma_g$ and $4s\sigma_u$ near degeneracy for Cu_2 /49/. The two configuration MCSCF is also used for the Cu_{24} cluster to have an equivalent treatment, even though the environmental Cu atoms lead to the formation of a 4sp conduction band which largely removes this particular near degeneracy. The adsorption site Cu atoms in Cu_2 are 1-fold coordinated, while they are 8-fold coordinated in Cu_{24}. Even though the cluster geometry is specific to a Cu(100) surface, the different coordination of the adsorption site Cu atoms in the two clusters is also relevant to the different atomic environment in different size Cu clusters.

The occupation of the 3d shell in the Cu atom and the Cu_2 and Cu_{24} clusters is estimated in two ways. In the first way, the 3d orbitals of the $3d^{10}4s^1(^2S)$ ground state

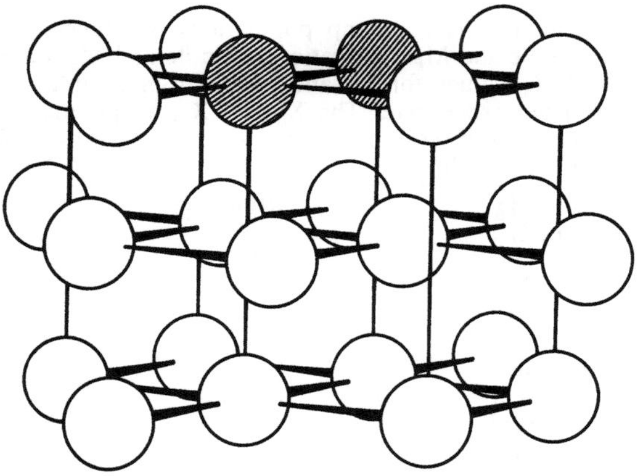

Figure 1. Schematic view of the Cu_{24} cluster which contains 8 atoms from each of the first three layers of a Cu(100) surface. The shaded atoms represent a bridge adsorption site; an adsorbate would be added directly above the center of these atoms. They are also the only atoms in the Cu_2 cluster

of the Cu atom are projected from the cluster wavefunctions as described in Sec. 2.3. In the second way, the 3d population is obtained from a Mulliken gross population analysis. These 3d occupations are shown in Table 2. By either measure, the 3d shell, or d "band", is almost completely filled for both the Cu_2 and Cu_{24} clusters. However, while for the Cu_2 cluster the 3d occupation is negligibly different from 10, the difference is significantly larger for Cu_{24}. The estimates of the 3d occupation from projection and from the Mulliken population are different. These differences are due to the different ways that projection /17/ and population analysis /44/ take into account the overlap between the 3d orbitals on a particular Cu atom and the 3d and 4sp orbitals on neighboring Cu atoms. It is entirely expected that the occupation given by projection is larger than that given by population analysis. But , both estimates give consistent trends; they both show that the Cu 3d occupation is smaller for Cu_{24} than for Cu_2.

Table 2. Occupations of the 3d shell of the Cu atom and of one of the shaded atoms of the Cu_2 or Cu_{24} clusters shown in Figure 1. Occupations are obtained both from projection and from population analysis

	Atom	Cu_2	Cu_{24}
Population	10.000	9.984	9.929
Projection	10.000	9.996	9.979

The change of the 3d occupation cannot arise simply from a direct covalent bonding between the 3d shells on different Cu atoms. The atomic d shells are fully occupied; bonding and anti-bonding levels would also have to be fully occupied and this would not lead to a change in the 3d occupation on a given atom. The reduction must arise from hybridization of the 3d shell to 4sp character. It is reasonable that the 3d $\rightarrow$ 4sp hybridization will be much larger for Cu_{24} than for Cu_2 since the coordination of the atoms is different in these two clusters. The atoms for which the 3d occupation is estimated have an 8-fold coordination in Cu_{24} but only a 1-fold coordination in Cu_2 It will be shown in Sec. 4 that this small reduction of 3d occupation contributes to a significant increase in the core level IP's of the Cu atom in the Cu_{24} cluster.

This section has focussed on the relation of hybridization to bonding in metal clusters. For alkaline earth clusters, in particular Be and Mg, it has been shown that there is a direct correlation between the extent of the ns $\rightarrow$ np hybridization and the bond energy.

Further, it has been shown (1) that atoms with higher coordination, more near neighbors, have a greater ns → np hybridization and (2) that the hybridization is greater for Be than Mg. These differences have been interpreted in terms of a balance between the energetic cost of the hybridization and the benefit from forming directed covalent bonds. For Cu clusters, it has been shown that the 3d → 4sp hybridization is small but that for Cu_{24} there is a greater reduction of the 3d occupation from the atomic value of 10 than there is for Cu_2. This has also been related to the different coordination of the Cu atom in the two clusters. It is quite possible that similar d → sp hybridization effects will be found for other transition metal clusters.

4. CLUSTER IONIZATION POTENTIALS

4.1 Core Level Ionization Potentials

There have been extensive X-ray photoemission spectroscopy, XPS, studies of the binding energies, BE's, or IP's, of the core level electrons of metal clusters on insulating supports /26/. (Note that in this paper, the terms BE and IP will be used interchangeably to describe ionization phenomena.) The general feature for several metals, e.q. Pd, Pt, and Ag, is that the core level BE's decrease by ~ 1eV as the size of the supported clusters increases towards a continuous film. In this section, the results of SCF calculations on Li, Al and Cu clusters will be analyzed and two important initial state effects which can contribute to cluster size dependent core level BE's will be identified.

It is standard, in theoretical analyses, to divide the contributions to the IP into initial and final state effects /26,50,51/. The initial state IP's reflect the chemical environment of the ionized atom and Koopmans' Theorem, KT, is used to give these IP's. IP(KT) = $-\varepsilon$ where ε is the SCF orbital energy for the electron which is removed. For open shell systems, where the ionic configuration may be coupled to give several multiplets, IP(KT) is for ionization to the weighted average of the ionic multiplets /52/. The KT IP is a frozen orbital, FO approximation since none of the orbitals are allowed to change from their form in the initial, unionized, system. This final state response, or orbital relaxation, due to the presence of the hole lowers the IP. The ΔSCF IP which includes both initial and final state effects is given as the difference of the energies for the initial and final, ionized, states:

$$\mathrm{IP}(\Delta\mathrm{SCF}) = E_{SCF}(\mathrm{Ion}) - E_{SCF}(\mathrm{Neutral}) \qquad (3)$$

$$= \mathrm{IP}(\mathrm{KT}) - E_R.$$

The relaxation energy, E_R, arises from intra-atomic orbital contraction and extra-atomic screening /53,54/. The final state effects for core ionization are normally quite large and are commonly described by Jolly's equivalent core model /55/. In the equivalent core model, the core ionized atom of nuclear charge Z is replaced by an atom of charge Z+1 where the cores are fully occupied. For example, consider the C_{1s} core ionized state of the CO molecule, $CO[C_{1s}\text{-}1]^+$; the equivalent core molecule for this core ionized state is NO^+. The 10 valence electrons of neutral CO, core-ionized CO, and the equivalent core NO^+ have the same configuration,

$$3\sigma^2 4\sigma^2 5\sigma^2 1\pi^4.$$

The difference of the SCF orbitals for neutral CO and for NO^+ from those for core ionized $CO[C_{1s}\text{-}1]^+$ may be seen by projecting the neutral CO and NO^+ orbitals onto the $CO[C_{1s}\text{-}1]^+$ SCF wavefunction; see Sec. 2.3. Projection operators for σ and π symmetries are defined by:

$$P(\sigma) = \Sigma_i \varphi_i \varphi_i^\dagger \quad i = 3\sigma, 4\sigma, 5\sigma \qquad (4)$$

$$P(\pi) = \varphi_{1\pi}\varphi_{1\pi}^\dagger,$$

where the φ used to construct $P(\sigma)$ and $P(\pi)$ are SCF orbitals for either CO^0 or NO^+, as appropriate. The expectation values, P_λ, for the $CO[CO_{1s}\text{-}1]^+$ SCF wavefunction are given in Table 3. The projection of the CO^0 valence orbitals, $P_\sigma + P_\pi$, is much smaller than 10 indicating that a considerable relaxation has taken place in response to the C_{1s} hole. The projection of the NO^+ orbitals shows that the equivalent core model gives a reasonable, although not entirely perfect, description of the valence charge distribution of $CO[C_{1s}\text{-}1]^+$. A large part of the relaxation is for the 1π orbital and this can be measured with $<z>_{1\pi}$. The 1π orbital for $CO[C_{1s}\text{-}1]^+$ has moved 0.25 bohrs closer to the C nucleus in order to screen the 1s core hole. Also the $<z>_{1\pi}$ for $CO[C_{1s}\text{-}1]^+$ and NO^+ are nearly equal.

Table 3. Projections of the neutral CO and NO^+ valence orbitals on the C_{1s} core ion SCF wavefunctions of CO, see Eq. 4. Also included is the expectation value of z for the 1π orbital; z is the CO or NO internuclear axis and C or N is taken as the origin. The bond distance for both CO and NO is 2.15 bohr.

Orbitals	P_σ	P_π	$P_\sigma+P_\pi$	$<z>_{1\pi}$
CO^0	5.90	3.89	9.79	1.66
NO^+	5.96	4.00	9.96	1.39
$CO[C_{1s}\text{-}1]^+$	---	---	---	1.41

Finally, it is appropriate to comment on the localized character of the ionic state when there are n equivalent atoms in the cluster which may be core ionized. It is possible to have a delocalized ion where 1/n of a core electron is removed from each of the n equivalent atoms. The SCF ionic state is delocalized if the full symmetry of the cluster or molecular point group is imposed on the orbitals. It is also possible to have a localized core ion where a single core electron is removed from one of the n equivalent atoms and the remaining n-1 have filled core levels. The SCF ionic state can become localized if the full point group symmetry is not imposed /53,56/. For SCF wavefunctions, the localized core hole provides the better representation of final state screening /53/. The core level ionic cluster final states discussed in this section will be for localized core holes.

The evidence obtained by NELIN and BAGUS /51/ about the dependence of the core level IP on the coordination of the ionized atom will be reviewed. The Li and Al clusters to be considered for this purpose have been chosen to have the geometry of the bulk crystal, bcc for Li and fcc for Al. However, within this geometric constraint, the bond distance or effective lattice constant, a_{eff}, of the cluster is varied and the equilibrium value for a given size cluster is obtained. The Li_9 cluster has a central atom, denoted C, and eight atoms in shell 1 which are the nearest neighbors of the central atom; thus C is 8-fold coordinated and the shell 1 atoms are 1-fold coordinated. The Li_{35} cluster is formed by adding 26 atoms to Li_9 so that the shell 1 as well as the center atom have the bulk 8-fold coordination; the cluster is shown in Fig. 2. Of the 26 added atoms, the 6 in shell 2a are 4-fold coordinated; the 12 in shell 2b are 2-fold coordinated; and the 8 in shell 2c are 1-fold coordinated. Clearly, clusters will not adopt the bulk geometry until they grow very large. The Li_9 and Li_{35} clusters are representative of real clusters in that they have atoms with different coordinations. This coordination will be shown to be very important for the core level IP; cluster atoms which have high coordination have lower IP's than cluster atoms with lower coordination. The generality of this coordination effect is shown by the Al_{13} cluster which is , for fcc Al, the analogous cluster to Li_9. The center atom has the bulk 12-fold coordination and the 12 shell 1 atoms are 5-fold coordinated. The details of the neutral ground state and core level ionic SCF cluster wavefunctions for Li_9, Li_{35}, and Al_{13} are given elsewhere /51/.

The IP(KT) for the Li and Al core levels are given in Table 4. The values of the ε's for orbitals associated with a 1s or 2s core level of atoms in the same shell or sub-shell are equal within ~0.001 eV. The 2p band of the edge, shell 1, Al atoms in Al_{13} is slightly wider, 0.035 eV, and an average IP(KT) is given for these 2p levels. There is a monotonic relation between the coordination of the core ionized atom and IP(KT); atoms with higher

Table 4 Koopmans' theorem IP's, in eV, for the differently coordinated shells of Li_9, Li_{35}, and Al_{13}. The coordination of the shell and the shift of the IP for that shell from the IP for the center atom are given. The IP(KT) are for a cluster a_{eff} near equilibrium

Cluster	Core	Atomic Shell (Coordination)		IP(KT)	Shift
Li_9[a]	1s	Center	(8)	65.4	---
		Edge	(1)	66.8	+1.39
Li_{35}[a]	1s	Center	(8)	65.2	---
		1	(8)	65.3	+0.04
		2a	(4)	66.0	+0.79
		2b	(2)	66.3	+1.10
		2c	(1)	66.8	+1.57
Al_{13}[b]	1s	Center	(12)	1589.6	---
		Edge	(5)	1591.8	+2.19
	2s	Center	(12)	131.7	---
		Edge	(5)	133.6	+1.83
	2p	Center	(12)	86.1	---
		Edge	(5)	87.8	+1.74

[a] a_{eff} = 6.6 bohrs [b] a_{eff} = 7.8 bohrs

Li_{35} Cluster

	Shell	Number	Coordination	Distance
Li	Center	1	8	
Li	1	8	8	$\sqrt{3}/2a$
Li	2a	6	4	a
Li	2b	12	2	$\sqrt{2}a$
Li	2c	8	1	$\sqrt{3}a$

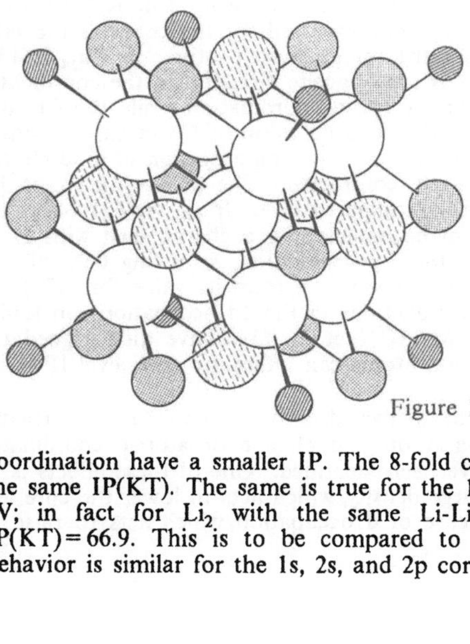

Figure 2. Schematic representation of the Li_{35} cluster.

coordination have a smaller IP. The 8-fold coordinated atoms of Li_9 and Li_{35} have nearly the same IP(KT). The same is true for the 1-fold coordinated atoms where IP(KT)=66.8 eV; in fact for Li_2 with the same Li-Li bond length as for the clusters, the 1s IP(KT)=66.9. This is to be compared to the 1s IP(KT)=67.5 for the Li atom. The behavior is similar for the 1s, 2s, and 2p core levels of the Al_{13} cluster; however, the shift between the center and edge atoms is somewhat larger than for the Li_n clusters. There is a simple explanation for the lower IP(KT) of the atoms with higher coordination. The cluster atoms are surrounded by charge arising from the valence shells of the neighboring atoms. This charge creates an electrostatic potential which lowers the core level IP. All of the Li cluster atoms have initial state IP's which are smaller than that for the free atom. Cluster atoms with a higher coordination are surrounded by the valence charge from more neighbors and hence have a lower IP(KT). The larger shifts for the Al core levels is related, at least in part, to the fact that an Al atom has 3 valence electrons, $3s^2 3p^1$, while a Li atom has only 1, $2s^1$.

These model calculations are related to the observed cluster size dependent shifts of the core level IP's of the metal clusters on insulating supports through the average coordination of the atoms of a given cluster. As the cluster size increases, the average coordination increases, until the cluster is very large and the bulk value is reached. The calculations predict a decrease of ~1-2 eV from low to bulk coordination. The similar behavior for Li_n and Al_n demonstrate the importance and generality of this initial state coordination effect. All metals have a diffuse sp electron conduction band which will contribute to the initial state effect that has been shown above.

It is possible that this initial state effect could be masked by different E_R for the differently coordinated atoms. Applying the equivalent core argument /55/ to a core ionized atom of Li_n, an edge Be impurity atom could be quite different from a central Be impurity atom. NELIN and BAGUS /51/ have investigated the different final state screening for the central and edge atoms of Li_9 and Al_{13}. The E_R is large, ~ 7 eV for Li_9 and ~ 15 eV for Al_{13}. However, there is virtually the same E_R for the two types of atoms; the edge and center E_R differ by 0.02 eV for Li_9 and by 0.2 eV for Al_{13}. Thus, E_R depends much more strongly on the total number of atoms in the cluster than on the coordination of the ionized atom. The coordination dependent, initial state shifts of the core level IP's are not masked by final state relaxation. The dependence of E_R with cluster size is not rapid and this argues strongly /26,51,57/ against the applicability to small clusters of the classical Coulomb mechanism for IP shifts proposed by WERTHEIM et al. /58/.

It was pointed out in Sec. 3.2, that the d electron occupation of the Cu atom is smaller in Cu_{24} than in Cu_2. Here, the shifts of the 3s IP's, or ε's, of Cu are examined for these clusters. The wavefunctions and clusters described in Sec. 3.2 are used. The 3s IP(KT) for the $3d^{10}4s^1$ configuration of the Cu atom is 137.1 eV; the value for Cu_2 is 136.9 eV or 0.2 eV smaller. This is exactly the direction of shift expected for the coordination effect discussed above. However, the 3s IP(KT) for the bridge site Cu atoms in Cu_{24} is 138.0 eV or 1.1 eV larger than the Cu_2 value. This is the opposite direction to the coordination shift described above; the Cu atom in Cu_2 has a 1-fold coordination while the Cu atom in Cu_{24} has an 8-fold coordination. However, the 3d occupation of Cu in Cu_{24} is smaller than that of Cu in Cu_2; see Table 2. The influence of removing a portion of a 3d electron comes from the fact that the 3d orbital is quite contracted. The SCF $\langle 1/r \rangle_{3d}$ is 1.39 $bohr^{-1}$ for the $3d^{10}4s^1(^2S)$ state of the Cu atom compared to $\langle 1/r \rangle_{4s}$ = 0.37 $bohr^{-1}$. A simple approximation is that the 3d shell can be modeled by a charge shell with 1/r = 1.4 $bohr^{-1}$ or r = 0.7 bohr = 0.38 Å. In this approximation, removing 0.02 of a 3d electron, see Table 2, would raise the 3s IP(KT) by 0.8 eV from its value for n(3d) = 10. This simple argument shows that very small reductions in the 3d occupation can lead to very large increases in the core level IP's; MASON et al. /59/ have also argued that changes in the d occupation of transition metal atoms can affect the core level IP's.

Further investigation of this effect of d to sp hybridization is required. In particular, we must understand how bond distance variation and change of atomic coordination influence the two competing effects that have been identified. One of these effects is the increase of diffuse "conduction band" sp charge density, which acts to lower the core level IP. The second of these effects is the decrease of d occupation, which acts to raise the core level IP.

4.2 Valence Level Orbital Energies

Knight and his collaborators /27/ have argued that a shell model for the valence charge of metal clusters may explain the special stability of clusters with particular numbers of atoms. In particular, it is claimed that "in free-electron-like clusters the origin of the discontinuities in the mass spectra is not related to the geometric packing pattern but to the electronic structure" /27/. Further, it is claimed that the electronic structure is well described by a simple spherical jellium model which leads to shells in the order 1s, 1p, 1d, 2s, 1f, ...When these shells are filled for a particular number of cluster atoms, that cluster is especially stable. Electronic structure results for the electron eigenvalues for Na_{13} and Na_{15} clusters forced to have fcc and bcc geometry are shown /27/ to have a pattern of degeneracies and near-degeneracies like that given by spherical jellium. This is taken to support the claim that the specific cluster geometry is not important.

Here, the valence level electron eigenvalues for 3 clusters are examined. The clusters are Li_{35} which has bcc geometry, Al_{13} which has fcc geometry (see discussion above) and Cu_{34} where the geometry is constrained to be that of a (100) surface of the fcc crystal. The Cu_{34} cluster has 9 atoms in the first layer and 16 and 9 atoms in the second and third layers respectively; the central atom of the first layer has been used to model an on-top site for chemisorption /7/. This cluster, denoted $Cu_{34}(9,16,9)$, is shown schematically in Fig. 4. Details of the calculation of the Cu_{34} SCF wavefunction are given in Ref. 7.

For Li_{35}, Al_{13}, and Cu_{34}, the SCF orbital energies, ε, of the occupied cluster "conduction band" orbitals with valence sp character, 2sp for Li, 3sp for Al, and 4sp for Cu, are given in Tables 5 and 6. Specifically, the differences of the orbital energies, $\Delta\varepsilon$, with respect to the lowest sp band ε are given. The Li_{35} and Al_{13} clusters have O_h point group symmetry and the orbital degeneracies are: A, 1-fold; E, 2-fold; and T, 3-fold. The Cu_{34} cluster has D_{4h} point group symmetry; the A and B orbitals are 1-fold and the E orbitals are 3-fold degenerate. The O_h and D_{4h} cluster orbital degeneracies are not those of the 3-dimensional rotation group orbitals given by the spherical jellium model. If the spherical jellium degeneracy and ordering are to be found from the Li_{35}, Al_{13}, and Cu_{34} cluster wavefunctions, there must be special patterns of degeneracy or near degeneracy for the cluster orbitals. In Tables 5 and 6, the shell ordering given by spherical jellium is also indicated and the cluster orbitals are grouped to show the extent of their correlation with the jellium shells.

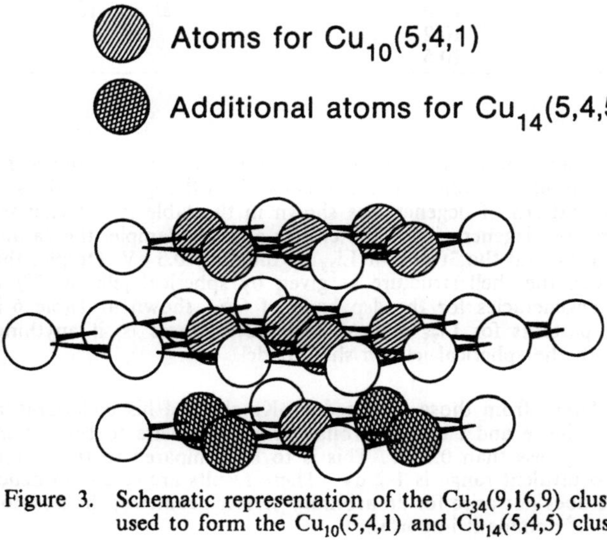

Figure 3. Schematic representation of the $Cu_{34}(9,16,9)$ cluster. The subset of atoms used to form the $Cu_{10}(5,4,1)$ and $Cu_{14}(5,4,5)$ clusters are indicated

Table 5. Valence orbital energy differences, $\Delta\varepsilon$, in eV, for the conduction band orbitals of bcc Li_{35} and fcc Al_{13}, the distribution of the cluster ε's is compared to that predicted by spherical jellium. The cluster orbital symmetry and occupation are indicated under the heading "Cluster Orbital". For Li_{35}, a_{eff} = 6.6 bohrs and for Al_{13}, a_{eff} = 7.8 bohrs; both values are close to equilibrium

Li_{35} Cluster Orbital	$\Delta\varepsilon$	Al_{13} Cluster Orbital	$\Delta\varepsilon$	Jellium Shells
1a(2)	0	1a(2)	0	1s
1t(6)	1.4	1t(6)	3.1	1p
2t(6)	2.8	2t(6)	6.4	
3a(2)	3.5	3a(2)	8.1	1d and 2s
1e(4)	3.8	1e(4)	8.2	
4a(2)	4.3	3t(6)	10.8	
3t(6)	4.8	4a(2)	11.5	1f and 2p
4t(6)	5.1	4t(6)	12.0	
5t(1)	5.9	5t(5)	13.3	

Table 6. Orbital energy differences, $\Delta\varepsilon$, in eV, for the 4sp conduction band orbitals of fcc Cu_{34}. The cluster a_{eff} = 6.8 bohrs is the value for bulk Cu

Cluster Orbital	$\Delta\varepsilon$	Jellium Shell
1a(2)	0	1s
1e(4)	2.1	1p
1b(2)	4.1	
2a(2)	4.6	
3a(2)	5.3	1d
2b(2)	5.4	
2e(4)	6.9	
3e(4)	7.4	
3b(2)	9.2	
4e(4)	9.3	
3a(2)	10.0	2s and 1f
4a(2)	10.2	
4b(2)	10.5	

The patterns of the bcc Li_{35} and the fcc Al_{13} orbital energies have been examined for a 5% range of bond distance about the values used in Table 5. Although the values of the ε's change somewhat, the pattern of degeneracies shown in the table is not changed significantly. There are some near degeneracies in the cluster ε's; for example, the 3a and 1e Al_{13} ε's are different by 0.1 eV and the 3t and 4t Li_{35} ε's differ by 0.3 eV. Despite this degeneracy, the correlation with the shell structure as given by spherical jellium /27/ is quite poor. The pattern of degeneracies for the 4sp band of Cu_{34} shown in Table 6 is somewhat different from the patterns for Li_{35} and Al_{13}. However, there is, if anything, an even poorer correlation with the spherical jellium shell model.

These results are quite different from those obtained by Knight and his collaborators /27/ for Na_{13} and Na_{15} where the e and t cluster orbitals which correlate to the jellium 1d shell have ε's which differ by less than 0.3 eV. This is to be compared to the results in Tables 5 and 6 where the equivalent range is 1-2 eV. These results are strong evidence that the jellium shell structure does not arise for an arbitrary cluster geometry; the position and number of the nuclei are of dominant importance.

5. BONDING OF MOLECULAR ADSORBATES, CO and CN

In this section, the nature of the bonding of the adsorbates CO and CN to a Cu cluster will be discussed. It will be shown that these two ligands represent examples of two very different kinds of bonding; the Cu-CN bond is ionic while the Cu-CO bond is covalent. It is reasonable that the same features of the metal-ligand interaction which are found to be important for Cu will also be important for other d-electron metals. While the precise value of various contributions to the interaction are different, the same general features have been found for the metal-CO bond for Cu, Ni, and Fe /24,39/; the similarity is expected for the metal-CN bond /16/. An understanding of the bonding is particularly important when this understanding can be related in a direct fashion to observables. This direct connection has been made for the shifts of the vibrational frequencies between the isolated CO and CN ligands and the ligands bonded to the Cu cluster /16,19,25,60/. It has also be made to explain the shifts of the valence level photoemission spectra, PES, peaks for adsorbed CO /18/.

The ligands will be placed at an on-top site for $Cu_{10}(5,4,1)$ and $Cu_{14}(5,4,5)$ clusters. These clusters are part of the $Cu_{34}(9,16,9)$ cluster shown in Fig. 3; the atoms which are included in Cu_{10} and Cu_{14} are indicated in the figure. The adsorbate, either CO or CN, is added directly above the central atom of the first layer of Cu atoms. The C atom is nearest Cu, and the ligand internuclear axis is normal to the layer. This adsorption geometry is known for CO on a Cu(100) surface and also for CO on other metals /61/. For CN, the adsorption geometry is accepted as holding in electrochemical environments /62,63/; using the same geometry for CO and CN makes possible a direct comparison of the metal ligand, ML, bonding and its consequences for these two ligands. The Cu_{10} and Cu_{14} clusters have been chosen as models for the Cu(100) surface. However, the description of the ML bonding obtained with these clusters is not restricted to this surface but will apply to a linear metal-ligand bond in general. This claim is supported by both computational /7,17,45/ and experimental /64/ evidence. The overall nature of the ML interaction is quite similar for both the Cu_{10} and Cu_{14} clusters as well as for other clusters both larger and smaller /7,17/. In some cases, the properties of the $Cu_{10}L$ and $Cu_{14}L$ clusters will be compared in order to examine the dependence of a property on the specific number of atoms in the cluster. Occasionally, comparisons will be made between $Cu_{10}CN$ and $Cu_{14}CO$; however, this will be done only for cluster size independent features. For the analysis of the nature of the bonding, metal-ligand and intra-ligand bond distances near the calculated /7,16,60/ equilibrium distances are used; in the case of CO, these are also near the experimentally determined /61/ equilibrium. Specifically for Cu_nCO, R(Cu-C) = 3.70 bohrs and R(C-O) = 2.15 bohrs are used; for Cu_nCN, R(Cu-C) = 3.90 and R(C-N) = 2.20 are used.

The traditional, and often cited, view of the metal-CO bond is that it is composed of two parts: (i) σ donation from CO to the metal; and (ii) π back-donation from the metal to CO /65/. Both the donation and back-donation give rise to dative covalent bonding between CO and the metal; the bonding is dative because it involves occupied orbitals on one component and unoccupied, or virtual, orbitals on the other. For the σ donation, the primary orbitals that are involved are the CO 5σ and empty metal σ orbitals; recall that the 5σ orbital is a C lone pair orbital. For the π back-donation, the metal d and sp conduction band both contribute /7/ to forming a covalent bond with the CO $2\pi^*$ orbital /66/; recall that the $2\pi^*$ is anti-bonding between C and O with dominant C character. This view of the bonding has also been used to analyze the metal-CN interaction /67/.

The projection of the ligand σ, 1π, and 2π orbitals on the $Cu_{10}L$ and $Cu_{14}L$ wavefunctions, Eqs. (2) and (4) is considered first; the results are shown in Table 7. To measure the 2π back-donation, the quantity $\Delta P_{2\pi}$ is used. This is the difference between the projection of the 2π orbital, $P(2\pi) = \varphi_{2\pi}\varphi_{2\pi}^{\dagger}$, on the bare Cu_n cluster and on the cluster plus adsorbate, Cu_nL. It is defined as:

$$\Delta P_{2\pi} = \langle \Psi(Cu_nL) \mid P(2\pi) \mid \Psi(Cu_nL) \rangle \qquad (5)$$
$$- \langle \Psi(Cu_n) \mid P(2\pi) \mid \Psi(Cu_n) \rangle .$$

Table 7. Projections of the ligand orbitals, P_σ, and $P_{1\pi}$ and $\Delta P_{2\pi}$, on the $Cu_{10}L$ and $Cu_{14}L$ clusters, L=CN or CO. For $Cu_{10}CN$, different total charge states of the cluster, $[Cu_{10}CN]^0$, $[Cu_{10}CN]^+$, and $[Cu_{10}CN]^-$, are used. For the occupied ligand orbitals, the orbitals from neutral $[CN]^0$ and anionic $[CN]^-$ are projected; but only the orbitals of neutral $[CO]^0$ are projected

System	Ligand Occupied Orbitals	P_σ	$P_{1\pi}$	$\Delta P_{2\pi}$ [a]
$[Cu_{10}CN]^0$	$[CN]^0$	9.91	3.98	---
$[Cu_{10}CN]^0$	$[CN]^-$	9.96	3.99	0.01
$[Cu_{10}CN]^+$	$[CN]^-$	9.96	3.98	0.00
$[Cu_{10}CN]^-$	$[CN]^-$	9.96	3.99	0.01
$[Cu_{14}CN]^-$	$[CN]^-$	9.96	3.99	0.015
$[Cu_{10}CO]^0$	$[CO]^0$	9.98	4.00	0.28
$[Cu_{14}CO]^0$	$[CO]^0$	9.98	4.00	0.45

[a] The ligand 2π orbital is taken from CO^- or $CN^=$

It is necessary to consider $\Delta P_{2\pi}$ because there is a significant overlap between the ligand 2π orbital and the orbitals of the bare Cu_n cluster, primarily the conduction band, 4sp, but also the 3d orbitals /16,17/. This difference provides a measure of the increase of ligand 2π character due to the metal to ligand π back-donation. For the projection of the $[CN]^0$ orbitals on $[Cu_{10}CN]^0$, $P_\sigma + P_{1\pi} = 13.89$ indicating that CN has nearly 14 electrons in these orbitals and that the adsorbed CN is CN^-. If the relaxation of the CN orbitals in CN^- is taken into account by projecting the $[CN]^-$ orbitals, $P_\sigma + P_{1\pi} = 13.95$, which is even closer to 14. The value of $P_\sigma + P_{1\pi}$ is also 14 for $[Cu_{10}CN]^+$ and $[Cu_{10}CN]^-$ showing that the CN bond is ionic independent of the cluster charge. When the cluster cation or anion is formed, the charge is removed from the Cu cluster conduction band; the CN remains an anion. Because the total cluster charge does not affect the nature of the Cu-CN bond /16,68/, the results for $[Cu_{10}CN]^-$ and $[Cu_{14}CN]^+$ projections are almost identical. There is a small σ donation from CN^- , P_σ = 9.96 electrons not 10. The Cu to CN π back-donation is negligible; CN^- does not want still another electron. The bonding of Cu to CO is quite different. The σ donation is even smaller in this case but the π back-donation is much larger. The difference in $\Delta P_{2\pi}$ for $Cu_{10}CO$ and $Cu_{14}CO$ is related to the π character of the Cu_n cluster 4sp conduction band /7/. However, the critical point is that there is a large 2π character of CO adsorbed on Cu in both cases! For CN, there is no 2π character and the bond is ionic. The 2π character for Cu_nCO leads to an increase in charge between Cu and the C atom, as is characteristic of a covalent π bond /66/.

The nature of the bond, ionic versus covalent, is clearly shown by the dipole moment curve for the frustrated translation of the ligand with respect to the Cu cluster /16,17/; for this translation the ligand bond distance is fixed. The dipole moment curves for this translation are quite linear /16/. The ionic CN bond has a large slope; -1.3 for $[Cu_{10}CN]^-$ and -1.4 for $[Cu_{14}CN]^-$. The slopes for $Cu_{10}CO$ and $Cu_{14}CO$ are <0.1; and this small value clearly indicates the covalent character of the Cu-CO π bond. Because the π bond is covalent, it is not proper to interpret the values of $\Delta P_{2\pi}$ as an effective ionicity of adsorbed CO. Rather, $\Delta P_{2\pi}$ shows that there is a major π involvement in the metal-CO bond which the small slope of the dipole moment curve indicates is a covalent bond.

The CSOV analysis, see Sec. 2.2, provides a definitive measure of the importance of the various charge rearrangements. While the projections in Table 7 do indicate that the Cu to CO π back-donation involves a larger charge rearrangement than the other inter-unit donations for either CN or CO, they do not indicate the magnitude of its energetic importance. In Fig. 4, the CSOV interaction energy, E_{Int}, for each step of the constrained variations is plotted for $[Cu_{10}CN]^-$ and $Cu_{14}CO$. The bottom of each numbered block in the figure gives E_{Int} at that step:

$$E_{Int}(Cu_nL; \text{step } n) = E(Cu_n) + E(L) - E(Cu_nL; \text{step } n); \tag{6}$$

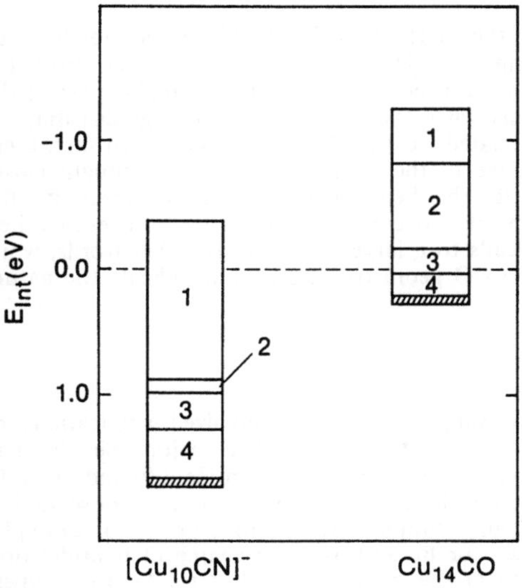

Figure 4. Block diagram for E_{Int} at the various CSOV steps

$E_{Int} < 0$ indicates repulsion while $E_{Int} > 0$ is an attractive interaction. For $[Cu_{10}CN]^-$, the interaction is between Cu_{10} and CN^-. The top of the block shows the repulsive FO interaction for the superposed L and Cu_n charge distributions. The FO repulsion is much larger for CO than for CN. For CO, there is a large Pauli repulsion due to the overlap and interpenetration of the CO and Cu_n charge distributions /7,13,39,44/. For CN, the repulsion is less because the electrostatic attraction of Cu_{10} and CN^- reduces the Pauli repulsion /13,69/. The size of block 1 measures the energetic importance of the Cu polarization in response to the presence of the ligand. It is modest for $Cu_{14}CO$ where the charge polarizes to reduce the Cu and CO interpenetration and consequent Pauli repulsion. It is large for $[Cu_{10}CN]^-$ because the Cu polarizes to form an image charge in response to CN^-; a characteristic of an ionic interaction. The Cu to CO π back-donation is responsible for the large size of block 2 for $Cu_{14}CO$; this block is negligible for $[Cu_{10}CN]^-$. Both of these results are fully consistent with the $\Delta P_{2\pi}$ values in Table 7. The ligand polarization is of modest importance, 0.1-0.2 eV contributions to E_{Int}. The ligand donation block arising from σ donation to Cu is larger for $[Cu_{10}CN]^-$ than for $Cu_{14}CO$. The hatched area indicates the difference between CSOV step 4 and the unconstrained SCF calculation; it is very small indicating that the CSOV steps have accounted for the major parts of the Cu-CO and Cu-CN interactions. The CSOV analysis confirms the characterization of the bonding given by the projection and the dipole moment curves. Most importantly, it makes it possible to assign energetic importance to the various charge rearrangements.

The consequences of the ligand to metal bonding have been studied for the ligand vibrations /16,19,25,60/ and for the PES spectra related to the ligand /18/. It is possible here to only briefly summarize the major conclusions of these studies. For the vibrations, two important effects have been identified. The first is an increase in the ligand stretch frequency at the FO charge superposition step. This increase is related to the FO repulsion shown in Fig. 4. It arises because the ligand is close to a "wall" of surface charge; stretching the ligand bond brings the ligand charge closer to the surface, a change which has a modest energetic cost which increases the CO stretch frequency by $\sim$ 200 cm^{-1} /19,25/; there are somewhat smaller, $\sim$100 cm^{-1}, increases in the CN stretch frequency /70/. The second effect is that the Cu back-donation into the ligand $2\pi^*$ reduces the frequency. This effect, which is important only for CO, cancels the increase due to the wall and leads to a small net change for the stretching frequency of CO adsorbed at an on-top site /19/. For CO adsorbed at an on-top site of a metal surface, the 5σ and 1π ionization energies are nearly degenerate and give rise to only a single PES peak /71/; in the gas phase, the 1π IP is 3eV larger than the 5σ IP. This shift had been incorrectly taken as evidence of a strong metal-CO σ bond /71/. However, as we have seen from the theoretical analysis above, the σ bond is not especially important. The major part of the differential 1π-5σ binding energy shift occurs when the metal and CO charge distributions are superposed /18,72/ and, hence, is not related to the chemical bonding. In fact, there is a monotonic increase of the 5σ IP relative to the 1π IP as the CO is brought closer to the metal /72/. The simple explanation for this behavior is that, at an on-top site, the 5σ lone pair screens the adsorption site metal and carbon atom nuclei from each other; removing an electron from the 5σ orbital leads to a large repulsion. In other words, while the differential 1π-5σ shift does indicate the CO geometry, it does not indicate the nature of the bond.

6. SUMMARY

This review has covered a wide range of subjects that have involved applications of rigorous quantum mechanical MO wavefunctions to clusters. The major objective has been to demonstrate that these wavefunctions, properly analyzed, can identify and quantify the importance of mechanisms that are responsible for the measured values of observables. Sometimes, the analyses give results that our intuition leads us to expect. For example, it is not at all surprising that, as was shown for Be and Mg, the degree of hybridization should be closely correlated to the strength of the bond in the cluster. At other times, the analyses give results that indicate limitations to the validity of simple models. For example, the jellium shell model is an extremely attractive way to explain the special stability of certain sizes of clusters. The jellium model (and its derivatives) discount the specific cluster nuclear geometry and shape as one of the important reasons for this special stability. Unfortunately, rigorous calculations indicate that the nuclear geometry may, in fact, be quite important for the cluster electronic structure. While it is necessary to understand the limitations of a model, it is even more important to develop models which provide correct views. Such models have been developed for the metal-CO and metal-CN bonds which explain a wide range of phenomena. These models have been obtained with the application of the CSOV method, orbital projection, and other techniques. They have been used to analyze properties which are quite different, as different as vibrational frequencies and photoemission spectra. New mechanisms have been identified and a broad unified view of the nature and consequences of the bonding of CO and similar molecules to metals, surfaces and clusters, has been obtained.

REFERENCES

1. For a survey of applications see, for example, (a) The Electronic Structure of Atoms and Molecules , H. F. Scheafer III (Addison-Wesley, Reading, 1972); see also (b) Applications of Electronic Structure Theory , H. F. Scheafer III editor (Plenum, New York, 1979).
2. H. P. Lüthi and J. Almlöf, Chem. Phys. Lett. 135 , 357 (1987); J. Almlöf and H. P. Lüthi in Supercomputer Research in Chemistry and Chemical Engineering , D. G. Truhlar and K. F. Jensen ed., American Chemical Society Symposium Proceedings (ACS, Washington, D. C., in press).
3. H. W. Krob, J. R. Heath, S. C. O'Brien, R. F. Curl, and R. E. Smalley, Nature 318 , 162 (1985).
4. C. W. Bauschlicher, P. S. Bagus, and H. F. Scheafer, III, IBM J. Res. Develop. 22 , 213 (1978).
5. K. Hermann, Phys. Bl. 36 , 227 (1980). 6. P. S. Bagus, in Plasma Synthesis and Etching of Electronic Materials , R. T. H Chang and B. Abeles, eds., Materials Research Society Symposia Proceedings, Vol 38 (Materials Research Society, Pittsburg, 1985), p. 179.
7. K. Hermann, P. S. Bagus, and C. J. Nelin, Phys. Rev. B 35 , 9467 (1987).
8. K. Hermann, and P. S. Bagus, J. Appl. Phys., in press.
9. W. A. Goddard, J. J. Barton, A. Redondo, and T. C. McGill, J. Vac. Sci. Technol. 15 , 1274 (1978).

10. P. S. Bagus and U. Wahlgren, Mol. Phys. 33 , 641 (1977).
11. S. J. M. Janssen and W. C. Nieuwpoort, Phil. Mag. B 51 , 127 (1985); Solid State Ionics 16 , 29 (1985).
12. P. S. Bagus, C. J. Nelin, and C. W. Bauschlicher, Surf. Sci. 156 , 615 (1985).
13. (a) P. S. Bagus, K. Hermann, and C. W. Bauschlicher, J. Chem. Phys. 80 , 4378 (1984); and (b) ibid., 81 , 1966 (1984).
14. P. S. Bagus, C. J. Nelin, and C. W. Bauschlicher, J. Vac. Sci. Technol. A 2 , 905 (1984).
15. L. G. M. Pettersson and P. S. Bagus, Phys. Rev. Lett. 56 , 500 (1986).
16. P. S. Bagus, C. J. Nelin, W. Müller, M. R. Philpott, and H. Seki, Phys. Rev. Lett. 58 , 559 (1987).
17. C. J. Nelin, P. S. Bagus, and M. R. Philpott, J. Chem. Phys. 87 , 2170 (1987).
18. P. S. Bagus and K. Hermann, Phys. Rev. B 33 , 2987 (1986).
19. P. S. Bagus and W. Müller, Chem. Phys. Lett. 115 , 540 (1985).
20. See, for example, Methods of Electronic Structure Theory , H. F. Scheafer III editor (Plenum, New York, 1977).
21. L. R. Kahn, P. Baybutt, and D. G. Truhlar, J. Chem. Phys. 65 , 3826 (1976).
22. J. C. Barthelat, P. K. Durand, and A. Serfini, Mol. Phys. 33 , 159 (1977).
23. O. Gropen, U. Wahlgren, and L. Pettersson, Chem. Phys. 66 , 453,459 (1982).
24. C. W. Bauschlicher, P. S. Bagus, C. J. Nelin, and B. O. Roos, J. Chem. Phys. 85 , 354 (1986).
25. W. Müller and P. S. Bagus, J. Vac. Sci. Technol. A 5 , 1053 (1987).
26. F. Parmigiani, E. Kay, P. S. Bagus, and C. J. Nelin, J. Elec. Sprectros. and Related Phenom. 36 , 257 (1985).
27. M. L. Cohen, M. Y. Chou, W. D. Knight, and W. A. de Heer, J. Phys. Chem. 91 , 3141 (1987); W. A. de Heer and W. D. Knight, paper in this volume.
28. R. K. Nesbet, Proc. Roy. Soc. London Ser. A 230 , 312 (1955).
29. R. Ahlrichs and P. R. Taylor, J. de Chimie Physique 78 , 315 (1981).
30. P. S. Bagus, B. Liu, A. D. McLean, and M. Yoshimine in Computational Methods For Large Molecules And Localized States in Solids F. Herman, A. D. McLean, and R. K. Nesbet, eds. (Plenum, New York, 1972).
31. C. C. J. Roothaan and P. S. Bagus, in Methods in Computational Physics , B. Alder, S. Fernbach, and M. Rotenberg, eds. (Academic Press, New York, 1963); C. J. J. Roothaan, Rev. Mod. Phys. 32 , 179 (1960); ibid., 23 , 69 (1951)
32. P. O. Löwdin, Adv. Chem. Phys. 2 , 207 (1959).
33. Y.-K. Kim and P. S. Bagus, J. Phys. B 5 , L193 (1972); Phys. Rev. A 8 , 1739 (1973).
34. B. O. Roos, P. R. Taylor, and P. E. M. Siegbahn, Chem. Phys. 48 , 157 (1980).
35. P. E. M. Siegbahn, J. Chem. Phys. 72 , 1647 (1980); 70 , 6391 (1979).
36. J. L. Whitten and T. A. Pakkanen, Phys. Rev. B 21 , 4357 (1980); T. A. Pakkanen and J. L. Whitten, J. Chem. Phys. 69 , 2168 (1978).
37. W. Müller and W. Meyer, J. Chem. Phys. 80 , 3311 (1984); W. Müller, J. Flesch, and W. Meyer, ibid., 80 , 3279 (1984).
38. P. S. Bagus and A. R. Williams, IBM J. Res. Develop. 25 , 793 (1981).
39. P. S. Bagus, C. W. Bauschlicher, C. J. Nelin, B. C. Laskowski, and M. Seel, J. Chem. Phys. 81 , 3594 (1984).
40. P. S. Bagus and L. G. M. Pettersson, to be published.
41. G. Pacchioni, P. S. Bagus, and M. R. Philpott, to be published.
42. P. S. Bagus and K. Hermann, to be published.
43. P.-O. Löwdin, Phys. Rev. 97 , 1474 (1955).
44. P. S. Bagus, C. J. Nelin, and C. W. Bauschlicher, Phys. Rev. B 28 , 5423 (1983).
45. C. W. Bauschlicher, and P. S. Bagus, J. Chem. Phys. 81 , 5889 (1984).
46. B. Liu and A. D. McLean, J. Chem. Phys. 72 , 3418 (1980).
47. C. W. Bauschlicher, P. S. Bagus, and B. N. Cox, J. Chem. Phys. 77 , 4032 (1982).
48. P. J. Hay and W. R. Wadt, J. Chem. Phys. 82 , 299 (1985).
49. G. del Conde, P. S. Bagus, and O. Novaro, Phys. Rev. A 26 , 3653 (1982).
50. K. Hermann, and P. S. Bagus, Phys. Rev. B 16 , 4195 (1977).
51. C. J. Nelin, and P. S. Bagus in Festkörperprobleme (Advances in Solid State Physics), Vol. 25, p. 135, P. Grosse, editor (Vieweg, Braunschweig, 1985)
52. P. A. Cox, Mol. Phys. 30 , 389 (1972).

53. P. S. Bagus and H. F. Scheafer, J. Chem. Phys. 56 , 224 (1972).
54. P. S. Bagus, K. Hermann, and M. Seel, J. Vac. Sci. Technol. 18 , 435 (1981).
55. W. L. Jolly in Electron Spectroscopy D. Shirley, editor (North-Holland, Amsterdam, 1972)
56. C. P. Keijzers, P. S. Bagus and J. P. Worth, J. Chem. Phys. 69 , 4032 (1978).
57. P. S. Bagus, C. J. Nelin, E. Kay, and F. Parmigiani, J. Elec. Sprectros. and Related Phenom., in press.
58. G. K. Wertheim, S. B. DiCenzo, and S. E. Youngquist, Phys. Rev. Lett. 51 , 2310 (1983).
59. M. G. Mason, L. J. Gerenser, and S.-T. Lee, Phys. Rev. Lett. 39 , 288 (1977); M. G. Mason, Phys. Rev. B 27 , 748 (1983).
60. P. S. Bagus, C. J. Nelin, K. Hermann, and M. R. Philpott, Phys. Rev. B, in press.
61. S. Andersson and J. B. Pendry, Phys. Rev. Lett. 43 , 363 (1979) and references therein.
62. K. Kunimatsu, H. Seki, W. G. Golden, J. G. Gorden II, and M. R. Philpott, Surf. Sci. 158 , 596 (1985).
63. R. Kötz and E. Yeager, J. Electroanal. Chem. 123 , 335 (1981).
64. C. D. Makowka, C. P. Slichter, and J. H. Sinfelt, Phys. Rev. B 31 , 5663 (1985); S. L. Rudaz, J.-P. Ansermet, P.-K. Wang, and C. P. Slichter, Phys. Rev. Lett. 54 , 71 (1985).
65. G. Blyholder, J. Chem. Phys. 68 , 2772 (1964).
66. P. S. Bagus, K. Hermann, W. Müller, and C. J. Nelin, Phys. Rev. Lett. 55 , 2618 (1986).
67. J. Billman and A. Otto, Surf. Sci. 138 , 1 (1984).
68. P. S. Bagus, C. J. Nelin, M. R. Philpott, and K. Hermann, to be published.
69. K. Hermann, W. Müller, and P. S. Bagus, J. Elec. Spectros. and Related Phenom. 39 , 107 (1986).
70. K. Hermann, P. S. Bagus, and C. J. Nelin, to be published.
71. C. L. Allyn, T. Gustafsson, and E. W. Plummer, Chem. Phys. Lett. 47 , 127 (1977); Solid State Commun. 24 , 531 (1977).
72. P. S. Bagus, C. J. Nelin, and K. Hermann, Aust. J. Phys. 39 , 731 (1986).

Clusters: Ionization, Reactions and Properties

A.W. Castleman, Jr.[1*] *and R.G. Keesee*[2]

[1]Physikalische Chemie-Elektrochemie,
Technische Hochschule Darmstadt,
D-6100 Darmstadt, Fed. Rep. of Germany
[2]Department of Chemistry, The Pennsylvania State University,
University Park, PA 16802, USA

1. INTRODUCTION

During the last decade or so, there has been a dramatic growth of activity in the field of cluster research. Reference to clusters permeates the literature of a very broad cross section of basic and applied science from that dealing with condensed matter and surfaces to interstellar space. In particular, clusters provide a unique way of investigating properties of matter at different degrees of aggregation and therefore yield information on changing properties and reactivity that serves to bridge the gas and condensed state.

Measurements of spectroscopic shifts upon successive clustering are especially interesting with regard to the onset of liquid or solid-like features in the spectra (1,2). Investigations of the processes of cluster ionization and dissociation are of particular interest since they contribute to a further understanding of the evolution of changes in ionization potentials as a system approaches the bulk work function.

Studies of the dynamics of formation and dissociation, and the changing properties of clusters at successively higher degrees of aggregation, enable an investigation of the basic mechanisms of nucleation to be probed at the molecular level. Work on systems of increasingly higher degrees of aggregation also has a direct bearing on the development of surfaces, and ultimately solvation phenomena and the formation of the condensed state. The progressive clustering of a molecule involves energy transfer and redistribution within the cluster system, with attendant processes of unimolecular dissociation taking place between growth steps.

Investigations of the molecular properties, reactions, and behavior of clusters generally require ionization in one of the steps as either a probe and/or a method of detecting clusters through mass spectrometry. Energy transfer and dissociation are operative during the reorientation of molecules about ions following the primary ionization event. Although ionization can be accomplished through electron impact as well as single

---

* U.S. Senior Scientist Alexander von Humboldt Awardee
Permanent Address: Department of Chemistry, The Pennsylvania State University, University Park, PA. 16802 USA

photon techniques, resonance enhanced multiphoton ionization often enables selective ionization of clusters in particular states. Recent advances in the field of molecular beam research, coupled with lasers and time-of-flight mass spectrometry, enable the details of these various processes to be investigated. Intracavity laser techniques, along with high pressure mass spectrometry, enable other aspects of dissociation and the thermochemistry of clusters to be determined.

The present manuscript surveys four interrelated aspects of this field, drawing largely on findings from the authors' own laboratory. Although various problems are discussed, emphasis is placed on the topic of ion clusters.

2. REACTIONS IN CLUSTERS, STUDIES OF CLUSTER DISTRIBUTIONS, AND THE ORIGIN OF MAGIC NUMBERS

Neutral clusters are often investigated by mass spectrometry, whereby they are detected as cluster ions. A subject of considerable interest is the relative abundance of the ensuing cluster ions as a function of the degree of aggregation. Many systems display anomalous abundances at certain cluster sizes which are invariant with the expansion conditions used to produce the neutral clusters; these features are commonly called magic numbers. In many cases these have been attributed to special features in the size distribution of the neutral clusters, but mounting evidence in most cases points to the importance of the processes which follow ionization and the properties of the cluster ions. Interest in the reactions in clusters following ionization also derives from the belief that the results will further elucidate the relationship of ion-molecule reactions in the gas phase to those in condensed media.

2.1 Experimental Techniques

Supersonic expansion techniques are commonly used to produce beams of neutral clusters (3). Cooling of the beam is accomplished through the conversion of thermal energy in the high pressure source gas into a directed beam velocity u. Taking C_p to represent the heat capacity of the expanding gas and T_s the absolute temperature in the expansion source, a simple energy balance, Equation (1), for an ideal gas of molecular weight m displays why the temperature T_b in the beam is colder.

$$C_pT_s = 1/2\ mu^2 + C_pT_b\ . \tag{1}$$

Translational temperatures below a few degrees Kelvin are routinely obtained. In the absence of clustering, simple molecules can also be substantially cooled rotationally and vibrationally. Cooling promotes condensation. Relationship (1) neglects the effect of the latent heat of condensation resulting from the formation of clusters, and clusters may be vibrationally "hot".

The time-of-flight (TOF) mass spectrometer has been in use for many years. The instrument is experiencing a resurgence in popularity due to the advent of pulsed lasers which supply efficient and short durations of ionization in a small volume. In a typical time-of-flight mass spectrometer, either a two element or alternately a single element accelerating field may be used in the region of ionization. This is followed by a field-free drift region at the end of which the arrival of the ions is detected. Using the conventional TOF method, dissociation which occurs with rates in the neighborhood of 10^6 to 10^8 sec^{-1} can be investigated by either of two methods. One involves analyzing the peak shape (arrival spectrum) of the ions under dual field acceleration conditions, in which case a knee is observed since the ions spend far more time in the first low-field region where ionization is initiated than in the second high-field region where the bulk of the acceleration occurs (4). An alternate method is to operate under single field conditions and deduce rates from the shape of the late arriving tail of the peak (5).

A third method of studying dissociation employs a reflecting field (reflectron) (3). Although originally designed to enhance the resolution of the TOF method, a reflectron can also be employed to investigate dissociation in the field-free drift region, so that slower processes may be observed. These experiments are performed by subjecting the cluster beam to multiphoton ionization using a Nd:YAG pumped dye laser with wavelength extension components that provide frequency doubling and mixing capabilities. Typical photon fluxes are about 10^{16} photons per pulse of approximately 5 ns duration. The ions are accelerated in the accelerating field to about 2 KeV, whereafter they enter a field-free region and then are electrically reflected and detected as shown in Figure 1. With appropriate potentials applied to the reflectron grids, nondissociating parent ions can be separated from those that dissociate while in the field-free region. A unique identification of these daughter ions can be accomplished by the time separation or by an energy analysis with the reflectron. The separation of the parent and daughter ions occurs as a result of the loss in kinetic energy, with essentially no change in velocity of the ion upon dissociation. Hence species with greater kinetic energy (parents) have a longer path to the detector than do the daughter (dissociation) products.

2.2 Internal Ion-molecule Reactions and Dissociation

2.2.1 Ammonia Clusters

Extensive investigations of the dissociation of ammonia clusters following multiphoton ionization at 266 nm were conducted in our laboratory (3). Interaction with the laser beam leads to ionization of one of the molecules in the hydrogen bonded neutral ammonia cluster and ejection of an electron from the cluster. Due to the large proton affinity of NH_3 (8.85 eV), and analogous to the well-known gas phase ion-molecule reaction between NH_3^+ and NH_3, an internal (intra-cluster) ion-molecule reaction rapidly forms NH_4^+ with the

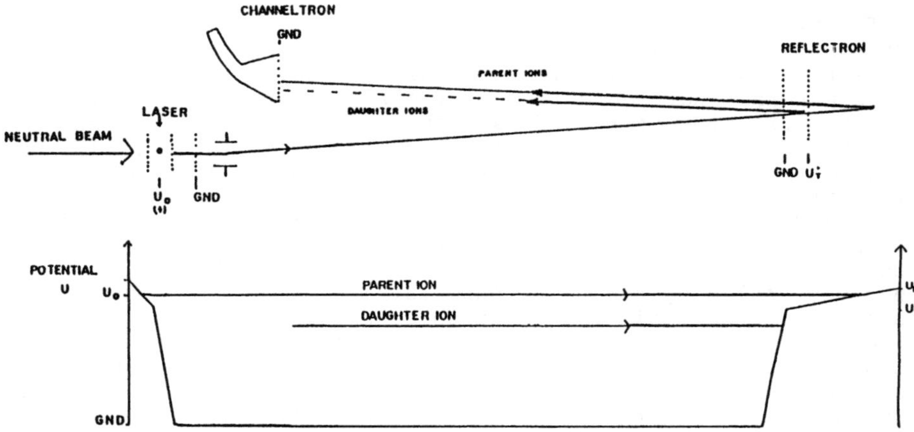

Figure 1. Scheme of time-of-flight mass spectrometer with reflectron where daughter ions are reflected before parent ions. Lower figure depicts the electrostatic potentials

ejection of NH_2. Thereafter the ammonia molecules reorient to accommodate the proton. Unimolecular (unicluster) dissociation may occur as a result of the excess vibrational energy created during the conversion from a neutral to an ionized system. Collision-induced dissociation is also generally observed as a result of the interaction of the cluster ions with residual molecules in the vacuum chamber.

$$(NH_3)_n + h\nu \rightarrow NH_3^+(NH_3)_{n-1} + e \quad (2)$$

$$NH_3^+ (NH_3)_{n-1} \rightarrow [NH_4^+(NH_3)_{n-2}]^* + NH_2 \quad (3)$$

$$[NH_4^+(NH_3)_{n-2}]^* \rightarrow NH_4^+(NH_3)_{n-2-m} + mNH_3 \quad (4)$$

The extent of fragmentation is determined through use of the reflectron.

The contribution of unimolecular, compared to collision induced, dissociation is determined through experiments in which the pressure in the field-free region is varied. An extrapolation of the data to a zero of pressure can be made in a linear fashion under the thin collision approximation, where, within the uncertainty of the true zero of pressure, the ordinate gives a direct measure of the fraction of dissociation by unimolecular (evaporative) loss. The component from unimolecular decay has been found to increase steadily with cluster size from n = 4 to 25, all being of order 10^5 sec^{-1}. Though at first glance the trends are surprising, they have been found to be consistent with simple considerations (3).

2.2.2 Methanol Clusters

Similar studies to the foregoing were undertaken using 266 nm light to ionize $(CH_3OH)_n$ clusters (6). Several processes were observed, the first of which has a direct analogy to the ammonia studies. Following ionization, methanol clusters undergo proton transfer with a concomitant loss of CH_3O. The exothermicity of the proton transfer reaction and subsequent relaxation of the alcohols around the protonated moiety also leads to excess internal energy and dissociation processes. The dissociation processes lead to unimolecular evaporation rates in a manner similar to the ammonia system and the rates are also in the order of 10^5 to 10^6 sec^{-1}.

A nonevaporative loss process was also observed, namely the elimination of H_2O from the protonated dimer. This process, however, was not detected for the larger cluster ions. This observation is in direct correspondence with observations (6) of gas-phase ion-molecule reactions:

$$CH_3OH_2^+ + CH_3OH \longrightarrow (CH_3)_2OH^+ + H_2O \qquad (5)$$

$$\xrightarrow{M} CH_3OH_2^+ \cdot CH_3OH \qquad (6)$$

and

$$CH_3OH_2^+ . CH_3OH + CH_3OH \longrightarrow (CH_3)_2OH^+ \cdot CH_3OH + H_2O \qquad (7)$$

$$\xrightarrow{M} (CH_3)OH_2^+(CH_3OH)_2 \quad . \qquad (8)$$

The rate for the H_2O loss from $CH_3OH_2^+ \cdot CH_3OH$ has been measured to be 5.5×10^5 s^{-1}, in good agreement with the range of 10^5 to 10^6 s^{-1} estimated for the gas-phase ion-molecule process.

Based on measurements and estimates for the thermodynamics of various dissociation channels in $H^+ \cdot (CH_3OH)_3$, one would expect the loss of H_2O from this and larger methanol clusters to be at least as energetically favorable as that for the CH_3OH. Two different structural arrangements are believed to be important in the reaction; it is thought that rearrangement from one conformation to the other is necessary for the water loss process and that this process can occur readily only in the protonated dimer, perhaps being impeded in higher order clusters due to structural hindrance. Rapid energy dissipation in larger clusters may reduce the possibility of the rearrangement. These data reflect the importance of structural rearrangements necessary for certain unimolecular processes to be operative, and thereby give further insight into differences which may be seen in the condensed phase compared to the gas for ion-molecule reactions.

2.2.3 Xenon Clusters

Recently, we conducted a multiphoton ionization study of xenon clusters (5). Figure 2 shows the cluster abundances for the system Xe_n^+. Mass spectra with high-energy electron impact ionization are virtually identical. This observation indicates the importance of the energy liberated following ionization of the cluster in effecting the magic numbers and size distribution of weakly bound clusters as compared to the energy

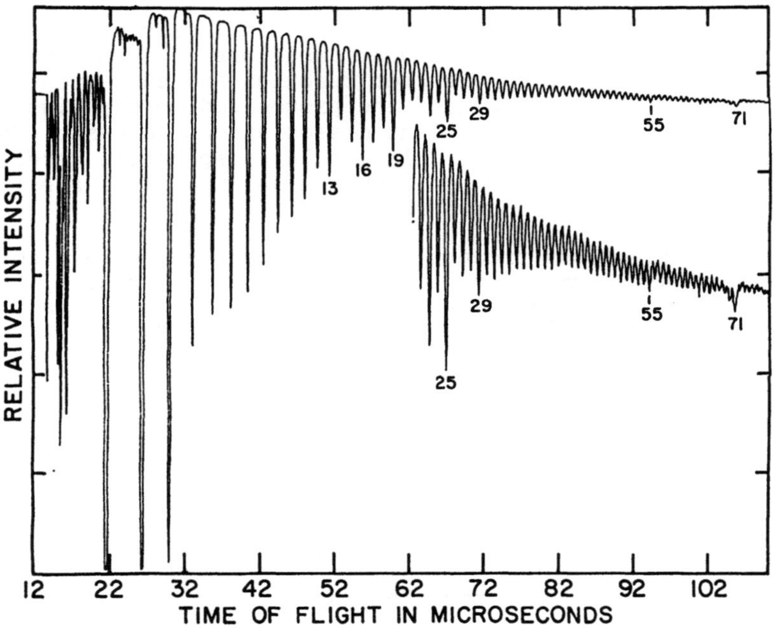

Figure 2. TOF mass spectra of Xe_n^+ obtained by multiphoton ionization of xenon clusters

deposited to initiate ionization. From a study of the peak shapes of the Xe^+, Xe_2^+ and Xe_3^+ in the time-of-flight spectra under single field acceleration, the dissociation rates could be estimated. A tail was found toward longer times, which is the result of rapid dissociation of larger clusters. The dissociation rates for Xe_5^+, Xe_4^+, and Xe_3^+ are found to lie in the neighborhood of 5×10^7 sec^{-1}. These are evidently the only species which are metastable in the time domain accessible by this method.

2.3 Origin of Magic Numbers

Substantial rearrangement occurs following the ionization of one of the moieties of a cluster; excess energy then arises within the cluster which leads to appreciable evaporative unimolecular dissociation (3,7). Consider a dissociation sequence

$$A_{n+1} \xrightarrow{A} A_n \xrightarrow{A} A_{n-1}. \tag{9}$$

Ascertaining the ratio of cluster concentrations at some (long) time following ionization and dissociation requires solution of the appropriate differential equations accounting for the rates of formation and loss due to dissociation, with due consideration of the initial concentrations. Nevertheless, the origin of the magic numbers can be seen simply from a consideration of the equation,

$$d[A_n] / dt = -k_{r,n}[A_n] + k_{r,n+1}[A_{n+1}], \quad (10)$$

and recognition that

$$k_{r,n+1}/k_{r,n} \approx \exp(-\Delta(E_{0,n})/RT) . \quad (11)$$

The latter follows from the fact that for a system with a large number of vibrational modes, the rate of unimolecular dissociation for both a microcanonical and canonical ensemble are nearly identical (8).

$$k_r = A\, e^{-E_{0,n}/\bar{E}} , \quad (12)$$

where $\bar{E}$ is the average energy per oscillator and $E_{o,n}$ is the energy required to remove a molecule from a cluster of size n. Whenever there are discontinuities in an otherwise smooth trend of decreasing $E_{o,n}$ with cluster size, a magic number will appear. A gool example, studied in some detail, is the case of ammonia where a magic cluster corresponding to the protonated pentamer, $(NH_3)_5H^+$, is well established (3,7).

3. CLUSTER IONIZATION AND SPECTROSCOPY: RELATIONSHIPS TO SOLVENT SHIFTS AND WORK FUNCTIONS

3.1 Spectral shifts

Resonance enhanced multiphoton ionization through the specific excitation of an electronic state of a chromophore contained within a cluster is a powerful method of ascertaining the properties of clusters and relating these to their counterparts in the condensed and isolated gas phases (1, 2, 9, 10). Generally, the clustering of atoms or molecules onto a chromophore results in a perturbation of the electronic states of that chromophore. The spectral shift of a given electronic transition from that of the isolated chromophore is a measure of the relative differences between the lower and upper states of the energetic perturbation induced by clustering.

We have employed both one- and two-color resonance enhanced multiphoton ionization to investigate the $S_1 \leftarrow S_0$ π-electron transition in phenylacetylene and p-xylene as clusters with various solvents. Single color multiphoton ionization studies of the perturbed $L_b(^1B_2)$ states of phenylacetylene (PA) bound with Ne, Ar, Kr, and Xe were all found to induce a lowering of the S_1 resonance with respect to the ground state. These observed red shifts have been attributed to dispersive interactions with solvent molecules (10-12). In general, a spectral shift is governed by three factors; (1) short-range electronic repulsive interactions which result in a blue shift, (2) electronic dispersive interactions which result in a red shift, and (3) differences of zero point energies between the excited and ground states.

A linear dependence of the spectral shift on the electrostatic polarizability of the rare gas atom has also been found. The sensitivity to atom polarizability is indicative of the important role of dispersive forces in the perturbation of the electronic states in the aromatic molecule. The spectral shift due to solute-solvent interactions, in the Onsager model, contains a term due to dispersion (13). Our results support the finding (10) that in aromatic molecule-rare gas atom systems, the spectral shifts are dictated by atom polarizability, i.e., the important role of dispersive forces in the perturbation of the S_1 excited state. The results of our study show a linear dependence of the spectral shift on the electrostatic polarizability of the rare gas atom, and conform to the Onsager model, but on a microscopic level.

Particularly interesting are the trends seen for larger clusters. Figure 3 shows a selected set of data for the spectral shifts relative to the S_1 electronic origin of phenylacetylene for clusters containing four to ten argon atoms. First it is interesting to note that the major feature asymptotically approaches a shift of approximately 50 cm^{-1}. Clearly the additivity rule does not apply. Secondly, the van

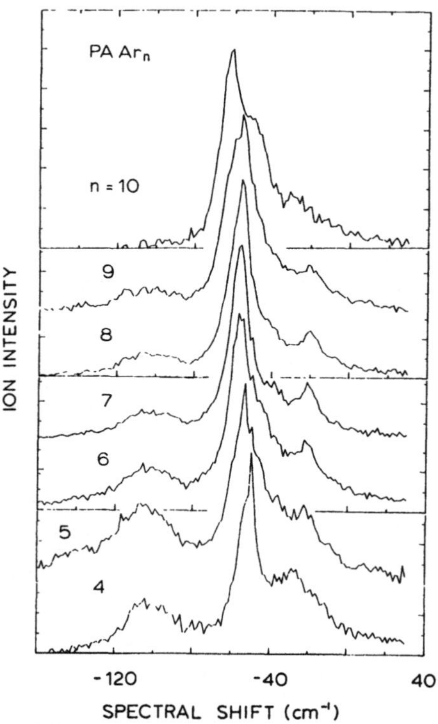

Figure 3. R2PI current versus one-photon energy. The ion currents are recorded at the m/e ratios corresponding to PA·Ar_n ($4 \le n \le 10$). The energy scale is relative to the S_1electronic origin of PA. The ion current scale is relative and different for each spectrum and P_o = 300 Torr

der Waals modes to the right side of the main resonance begin to fill in for large cluster sizes. The spectra are broadening in analogy to those seen in the condensed phase, and the features to the right resemble phonon modes for a system of infinite lattice.

Other interesting spectral shifts have been observed for the clustering of N_2, O_2, N_2O, NH_3, H_2O, CCl_4, and CH_4 to phenylacetylene. In most cases the main resonance is also red-shifted, although in a few a substantial blue shift is observed, most notably for H_2O. The striking difference between the isoelectronic molecules H_2O and NH_3 can be rationalized in terms of the excitation of the π system leading to a repulsive interaction with the two lone-pair electrons of the H_2O molecule (9).

3.2 Shifts in Ionization Potential with Degree of Aggregation

3.2.1 Organic Chromophores Solvated with Rare Gas Atoms

It is interesting to compare the behavior of the ionization threshold of neutral clusters (or the appearance potential for the ionized products) as a function of the degree of aggregation. The ionization potentials of p-xylene bound with argon ($PX \cdot Ar_n$) were determined through studies in which the energy of one photon was fixed at the L_b resonance and the wavelength of the second laser was scanned.

It is well known that the Stark effect leads to a shift in ionization potential when measured in an electric field, and correction is necessary to account for shifts of the order of 50 cm^{-1}. The ionization potentials are found to vary with the square root of the electric field present in the region of ionization in accordance with expectations and findings of others (14). Extrapolation to zero field is readily accomplished in view of the linear dependence and the fact that various cluster systems display lines of identical slopes in these weakly perturbed rare-gas aggregates.

The shifts (10) in ionization potential of p-xylene in the rare gas aggregates are shown in Figure 4 for clusters with one to six argon atoms. The shift in relative ionization potential is observed to display a broadly linear dependence on the number of argon atoms. The largest deviations from this trend are observed for the dimer and pentamer. The observed total shift of about 750 cm^{-1} for the hexamer is to be contrasted with the matrix isolated value which is about 6000 cm^{-1} for a similar molecule (benzene) in an argon matrix (15). Evidently, the "local environment" with which a molecule interacts is relatively large in such a matrix, and the observed shift is far from the expected bulk value. This is in interesting contrast to the metal systems discussed in the next subsection. When corrected for ion image potential effects, the ionization potential of aggregates composed of only a few alkali metal atoms display nearly the bulk ionization potential of the polycrystalline metallic system (1, 16, 17).

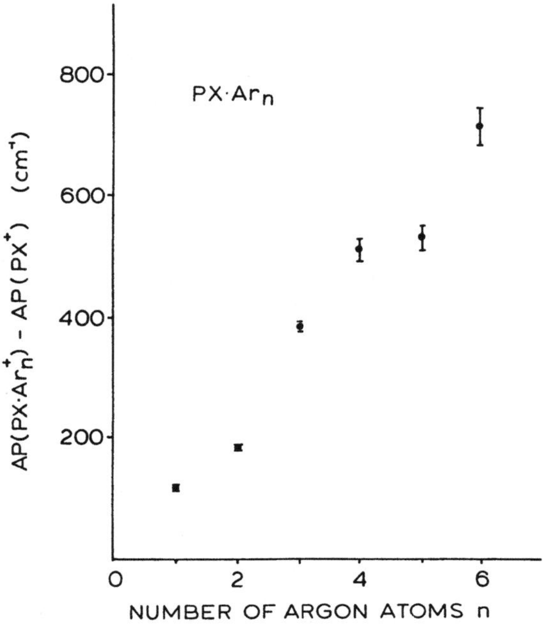

Figure 4. Relative appearance potentials. Photo ionization of $PX \cdot Ar_n$ (n=0-6) in a 150V/cm dc field. $AP(PX \cdot Ar_n^+)$ increases with the coordination number n

3.2.2 Comparison with Metallic Systems

Due to their simple one-electron nature, and the ease with which they can be compared with theoretical calculations, systems comprised of alkali metals are particularly interesting for study. Photoionization spectra for Na_x, with x ranging from 1 to 8, have been obtained in our laboratory using a molecular beam coupled with a UV light source and a quadrupole mass spectrometer detection system (16). Sodium clusters are produced by vaporizing the metal and expanding it with argon gas at pressures ranging between 70 and 100 torr through a 300 µm diameter nozzle. Once formed, the clusters pass through a skimmer into a detection chamber where they are photoionized and mass analyzed. Photoionization is accomplished by use of a 500 watt xenon arc lamp. The desired wavelength is selected with a monochromator. The light intensity of the monochromator is continually monitored using a thermopile detector, and the collected spectra are corrected for variations in power and the monochromator slit function.

Data from our laboratory (16,17) agree extremely well with measurements reported by Schumacher and coworkers (18) for all clusters with the exception of x = 5, 6, and 8, where our findings point to slightly lower values.

Perhaps of most interest are the trends observed in the ionization potentials with cluster size. The experimental data

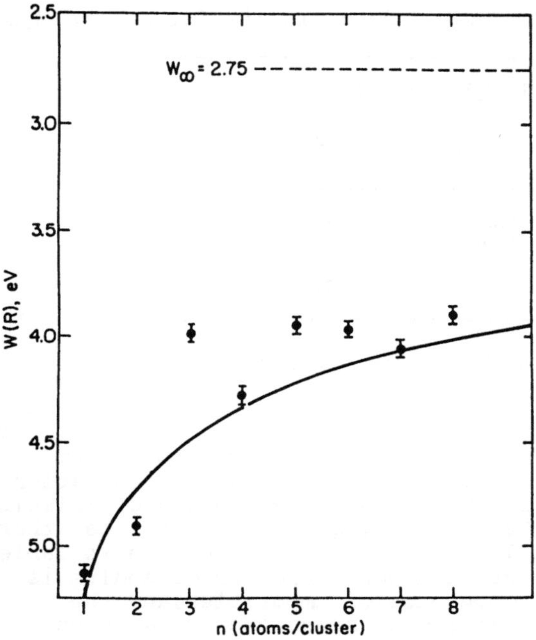

Figure 5. The ionization potential or work function of sodium clusters as a function of size. The curve is based on Eq. (13) where radius R is related to the number of atoms by assuming the bulk density for the clusters

correlate resonably well with the classical electrostatic expression relating the work function of a system of radius r to the bulk work function, W_∞ (see Figure 5). A classical equation relates the influence of particle size on the ion image potential contributions to the work function of a system with spherical symmetry. The equation is derived from classical equations of electrostatics:

$$W(R) = W_\infty + 3e^2/8R \,, \qquad (13)$$

where W(R) represents the work function of a cluster with radius R of the equivalent sphere supporting elementary charge e, and W_∞ the bulk work function of the polycrystalline metal. The photoionization thresholds of sodium clusters are in good agreement at all but the smallest sizes. A number of other metallic systems also correlate well with this classical theory.

Recently, limited data have become available for the photoionization of sub-oxides of alkali metals for further comparison with theory (19,20). Table I lists the difference in ionization potential between Na_xO and Na_x, where measurements are available from our studies for x up to 4. The ionization potential for NaO is much greater than for Na. The difference between the oxide and pure species becomes diminishingly small

Table I. Difference in Ionization Potential[a] for M_x and M_xO

$\Delta \equiv IP(M_x) - IP(M_xO)$

x	Na Δ, (eV)	K Δ, (eV)
1	- 1.36	——
2	- 0.20	- 0.65
3	0.08	- 0.35
4	0.33	- 0.02

[a]results calculated from data given in ref. 19

for x = 3 and the trend reverses for x = 4. The potassium system (19), while not displaying this reversal in ionization potential between oxide and pure metal, does show a diminishing trend with cluster size. It is tempting to speculate that there would be a reversal for a larger cluster (see results in Table I). It has been reported that the work function of sodium is appreciably lowered by the presence of impurities,and the discrepancy between early and recent measurements has been traced to surface contamination.

3.3 Intracluster "Penning Ionization" and Evidence for Slow Ionizaton Processes in Clusters

Clusters provide interesting systems for comparing ionization and concomitant electron transfer processes for bimolecular processes in the gas phase, including Penning ionization, with analogous ones in the condensed phase. Toward this goal, ionization of clusters comprised of p-xylene (PX) bound to NH_3 and $N(CH_3)_3$ were studied (21) following the absorption of photons through the perturbed S_1 state of p-xylene. An interesting comparison is provided by results of studies involving adducts of p-xylene bound to NH_3 and trimethylamine, since the ionization of p-xylene is less than that of ammonia but greater than that of trimethylamine. In the case of $PX \cdot NH_3^+$, ionization by adsorption of a second photon,which is adsorbed by the perturbed S_1 state of p-xylene,begins near the ionization threshold of p-xylene and leads to the expected cluster ion $PX \cdot NH_3^+$. Two other channels are possible at higher photon energies, namely the formation of NH_4^+ at 0.1 eV above the ionization potential of p-xylene,and NH_3^+ at 1.8 eV above; NH_4^+ is observed in the two-color experiments at high fluence of the ionizing laser,where two photons are absorbed by the S_1 state.

By contrast, absorption into high Rydberg states of p-xylene below its ionization potential in $PX \cdot N(CH_3)_3$ leads to the production of predominantly $N(CH_3)_3^+$, with $H^+N(CH_3)_3$ as a minor product. No $PX \cdot N(CH_3)_3^+$ ion is detectable. One conclusion is that photoexcitation of p-xylene leads to an intracluster ionization process bearing analogy to Penning ionization where the perturbed high Rydberg states of p-xylene interact with the partner molecule $N(CH_3)_3$. A second, and more startling observation, was the finding of a slow ionization process. Since the laser interacts with the molecules in the first of a two-field acceleration region, a long tail is only possible when the ionization process is slow. Fragmentation leads to a knee in the peak shape and not a long tail as observed in Fig. 4. Interestingly, the process is substantially slower with a decrease in the energy of the ionizing photon. Questions arise as to whether the slow step is associated with the proton transfer channel (i.e., the $(CH_3)_3NH^+$ product) or an electron transfer process (i.e., the $(CH_3)_3N^+$ product). Careful measurements with deuterated species reveal that the tail is largely associated with the electron transfer process. Interestingly, Hatano (22) has found that orientational effects in the liquid phase, where motion is restricted, can lead to a significant reduction in the rate of Penning ionization. Likewise, Harris (23) has evidence for long delays in ionization of NH_3 on silver electrode surfaces. Whether there is some analogy to the foregoing observations is currently unknown. A plausible explanation for the present findings is that a large geometry change is involved in the formation of the trimethylamine ion during the Penning-like ionization process.

4. CLUSTER IONS: DYNAMICS OF DISSOCIATION

Study of the dynamics of photodissociation of clusters provides a way of investigating intramolecular energy transfer for a series of nearly identical molecules having varying degrees of freedom. Furthermore, the interesting question of coupling between modes of widely differing frequency can be addressed. Effects of solvation on the dynamics can also be directly investigated. Studies of ion clusters are particularly interesting, since investigations can then be done on mass selected systems, and energy release upon dissociation can be determined in a relatively straightforward manner.

Vibrational predissociation can occur when one of the molecular subunits absorbs an amount of energy larger than a bond dissociation energy of the cluster. Many photodissociation experiments have involved neutral van der Waals molecules, where vibrational predissociation proceeds by the transfer of a small amount of energy from the initially excited mode of a molecule to the intermolecular bonds of the cluster. The energy associated with one vibrational quantum in a molecular entity of a neutral cluster generally exceeds that required to dissociate the cluster. In the case of cluster ions, an energy corresponding to a change of several vibrational quanta is required for dissociation to occur.

This section deals with results from our laboratory of studies of the dissociation dynamics of a homomolecular system, the anion $(SO_2)_2^-$, and a heteromolecular system, CO_3^- and its hydrates. The bonding of all of these cluster anions exceeds the energy associated with one quantum of vibration in the core ion.

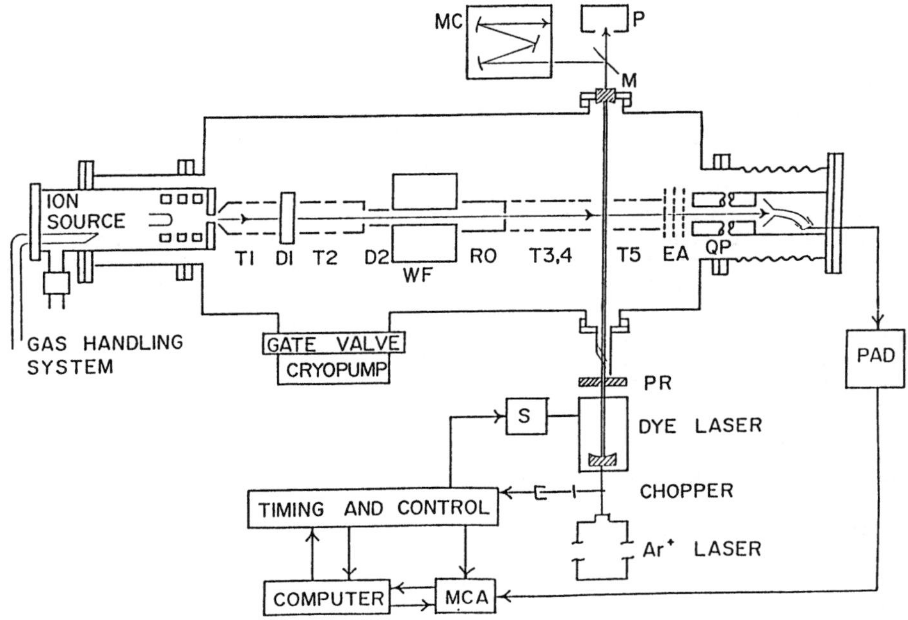

Figure 6. Schematic diagram of photodissociation spectrometer.

4.1 Experimental Technique and Data Analysis

The apparatus (Figure 6) used for energy resolved photodissociation spectroscopy consists of an ion source, a system of electrostatic ion optics to control the position and energy of the ion beam, a Wien velocity filter to mass select a particular ionic species, a tunable dye laser (with the ion beam traversing intracavity) pumped by an argon ion laser, a triple screen retarding field energy analyzer, a quadrupole mass spectrometer, and associated signal acquisition and control electronics (24). The ion source is of a type used for high pressure mass spectrometry in our laboratory (25). Photodissociation occurs in the interaction region, which is surrounded by a cylindrical screen held at a well defined electrical potential. The intensities of parent ions and all photofragment ions are monitored by the quadrupole mass spectrometer as a function of both laser frequency and ion kinetic energy in order to determine the spectroscopic structure of the parent ions and the dynamics of the dissociation process.

Energy analysis is performed with the two outer screens of the energy analyzer held at the same potential as the surrounding ion optical elements, while the center screen is scanned over a range of repulsive potentials (26,27). Only ions whose energies are equal to or greater than the center potential are able to pass through the analyzer and be detected; all other ions are repelled. This experiment yields the integral of the lab frame energy distribution. The actual energy distribution is generated by numerically differentiating the raw data. The amount of energy released by the fragmentation process into relative translation is calculated from the location and the width of both parent and fragment peaks (24,26).

The energy released into relative translational energy of the fragments, E_r, is obtained from

$$E_r = [E_b m_1 + E_{1x} M \pm 2(m_1 M E_b E_{1x})^{1/2}]/m_2 \cos^2\Theta_1 \,, \qquad (14)$$

where Θ_1 is the angle that the fragment ion makes with the beam axis, M the parent mass, m_1 the fragment ion mass, and m_2 the neutral fragment mass. The value of E_{1x} corresponds to the center of the measured fragment energy distribution. The probability of a given E_{1x} is then given by

$$P(E_{1x}) = P(E_b)P(E_r)P(\Theta) \,, \qquad (15)$$

where $P(E_b)$ is the probability of a given parent energy (measured in an experiment), $P(E_r)$ is the probability of a given kinetic energy release (calculated from either the isotropic distribution or phase space), and $P(\Theta)$ is the probability of a given recoil angle.

The probability of product recoil per solid unit angle is (28)

$$P(\Phi) = (1/4\pi)\ [1 + \beta P_2(\cos \Phi)] \,, \qquad (16)$$

where Φ is the recoil angle with respect to laser polarization, $P_2(\cos \Phi)$ is the second Legendre polynomial, and β is the anisotropy parameter. The value of β ranges from 2 for a pure parallel transition, 0 for an isotropic transition, to -1 for a pure perpendicular transition. Physically, the anisotropy parameter can be thought of as the product of two angle dependencies; one relating the angle between the internuclear axis and transition dipole moment and the second between the internuclear and dissociation axes (28, 29). The second arises as a result of molecular rotation which occurs after photon absorption but before dissociation.

4.1.1 The $(SO_2)_2^-$ System

The SO_2 dimer anion has been observed in our laboratory to photodissociate to SO_2^- as the sole ionic photofragment (26); its photodissociation cross section has been reported (30) to be a smooth peak displaying a maximum near 2.1 eV and having a total cross section of 1.9×10^{-17} cm^2. Power studies made in our laboratory at 600 nm (2.07 eV) indicate that the dissociation is a one-photon process. We have measured the bond energy for SO_2^- to be 1.04 eV (31) and that for $SO_2 \cdot SO_2$ to be 0.19 eV (32). Based on the reported electron affinity for SO_2 of 1.1 eV (33), the results lead to 1.95 eV as the electron affinity of the dimer ion.

With the above relationships and the isotopic distribution, computations of relative intensity as a function of retarding

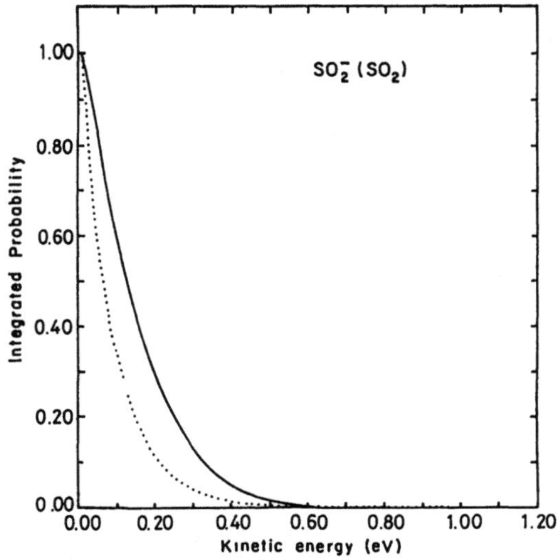

Figure 7. The experimentally determined (solid line) and statistical (dotted line) integrated probability of kinetic energy release for photodissociation of $SO_2^- \cdot SO_2$

potential can be made and compared with the original experimental data. Our experimental laboratory energy distributions of the SO_2^- photofragments are best fit by an anisotropy parameter of 1.2. This result suggests that the transition is more parallel than isotropic and that the dissociation is rapid with respect to rotation.

Figure 7 shows the probability of having a kinetic energy release exceeding a certain energy. The experimental results indicate that 50% of the fragments have a kinetic energy release of 0.2 eV or greater. The integrated probability expected if the process were statistical is also shown in Figure 7. This calculation is based on a statistical phase space theory developed by Bowers and coworkers (34). Although the calculated result would vary slightly, depending upon the vibrational frequencies employed in the calculations, the difference between phase space theory and the experiment is significant and further points to dissociation on a repulsive surface.

4.1.2 CO_3^- and Its Hydrates

Studies of the photodissociation of CO_3^- and its hydrates have also been conducted in our laboratory (24,35). In the photon energy range 1.95 to 2.2 eV, photodissociation of CO_3^- leads to O^- and CO_2 fragments. The photodissociation spectrum measured in our laboratory shows a well-defined, sharp structure. Excitation to a bound intermediate excited state ion has been confirmed by our work. One involves two-photon excitation to a repulsive surface and the other results from collisional dissociation of the excited anion which has undergone internal conversion from the excited intermediate state to high vibrational levels of the ground state. Photodissociation is initiated by a $^2A_1 \leftarrow {}^2B_1$ bound-bound transition that promotes the ion from the ground state to the intermediate weakly bound excited. Analysis of the kinetic energy for the two-photon mechanism indicates that a large fraction (about 20%) of the available energy is deposited into relative translation of the photofragments.

The bond energies for water with CO_3^- have been measured in our laboratory and found to be approximately 0.6 eV each (25). The only photodissociation channel observed for the CO_3^- hydrates is loss of all attached water molecules. It is an interesting observation that, in the larger clusters where partial dehydration is possible, the loss of all ligands is the only observed channel. The kinetic energy analysis of the CO_3^- photofragment ions from all three hydrates indicates that more energy is partitioned into translation than is predicted by phase space. The overall percentage of available energy that is channeled into relative translation is 27%, 18%, and 27% for the first, second, and third hydrates, respectively.

In the case of the hydrates, no O^- photofragment ion is detected. Conversion of energy from the core ion into the vibration of ion-water bonds leads to dissociation of the cluster. In the case of the CO_3^- hydrates neither a $\Delta v = -1$ nor -2 transition would be sufficient to result in bond dissociation. Although vibrational predissociation would be expected to pertain to these systems, a surprisingly large amount of energy is partitioned into relative translation and we cannot rule out the possibility that dissociation to some repulsive surface is operative in this cluster system.

5. CLUSTER IONS: THERMOCHEMISTRY, SOLVATION, AND NUCLEATION PHENOMENA

Research on the properties of ion clusters has been particularly valuable and revealing (1-3) in the field of interphase physics, which is concerned with the molecular details of phase transitions (nucleation phenomena), the development of surfaces, and ultimately solvation phenomena and formation of the condensed state. Clusters are often formed following the bombardment of surfaces with ions, and studies of the factors influencing their size, stability, and mechanisms of formation also comprise a related active area of research in surface science.

5.1 Experimental Methods for Deriving Thermochemical Data

The formation of ion clusters in the gas phase proceeds via a sequence of association reactions between ions and the molecules with which they interact:

$$A^{\pm}\cdot B_{n-1} + B + M == A^{\pm}\cdot B_n + M \quad . \qquad (17)$$

Here, $A^{\pm}$ designates the ion (of either sign) about which molecule B is clustered; M is the third body which serves to stabilize the cluster during the formation process.

High pressure mass spectrometry has been particularly valuable in quantitatively determining the stability of ion clusters. In this method, ions effuse from a high pressure source (typically a few tenths to a few torr) through a small aperture into a mass spectrometer where the distribution of ion clusters is determined(25,36). Ions are created with a heated filament, via thermionic emission in the case of positive metallic ions, and via emission of electrons with subsequent attachment to form negative ions. The resulting ions are focussed into a region which is thermally controlled by either resistive heaters or a circulation liquid. The pressure of the ion source is maintained sufficiently high, such that ions reside in a region of well-defined temperature for a time adequate to ensure attainment of equilibria among the various ion cluster species of interest. A sample of these ions effuses through an orifice and is mass analyzed with a quadrupole mass filter using sequential pressure of the clustering neutrals in the ion source; equilibrium constants for the reactions are evaluated. By changing the temperature of the reaction region, one can construct van't Hoff plots; the enthalpy and entropy changes are calculated from the slope and the extrapolated infinite temperature-intercept, respectively.

5.2 Ion Induced Nucleation

In order for a phase transition between the gaseous and the condensed state to take place, molecules must first cluster together and subsequently grow to critical size (37-39). In principle, condensation is possible whenever the partial pressure of a vapor exceeds the vapor pressure of the condensed phase. Based on a formation involving kinetics of association reactions, a critical-size cluster, designated n^*, is assumed to exist. It is frequently assumed that subsequent collision of the critical-size cluster with another molecule leads irreversibly to a transition from the gaseous to the condensed state. A major problem of current interest concerns the development of theoretical formulations to quantitatively determine the size and concentration of critical clusters.

The classical liquid drop formulation of Thomson (40),

$$\Delta G_{0,n} = -nRT \ln S + 4\pi N r^2 \sigma + (q^2 N/2)(1 - 1/\varepsilon)(1/r - 1/r_i), \quad (18)$$

is commonly used to evaluate the free energy for formation of the nth cluster. The cluster size at which $\Delta G_{0,n}$ obtains a local maximum defines the critical cluster. The first term accounts for the change in free energy due to the condensation of n molecules at a saturation ratio S (where S is the ratio of the partial pressure of the ligand to its normal bulk vapor pressure at the same temperature). For the standard state, the partial pressure is taken to be 1 atm. The second represents the work done in forming a droplet of radius r (where σ is the surface tension and N is Avogadro's Number). The final term gives the change in field energy due to the condensation of a dielectric about the ion (where ε is the dielectric constant of the solvent, q the ionic charge, and r_i the radius of the ionic cavity). It should be noted that, in the Thomson equation, the nature of the ion is accounted for only by its radius and total charge. The stepwise change is given simply by $\Delta G_{0,n} - \Delta G_{0,n-1}$.

Differentiation of eq.18 with respect to temperature enables an evaluation of entropy changes; enthalpy changes are then deduced by substitution into well-known thermodynamic relationships. Studies of gas-phase clustering equilibria provide experimental data for the small size domain and also direct comparison to the theoretical treatment (41,42).

5.3 Energetics of Ion Solvation

Data on the clustering of molecules to individual ions can be used to determine the energies of solvation of individual ions in the condensed phase. In the former case, the overall reaction is

$$A^{\pm q}(g) + nB(g) \rightarrow A^{\pm q} \cdot nB \quad , \tag{19}$$

with a standard free energy change designated by

$$\Delta G^\circ_{0,n} = \sum_{i=1}^{n} \Delta G^\circ_{i-1,i} \quad , \tag{20}$$

whereas in the latter case

$$A^{\pm q}(g) + nB(l) \rightarrow A^{\pm q} \cdot nB(l) \quad , \tag{21}$$

with a single free energy of solvation of $\Delta G^\circ_{(solv)}$ (at infinite dilution for large n). When n is sufficiently large, the product of reaction 19 is also a liquid droplet and so the free energies are related by

$$\Delta G^\circ_{0,n} = \Delta G^\circ_{solv} + \sum_{i=2}^{n} \Delta G^\circ_{i=1,i}\,(B) \quad . \tag{22}$$

Hence, the last term is the free energy change for forming a droplet through the stepwise clustering of the solvent B. At sufficiently large n, $\Delta G^\circ_{i-1,i}(B)$ becomes the free energy of condensation for neutral B. Corresponding equations can be written for the enthalpy and entropy changes.

One of the first attempts to predict solvation energies in the condensed state was based on the Born relationship (43) where the ion is assumed to be a rigid sphere of radius r_i and charge q immersed in a structureless continuum with dielectric constant ε. ΔG°_{solv} is taken as the electrostatic potential difference between the ion in the solvent and in a vacuum.

An approach for relating the gas-phase data to solvation is to consider the ratio $\Delta H^\circ_{solv}/\Delta H^\circ_{0,n}$ as a function of cluster size. Figure 8 shows the ratio, as calculated by the Born and Thomson equations, compared to experimental data for the hydration of positive ions. The ratio is seen to converge for a range of ionic radii for clusters of about five solvent molecules. Similar results are also obtained for the

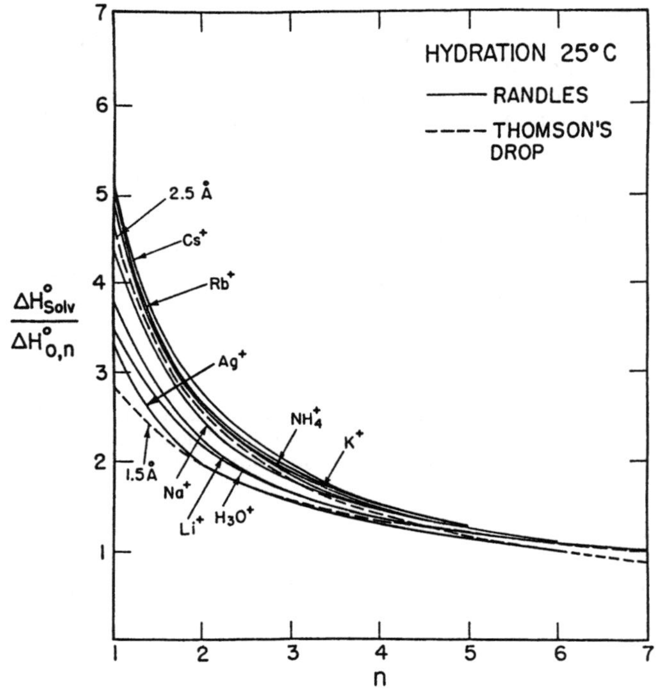

Figure 8. Ratio of Randles' total enthalpy of solvation to the partial gas-phase enthalpy of hydration for positive ionic cluster size n

ammoniation of positive ions and hydration of negative ions (41,42). The findings suggest that solvation data can be deduced from the gas-phase experiments.

6. SUMMARY AND FUTURE PROSPECTS

The field of cluster research is rapidly growing and the conceptual gap between the gaseous and condensed phases is closing. Hopefully, an appreciation of the importance of cluster research, the advances which are being made, and the exciting opportunities for fundamental studies pertaining to the dynamics and properties of what is sometimes termed a "fifth state of matter" has been gained.

Support by the Department of the Army, Grant No. DAAG29-82-K-0160 and DAAG29-84-K-0087, the Department of Energy, Grant No. DE-AC02-82-ER60055, and the National Science Foundation, Grant No. ATM-82-04010, is gratefully acknowledged. I wish to especially acknowledge a collaborative interaction with Professor T. D. Märk (Innsbruck) made possible though a joint USA-NSF/Austrian Science Foundation Grant No. INT 8420823. AWC Jr. thanks the Alexander von Humboldt Foundation for the award and the hospitality of Prof. K.G.Weil (Technische Hochschule Darmstadt) during the time which this manuscript was prepared.

1. A. W. Castleman, Jr. and R. G. Keesee, "Clusters: Bridging the Gas and Condensed Phases,.." Accts. Chem. Res., 19, 413 (1986).
A. W. Castleman, Jr. and T. D. Mark, in: Gaseous Ion Chemistry/Mass Spectrometry (F. H. Futrell, Ed.) John Wiley and Sons, pp. 259-303(1986).
2. A. W. Castleman, Jr. and R. G. Keesee, Chem. Rev. 86, 589 (1986).
3. O. Echt, P. D. Dao, S. Morgan, and A. W. Castleman, Jr. J. Chem. Phys. 82, 4076 (1985). See also O. Echt, S. Morgan, P. D. Dao, R. J. Stanley, and A. W. Castleman, Jr. Ber. Bunsenges. Phys. Chem. 88, 217 (1984).
4. J. L. Durant, D. M. Rider, S. L. Anderson, F. D. Proch, and R. N. Zare, J. Chem. Phys. 80, 1817 (1984).
5. O. Echt, M. Cook and A. W. Castleman, Jr., "Multiphoton Ionization of Rare Gas Clusters: Xe_n," Chem. Phys. Lett. 135, 229 (1987).
6. S. Morgan and A. W. Castleman, Jr., "Evidence of Delayed 'Internal' Ion Molecule Reactions Following the Multiphoton Ionization of Clusters: Variation in Reaction Channels in Methanol with Degree of Solvation," J. Am. Chem. Soc. 109, 2867 (1987).
7. A. W. Castleman, Jr., S. Morgan, O. Echt, P. D. Dao, and R. G. Keesee "Considerations of the Origin of Magic Numbers in Hydrogen Bonded Clusters," (to be published).
8. R. M. Stratt and J. E. Adams , J. Chem. Phys. 78, 2368 (1983)
9. P. D. Dao, S. Morgan, and A. W. Castleman, Jr., Chem.Phys. Lett. 111, 38 (1984).
10. P. D. Dao, S. Morgan, and A. W. Castleman, Jr., Chem.Phys. Lett. 113, 219 (1985).
11. S. Leutwyler, U. Even and J. Jortner, J. Chem. Phys. 79, 5769 (1983).
12. D. H. Levy, Ann. Rev. Phys. Chem. 31, 197 (1980); A. Amirav, U. Even and J. Jortner, J. Chem. Phys. 71, 2319 (1979); 75, 3770 (1981); A.M. Griffiths and P. A. Freedman, Chem. Phys. 63, 469 (1981); A. Amirav and J. Jortner, Chem. Phys. 85, 19 (1984); and K. Rademann, B. Brutschy and H. Baumgartel, Chem. Phys. 80, 129 (1983).
13. S. Basu, Advan. Quantum Chem. 1, 145 (1964).
14. K. H. Fung, H. L. Selzle and E. W. Schlag, Z. Naturforsch 36a, 1257 (1981).
15. J. Jortner, in: Vacuum Ultraviolet Radiation Physics (E.E. Koch, R. Haensel, and C. Kunz, Eds.) Pergamon Press, Oxford, p. 291 (1974).
16. K. I. Peterson, P. D. Dao, R. W. Farley, and A. W. Castleman, Jr. J. Chem. Phys. 80, 1780 (1984).
17. A. W. Castleman, Jr. and R. G. Keesee, "Metallic Ions and Clusters: Formation, Energetics, and Reaction," Zeitschrift für Physik, 3, 167 (1986).
18. A. Herrmann, S. Leutwyler, E. Schumacher, and L. Woste, Helv. Chem. Acta 61, 453 (1978).
19. P. D. Dao, K. I. Peterson, and A. W. Castleman, Jr., J. Chem. Phys. 80 563 (1984).
20. K. I. Peterson, P. D. Dao and A. W. Castleman, Jr., J. Chem. Phys. 79, 777 (1983). 21. P. D. Dao and A. W. Castleman, Jr., J. Chem. Phys. 84, 1435 (1986).

22. Y. Hatano (personal communication); see also T. Wada, K. Shinsaka, H. Namba, and Y. Hatano, Can. J. Chem. 55, 2144 (1977).
23. C. Harris (personal communication); Univ. of California, Berkeley.
24. D. E. Hunton, M. Hofmann, T. G. Lindeman, and A. W. Castleman, Jr., J. Chem. Phys. 82, 134 (1985).
25. R. G. Keesee, N. Lee, and A. W. Castleman, Jr., J. Am. Chem. Soc. 101, 2599 (1979).
26. A. W. Castleman, Jr., C. R. Albertoni, K. O. Marti, D. E. Hunton, and R. G. Keesee, "Photodissociation of Negative Ions and Their Clusters", Faraday Discuss. Chem. Soc. 82 (1986).
27. A. W. Castleman, Jr., and R. G. Keesee, "Dynamics of Dissociation and Reactions of Cluster Ions", Structure/Reactivity and Thermochemistry of Ions, (P. Ausloos and S. G. Lias, Eds.) D. Reidel Publishing Company, pp. 185-217 (1987).
28. R. N. Zare, Mol. Photochem. 4, 1 (1972); G. E. Busch and K. R. Wilson, J. Chem. Phys. 56, 3638 (1972).
29. S. -C. Yang and R. Bersohn, J. Chem. Phys. 61, 4400 (1974).
30. R. V. Hodges and J. A. Vanderhoff, J. Chem. Phys. 72, 3517 (1980).
31. R. G. Keesee, N. Lee, and A. W. Castleman, Jr., J. Chem. Phys. 73, 2195 (1983).
32. J. J. Breen, K. Kilgore, K. Stephan, R. Sievert, B. D. Kay,
R. G. Keesee, T. D. Mark, J. van Doren, and A. W. Castleman, Jr., Chem. Phys. 91, 305 (1984).
33. R. J. Celotta, R. A: Bennett, and J. L. Hall, J. Chem. Phys.60, 1740 (1974).
34. W. J. Chesnavich and M. T. Bowers, Prog. React. Kinet. 11, 137 (1982); W. J. Chesnavich and M. T. Bowers, in Gas Phase Ion Chemistry, Vol. 1, ed. M. T. Bowers (Academic, New York, 1979) pp. 119-151; computational program supplied by M. T. Bowers.
35. D. E. Hunton, M. Hofmann, T. G. Lindeman, C. R. Albertoni, and A. W. Castleman, Jr., J. Chem. Phys. 82, 2884 (1985).
36. A. W. Castleman, Jr. P. M. Holland, D. M. Lindsay and K.I. Peterson, J. Am. Chem. Soc., 100, 6039 (1978).
37. A. W. Castleman, Jr., P. M. Holland and R. G. Keesee, J. Chem. Phys. 68, 1760 (1978).
38. A. W. Castleman, Jr., Adv. Colloid Interface Sci. (1979), 10, 73.
39. A. C. Zettlemoyer, Nucleation; Dekker: New York (1969).
40. J. J. Thomson, Application of Dynamics to Physics and Chemistry. 1st ed: Cambridge Cambridge 1888.
41. N. Lee, R. G. Keesee and A. W. Castleman, Jr., "On the Correlation of Total and Partial Enthalpies of Ion Solvation and the Relationship to the Energy Barrier to Nucleation," J. Colloid Interface Sci., 75, 555 (1980).
42. P. M. Holland and A. W. Castleman, Jr., "The Thomson Equation Revisited in Light of Ion Clustering Experiments," J. Phys. Chem., 86, 4181, (1982).
43. M. Born, Z. Phys. 1, 45 (1920).

Chemistry on Molecular Surfaces: Reactions of Gas Phase Clusters

D.M. Cox, M.R. Zakin, and A. Kaldor

Exxon Research and Engineering Company, Clinton Township, Route 22 East, Annandale, NJ 08801, USA

1. Introduction

Studies of gas phase metal cluster chemistry start with a fully coordinatively unsaturated, "naked" cluster free of support interactions or ligands. This offers a unique opportunity to explore a number of key scientific issues in chemistry and physics as the number of constituent atoms is increased and bulk properties as well as stable surface structures evolve. These include questions such as: What are the structures of the various clusters? Are they unique? How do the electronic, optical and chemical properties of the clusters change as a function of the number and type of constituent atoms? How large does the cluster have to be before solid state theoretical description applies? What are the magnetic properties of "naked," and "dressed" clusters? Is catalytic chemistry possible on such clusters, etc?

This area lies between traditional inorganic cluster chemistry and surface chemistry. In the former, the clusters are "clothed", i.e. the fully ligated or even partially unsaturated clusters are stable. In surface chemistry long range order, subsurface atoms, as well as support interactions play a significant role. This approach may be able to connect two areas where the common thread is the identification of the localized surface bond. While clusters mimic the surface where localized bonding is prevalent, small clusters, in addition, mimic the surface with the greatest disorder and the least amount of long range order.

Next we note that for clusters containing less than 100 atoms most atoms of the cluster are "surface atoms". For instance, if one simply assumes bulk densities and spherical clusters, a 100 atom cluster would have about 72 surface atoms. A 20 atom cluster, on the other hand, would have about 18-19 surface atoms. Obviously, if the cluster structure is planar or chain-like, all atoms should be considered as surface atoms. Thus, probably with reasonable justification, it has been suggested that the study of clusters should be viewed as the study of "molecular surfaces"[1]. While the structure of these clusters is not known, it may be reasonable to assume that many structural isomers coexist, with a range of energy barriers among them. One could characterize them as fluxional; a surface analogue would be surface reconstruction and faceting. Such restructuring during exothermic chemical reactions is very probable. Beyond the obvious implications of bringing together two important areas of chemistry, and the theoretical foundations underlying them, equally important is that developments in this field may force solid state theory and molecular electronic structure theory to resolve conflicts, whether in language or in fact.

Recent advances in experimental techniques have made it possible to study in the gas phase the physical and chemical properties of clusters of virtually any material[2]. Although clusters consisting of hundreds of atoms have been synthesized, the majority of studies to date have concentrated on clusters of fewer than 30 atoms[3]. In this paper we briefly review selected results from our laboratory on a few systems: The nature of large carbon clusters, the remarkably selective reactions of aluminum clusters with a number of reagents, the reactivity of various transition metals with H_2, CO and C_6H_6, and infrared multiple photon dissociation spectroscopy of transition metal clusters containing adsorbates.

2. Experimental Technique

Only a general overview of the experimental techniques will be given here. For the interested reader further details are readily available from the references[4-11]. Figure 1 shows, schematically, the chemical reactor addition to the cluster source used in the gas phase reaction studies. In operation a high pressure pulse of helium gas is present at the metal rod when a high power vaporizing laser is fired. The metal vapor is entrained in the helium carrier and confined to flow through a narrow channel, where cluster growth occurs. The clusters then expand into the reactor tube where a second pulsed valve injects either helium or a mixture of helium plus reactant. The reaction is quenched when the cluster/helium/reactant mixture expands into a vacuum at the exit of the reaction tube. The resulting beam is collimated and subsequently interrogated downstream by photoionization time of flight mass spectrometry.

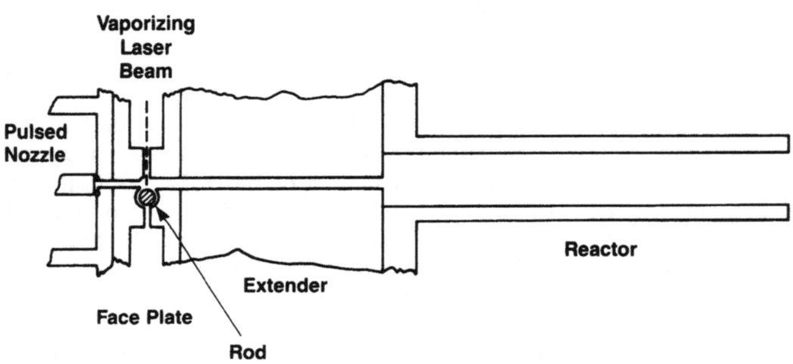

Fig. 1. Scale-drawn scheme of the essential features of the cluster source and chemical reactor system. The channel from the nozzle to the rod is 0.1 cm diameter, opening to 0.2 cm diameter from the rod to end of the 1.27 cm wide faceplate. The extender and reactor tubes as shown are 0.2 cm diameter by 3.81 cm long and 1 cm diameter by 7.6 cm long, respectively. Different length and diameter extender and reactor tubes are easily installed.

The clusters are ionized by either an excimer laser operating with a variety of gas mixtures to yield 7.87, 6.42 or 5.0 eV photons, or by a tunable UV laser source which is able to produce radiation at photon energies lower than 6.5 eV. We take care, in the experiments, to minimize the laser fluence used, to avoid fragmentation of the clusters. This means that we attempt to avoid multi-or multiple photon processes by using low fluence, highly efficient single photon ionization.

3. Carbon Clusters

Rohlfing et.al. reported the first mass spectrum of carbon clusters which exhibited a unique bimodal distribution[12]. Only even atom clusters were observed for those containing more than forty atoms. Certain mass peaks were observed to be significantly more intense than others, and Kaldor et.al. [1c] wondered about the "magic" nature of certain mass peaks in the spectrum, specifically C_{50}^+, C_{60}^+, and C_{70}^+. More recently, at Rice University, Smalley, Curl, Kroto and their collaborators performed a series of experiments which indeed brought forth the magic [13-16]. They reported that by providing more efficient cooling during cluster production they were able to shift conditions to maximize C_{60}^+ ion yield[13]. They achieved a contrast of about 40 compared to other ion peaks in the distribution.

They reported that more than 50% of the total carbon cluster yield was C_{60}^+. They also observed odd carbon species in this large cluster range. In reactivity studies of the large carbon clusters they showed that C_{60}, as well as all even C_n clusters, are inert, while the odd clusters react away[14]. From these observations they concluded that large C_n clusters are especially stable, and specifically C_{60} is an ultrastable cluster. They proposed that a "soccerball"-like structure could explain the ultrastability. In further experiments they argued that the interior space of such a cluster would be large enough to accommodate only one metal atom, such as lanthanum, or barium, and argued that the observation of only C_nLa^+, for n>40, and specifically of only the monoadduct $C_{60}La^+$, was further demonstration of the validity of the proposed "soccerball" structure[15].

In a series of experiments we have re-examined the carbon cluster system, C_n, and have observed that C_{60}^+ is only a minority constituent of the total carbon cluster ion signal in the beam; it is less than 0.1%[17-18]. We find two sources for production of carbon clusters, one which produces mostly smaller (n<35) clusters and one which preferentially produces larger (n>35) clusters, thus accounting for the bimodal distribution. In contrast to what we reported earlier[12], C_n^+, $n > 40$ exhibit a quadratic laser fluence dependence for 6.42 eV photoionization when the more efficient cooling approach, demonstrated by Smalley et al [13] is used. The ionization potential for these clusters lies above 6.42 eV. This last observation has an important impact on the metal doping experiments as well. As shown in Figure 2, the sensitivity for the detection of the naked carbon clusters is quite different from that for the lanthanum doped ones. The ionization potential for the doped clusters is reduced substantially when complexed with a low ionization potential metal; falling below 6.42 eV for K and La, C_nK and C_nLa are single photon ionized. With 7.87 eV photons, all the large clusters are single photon ionized and a more realistic measure of their absolute concentration is achieved. The lower signal of $C_{60}La^+$ relative to C_{60}^+, obtained with 7.87 eV ionization alluded to above, suggests a low concentration of $C_{60}La^+$, and an undetectably low concentration of $C_{60}La_2^+$.

In experiments where the graphite target was treated with KOH to introduce potassium onto the graphite, we were able to detect carbon clusters containing multiple potassium atoms [17,18], including $C_{60}K_3^+$, shown in Figure 2. This suggests that the metal loading is not on the interior of the "soccerball"-like structure, rather it is on the exterior. It appears likely that an edge site of an open structure is being decorated, something often observed for potassium loaded carbon

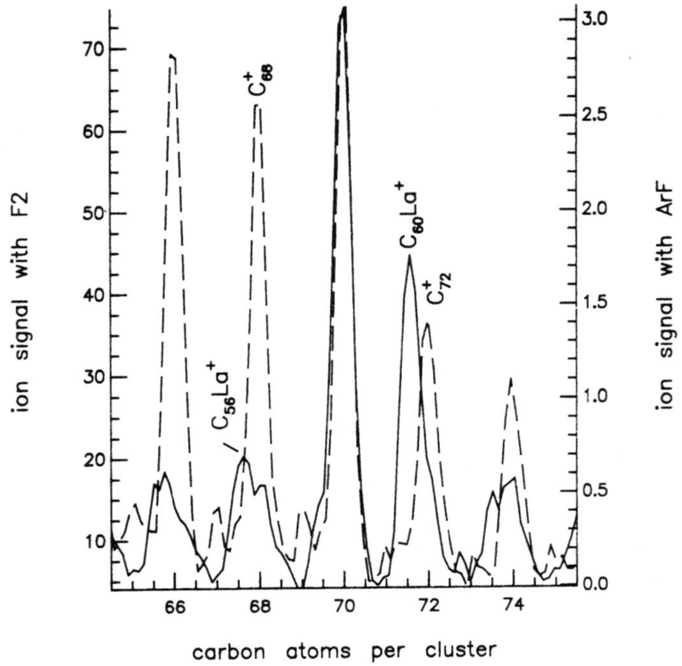

Fig. 2 Photoionization time-of-flight mass spectra obtained from pulsed-laser vaporization of a lanthanum-impregnated graphite rod. The solid line and right-hand ordinate refer to the spectrum taken with 6.42 eV ionization and the dashed line and left-hand ordinate refer to the spectrum taken with 7.87 eV ionization.

surfaces[19]. It is interesting to note in Figure 3 that the observed loading shows strong fluence dependence. At higher fluence level there are fewer metal atoms observed on the carbon clusters, strongly suggesting that significant laser "cooking" is taking place, something that the Rice group [15] has also noted in lanthanum experiments, but attributed to the generation of stable species. The metal loading experiments with K suggest that the structure of the carbon clusters is not a <u>closed</u> shell.

Time of arrival studies of different carbon clusters point to two different sources for the production of the clusters[18]. One cluster source is the vaporization zone and produces mostly small clusters (n<32), and the other source produces clusters from deposits formed during the experiment on the walls of the nozzle system and gives rise to mostly larger clusters (n>32). This latter source was discovered by accident when LaC_n^+ signal continued to be observed after a lanthanum doped graphite target was replaced with one that contained no lanthanum. By adjusting the experimental conditions in the source the relative ion signal of different clusters can be dramatically altered. Similarly the relative ion signal depends on the arrival time chosen for detection, and on the ionization conditions.

We suggest that the "magic" of C_{60} can be explained by <u>assuming</u> that the enhanced C_{60}^+ signals observed in the photoionization of neutrals arise from highly selective differences in ionization processes (single

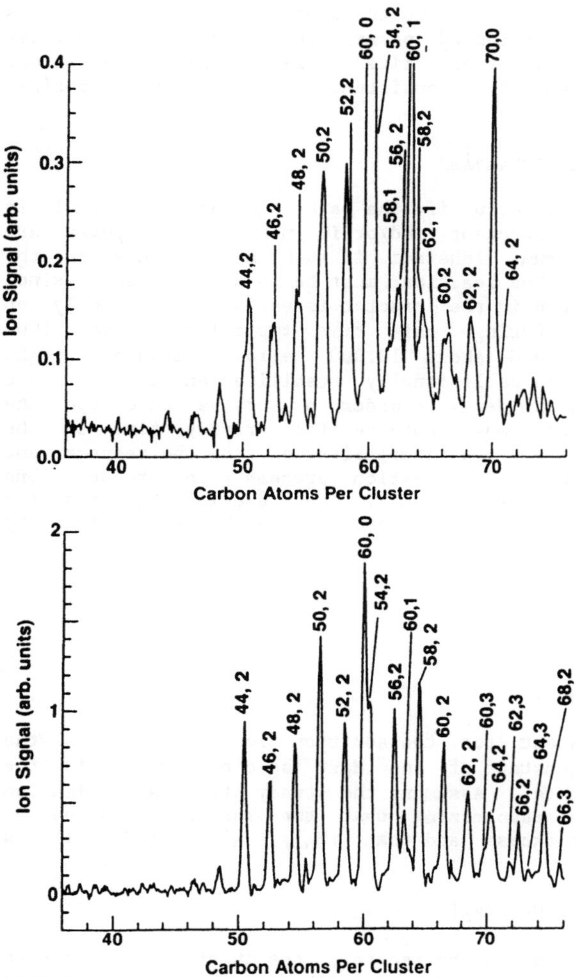

Fig. 3 Time-of-flight mass spectrum obtained when a potassium-treated graphite rod is vaporized. Ionization is with 6.42 eV (ArF) photon energy at 0.2 mJ/pulse (lower panel) and 1.3 mJ/pulse (upper panel). The peak labels n,m indicate C_nK_m

photon near threshold ionization with 7.87 eV photons, near-resonant two photon ionization with 6.42 eV and non-resonant two photon ionization with 5 eV photons) and that C_{60} has a somewhat lower ionization potential and higher electron affinity than other nearby clusters. The enhanced C_{60}^- and C_{60}^+ ion signals observed when positive or negative cluster ions are detected directly may also be explained if C_{60} possesses a relatively lower ionization threshold and higher electron affinity than other nearly C_n, properties which would affect the relative ion concentrations in the plasma.

We note that although there has been a groundswell of theoretical support for the proposed spheroidal structure[20] direct experimental confirmation of any structure for C_{60} and the other large carbon clusters is still lacking. If C_{60} does indeed possess the electronic properties proposed above, we now have a set of criteria against which theoretical calculations can be tested for consistency with various possible structures.

4. Kinetic Model of Cluster Reactivity

Reaction is presumed to occur if the bare cluster ion signal is depleted upon addition of reactant and/or if new product peaks are detected in the mass spectrum. Inherent in such measurements are the following major assumptions: The reaction must be relatively fast, since the residence time of clusters in the reactor is estimated to be only on the order of 100 usec. During this time typically 10 to 1000 cluster-reactant collisions and about 10,000 helium collisions might occur. The products must be collisionally stabilized on a time scale rapid compared to unimolecular decay in order to survive and escape the reactor. Similarly products must survive the transit time to the detection zone (~500 usec in our apparatus) after exiting the reactor, and in addition must survive the photoionization process (i.e. product ions observed are assumed to be parent ions). The reaction kinetics for chemisorption of a reagent molecule R onto a cluster M_n is described by equ. 1 and 2,

$$M_n + R \underset{k_n'}{\overset{k_n}{\rightleftharpoons}} M_nR^* \quad (1)$$

$$M_nR^* + He \xrightarrow{k_s} M_nR + He \ , \quad (2)$$

where k_n is the rate constant for the addition reaction, k_n' is the unimolecular decay rate constant of the M_nR^* species and k_s is the collisional stabilization rate. Assuming the steady state approximation can be applied to M_nR^*, recognizing that the reactant and helium concentrations are both in excess and that $[M_n]_f/[M_n]_o = I_f/I_o$, it is easily shown that

$$-\ln(I_f/I_o) = k_n\ [R]\ t\ (k_s[He]/(k_s[He] + k_{n'}) \ , \quad (3)$$

where t is the residence time in the reactor, the quantities in square brackets are the concentrations of the respective species in the reactor and I_f and I_o are the bare cluster ion signals after and before reaction, respectively. The last expression in parentheses on the right hand side of equ. 3 represents the competition between unimolecular decay and collisional stabilization of M_nR^*. If collisional stabilization is dominant then equ. 3 simply becomes

$$-\ln(I_f/I_o) = k_n\ [R]\ t \ , \quad (4)$$

and we find that an experimentally determined quantity, $-\ln(I_f/I_o)$, is directly proportional to the pseudo first order rate constant, k_n. Inherent in the derivation of equ. 4 are the assumptions that the reactant concentration [R] is sufficiently high that it is little changed in the reaction, and that all clusters have the same residence time, i.e. the proportionality constant [R]t in equ. 4 is the same for all n, so the relative rate constant for different cluster sizes can be obtained from measurement of the survival fraction for the n-atom cluster, I_f/I_o. A further assumption is that although the effective temperature for all bare clusters is not well known, it is the same for each cluster prior to reaction. For our present pulsed reactor experimental setup, absolute rate constants are relatively uncertain since reactant concentration and cluster residence time can only be crudely estimated. Absolute rate constants have been reported by one research group [21] through use of a continuous flow system for the reactant and helium buffer but still using pulsed laser vaporization of metal. A comparison of H_2 chemisorption on iron clusters measured independently by three different research groups shows good qualitative agreement[7,21-22].

Keep in mind that the rate constant measurements reported here reflect only the initial reaction step, i.e. addition of the first reactant molecule to the cluster $M_n + R \longrightarrow M_nR$. Although observation of product ions is taken as additional confirmation that reaction has occurred, we do not use these ion signals to estimate the reactivity. This is because the ionization cross section for the bare cluster and the same cluster with a reactant molecule chemisorbed onto it may differ and thus give rise to different detection efficiencies. We have shown that the ionization threshold of a cluster with chemisorbed hydrogen increases substantially over that of the bare cluster, and that the direction of this change (increase for hydrogen [23] and decrease for ammonia [24]) can be predicted from simple frontier orbital considerations[23]. Similarly the rates for chemisorption of a second, third, etc. molecule onto an already reacted cluster are not yet known. Further experiments designed to probe the ionization thresholds, cross section variation, and subsequent reactivity for clusters containing adsorbates are in progress.

5. Chemical Reactivity of Aluminum Clusters

Many classes of cluster reactions are just now beginning to be studied. For instance we have performed initial survey studies of oxidation reactions on clusters of Al, Fe, Pt, Pd and Ag; H-H, N-N, C-O and C-H bond activation reactions on many different metal clusters, sulfurization reactions on Al and Fe clusters, and dehydrogenation reactions on Nb, Pt and Rh clusters.

One of the most interesting systems we have explored is the reactivity of aluminum clusters with a number of gaseous species[3b,10,25]. The chemisorption of deuterium on aluminum clusters is the most cluster size-specific of any system we have studied. The six atom cluster is the most reactive, and no reaction is observed for clusters containing fewer atoms. Reactivity falls off rapidly for Al_7 and Al_8. Under the most severe conditions reaction has been observed up to Al_{20}, but not beyond. Similar results are observed for H_2, but the relative rate constants are about 1.7 ± 0.4 higher than for D_2. The primary products of the reaction are mono-molecular adducts. Unlike other hydrogen chemisorption systems we and others have studied, there is no evidence of further hydrogen addition. Saturation coverage appears to be achieved with only one molecule of hydrogen.

This does not appear to be a detection problem. If one adds the reactant to the carrier gas, atomic deuterium is produced which can be incorporated into the cluster in a variety of different mechanisms, including reaction. Under these conditions and using 7.87 eV photons to ionize the products for mass analysis, stable products are produced and detected, of the generic formula Al_nD_y, with n=6, y=1,2; n=7, y=1-3; n=8-12, y=1-5; n=13-21, y=1-3. These observations suggest that the detection of the chemistry products is a reliable indicator of neutral stabilities.

The reactivity pattern for chemisorption of different molecules by aluminum clusters in each case exhibits a different and unique size dependence. It is interesting that Al_6 is also the most reactive cluster with CO, but in contrast to the hydrogen case both smaller and larger clusters will react with CO at high enough pressure. For O_2 the most reactive cluster is the dimer. The reactivity drops dramatically for Al_3-Al_6 followed by a slow, monotonic rise towards larger clusters. Above 20 atoms, the rate is comparable to that of the dimer.

For water, Al_{10} is found to be the most reactive cluster, while smaller clusters are relatively inert and larger ones are reactive. Methanol exhibits a surprising reactivity pattern which suggests that there may be "magic numbers" for the minimum in reactivity at Al_7, Al_{14}, and Al_{20-21}. The primary reaction of H_2S is dissociative sulfidation, producing a species Al_nS as the product. This species is quite facile toward further adsorption of H_2S, but the system is poisoned towards further dissociative adsorption as evidenced by lack of dehydrogenation. Methane is found to be inert with Al_n under all reaction conditions. The relative reactivity of these molecules is ordered as follows:

$$O_2 > CH_3OH > CO > D_2O > D_2 > CH_4 \quad ,$$

with O_2 about a thousand times more reactive than D_2.

There is significant theoretical activity to understand the electronic structure and reactivity of aluminum clusters[26]. Calculations of the ionization potential and the stability of aluminum clusters using the jellium model have been performed[27]. The results suggest that a shell model may be useful in explaining some of the observations. Upton has performed extensive calculations on Al_n and Al_nH_2 $n \leq 6$ to obtain relative stabilities for several geometrical structures[28]. While he did not do an exhaustive search of all possible structures, he did find that the heat of adsorption of H_2 increases with cluster size and that there is only one structure of Al_6 among those considered for which chemisorption is exothermic. His findings are consistent with the observations reported above for $n \leq 6$. Why adsorption of D_2 terminates abruptly beyond Al_6 remains an open question.

6. Hydrogen Chemisorption on Transition Metals

Recent studies of transition metal clusters also revealed that the chemisorption of hydrogen is remarkably sensitive to both the number of metal atoms in the cluster as well as the metal type[7,8,21,22]. For example Nb_8, Nb_{10}, and Nb_{16} are significantly less reactive toward molecular hydrogen than other niobium clusters[8,9,22]. For cobalt, Co_6 and Co_7 are the least reactive[22].

In Figure 4 we compare our results for the reactivity of iron[7], vanadium[8] and niobium[8] clusters toward molecular hydrogen. In all cases the reactivity is observed to vary dramatically and non-monotonically with cluster size, and to exhibit a unique size-dependent reactivity for each metal. For instance, note the oscillatory behavior for iron clusters in which a local maximum in reactivity is observed near Fe_{10} followed by a local minimum near Fe_{16-17}.

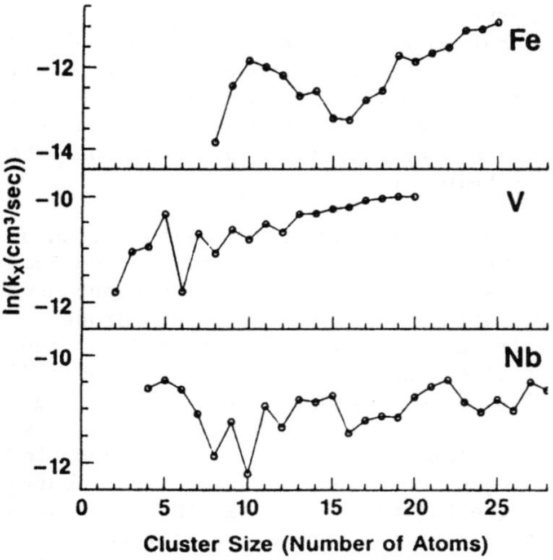

Fig. 4. The logarithm of the rate constant, $\ln(k_n)$, for reaction toward deuterium is plotted as a function of cluster size for iron, vanadium and niobium clusters. The reactivity data (note the log scale for the abscissa) is presented in this fashion in order to better observe the correlation with the ionization threshold energies shown in fig. 4 and predicted from equ. 5. The absolute rate constants for our data are determined by normalizing our value at Fe_{10} with that measured by Richtsmeier et al.[21]

For V and Nb we note that, as for iron, certain clusters are considerably less reactive toward hydrogen than others. For instance most vanadium clusters containing an even number of atoms are less reactive than their odd atom neighbors, ie. V_6, V_8, V_{10} and V_{12} exhibit considerably reduced reactivity compared to V_5, V_7, V_9, V_{11} and V_{13}. For niobium Nb_8, Nb_{10}, Nb_{16}, Nb_{24} and Nb_{26} exhibit considerably reduced reactivity toward hydrogen than neighboring clusters. As a contrast note that the small niobium clusters exhibit an enhanced reactivity.

In addition, several orders of magnitude variation in reactivity occurs between the least and most reactive clusters, with the most dramatic variation in reactivity occurring for clusters containing fewer than about 30 atoms. For instance about three orders of magnitude variation in reactivity is found between Fe_8 and Fe_{25} and about 1.5 orders of magnitude variation between Fe_{11} and Fe_{15}. In contrast the reactivity is high and changes little for iron clusters larger than about 23-25 atoms[21]. It has been suggested that this is the cluster size for which the onset of bulk- like behavior occurs, since hydrogen chemisorption on bulk iron is facile (and dissociative). Applying a similar interpretation to our results on vanadium and niobium would suggest that bulk-like behavior is achieved about V_{17} or V_{18}, while for niobium it must occur for clusters containing greater than 28 atoms.

Above we have shown that the chemisorption of the first hydrogen molecule on metal clusters exhibits a pronounced dependence on cluster size. In this section we briefly address the question of how much hydrogen can be chemisorbed upon a given size cluster, i.e. what is the saturation coverage which can be obtained and how does it vary with cluster size? The saturation coverage of iron clusters has been investigated in detail by Parks et al [29]. Their results suggest that for iron clusters dissociative chemisorption terminates when each metal SURFACE atom has chemisorbed one hydrogen atom. For clusters containing less than 30 atoms, essentially all metal atoms are surface-like with at most one or two atoms possibly in the "interior" or "bulk-like". Thus for Fe_{30} and smaller clusters, essentially a saturation coverage of one hydrogen atom per metal atom is observed. For clusters ranging in size from 30 to 80 atoms, Parks observes that the 1:1 ratio is achieved only if one counts surface atoms. We have measured the hydrogen uptake for small vanadium clusters $n<30$, and find that the small vanadium clusters take up about 1.4 hydrogen atoms per metal surface atom, somewhat larger than that measured for the small iron clusters. Hydrogen uptake experiments on the larger vanadium clusters is presently being investigated.

To this point, we have presented results for hydrogen chemisorption on Fe, V, and Nb clusters and have shown that there is a very pronounced variation with cluster size and metal type. Such a pronounced size-selectivity in itself is quite exciting. However another property, the ionization threshold energy, is also found to exhibit a strong non-monotonic variation with cluster size. Interestingly, it was found[7] that for Fe clusters larger than eight atoms, clusters with low (high) ionization threshold energies are those which exhibit high (low) reactivity toward di-hydrogen. The explanation put forth to account for this anticorrelation was that hydrogen chemisorption required charge transfer from the metal cluster for the activation of the H-H bond, with charge transfer being more facile the lower the ionization threshold[7]. A simple empirical model was proposed in which the IP and chemisorption rate constant k_n of the n-atom cluster are related according to

$$k_n = A \exp\left[-\epsilon(IP_n - E_o)/kT\right] , \qquad (5)$$

in close analogy to an Arrhenius expression for an activated process. The term $\epsilon(IP_n - E_o)$ may then be interpreted as something akin to an activation energy E(act). The relative rate between two different size clusters n and n* is then expressed as

$$k_n/k_n^* = \exp\left[-\epsilon(IP_n - IP_n^*)\right] , \qquad (6)$$

which allows scaling of ratios of rate constants to differences in IPs. The IP-ln(rate) empirical fit of the experimental IP and relative reaction rate constants is found to be quite good for iron clusters containing more than 8 atoms[7]. Ionization threshold measurements have now been extended to V [8] and Nb [8,9] clusters, and these results together with those for iron are compared with their corresponding rate constants in fig. 5. We note a striking correspondence between high ionization threshold and low reactivity for clusters larger than about eight atoms in size for all three metals.

However, we also observe that certain smaller ($n<8$) clusters which have higher ionization thresholds also exhibit considerable reactivity, exactly opposite to the behavior observed in the larger clusters. This is particularly dramatic for the small niobium clusters $n<7$, but Fe_4 and V_5 also have considerably higher reactivity than would be predicted by the simple empirical model of metal to H_2 charge transfer originally proposed to explain the IP-ln(rate) correlation observed for the larger clusters. In the next section we outline the general features of a simple model which can account for the activation of hydrogen on both large and small metal clusters.

7. Model for Hydrogen Activation by Metal Clusters

In this section we will discuss the various factors believed to be most important in the H-H bond activation by metal clusters. We will present a simple frontier orbital model whose primary features are:

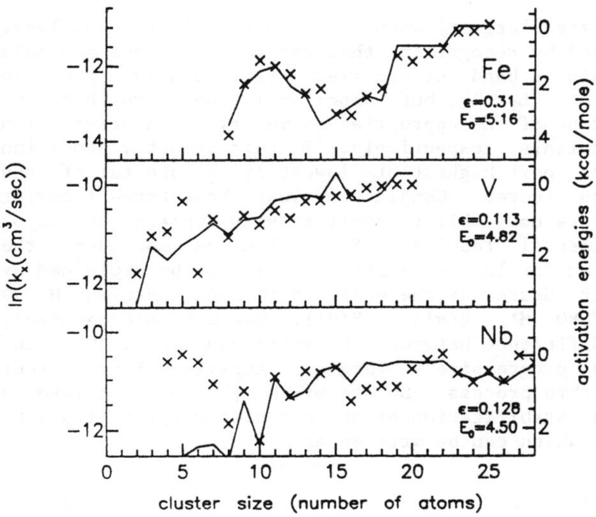

Fig. 5 Comparison of rate constants (crosses) and bare cluster ionization thresholds (solid line) for iron, vanadium and niobium. The ionization potential is given by IP=Eo+E(act)/ε (see text). The constant A is estimated by assuming the large clusters have no barrier toward hydrogen chemisorption and setting it equal to the rate constant measured for large iron clusters[21]. E_0 and ϵ are determined by least squares fitting the ionization threshold energies to the rate constants, assuming the constant A is independent of cluster size. The values of E_0 and ϵ for each metal are given on the plot. The same value of the constant A is assumed for iron, vanadium and niobium

(1) Attractive metal cluster- donor interactions between the high lying unoccupied H_2 σ^* antibonding molecular orbital and the highest lying occupied metal orbital M'(HOMO) at the Fermi energy.

(2) Attractive metal-acceptor interactions between the occupied H_2 σ bonding orbital and the lowest unoccupied metal orbital M(LUMO).

(3) Repulsive (Pauli repulsion) interactions between the filled σ H_2 orbital and filled cluster valence orbitals.

Interactions (1) and (2) are attractive and lead to H-H bond weakening as well as increased M-H_2 bonding. This is because in interaction (1) charge density is transferred into the antibonding H_2 σ^* orbital, whereas in (2) charge density is transferred out of the bonding H_2 σ orbital. The repulsive interaction is directly responsible for the barrier to hydrogen chemisorption and dominates the energetics in the approach to the transition state which leads to dissociative chemisorption. It develops at large interaction distances. Elongation of the H-H bond occurs only at shorter distances when it is possible to transfer charge either into the H_2 σ^* or from the H_2 σ. Without (3) the combination of (1) and (2) would lead to facile (non-activated) dissociative chemisorption with the formation of two new low-lying (relative to the Fermi energy) metal-hydride bonds. The energetic cost in overcoming the activation barrier (3) and in enhancing metal to hydrogen charge transfer (1) is expected to be smallest (and thus the dissociative adsorption rate largest) when the cluster IP is smallest. This is easily rationalized by recognizing that the metal valence orbitals will primarily fall within a band of energies significantly higher in energy than that of H_2 σ. Lastly, but importantly, we note that the M(LUMO) and M'(HOMO) must be of the appropriate symmetry to interact with the H_2 σ and the H_2 $\sigma*$ orbitals, respectively. If this symmetry condition is not satisfied, then the next highest or lowest lying orbital of the proper symmetry must be considered. Considering just these three frontier orbital interactions, we are now able to qualitatively explain the major features of the experimental results. First let us consider the IP-ln(rate) anticorrelation for larger clusters. This can be explained by recognizing that as the IP decreases, the energy difference between H_2 $\sigma*$ and the metal cluster HOMO M', $E(\sigma*)$ - $E(M')$, becomes progressively smaller and the energy difference between the metal cluster LUMO M and H_2 σ, $E(M)-E(\sigma)$, becomes progressively larger. Applying first order perturbation theory to this process, Shustorovich [30] has derived a simple expression for the stabilization of an occupied metal orbital M' due to mixing with the $\sigma*$ which can be written as

$$Q'_M \sim (\beta*)^2 / [E(\sigma*) - E(M')] \quad . \tag{7}$$

Similarly, stabilization of the H_2 σ due to mixing with an unoccupied metal orbital M is written as

$$Q_H \sim (\beta^2)/[E(M)-E(\sigma)] \quad ; \tag{8}$$

β and $\beta*$ are the interaction matrix elements between a metal orbital (M or M') and the H_2 σ and $\sigma*$ respectively. In general it is assumed that E(M) and E(M') are near the Fermi level with E(M)>E(M'). Donation to $\sigma*$ will dominate when $Q_M'/Q_H > 1$. This can occur even though the energy denominator $E(\sigma*)-E(M')$ may be greater than $E(M)-E(\sigma)$ if $\beta*$ is sufficiently greater than β. This appears to be true for metal surfaces [31], but such conditions will become progressively harder to satisfy for small clusters with high IPs. Equ. 7 shows that the stabilization of the occupied metal orbital M' depends inversely on the energy separation between $E(\sigma^*)$ and E(M'). Recognizing that E(M') will vary with cluster size and is related to the ionization threshold energy, we predict a direct correlation between the ability of the cluster to donate charge to the H_2 σ^* (hydrogen activation) and the energy of M' (the cluster IP), thus an anticorrelation with ionization threshold energy.

For the smallest clusters the ionization threshold energy may become sufficiently large that donation from H_2 σ to M can dominate. The stabilization due to this interaction will increase as the IPs increase (see Equ. 8), and in the simplest picture the chemisorption rate should increase with increase in IP. However, symmetry considerations are expected to play an increasingly important role as the cluster size decreases. This is simply due to the fact that the density of states near the Fermi energy becomes increasingly sparse as the cluster size decreases. Thus, finding a metal orbital of the appropriate symmetry near the Fermi energy becomes increasingly less probable. Thus it may not be surprising that small clusters (≤8 atoms) may be perverse in their reactivity pattern as a function of cluster size and metal type. For example, the pentamer of vanadium exhibits pronounced reactivity, whereas the iron pentamer is highly unreactive. Just the opposite behavior is observed for their respective tetramers. For niobium, on the other hand, the small $4<n<8$ clusters are nearly as reactive as the larger clusters.

In each instance the IPs are sufficiently high that little reactivity would be predicted based only on the metal to hydrogen charge transfer picture used to explain the behavior of the larger clusters. Small clusters, in particular, should be ideally suited for in depth theoretical studies, as has recently been carried out for aluminum clusters [26,28].

The hydrogen chemisorption rate as a function of cluster size reaches a maximum and levels off for the transition metal clusters studied to date. This is precisely what would be expected for a non-activated process, where the barrier is non-existent or extremely small and results in facile chemisorption. Harris and Anderson [32] have examined the conditions whereby the activation barrier may be alleviated. They point out that when H_2 closely approaches a metal surface, destabilized metal orbitals may relax by populating entirely new metal orbitals which interact less repulsively with H_2. In particular, they presented a case for 4s orbital occupancy shifting into 4d orbitals, and found a resultant decrease in the repulsive barrier. Theirs is one of a large number of theoretical studies of the activation of hydrogen by metal clusters and surfaces.

Upton has studied the electronic factors important in the dissociative chemisorption of hydrogen on aluminum clusters, and specifically on Al_6 in some detail [26,28]. He explored the role of the charge state of the cluster, be it anion, neutral, or cation and found that the charge had only a small effect on the activation barrier. Similar size selective behavior might be expected from charged and neutral clusters, particularly if the orbital ionized does not participate directly in the reaction. While the calculations were for aluminum, conceptually they explain why the size selective reactivity patterns [33] for Nb_7^+, Nb_8^+, and Nb_9^+ with H_2 are similar to their neutral counterparts.

8. Chemisorption of Carbon Monoxide

The chemisorption of CO has been measured [34] in our laboratory for first row transition metals V, Fe, Co, Ni and Cu; second row metals Nb, Mo, Ru, and Pd; and third row metals W, Ir, and Pt as well as for Al. Our results show that CO is readily chemisorbed on most transition metal clusters containing three or more atoms and that the reactivity of the larger ($n > 4$) clusters is comparable, nearly independent of cluster size or metal type. This is in direct contrast to the very pronounced size selectivity discussed above for hydrogen chemisorption and suggests that CO chemisorption is a non-activated reaction.

The CO chemisorption on niobium and cobalt clusters has been reported previously [22]. The two sets of measurements are in reasonable agreement for clusters containing 3 or more atoms. However, under our particular experimental conditions, the cobalt atom and dimer also exhibit depletion whereas no depletion was observed on these two species in the earlier measurements.

In these studies we have purposely held all detection, reaction, and vaporization conditions as constant as possible and changed only the metal rod being vaporized. Dependent on the metal, the onset of reactivity may begin with the atom, the dimer, the trimer or even a larger size cluster. Why should some metals show reaction for all size metal clusters (including the atom) whereas others only show reactivity above a certain minimum cluster size? By appealing to the the kinetic model described in section IV, we explain such results as a competition between collisional stabilization versus unimolecular decay in the reactor, and from this to infer an ordering of metal-CO bond strengths as a function of the metal type.

To see this let us examine equ. 3 in more detail. The term in brackets on the right hand side of equ. 3 expresses the relationship between the collisional stabilization rate, $k_s[He]$, and the unimolecular decay rate k_n'. For a fixed set of pressure conditions in the reactor we may reasonably expect the collisional stabilization rate to be nearly independent of cluster size (ignoring the slow increase associated with growth in the physical size of the clusters). However, the unimolecular decay rate depends very strongly on cluster size, decreasing rapidly with increasing number of atoms. This is most easily seen by considering the simple RRK prediction for k_n',

$$k_n' = A[(E-E_o)/E]^{s-1} \qquad E > E_o \quad , \qquad (9)$$

where A is the frequency factor (typically around 10^{14} sec^{-1}), s is the number of modes (3n for the M_nCO^* species), E is the average vibrational energy in the M_nCO^* species and E_o is the critical energy needed for decomposition, which we shall assume for this simple argument is the cluster-CO bond strength assumed equal to the heat of adsorption. Assuming A does not change dramatically with cluster size, we see the term in brackets rapidly becomes smaller as n increases, quickly reducing the unimolecular decomposition rate for larger clusters. For $E_o = 0.5$ eV, T=400K and n=3, the unimolecular decomposition rate as predicted from equ. 9 is about 2×10^8 sec^{-1} or of the order of the collision frequency. Thus trimers with metal-CO bond strengths ≤ 0.5 eV should decompose in the reactor and no reaction would be observed. On the other hand, if E_o is 1.0 eV, the trimer decomposition rate would be only about 10^6 sec^{-1} and products should be stabilized, i.e. reaction will be observed. Thus for a given bond strength E_o, a minimum cluster size is predicted to exist, below which the cluster-reactant complex will unimolecularly decompose and above which it will be stabilized.

In applying this model to CO chemisorption studies we have assumed that the metal-CO bond strength varies little with cluster size and the CO chemisorption is non-activated. For a fixed set of experimental conditions we find that reactivity is first observed for atoms of Pt and Co; dimers of Pd, trimers of V, Ni, Nb, Ru, and Ir; tetramers of W; pentamers of Cu and Mo, hexamers of Fe and Al. Thus we propose that the metal-CO bond strengths be ordered according to: Co, Pt, Pd > V, Ni, Nb, Ru, Ir, W > Mo, Fe, Cu, Al. Interestingly this ordering is consistent with heats of desorption of CO from metal crystals and surfaces[35]. By carefully controlling the reactor buffer gas pressure, i.e. collisional stabilization rate, a much less coarse ordering of bond strengths should be possible.

9. "Dehydrogenation" Reactions

Before discussing specific reactions of hydrocarbons with metal clusters, we first describe several effects which are important to consider in interpreting these results. First, the rate limiting step is assumed to be the initial chemisorption, and depletion of a bare metal cluster ion signal accurately reflects its change in concentration. Second, in identifying products we have assumed in the prior sections that the photoionization "detector" monitors the chemistry faithfully and produces only parent ions, i.e. photodissociation is not occurring so daughter ions are not produced. This is supported by the observation of product peaks of the generic form $M_n(ABC)y$ where M_n is the metal cluster and ABC is the mass of the chemisorbed molecule, i.e. no mass loss is detected. Third, the chemical identity of molecule ABC on the cluster may or may not be preserved; whether the chemisorption of ABC onto M_n is molecular or dissociative must be argued on grounds other than the mass of the product ions.

In the experiments discussed below, there are instances in which product ions are observed for which loss of hydrogen from the cluster-adduct has occurred. A key issue to be resolved is whether the hydrogen loss occurs in the reaction as a result of chemistry, or if it is induced by the ionization process. If it takes place in the reactor the observed mass spectrum may reflect the true parent ions. The situation is more complex in the latter case, because several processes must be considered. First, dehydrogenation may have already taken place prior to ionization, but the hydrogen has remained bound to the cluster. In this instance any excess energy deposited into the cluster-adduct complex over that needed for ionization may volatilize the hydrogen. Second, dehydrogenation may not have occurred but C-H activation may be photo- or thermally-induced by the ionization laser with subsequent desorption of the liberated hydrogen. Third, the ionizing conditions may be sufficiently severe (e.g. high ionizing intensity leading to multiphoton processes, sufficiently high photon energies leading to single photon dissociative ionization) that cluster fission may result, producing fragments. Thus whether "dehydrogenation" is observed may depend strongly on ionizing photon energy, intensity, and fluence.

Benzene reacts readily with platinum[36], niobium[37,38] and rhodium[39] clusters to form a rich spectrum of products, some with significantly lower ionization potentials than either the bare metal clusters or benzene. Complexes containing up to four benzene adducts have been detected for all Pt clusters studied ($n \leq 9$)[36,40]. The observed product distribution is particularly sensitive to the ionizing laser photon energy and intensity. Early studies of the reaction of platinum clusters with benzene[36], using high fluence 6.42 eV ionization, indicated extensive "dehydrogenation", but re-examination of the Pt_n-benzene system using low-fluence 7.81 eV ionization indicated a product distribution dominated by "molecular" addition[10,40].

The effect of ionizing fluence and photon energy on the product distribution formed in the reactions of Nb_n with normal- and perdeutero-benzene has been examined[37]. The products obtained not only depend dramatically on both cluster size and benzene isotope, but also on the photoionization conditions. With three notable exceptions, the major reaction channel observed with gentle, low-fluence (< 80 $\mu J/cm^2$), single-photon 6.42 eV ionization is "molecular" addition to the metal clusters. Clusters containing 5,6, and 11 niobium atoms, however, also give rise to a significant yield of products exhibiting loss of hydrogen for the reaction with benzene-h_6, but not for benzene-d_6. This is interpreted as a kinetic isotope effect occurring upon C-H bond activation on these specific clusters, suggesting that dehydrogenation of benzene has occurred in the fast-flow reactor, and that the hydrogen (deuterium) desorbs from the clusters in the reactor. Desorption of hydrogen can also be induced during the ionization process at high ionizing intensities, except for complexes containing less than three niobium atoms. Whether dehydrogenation (C-H bond activation) has already occurred in the reactor but the hydrogen remained bound to the clusters and is desorbed at higher intensities, or whether both dehydrogenation and desorption occur upon ionization at high intensities, is not certain. However, the shift towards increasing hydrogen loss and/or fragmentation with increasing ionizing photon energy and fluence mandates that gentle, low-energy, single-photon ionization near the ionization threshold is necessary for obtaining the "true" parent ions of the products.

10. Vibrational Spectroscopy of Metal Cluster-Adducts

Vibrational spectroscopy is an important probe used to determine bonding and structural properties of molecules. Powerful techniques such as electron energy loss spectroscopy (EELS) have been developed which allow one to obtain the vibrational properties of molecules chemisorbed upon surfaces. Due to low concentration, the highly reactive nature of the clusters, and the large number of possible species which typically are present in the cluster beams used to date, unconventional techniques are required in order to obtain spectroscopic information. One unconventional but powerful technique, infrared multiple photon dissociation (IRMPD), has recently been applied to the study of the vibrational properties of gas phase metal clusters upon which one or more molecules have been chemisorbed[41]. This same technique, IRMPD, has previously been used to obtain the vibrational spectra of ions[42], species for which it is difficult to apply conventional absorption techniques.

The basic feature of these experiments consists first of producing a metal cluster-adduct species as described earlier. Second, a high power infrared laser is aligned spatially with the cluster beam and fired at a predetermined time so as to intersect the clusters and cluster-adducts prior to their arrival at the ionization zone. The effect of the infrared laser is then monitored as a change (loss) in ion signal of the species which have undergone infrared multiple photon dissociation. From measurements of depletion as a function of laser frequency and fluence, an infrared dissociation spectrum of the complex is obtained which can be related to its infrared absorption spectrum[43]. Experiments are primarily limited by the fact that tunable, high-power infrared lasers are readily available only for limited frequency ranges.

In the initial experiments, measurements of the IR decomposition of methanol chemisorbed on iron clusters demonstrated the feasibility of this technique. Figure 6 shows the change in the mass spectrum when a high power CO_2 laser operated at 984.4 cm^{-1} irradiates the cluster beam. Note that the $Fe_n(CH_3OH)_y$ adducts are decomposed as evidenced by the loss of nearly all ion signal on these mass peaks, whereas the bare iron clusters are unaffected. The bare iron clusters are transparent at all CO_2 laser frequencies examined between about 930 and 1085 cm^{-1}. The methanol-cluster adducts exhibit strong depletion near 984 and 1075 cm^{-1}. Based on high resolution EELS of methanol on iron single crystal surfaces[44], and IRMPD of other methanol isotopes (CH_3OD and CD_3OH), the 984 cm^{-1} feature is assigned to the methoxy stretch. Table I summarizes results from experiments with isotopically substituted methanol. These results are consistent with the assignment of the 984 cm^{-1} feature as the methoxy stretch. Initially, from experiments with CH_3OH and CH_3OD, we proposed[41] that the 1075 cm^{-1} feature might be assigned to a bridge-bonded hydrogen stretch[45]. However, the non-observation of IR decomposition near 1075 cm^{-1} with the CD_3OH species now makes this assignment suspect unless isotopic scrambling occurs.

The IR laser-induced decomposition of iron cluster-methanol complexes exhibits a threshold behavior on laser fluence consistent with a multiple

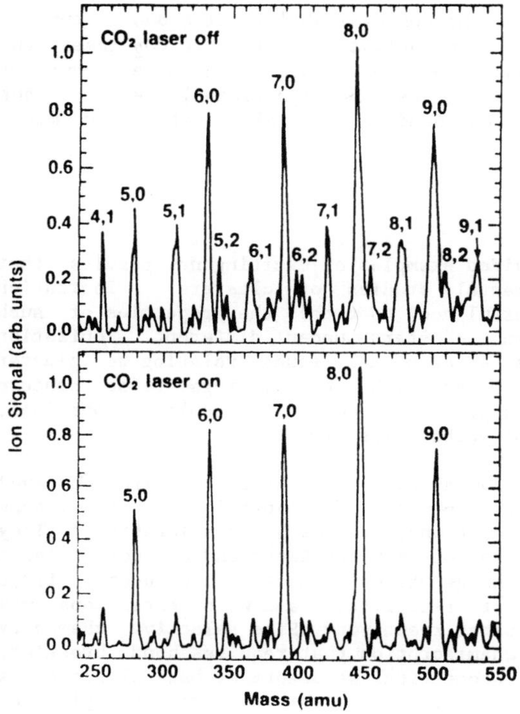

Fig. 6 Time-of-flight mass spectra for the iron cluster-methanol complexes. The upper trace shows the reference spectrum obtained from the reaction of methanol with iron clusters. The peaks are labeled according to (n,y), corresponding to masses equivalent to the chemical composition $Fe_n(CH_3OH)_y$. The upper trace is taken with the CO_2 laser blocked. The lower trace is taken with the CO_2 laser unblocked. The CO_2 laser frequency is 984.4 cm^{-1}

Table I. IR frequencies for which decomposition of methanol isotopes chemisorbed on iron clusters is observed

Isotope	Frequency (cm-1)	
CH_3OH	984	1075
CH_3OD	984	----
CD_3OH	955	----

photon absorption process. This is similar to results obtained on large molecules which have a large number of low-lying vibrational levels[43]. As IR laser fluence is increased the dissociation spectra exhibit red-shifted broadening expected in multiple photon dissociation processes. The IR dissociation spectra of iron clusters containing one or two chemisorbed methanols are found to be nearly identical,suggesting that the first two chemisorbed methanols are bonded in a similar fashion to the cluster. In addition,the IR multiple photon dissociation spectrum is independent of n, the number of Fe atoms in the cluster, in the range $5 \leq n \leq 12$. This suggests that the nature of the CH_3OH chemisorbed state is independent of cluster size.

These results show that the vibrational spectra and thus the structure of molecules chemisorbed on metal clusters can be probed through use of infrared multiple photon dissociation spectroscopy. These early experiments have been limited by the non-availability of broadly tunable high power infrared lasers, but with only a line tunable CO_2 laser the potential of such an approach has been demonstrated. The development of new tools such as this will allow us to significantly advance our understanding of the structure and bonding of metal clusters and metal clusters containing adsorbates.

Summary

In this paper we have provided examples of exciting new results that are emerging in the experimental studies of clusters. Ionization threshold studies offer a powerful tool to probe the properties of such systems. Detection of carbon clusters under different ionization conditions was used to resolve experimental issues relating to cluster stability. The results, when combined with other findings, raise concern about the experimental "support" for ultrastable spherical, "soccerball"-like structures for carbon clusters.

Ionization thresholds are found to be useful to explain observed reaction rate measurements for a number of transition metal systems undergoing demanding activated reactions, such as H_2 chemisorption. They are probes of electronic structure, of both reactants and products. There is an emerging picture that charge transfer models are useful first approximations to explain cluster reactivity, and with extensions may provide a general model for cluster reactivity. Size selective chemistry appears to be associated with chemisorption that requires bond breaking, i.e. with an activated process. Geometrical structure of bare clusters is an important piece of information missing from our general picture. Chemical probes are useful first tools to give insight and information about structure using surface adsorption techniques.

On the other hand we show that for a facile non-activated reaction such as CO chemisorption, a kinetic model can be developed from which relative CO-metal cluster bond strengths can be inferred, and that such bond strength ordering agrees well with heats of adsorption on the corresponding metal surface. Finally, we show that IR spectroscopy of molecules chemisorbed onto clusters can be probed using the technique of infrared multiple photon dissociation. Such a technique, by probing cluster and cluster-adsorbate vibrational frequencies, has general applicability for helping to advance our understanding of metal-metal bonding in metal clusters themselves as well as bonding of adsorbates to clusters. It is limited at present by the fact that only in limited frequency ranges are sufficiently powerful IR lasers readily available.

The field of gas phase cluster research is expanding rapidly. We have only seen the tip of the iceberg. Strong interaction between experiment and theory is essential for rapid progress in the field, especially in light of the fact that only limited kinds of experimental data are currently available. It is important to focus the effort where theory and experiment can interact effectively.

References

1. (a) A. Kaldor, E. Rohlfing, and D. M. Cox, Laser Chem. 2, 185 (1983). (b) D. M. Cox, E. A. Rohlfing, D. J. Trevor, and A. Kaldor, J. Vac. Sci Technol. A2, 812 (1984).

(c) A. Kaldor, D. M. Cox, D. J. Trevor and R. L. Whetten, ACS Symposium Series No. 288, "Catalyst Characterization Science", Eds. M. L. Deviney and J. L. Gland (1985).
2. (a) T. G. Dietz, M. A. Duncan, D. E. Powers and R. E. Smalley, J. Chem. Phys. 74, 6511 (1981).
(b) S. J. Riley, E. K. Parks, C. R. Mao, L. G. Pobo and S. Wexler, J. Phys. Chem., 86, 3911 (1982).
(c) E. A. Rohlfing, D. M. Cox, R. Petkovic-Luton and A. Kaldor, J. Phys. Chem., 88, 6229 (1984).
3. For recent reviews, see: (a) R. L. Whetten, D. M. Cox, D. J. Trevor and A. Kaldor, Surf. Sci., 156, 8 (1985).
(b) A. Kaldor, D. M. Cox and M. R. Zakin, Adv. Chem. Phys., in press
4. E. A. Rohlfing, D. M. Cox, and A. Kaldor, Chem. Phys. Lett. 99, 161 (1983).
5. E. A. Rohlfing, D. M. Cox, and A. Kaldor, J. Phys. Chem. 88, 4497 (1984)
6. E. A. Rohlfing, D. M. Cox, A. Kaldor and K. H. Johnson, J. Chem. Phys. 81, 3846 (1984)
7. R. L. Whetten, D. M. Cox, D. J. Trevor, and A. Kaldor, Phys. Rev. Lett. 54, 1494 (1985).
8. D. M. Cox, R. L. Whetten, M. R. Zakin, D. J. Trevor, K. C. Reichmann, and A. Kaldor, AIP Conference Proceedings no. 146, Adv. in Laser Science I, Eds. W. C. Stwalley and M. Lapp, American Institute of Physics, (1986)
9. R. L. Whetten, M. R. Zakin, D. M. Cox, D. J. Trevor, and A. Kaldor, J. Chem. Phys. 85, 1697 (1986)
10. A. Kaldor, D. M. Cox, D. J. Trevor and M. R. Zakin, Z. Physik. D3, 195 (1986).
11. R. L. Whetten, D. M. Cox, D. J. Trevor, and A. Kaldor, J. Phys. Chem. 89, 566 (1985)
12. E. A. Rohlfing, D. M. Cox and A. Kaldor, J. Chem. Phys. 81, 3322 (1984).
13. H. W. Kroto, J. R. Heath, S. C. O'Brien, R. F. Curl and R. E. Smalley, Nature 318, 162 (1985).
14. Q. L. Zhang, S. C. O'Brien, J. R. Heath, Y. Liu, R. F. Curl, H. W. Kroto and R. E. Smalley, J. Phys. Chem., 90, 525 (1986).
15. J. R. Heath, S. C. O'Brien, Q. Zheng, Y. Liu, R. F. Curl, H. W. Kroto, F. K. Tittel and R. E. Smalley, J. Am. Chem. Soc. 107, 7779 (1985).
16. Y. Liu, S. C. O'Brien, Q. Zhang, J. R. Heath, F. K. Tittel, R. F. Curl, H. W. Kroto and R. E. Smalley, Chem. Phys. Lett., 126, 215 (1986).
17. D. M. Cox, D.J. Trevor, K. C. Reichmann and A. Kaldor, J. Am. Chem. Soc., 108, 2457 (1986).
18. D. M. Cox, K. C. Reichmann and A. Kaldor, J. Chem. Phys., submitted.
19. T. R. Baker, private communication.
20. See for example:
(a) A. D. J. Haymet, J. Am. Chem. Soc., 108, 319 (1986); Chem. Phys. Lett 122, 421 (1985).
(b) R. C. Haddon, L. E. Gries and D. Raghavachari, Chem. Phys. Lett., 125, 459 (1986).
(c) M.D. Newton and R. E. Stanton, J. Am. Chem. Soc., 108, 2469 (1986).
(d) M. Osaki and A. Takahasi, Chem. Phys. Lett. 127, 242 (1986).
(e) D.J. Klein, T. G. Schmalg, G. E. Hite and W. A. Sertz, J. Am. Chem. Soc., 108, 1301 (1986).
(f) R. L. Disch and J. M. Schulman, Chem. Phys. Lett. 125, 465 (1986).

21. S. C. Richtsmeier, E. K. Parks, K. Liu, L. G. Pobo, and S. J. Riley, J. Chem. Phys. 82, 3659 (1985).
22. M. D. Morse, M. E. Geusic, J. R. Heath, and R. E. Smalley, J. Chem. Phys. 83, 2293 (1985).
23. M. R. Zakin, D. M. Cox, R. L. Whetten, D. J. Trevor, and A. Kaldor, Chem. Phys. Lett. 135, 223 (1987).
24. (a) S. J. Riley, E. K. Parks, K. Liu, and L. G. Pobo, Am. Chem. Soc. 191st annual meeting, paper 160, Div. of Coll. and Surf. Chem., April, 1986.
(b) S. J. Riley and E. K. Parks, Int'l Symp. on Physics and Chem. of Small Particles, Richmond, VA. Oct. 28-Nov. 1, 1986.
25. (a) D. M. Cox, D. J. Trevor, R. L. Whetten and A. Kaldor, submitted.
(b) T. H. Upton, D. M. Cox and A.Kaldor, Int'l Symp. on Physics and Chem. of Small Particles, Richmond, VA, Oct. 28-Nov. 1 (1986).
26. See, for example, T. H. Upton, Phys. Rev. Lett. 56, 2168 (1986).
27. M. L. Cohen, Proceedings of the NEC symposium on Clusters, "Microclusters", Eds. S. Sugano, Y. Nishina and S. Ohnishi, Spinger Series in Materials Science 4, Springer-Verlag, Toyko, Japan, 1986.
28. T. H. Upton, J. Chem. Phys., submitted, and references therein.
29. E. K. Parks, K. Liu, S. C. Richtsmeier, L. G. Pobo, and S. J. Riley, J. Chem. Phys. 82, 5470 (1985).
30. (a) E. Shustorovich and R. C. Baetzold, Science 227, 876 (1985).
(b) E. Shustorovich, J. Phys. Chem. 87, 14 (1983).
(c) E. Shustorovich, R. C. Baetzold and E. L. Muetterties, J. Phys. Chem. 87, 1100 (1983).
31. J. Y. Saillard and R. Hoffmann, J. Am. Chem. Soc. 106, 2006 (1984).
32. J. Harris and S. Andersson, Phys. Rev. Lett. 55, 1583 (1985)
33. (a) P. J. Brucat, C. L. Pettiette, S. Yang, L-S. Zheng, M. J. Craycraft and R. E. Smalley, J. Chem. Phys., 85, 4747 (1986).
(b) J. M. Alford, F. D. Weiss, R. T. Laaksonen and R. E. Smalley, J. Phys. Chem., 90, 4480 (1986).
34. D. M. Cox, K. C. Reichmann, D. J. Trevor and A. Kaldor, submitted.
35. (a) H. Papp, Ber. Bunsenges. Phys. Chem. 86, 555 (1982).
(b) P. Imnbihl, M. P. Cox, G. Ertl, H. Muller, and W. Brenig, J. Chem. Phys. 83, 1578 (1985).
(c) P. W. Davies and R. M. Lambert, Surf. Sci. Lett. 111, L671 (1981).
(d) D. Brennen and F. H. Hayes, Phil. Trans. Roy. Soc. A258, 347(1965).
(e) G. A. Samorjai, Chemistry in Two—Dimensions: Surfaces, Cornell University Press, Ithaca, NY (1981).
(f) H. Pfnur, P. Feulner, H. A. Engelhardt and D. Menzel, Chem. Phys. Lett. 59, 481 (1978); E. D. Williams and W. H. Weinberg, Surf. Sci. 82, 93 (1979).
(g) J. L. Taylor, D. E. Ibbotson, and W. H. Weinberg, J. Chem. Phys. 69, 4298 (1978); B. E. Nieuwenhuys and G. A. Samorjai, Surf. Sci. 72, 8 (1978); M. E. Thomas, H. Poppa, and G. M. Pound, Thin Solid Film, 58, 273 (1979).
(h) G. Wedler, H. Papp and G. Schroll, Surf. Sci. 44, 463 (1974); C. R. Helms and R. J. Madix, Surf. Sci. 52, 677 (1975).
(i) T. J. Vink, O. L. J. Gijzeman and J. W. Geus, Surf. Sci. 150, 14 (1985).
(j) J. W. Erickson and P. J. Estrup, Surf. Sci. 167 519 (1986).
(k) J. C. Tracy, J. Chem. Phys. 56, 2748 (1972); K. Horn, M. Hussain and J. Pritchard, Surf. Sci. 63, 244 (1977).
(l) C. B. Bargeron and B. H. Nall, Surf. Sci. 119, L319 (1982).
36. D. J. Trevor, R. L. Whetten, D. M. Cox and A. Kaldor, J. Am. Chem. Soc., 107, 518 (1985).

37. M. R. Zakin, D. M. Cox and A. Kaldor, J. Phys. Chem., accepted.
38. R. J. St. Pierre and M. A. El Sayed, J. Phys. Chem., 91, 763 (1987).
39. M. R. Zakin, D. M. Cox and A. Kaldor, in preparation.
40. D. J. Trevor and A. Kaldor, Proceedings of the American Chemical Society, Annaheim, CA, September 1986.
41. M. R. Zakin, R. O. Brickman, D. M. Cox, K. C. Reichmann, D. J. Trevor and A. Kaldor, J. Chem. Phys., 85 1196, (1986).
42. D. S. Bomse, R. L. Woodin and J. L. Beauchamp, J. Am. Chem. Soc., 101, 5503, (1979).
43. T. G. Dietz, M. A. Duncan, R. E. Smalley, D. M. Cox, J. A. Horsley and A. Kaldor, J. Chem. Phys., 77, 4417, 1982.
44. P. H. McBreen, W. Erley and H. Ibach, Surf. Sci., 133, L469, (1983).
45. A. M. Baro and W. Erley, Surf. Sci., 112, 1759 (1981).

Compound Clusters

T.P. Martin and H. Limberger

Max-Planck-Institut für Festkörperforschung, Heisenbergstr. 1, D-7000 Stuttgart, Fed. Rep. of Germany

1. INTRODUCTION

The cluster beam technique offers promise of becoming a useful tool for chemists in their search for new compounds. Using this technique elemental vapors can be made to react under controlled conditions to form units of condensed matter (clusters) small enough to be analyzed in a mass spectrometer,but still large enough to contain several structural building blocks. Periodic patterns in cluster mass spectra often allow these building blocks to be uniquely identified. Of particular interest is the identification of building blocks unknown in the crystalline state. The mass spectrometer has been used primarily to measure the number of atoms in elemental clusters; $Xe_{55}[1-3], Na_{20}[4-5], Si_{10}, Ge_{10}[6-9], C_{60}[10]$, to mention a few. In this paper we will emphasize a slightly different aspect of mass spectrometry. The technique can be used to chemically analyze compound clusters.

If the building blocks of a chemical compound can be identified, the character of its bonding and its crystalline structure are quick to follow. Building blocks in clusters are often easily identified by means of mass spectroscopy. Since the same building blocks can be expected to form solid materials, cluster beam techniques should prove to be useful in the preparation and characterization of new chemical compounds. This paper presents some recent results in our search for new compounds.

What happens when two metallic vapors are mixed to form clusters? The results of such an experiment depend on whether the two metals are chemically similar [12,13] or whether they lie far apart in the periodic table [6,14]. Chemically similar metals mix in all concentrations to form alloys with properties intermediate to the pure elements. However, when Cs and Au are mixed together something quite surprising happens [15-17]. Although Cs and Au are both model metallic elements, if they are mixed in the ratio of 1:1, the result is a non-metallic compound. The reason for this can be found in the large difference in the electronegativity of these two materials, its almost as large as that between cesium and iodine. In fact Cs and Au combine to form a crystal with the CsCl structure, a structure characteristic of ionic materials. Cesium transfers an electron to gold,resulting in a bond with a strong ionic contribution.

In this paper we describe the results of experiments on the systems Cs-Pb and Cs-Sn, for the practical reason that gold has a relatively low vapor pressure. However, the electronegativity difference between the components of these two systems is almost as large as that between Cs and Au, and unusual intermetallic clusters can be observed in the mass spectra.

2. EXPERIMENT

Compound clusters, i.e. clusters composed of more than one element, can be produced by allowing the elements to react in the vapor phase [6,11]. It is then possible to determine the composition of the clusters using a mass spectrometer. By varying the partial pressures of the two elements, the stability of compounds of all conceivable compositions can be tested in just one experiment. Various combinations of elemental vapors (Ge, Sn, Pb, Ga, In, Cs, Rb, Sr, Tm, Eu, Cl, O, S and P) have been allowed to react in the vapor phase and simultaneously thermally quenched in cold He gas. Compound clusters condensed out of the vapor and were identified by using a quadrupole mass spectrometer.

Elemental vapors were created in ovens of various design and allowed to react in a chamber containing from 0.7-5 mbar He gas. The outer mantel of the reaction chamber was cooled with liquid nitrogen. The reaction chamber contained two of these ovens. The vapors from the ovens were mixed and cooled. Compound clusters condensed out of the supersaturated vapor and passed through a 3 mm diameter hole into an intermediate chamber where the excess He gas was pumped off with a 500 l/s turbopump. The cluster beam then passed through a 1 mm diameter hole into a high vacuum chamber containing a quadrupole mass spectrometer pumped with a 360 l/s turbopump, where the clusters were ionized with 80 eV electrons. The spectrometer could be operated in the mass range from 1-2000 amu.

3. GROUP IV ELEMENT CLUSTERS

Before combining Cs with elements from group IV, it will be useful to first consider the properties of pure group IV clusters. All of these elements, carbon [18], silicon [19], germanium [20], tin [21], and lead [22], cluster very easily. In fact, polyatomic species can be identified even in equilibrium vapors. As we will see, this strong tendency for self-aggregation will also affect in a crucial way the structure of intermetallic compound clusters.

A mass spectrum of clusters formed by quenching Ge vapor in He gas is shown in Fig.1. Peaks corresponding to clusters containing 6, 10, 14, 15 and 18 Ge atoms are particularly strong. The dominant feature in the spectrum of Sn clusters is the weakness of the peak Sn_{14}^{+}. In this respect Sn resembles Pb [23]. That Sn should more resemble Pb than Ge can be seen in the solid state. Although Sn does have a low temperature semiconducting crystallographic modification with the Ge structure,

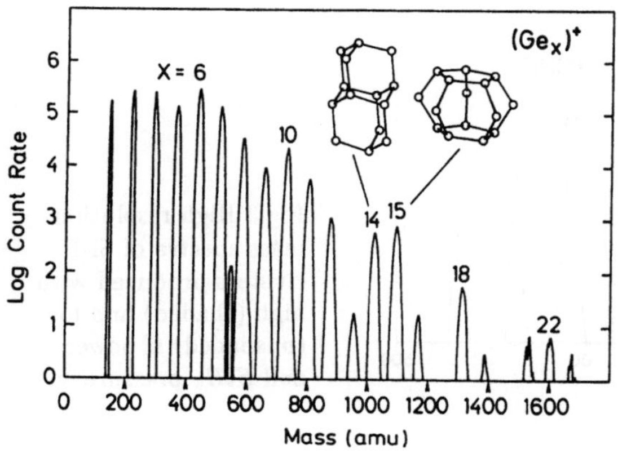

Fig.1 Mass spectrum of clusters formed by quenching Ge vapor in He gas. Peaks corresponding to clusters containing 6,10,14,15 and 18 atoms are particularly strong. Two structures with bond angles of nearly 109 degrees are indicated

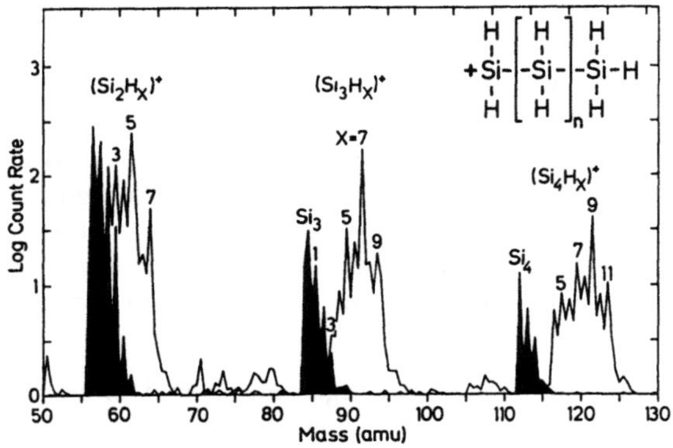

Fig.2 Mass spectrum of Si-H clusters formed by decomposition and ionization of SiH_4 with an rf discharge. Notice that the peaks separate into almost pure Si (shaded) clusters and clusters with approximate composition $(SiH)_n$

at room temperature the properties and structure are metallic. Because of silicon's low vapor pressure and high reactivity, it is not easily evaporated from an oven. However, clusters can be produced by rf discharge dissociation of SiH_4 gas [8]. The composition of the clusters can be changed from almost pure $(Si)_n$ to $(SiH)_n$ by varying the discharge conditions. A portion of a mass spectrum of such clusters is shown in Fig. 2.

Each peak is broadened by a distribution of unresolved hydrogen atoms. The hydrogen content clearly changes with cluster size. In particular, there is a well defined transition from silicon-rich clusters to hydrogen-rich clusters near n=10. Notice also that the peak corresponding to pure Si_{10}^+ is particularly strong, just as it was for Ge. The flux of small Si-H clusters is large enough to allow the mass spectrometer to be operated under conditions for high mass resolution, Fig.3. Each peak corresponds to the successive addition of a hydrogen atom. Low SiH_4 pressures lead to pure Si clusters, high pressures produce predominantly $[(SiH_2)(SiH_2)_n(SiH_3)]^+$ clusters. Clusters containing an odd number of hydrogen atoms seem to be particularly stable. The reason for this will be discussed later in the paper.

Excellent theoretical work is beginning to appear on the structure of group IV element clusters [24-33]. It would appear that these clusters, in particular Si, will become test objects for calculational methods.

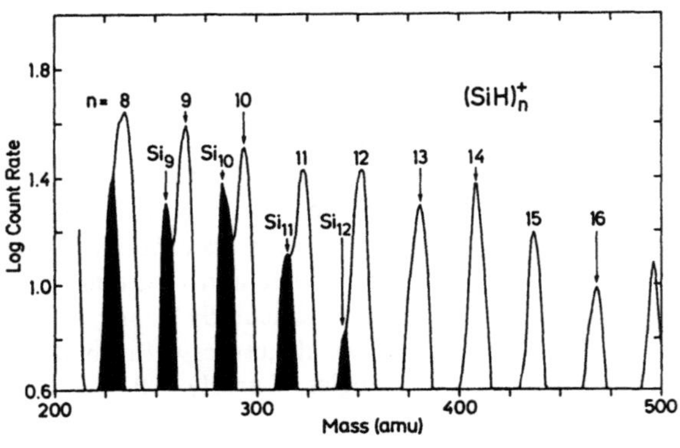

Fig.3 High resolution mass spectra of Si-H clusters produced with high (shaded) and low (unshaded) rf power and SiH_4 pressure

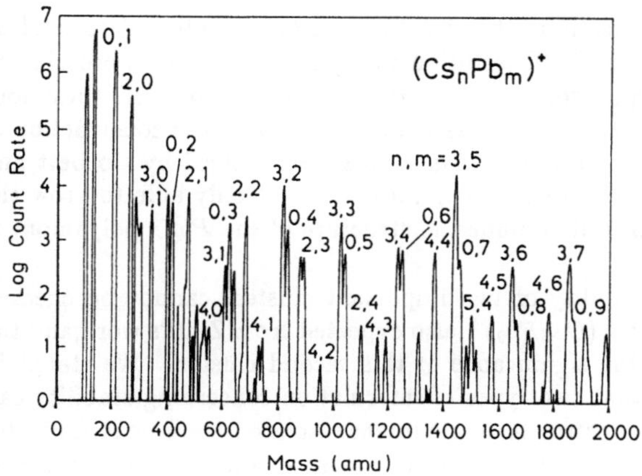

Fig.4 Mass spectrum of Cs-Pb clusters. Peak corresponding to $(Cs_3Pb_5)^+$ is particularly strong

4. INTERMETALLIC CLUSTERS

Figure 4 shows a mass spectrum of clusters formed by mixing a small amount of Cs vapor with Pb vapor. Mass peaks due to pure Pb_n^+ are plainly evident. Of the mass peaks corresponding to clusters containing both Pb and Cs, two stand out particularly strongly, $(Cs_3Pb_2)^+$ and $(Cs_3Pb_5)^+$. Why are these peaks so strong? What do these clusters look like? In order to answer these questions it is necessary to make a digression and further discuss the properties of pure group IV clusters. Figure 5 shows the results of an extended Hückel calculation assuming specific geometries. The matrix element parameters have been fitted to the atomic ionization potential and the well-known band structure of Si. Notice that for a pentamer with the trigonal bipyramid structure it is possible to add two bonding electrons to form a very stable M_5^{2-} polyanion. In addition, the calculation indicates that the tetrahedral M_4^{4-} and the M_2^{2-} polyanions should also demonstrate unusually high stability.

How can these results concerning pure group IV element clusters be applied to the Cs-Pb system and in particular to the cluster $(Cs_3Pb_5)^+$? Could Pb_5 be a trigonal bipyramidal polyanion? If so, where does it get its electrons? We believe that two Cs atoms transfer their electrons to Pb allowing it to condense into a stable, five-atom polyanion. The cluster needs the third Cs ion to provide an overall positive charge.

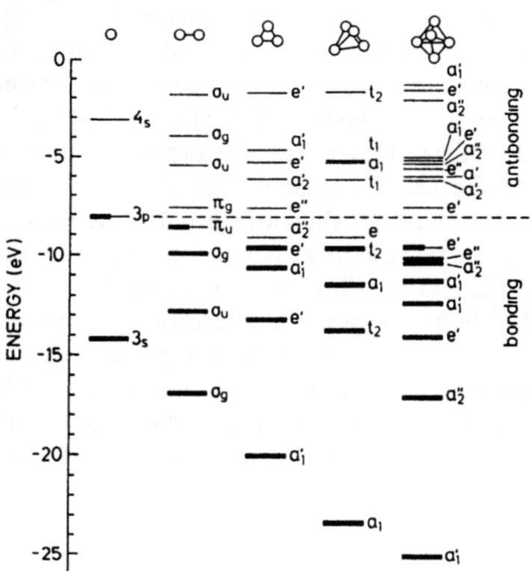

Fig.5 Calculated electronic energy levels of Si clusters. Notice that two bonding electrons can be added to Si_2 and Si_5 and four bonding electrons to Si_4

Additional evidence exists concerning the stability of Pb_5^{2-}. Zintl and co-workers [34,35] made an early study of a Cs-Pb solution in liquid NH_3 They were able to identify many polyanions in liquid solution, for example Pb_5^{4-}. However, these ions did not prove to be sufficiently stable to act as building blocks in the construction of crystals when the solvent was evaporated. This problem was solved by Corbett and co-workers with 2,2,2-crypt [36,37]. In this way they were not only able to grow the salt [crypt $Na^+]_2Pb_5^{2-}$ but also to determine the structure of the Pb_5^{2-} polyanion: it is a trigonal bipyramid.

Lead is so heavy that the stability of two important clusters cannot be checked with our spectrometer: the cluster $(Cs_5Pb_9)^+$,also expected from Zintl's work,and the cluster $(Cs_5Pb_{10})^+$. What is the significance of this second cluster? We stated in the introduction of this paper that two types of information concerning stability can be obtained from cluster mass spectra. A single intense isolated mass peak indicates a highly stable cluster; however, a periodicity in the mass spectrum means that this cluster can be used as a building block to constuct larger clusters. We have noticed that $[Cs(Cs_2Pb_5)]^+$ is stable. Is the next cluster in the series,$[Cs(Cs_2Pb_5)_2]^+$, stable ? In order to test the stability of these two clusters we turn to the system containing the lighter element Sn. This change in material has, however, the disadvantage that Sn has many more isotopes and the mass peaks will be broader.

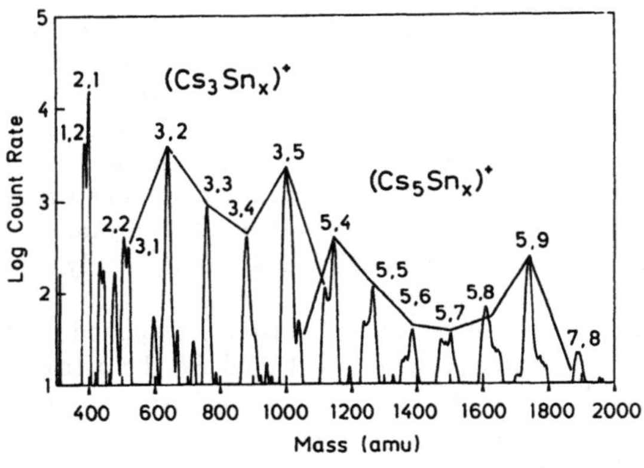

Fig.6 Mass spectrum of Cs-Sn clusters. The peaks are broad due to the large number of isotopes

A mass spectrum of Cs-Sn clusters is shown in Fig.6. Two series of peaks emerge, one series for clusters containing three Cs atoms and the other series for clusters containing five Cs atoms. This selectivity reflects the strong tendency to build clusters with paired electrons only. In Fig. 6,peaks corresponding to the clusters $(Cs_3Sn_5)^+$, $(Cs_5Sn_4)^+$ and $(Cs_5Sn_9)^+$ are particularly strong. The first cluster was already observed and discussed for the Cs-Pb system. In order to discuss the structure of $(Cs_5Sn_4)^+$ it is useful to return to the results of the Hückel calculation shown in Fig.5. If a neutral group IV element tetramer is assumed to be a tetrahedron, it can be seen that the unoccupied orbital offers a place for four additional bonding electrons, resulting in the formation of a highly stable polyanion M_4^{4-}. Four Cs atoms are necessary to provide the electrons. In our case a fifth Cs ion is necessary to provide the net positive charge on the cluster, $(Cs_5M_4)^+$. Actually, it is quite plausible that such a polyanion should be tetrahedral. Si_4^{4-} and Ge_4^{4-} are isoelectronic to neutral P_4 and As_4, both well-known tetrahedral molecules. In fact, tetrahedral Ge_4^{4-} has been observed in Li- and Na-germanide crystals [38-41].

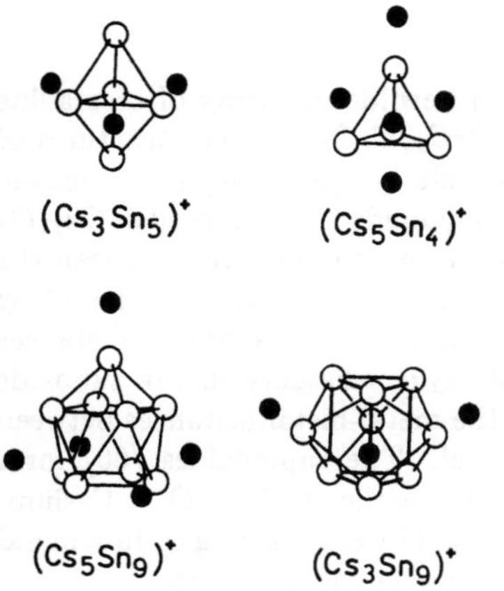

Fig.7 The known structure of several polyanions and the probable distribution of alkali metal ions (black) in several Cs-Sn clusters

Clearly, Zintl clusters exist in the vapor phase. These clusters can probably be described as a core-cluster within a cluster, Fig. 7.

5. PHOTOIONIZATION OF Cs-O CLUSTERS

Introduction

Does the ionization energy of an atom transform continuously into the work function of a metal with increasing cluster size? Early experiments indicated that the transition should not be smooth/42/. In addition, calculations /43,44/predicted that clusters having closed electronic shells with orbital angular momentum ℓ should be very hard to ionize; open shell configurations should give up their electrons easily. These predictions were based on the assumption that clusters have spherical symmetry, an assumption which could not be supported by structure calculations. There the matter stood until 1984,when mass spectra of Na clusters were published showing unusually strong mass peaks for clusters containing 8,20,40,58,92 valence electrons, i.e., closed shell configurations/45/. In the meantime mass spectra have been published by many groups /46,47/ apparently confirming the conclusion that closed-shell clusters have an enhanced stablity. More direct evidence concerning the electronic structure can be obtained by examining the ionization energy of clusters/48-52/.

In this paper previous photoionization investigations will be complemented in two ways. The first way is trivial. The experiments will be extended to cover the metal Cs. However, we have investigated not only pure metal clusters, but the entire binary system Cs_nO_m. Compound clusters composed of alkali metal atoms have received little attention in the past. Notable exceptions are dimers with composition M_2X. Both sodium and potassium halide and oxide dimers have been produced using two quite different techniques. Peterson et al./53/ coexpanded sodium vapor and an inert gas out of a tubular nozzle. The reactant gas O_2 was added during the expansion. Kappes et al./54/ expanded the vapor from an oven containing a mixture of alkali metal and alkali halides. In both these careful studies well-defined ionization thresholds were observed. Dao et al./55/ extended this work on dimers to trimers and tetramers with composition M_3O and M_4O.

The Cs-O System

The elements cesium and oxygen combine to form a bewildering array of crystalline compounds ($Cs_7O, Cs_4O, Cs_3O_{11}, Cs_2O_2, Cs_2O_3, CsO_2, CsO_3$). The structures of the suboxides Cs_7O , Cs_4O and Cs_3O_{11} are particularly interesting and unusual. Crystallographic investigations indicate that these compounds are composed of $Cs_{11}O_3$ clusters having strongly ionic character /56-59/. However, the bonding between the clusters is metallic. The most immediate support of this bonding model is found by comparing the interatomic distances in the suboxides with the corresponding distances in the normal oxides and pure metals. The metal-oxygen distance in the suboxide clusters are nearly the sum of the ionic radii, while the metal-metal distances between clusters are nearly those found in a pure alkali metal. The superoxides /60/ form because oxygen possesses the capability of appearing as either O_2^{2-} or O_2^-. Cesium-oxygen clusters of all possible compositions were produced by evaporating cesium metal into a helium-oxygen gas mixture containing increasing amounts of oxygen.

Experiment

Photoionization experiments on alkali metal clusters have been carried out successfully in the past using other types of light sources; tunable pulsed dye lasers and Hg lamp radiation monochromatized by gratings. We have chosen to use a continuous dye laser because it provides a highly monochromatic beam of light without producing high peak power levels which could promote multiphoton ionization and fragmentation. However, a continuous laser also has a distinct disadvantage. Usable dyes exist only for wavelengths less than 3 eV, and frequency doubling is less efficient than with a pulsed laser. Since we are limited to studying photoionization thresholds below 3 eV, the choice of the materials to be investigated is also severely limited. Cs has the lowest ionization energy of all elements, 4.0 eV. As we shall see, Cs compound clusters have even lower ionization thresholds.

Cesium vapor created in a heated quartz crucible was quenched in He gas having a pressure of 1 mbar. The reactant, O_2, was introduced as a gas mixed with the He. The outer mantel of the reaction chamber was cooled to liquid nitrogen temperatures. Clusters and microcrystals condensed out of the quenched vapor and were transported by the gas stream through a 2 mm nozzle into a chamber of intermediate, $1x10^{-4}$ mbar, pressure. After passing through a 1 mm diameter skimmer, the clusters entered a high vacuum, $2x10^{-6}$ mbar, chamber as a well-collimated beam. This beam intersected the axis of a quadrupole mass spectrometer where the clusters were ionized with light from a tunable dye laser. The mass spectrometer was operated in the mass range from 0 to 2000 amu. The composition of the Cs clusters was controlled by changing the partial pressures of the reactant and of the Cs vapor.

Results and Discussion

Photoionization spectra were recorded for 93 distinct Cs-O clusters. For each of the clusters seven different dyes were used in order to scan the wavelength over the entire range 15500 - 25000 cm^{-1}. Background signal caused by ionization from photoelectrons or from broad band fluorescence was so low that ion signals could be recorded over five orders of magnitude. An example of such a photoionization spectrum (for Cs_9O_4) is shown in Fig.8. Since the ion signal has been normalized to the dye laser intensity, the noise increases at the extreme wavelengths attainable with each dye.

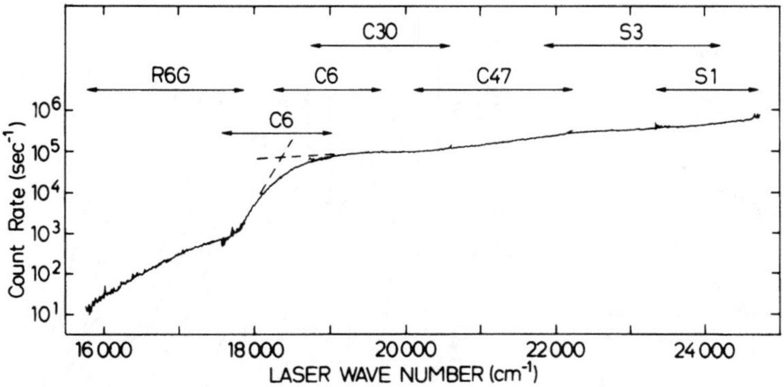

Fig.8 Laser frequency dependence of the ion current for Cs_9O_4 clusters. The partially overlapping curves represent data collected using seven different dyes

Photoionization spectra that consist of one sharp step can be straightforwardly interpreted. However, in practice the spectra are usually much more complicated, Fig.9. For the optimist, this means that the spectra are very rich in information.
Ion counting experiments have a very large dynamical range; usable signals vary from 10 cps to 10^5. Therefore, data must be presented on a logarithmic scale if information is not to be lost, even though such a presentation tends to suppress detailed but reproducible structure. It is not exceptional to find that the photoionization spectra have two or more well-defined steps, Fig. 9. Possible explanations are:

1) Different electrons within the cluster require different energies for ionization.
2) The cluster has several isomeric forms, each with its own threshold.
3) The cluster ion observed can also be produced as a fragment resulting from ionization of a larger cluster.

Clearly, the interpretation of photoionization threshold spectra is far from straightforward. Assumptions must be made concerning, the temperature of the cluster, the electronic-vibrational coupling (Franck-Condon factors), the form of the vibration-free photoionization efficiency curve, and the extent and the temperature dependence of isomerization and fragmentation. All of these factors will be discussed in a forthcoming detailed report on these measurements. For this qualitative presentation we will merely interpret breaks in a semilogarithmic plot of the photoionization spectra as defining ionization thresholds. This assumption, while leading to consistently high threshold values, does not affect the qualitative discussion which will now be presented.

The results of the experiments are summarized in Fig.10 Here the ionization energies of Cs_nO_m are plotted as a function of n and m. No experimental information could be obtained above about 3.0 eV. All points plotted at this limiting value correspond to thresholds outside the experimentally accessible spectral region.

In discussing these results it is useful to speak of four types of cluster; "metal-rich" clusters with composition from Cs_n to $Cs_{2n+2}O_n$, "doped oxide" clusters containing just one excess Cs atom $Cs(Cs_2O)_n$, "oxide" clusters with the composition $(Cs_2O)_n$, and "oxygen-rich" clusters containing more oxygen than oxide clusters. Using this terminology it is now possible to formulate several experimental observations. Three of these observations are qualitative.

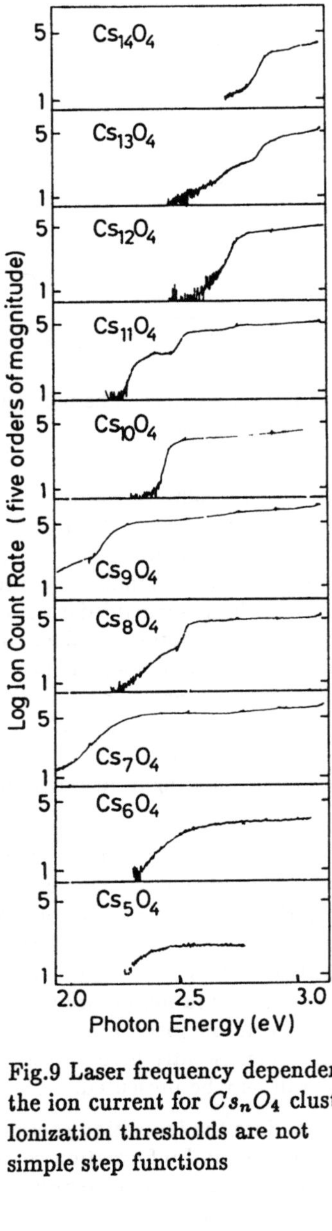

Fig.9 Laser frequency dependence of the ion current for Cs_nO_4 clusters. Ionization thresholds are not simple step functions

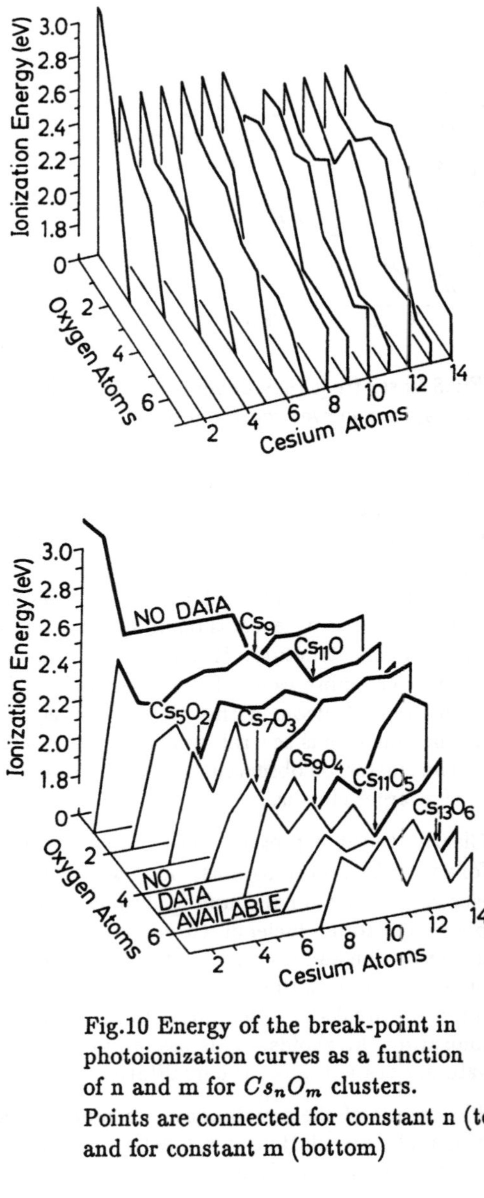

Fig.10 Energy of the break-point in photoionization curves as a function of n and m for Cs_nO_m clusters. Points are connected for constant n (top) and for constant m (bottom)

1) The ionization energies of metal-rich clusters *decrease* with increasing oxygen content.
2) "Doped oxide" clusters have unusually low ionization energies.
3) The ionization energies of oxygen-rich clusters Cs_nO_m alternate; high for even values of n, low for odd values of n.

In addition to these general observations several specific results are worth mentioning.

4) Cs_9,Cs_4O, $Cs_{11}O$ have particularly low ionization energies.
5) Cs_2O, $Cs_{10}O$ have particularly high ionization energies.
Each of these results will now be discussed in turn.

Will the ionization energy of a Cs cluster increase or decrease when an oxygen atom is added? Intuitively one might think that a metal should be easier to ionize than an oxide. Fig.10 clearly shows that this is not true for Cs. In fact, oxidized Cs has perhaps the lowest work function of any known material, a property utilized for decades in the construction of photocathodes sensitive in the infrared. Many models have been presented over the years to explain this effect on solid surfaces. Perhaps the effect can be made plausible in a cluster in the following way. Among the alkali metal oxides, the Cs-O bond has the highest ionicity. An electron removed from a Cs s-orbital will be given an additional push by the nearby negative oxygen ion. Clearly, realistic ab initio calculations for Cs_nO clusters are needed to clarify this issue.

"Doped oxide" clusters containing one extra cesium atom have very low ionization energies. Optical transmission measurements /61/ on thin films of Cs_2O indicate that the normal oxide is a semiconductor with a band gap of about 2.0 eV. If the material is prepared under conditions which result in an excess of cesium, measurements of the activation energy of electrical conductivity indicate the presence of donor states 0.25 eV below the conduction band /62/. In analogy with the bulk properties, the excess cesium in an oxide cluster may act as a donor state well above the highest occupied orbital (top of the valence band) of the normal oxide, resulting in an appreciable lowering of the ionization energy (work function).

The third effect listed above can also be interpreted as being due to doping by an excess Cs atom. However, in this case the doped material is not a simple oxide but a mixture of normal and super oxides. For example, Cs_8O_5 can be thought to contain of three different building units; Cs^+, O^{2-}, and O_2^{2-}. Using this scheme Cs_8O_5 becomes $(Cs_2O)_3(Cs_2O_2)$, i.e., a mixed oxide with no excess cesium atoms. On the other hand, Cs_9O_5 can be written as $Cs(Cs_2O)_3(Cs_2O_2)$, i.e., this cluster is doped with an excess cesium atom and has, therefore, a low ionization energy. This results in the even-odd alternation in the ionization energy of oxide-rich clusters. Even-odd alternations have been observed previously in the mass spectra of metal clusters and metal compound clusters. That pure alkali metal clusters containing an odd number of atoms are easily ionized has also been established. Now we see that such an alternation also exists in the ionization potential of metal-oxygen clusters. The most obvious explanation is that a paired electron is difficult to remove.

Both Cs_9 and $Cs_{11}O$ have unusally low ionization energies. Both clusters contain nine free electrons. As explained in the introduction to this section, both theory and experiment give support to the view that the ninth electron in a pure, spherical metal cluster opens a new angular momentum shell and that this electron is weakly bonded. Here we see that apparently such considerations qualitatively explain the electronic properties of not only pure alkali metal clusters, but also of clusters containing an oxygen atom.

It is now a well established fact that ionization is often accompanied by fragmentation. Although fragmentation should be considerably reduced near the photoionization threshold, we must consider the possiblity that some of the spectral features observed in these photoionization experiments reflect the properties of a larger parent cluster.

Common features observed in the spectra of different clusters indicate that fragmentation might be playing a role. Of course,it is not difficult to find correlations when comparing 90 spectra with one another. Therefore, this discussion will be limited to the three most likely decay channels:

$$Cs_mO_n \longrightarrow (Cs_{m-1}O_n)^+ + Cs + e$$

$$Cs_mO_n \longrightarrow (Cs_{m-2}O_{n-1})^+ + Cs_2O + e$$

$$Cs_mO_n \longrightarrow (Cs_{m-4}O_{n-2})^+ + Cs_4O_2 + e.$$

Much to our surprise the spectra show no evidence for the decay by formation of neutral Cs or Cs_2O. The data are, in a few instances only, consistent with the third decay channel characterized by the splitting off of a neutral Cs_4O_2. One example is shown in Fig.11.

Some information concerning small alkali-metal-oxygen clusters is available from calculations. For example, LCAO-MO SCF calculations /63/ indicate that Li_2O_2 is planar with D_{2h} symmetry. The most stable form of Li_4O has tetrahedral symmetry /64/. Recent SCF pseudopotential calculations /65/ offer an explanation for the low ionization energy of Cs_4O. The a_1(s) orbitals on the oxygen atom are destabilized by the large Cs_4-orbital charge density within the tetrahedron.

The stucture of positively charged clusters with composition $[Cs(Cs_2O)_n]^+$ can be determined by minimizing the total energy which is given by the sum of all two-body interactions.

$$V_{ij} = Z_iZ_j/r_{ij} + Aexp(-r_{ij}/\rho) \quad .$$

Each of the ions in these clusters has a rare gas electron configuration. In this respect they might be referred to as normal oxides. For such cases there is no ambiguity as to how the charge should be distributed among the ions, namely, Cs^+ and O^{2-}. The parameters used, A = 4000eV and $1/\rho$ = 303pm, were chosen so that the Cs-O distance in large clusters is that observed for crystalline Cs_2O. $(Cs_3O)^+$ has only one stable configuration, a planar structure with 119.47eV binding energy with respect to dissociation into ions. The dimer has two stable configurations. The symmetrical dimer with a threefold symmery axis (Fig.11) has 0.6eV more binding energy than its nearest competitor. Larger clusters with a threefold axis are also energetically favored over less symmetrical structures.

6. CONCLUDING REMARKS

The technique of producing compound clusters by allowing the component elements to react in the vapor phase is a convenient way of studying binary systems. In this way it is easy to scan through the entire range of compositions and to search for clusters with unusually high or low stability or to measure the photoionization thresholds.
Adding oxygen to cesium clusters lowers the ionization potential. Two types of Cs-O clusters have particularly low ionization energies; those with composition $Cs(Cs_2O)_n$, i.e. "doped" oxide clusters, and those containing nine "free" electrons. The photoionization spectra often show considerable structure near the threshold energy. This structure is only partly explained by fragmentation and may be due to the presence of structural isomers among clusters of a given size. Molecular dynamics calculations are now being carried out to test this explanation.

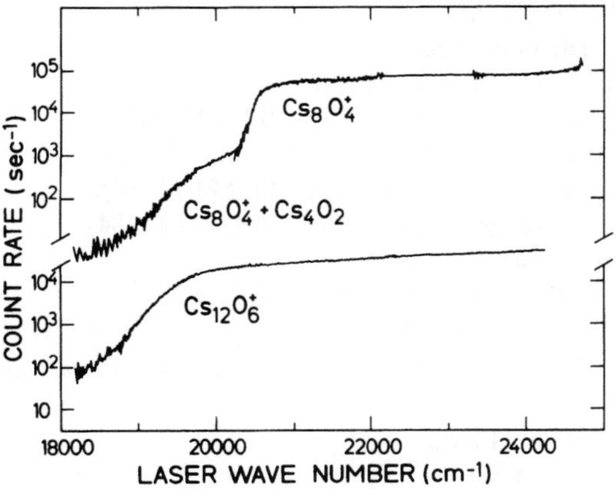

Fig.11 Photoionization spectra indicating possible fragmentation

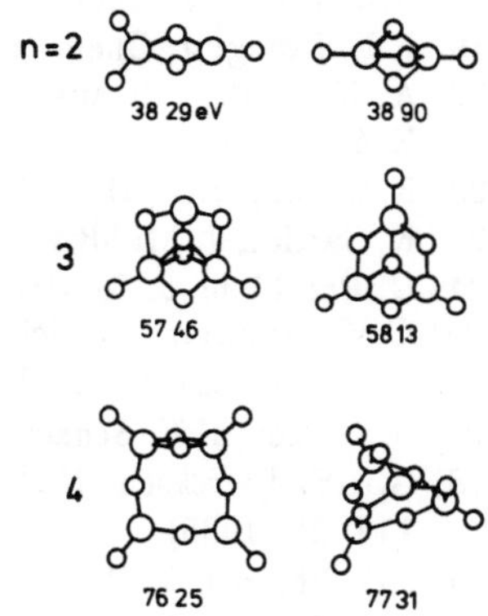

Fig.12 The most stable forms of $(Cs_{2n+1}O_n)^+$ clusters

REFERENCES

1. O. Echt, K. Sattler, E. Recknagel: Phys. Rev. Lett. **47**, 1121 (1981).
2. A. Ding, J. Hesslich: Chem. Phys. Lett. **94**, 54 (1983).
3. H.P. Birkhofer, H.P. Haberland, M. Winterer, D.R. Worksnop: Ber. Bunsenges. Phys. Chem. **88**, 207 (1984).
4. W.D. Knight, K. Clemenger, W.A. de Heer, W.A. Saunders, M.Y. Chou, M. L. Cohen: Phys. Rev. Lett. **52**, 2141 (1984).
5. M.M. Kappes, R.W. Kunz, E. Schumacher: Chem. Phys. Lett. **91**, 413 (1982).
6. T.P. Martin: J. Chem. Phys. **83**, 78 (1985).
7. L.A. Bloomfield, R.R. Freeman, W.L. Brown: Phys. Rev. Lett. **20**, 2246 (1985).
8. T.P. Martin, H. Schaber: J. Chem. Phys. **83**, 855 (1985); Z. Phys. B-Condensed Matter **35**, 61 (1979).
9. J. Van de Walle and P. Joyes: Phys. Rev B **32**, 8381 (1985); J.R. Heath, Liu Zuan, S.C. O'Brien, Qing-Ling Yhang, R.F. Smalley, F.K. Tittle: J. Chem. Phys. **83**, 5520 (1985).
10. H.W. Kroto, J.R. Heath, S.C. O'Brien, R.F. Curl, R.E. Smalley: Nature **318**, 162 (1985).
11. T.P. Martin: J. Chem. Phys. **81**, 426 (1984); S. Becker, H.-J. Dietze: Internat. J. of Mass Spectr. and Ion Processes **73**, 157 (1986); D. Schild, R. Pflaum, K. Sattler, E. Recknagel: J. Phys. Chem. **91**, 2649 (1987); R.G. Wheeler, K. LaiHing, W.L. Wilson, M.A. Duncan: to be published; R.B. Freas, B.A. Waite, B.I. Dunlap, J.E. Campana: to be published.
12. E.A. Rohlfing, D.M. Cox, R. Petkovic-Luton, A. Kaldor: J. Phys. Chem. **88**, 6227 (1984).
13. M.M. Kappes, P. Radi, M. Schär, E. Schumacher: Chem. Phys. Lett. **119**, 11 (1985).
14. T. Scheuring, K.G. Weil: Surf. Sci. **156**, 457 (1985).
15. W. van der Lugt, W. Geertsma: J. NonCryst. Solids **61**, 187 (1984).
16. H. Hoshino, R.W. Schmutzler, F. Hensel: Phys. Lett. A **51**, 7(1975).
17. N.E. Christensen, S. Kollar: Solid State Commun. **46**, 727 (1983).

18. R.E. Honig: J. Chem. Phys. **22**, 126 (1954).
19. R.E. Honig: J. Chem. Phys. **22**, 1610 (1954).
20. A. Kant, B.H. Strauss: J. Chem. Phys. **45**, 822 (1966).
21. K.A. Gingerich, A. Desideri, D.L. Cocke: J. Chem. Phys. **62**, 731 (1976).
22. K.A. Gingerich, D.L. Cocke, F. Miller: J. Chem. Phys. **64**, 4027 (1976).
23. K. Sattler, J. Mühlbach, E. Recknagel: Phys. Rev. Lett. **45**, 821 (1980).
24. G. Pacchioni, J. Koutecky: Ber. Bunsenges. Phys. Chem. **88**, 242 (1984).
25. C.M. Weinert: Surf.Sci. **156**, 641 (1985).
26. K. Raghavachari: J. Chem. Phys. **83**, 3520 (1985); **84**,5672 (1986).
27. R.S. Grev, H.E. Schaefer: Chem. Phys. Lett. **119**, 111 (1985).
28. G.H.F. Diercksen, N.E. Grüner, J. Oddershede, J.R. Sabin: Chem. Phys. Lett. **117**, 29 (1985).
29. J.C. Phillips: J. Chem. Phys. **83**, 3330 (1985).
30. R.O. Jones: Phys. Rev. A, **32**, 2589 (1985).
31. E. Blaisten-Barojas, D. Levesque: Phys. Rev. B **33**, 3910 (1986).
32. D. Tomanek, M.A. Schluter: (to be published).
33. S. Saito, S. Ohnishi, S. Sugano: Phys. Rev. B **33**, 7036 (1986).
34. E. Zintl, G. Brauer: Y. Elektrochem. **41**, 297 (1935).
35. E. Zintl, H. Kaiser: Y. Anorg. Allg. Chem. **211**, 113(1933).
36. P.A. Edwards, J.D. Corbett: Inorg. Chem. **16**, 903 (1977).
37. C.H.E. Belin, J.D. Corbett, A. Cisar: J. Am. Chem. Soc.**99**, 7163 (1977).
38. H.-G. von Schnering: Nova Acta Leopoldina **59**, 165 (1986).
39. J. Witte, H.-G. von Schnering: Z. Anorg. Chem. **327**, 260 (1964).
40. J. Llanos, R. Nesper, H.-G. von Schnering: Angew. Chem. **95**, 1016 (1983).
41. E. Niecke, R. Rüger, B. Krebs: Angew. Chem. **94**, 553 (1982).
42. A. Herrmann, S. Leutwyler, E. Schumacher and L. Wöste: Helv. Chim. Acta **61**, 453 (1978).
43. J.L. Martins, R. Car and J. Buttet: Surf. Sci. **106**, 265 (1981).
44. W. Ekardt: Ber. Bunsenges. Phys. Chem. **88**, 289 (1984).
45. W.D. Knight,K. Clemenger, W.A. de Heer, W.A. Saunders, M.Y. Chou and M.L. Cohen: Phys. Rev. Letts. **52**, 2141 (1984).
46. W. Begemann, S. Dreihofer, K.H. Meiwes-Broer and H.O. Lutz: Z. Physik D **3**, 183 (1986).
47. I. Katakuse, I.Ichihara, Y. Fujita, T. Matsuo, T. Sakurai and H. Matsuda: Intern. J. Mass Spectrom. Ion. Proc. **67**, 229 (1985).
48. A. Hermann, E. Schumacher and L. Wöste: J. Chem. Phys. **68**, 2327 (1978).
49. M.M. Kappes, R.W. Kunz and E. Schumacher: Chem. Phys. Letters **91**, 413 (1982).
50. K.I. Peterson, P.D. Dao, R.W. Farley and A.W. Castleman, Jr.: J. Chem. Phys. **80**, 1780 (1984).
51. C. Brechignac, Ph. Cahuzac and J. Ph. Roux: Chem. Phys. Letters **127**, 445 (1986).
52. W.A. Saunders, K. Clemenger, W.A. de Heer, and W.D. Knight: Phys. Rev. B **32**, 1366 (1985).
53. K.I. Peterson, P.D. Dao and A.W. Castleman, Jr.: J. Chem. Phys. **79**, 777 (1983).
54. M.M. Kappes, P. Radi, M. Schär and E. Schumacher: Chem. Phys. Letters **113**, 243 (1985).
55. P.D. Dao, K.I. Peterson and A.W. Castleman, Jr.: J. Chem. Phys. **83**, 563 (1984).

56. A. Simon, in *Structure and Bonding, Vol. 36* (Springer-Verlag, Berlin, 1979).
57. A. Simon, W Brämer, and H.-J. Deiseroth: Inorg. Chem. **17**, 875 (1978).
58. A. Simon: Z. Anorg. Allg. Chem. **431**, 5 (1977).
59. T.P. Martin, H.-J. Stolz, G. Ebbinghaus and A. Simon: J. Chem. Phys. **70**, 1096 (1979).
60. H. Föppl: Z. Anorg. Allg. Chem. **297**, 12 (1957).
61. P.G. Borzhak, V.F. Bibik and G.S. Kromerenko: Bull. Acad. Sci. USSR, Phys. Ser. (English Transl.) **20**,939 (1956).
62. J.E. Davy: J. Appl. Phys. **28**, 1031 (1957).
63. M. Allavena, E. Blaisten-Barojas and B. Silva: J. Chem. Phys. **75**,787 (1981); G. Bravo and E. Blaisten-Barojas: Chem. Phys. Lett. **108**, 237 (1984).
64. E.U. Würthwein, P.v.R. Schleyer and J.A. Pople: J. Am. Chem. Soc. **106**, 6973 (1984).
65. A. Savin, H. Preuss and H. Stoll: Rev. Rumanian Chem., to be published.

Molecular Metal Clusters: Structures and Bonding

G. Pacchioni

Dipartimento di Chimica Inorganica e Metallorganica,
Università di Milano, Via Venezian 21, I-20133 Milano, Italy

1. INTRODUCTION

Molecular metal clusters have been defined as a group of three or more metal atoms connected by chemical bonds and surrounded by ligands. Actually, the not well defined concept of metal-metal bond in molecular metal clusters makes it difficult to distinguish between "real" clusters and other metal-containing inorganic compounds.

The molecular metal clusters experimentally identified today fall in the range of aggregation between polynuclear molecular metal complexes and small metallic crystallites, and the ligands include halides, carbonyls, isocyanides, phosphines, chalcogenide atoms, and many other kinds of organic and inorganic molecules |1|. The cluster chemistry on the left of the d block is dominated by the halide clusters of general formula M_6X_8 and M_6X_{12}, but the variety of nuclearity, geometries and ligands increases towards the right of the transition series.

Structural and chemical investigations on molecular metal clusters can increase understanding of the geometrical and electronic properties of very small metallic particles |2|, and also of their surface |3| and catalytic chemistry |4|. In addition, molecular metal clusters can be potential starting material for the preparation of very small metallic particles, which can be characterized by specific and sometimes unusual physical and chemical properties.

However, several aspects of molecular cluster chemistry and physics are largely unexplored: a) are large homonuclear or mixed molecular clusters analogs of small metallic aggregates (or alloys) from the theoretical, geometrical and structural point of view? b) may the chemistry of molecular clusters give further information on the mechanism of nucleation of small metallic particles? c) can the stereochemistry of the ligands of the coordination sphere be an acceptable model for the mechanism of chemisorption on metal-surfaces? d) is the bonding in small bare transition metal clusters of the same kind as in their molecular counterparts?

2. STRUCTURE OF MOLECULAR METAL CLUSTERS

More than a thousand metal cluster structures have been reported in the literature up to now. They cover a wide range of possible geometrical packing, even if the majority has three-dimensional shapes resembling fragments of metallic structures |1c|. Typically, some of the largest cluster structures resolved by X-ray crystallography show hcp or ccp arrangements of the metal atoms (e.g. $|Pt_6Ni_{38}(CO)_{44}|^{6-}$ |5|, $|Pt_{38}(CO)_{44}|^{2-}$ |1c|, $|Rh_{22}(CO)_{37}|^{4-}$ |6|,

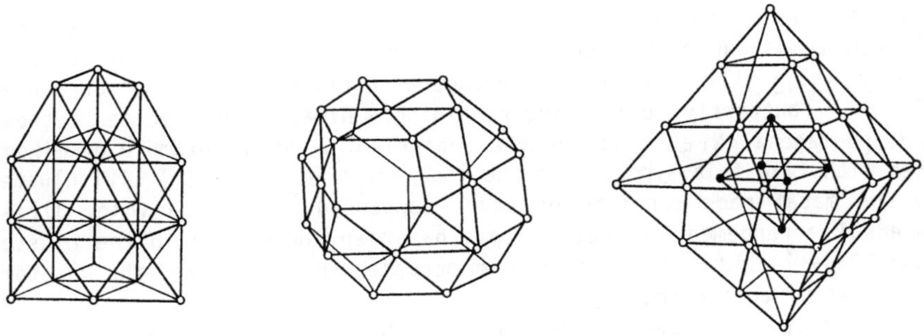

I II III

Fig. 1 - Compact packings in molecular metal clusters: (I) $|Rh_{22}(CO)_{37}|^{4-}$, ccp-hcp |6|; (II) $|Pt_{38}(CO)_{44}|^{2-}$, ccp |1c|; (III) $|Pt_6Ni_{38}(CO)_{44}|^{6-}$,ccp|5|.

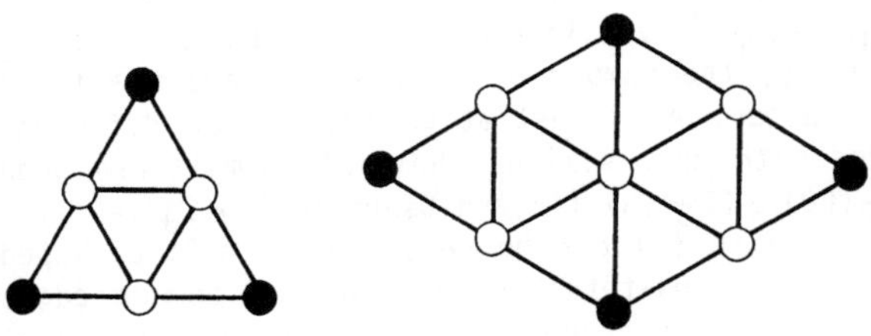

I II

Fig. 2 - Planar packing in molecular metal clusters. (I) $|Fe_3Pt_3(CO)_{15}|^{7-}$ |7| and $|Fe_3Cu_3(CO)_{12}|^{3-}$ |8|; (II) $|Fe_4Cu_5(CO)_{16}|^{3-}$ |8|.

Fig. 1). However, the tendency to assume compact structures is not the driving force determining the geometry of the metal core. In fact, some relevant examples of planar structures are known (Fig. 2); in the $|Fe_3Pt_3(CO)_{15}|^{-}$ |7| and $|Fe_3Cu_3(CO)_{12}|^{3-}$ |8| compounds the metal framework may be regarded as an equilateral triangle of Pt atoms bridged on each edge by a $Fe(CO)_4$ group; in the $|Fe_4Cu_5(CO)_{16}|^{3-}$ anion |8|, the largest planar cluster synthesized up to now, the metal atoms describe a regular rhombus (Fig. 2).

Pentagonal or icosahedral packings have also been reported for a series of clusters (Fig.3). Among homonuclear clusters we recall here the

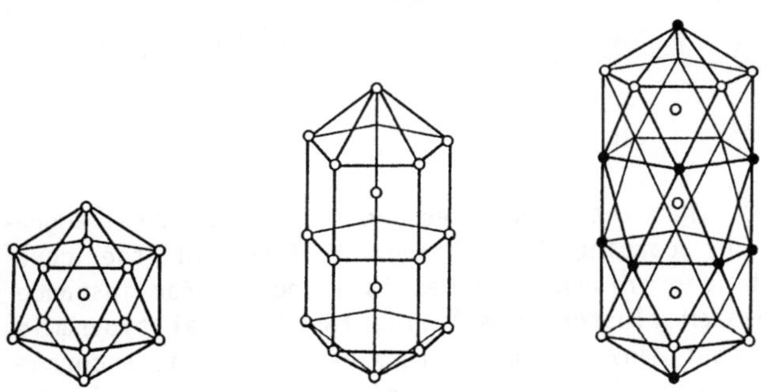

I II III

Fig. 3 - Icosahedral packings in molecular metal clusters. (I) $|Au_{13}(PR_3)_{10}Cl_2|^{3+}$ |9|; (II) $|Pt_{19}(CO)_{22}|^{4-}$ |10|; (III) $|Au_{13}Ag_{12}(PR_3)_{12}Cl_6|^{m+}$ |12|.

icosahedral $|Au_{13}(PR_3)_{10}Cl_2|^{3+}$ |9| and the $|Pt_{19}(CO)_{22}|^{4-}$ |10| clusters; this latter system can be considered as obtained by condensing three pentagonal bipyramids. A rather complicated structure is that of the cluster $Ni_{34}Se_{22}(PR_3)_{10}$ |11| containing a pentagonal antiprism Ni_{10} core surrounded by five Ni_4 "butterfly" units; the remaining 4 Ni atoms are located between the two pentagonal Ni_5 rings of the antiprism. A recent example of mixed-cluster with five-fold symmetry is the $|Au_{13}Ag_{12}(PPh_3)_{12}Cl_6|^{m+}$ |12| cluster where the metal cage consists of three condensed icosahedra (Fig. 3). An icosahedral arrangement of metal atoms has been found also in the cluster $|Rh_{12}Sb(CO)_{27}|^{3-}$ |13| where the Sb atom occupies the center and the Rh atoms the vertices of an icosahedron.

In transition molecular metal clusters the symmetry of the metal cage depends on a variety of factors including number, type and coordination mode of ligands, number of electrons, total molecular charge, etc. Often an interchange from one kind of packing to another has been observed in clusters of similar nuclearity. In a series of Rh carbonyl clusters it has been found |14| that the geometry of the metal cage changes from hcp in $|Rh_{13}(CO)_{24}|^{5-}$ to ccp in $|Rh_{15}(CO)_{30}|^{3-}$ via intermediate mixed hcp-ccp structures by changing the nuclearity, the number of CO ligands, and the total number of electrons of the cluster. An illustrative example of different cluster geometries for systems with very similar chemical formula has been reported |15| for the $Os_6Pt_2(CO)_{16}(C_8H_{12})$ (I) and $Os_6Pt_2(CO)_{17}(C_8H_{12})$ (II) clusters (Fig. 4). The metal core in (I) can be described as a Pt bicapped Os-tetrahedron condensed to a second Os-tetrahedron through a common edge, while the metal cage in (II) can be regarded as a regular octahedron of Os atoms with two face capping $Pt(C_8H_{12})$ units. However, it has been found that, with few exceptions, clusters with the same formal number of skeletal electrons assume similar geometries of the metal polyhedron (see § 3.2). Small changes of this number (e.g. by adding a CO ligand which is assumed to be a two-electron donor group) result in a substantial rearrangement of the metal cage, a sign of easy deformability of the system.

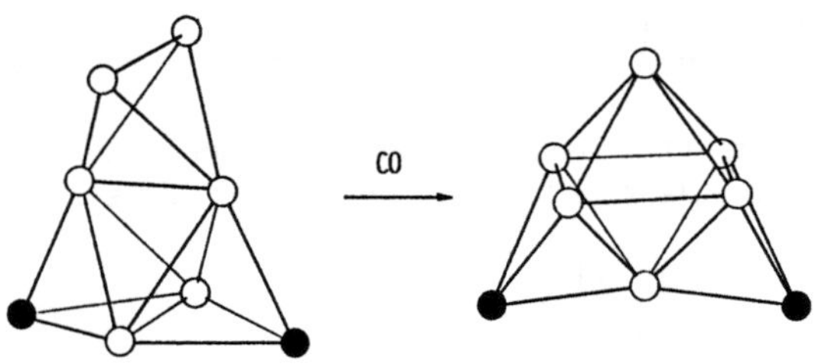

I II

Fig. 4 - Metal core structure of
(I) $Os_6Pt_2(CO)_{16}(C_8H_{12})_2$
and
(II) $Os_6Pt_2(CO)_{17}(C_8H_{12})_2$
|15|.

It has been proposed |2| that the growth sequence of simple metal aggregates proceeds initially by capping the triangular faces of the core polyhedron. The tendency to cap triangular faces is supported for instance by the high number of 5-atom clusters displaying the trigonal bipyramid structure and by the fact that many 7-atom clusters have a structure corresponding to a capped octahedron |16|. There is only one example of a 7-atom cluster with real pentagonal arrangement, namely a pentagonal bipyramid Au_7 cluster |17|.

Many clusters incapsulate hetero-atoms such as H, C, N, or even As, Sb, P, S, ets. inside their metallic cages. Usually the metal core arrangement is such to produce a hole in which the heteroatom can be accommodated. They represent useful models of interstitial materials |18|.

The metallic core can be fundamentally cubic or pentagonal. The number of interstitial atoms embedded in the metal cage depends on the geometry and on the dimensions of the cluster. In some cases a small molecule can be encapsulated into the metal core. For instance, the $|Co_6Ni_2C_2(CO)_{16}|^{2-}$ |19| and the $|Ni_{16}(C_2)_2(CO)_{23}|^{4-}$ |20| clusters contain one and two C_2 molecules, respectively. The heteroatoms do not always occupy interstitial positions. One of the classical examples is the $|Fe_4C(CO)_{12}|^{2-}$ complex where the C atom is exposed and more suitable for chemical attack |21|. Recently, an example of peripheral dicarbide carbonyl cluster has been reported |22|. In this molecule the metal cage is an open Co_6 "boat" framework where the two apical Co atoms are connected through a Co-C-C-Co array.

Many other group IV atoms, beside carbon, can be included in metal clusters |18|. One of these systems, the $|Co_9Si(CO)_{21}|^{2-}$ cluster with a Si atom in the center of a capped square antiprism of Co atoms |23| is a rare example of kinetically stable paramagnetic cluster.

The great variety of geometrical arrangements reported in the literature suggests that interstitial compounds may be structurally rather complex since they do not follow simple rules. This can be an important indication for hypothesis on the investigation on materials which are becoming technologically very important (in particular hydrides, carbides and nitrides).

Of particular interest are also mixed metal clusters with two or more metals. Structures of large clusters containing two metals have been solved; we have already mentioned the examples of Pt_6Ni_{38} |5| and $Au_{13}Ag_{12}$ |12| systems (Figs. 1 and 3). In both cases we have islands or layers of the same metal surrounded or followed by shells or layers respectively of the other metal. This trend would suggest that metal alloys might form similar homogeneous blocks. The effect should be that of diluting the two metals, whithout changing very much their individual properties.

This statement would support the "minimum polarity model" of alloys, according to which each component of an alloy maintains basically an electronic structure similar to that of its pure phase. In addition it is worth mentioning that many mixed clusters easily (for instance thermally or photochemically) transform into mixtures of simple homonuclear clusters, suggesting that despite the complexity of the molecular structures of mixed clusters there is a tendency or a driving force for the total dissociation of the original cluster into homonuclear metallic structures.

3. BONDING IN MOLECULAR METAL CLUSTERS

The understanding and interpretation of the chemical bonding in molecular metal clusters represents a big challenge to quantum-chemists. One of the fundamental questions is whether the bonding between metal atoms resembles that of the corresponding bulk metals or is typically molecular. Unfortunately, the dimensions of these systems are so large that accurate quantum-mechanical calculations are not feasible at the present. Many attempts to rationalize the structural behaviour of molecular clusters on the basis of few simple phenomenological rules have been done, although these qualitative approaches do not provide any insight into the nature of the bonding as such.

In the following we briefly discuss the basic ideas of various approaches of increasing complexity which have been outlined as interpretative tools of the structure and reactivity of molecular metal clusters.

3.1 Hard-Sphere Packing Model

One of the theories proposed for a systematic classification of polynuclear metal complexes is based on the assumption that an important factor governing the structure of inorganic clusters, and in particular of carbonyl clusters, is the interaction between the ligand groups. According to this model |24|, which in the last analysis can be considered as an extension of the Valence Shell Electron Pair Repulsion (VSEPR) theory |25|, the carbonyl ligands form a polyhedron which minimizes the non-bonded (repulsive) interactions between themselves thus creating a cavity large enough to accomodate the metal core. This latter orients itself inside the polyhedron in order to maintain an "optimal" distance from the ligands. The observed fluxionality of carbonyl clusters in solution |24| is therefore the result of the reorientation of the metal core within the ligand polyhedron.

Benfield and Johnson |24| have calculated the most stable arrangements of points on a spherical surface modeling the polyhedral structure of carbonyl ligands in a cluster compound of the type $M_n(CO)_m$. The predicted polyhedron is that with the smallest non-bonded repulsion energy, giving also a possible explanation of the ligand structure of several clusters.

The validity of this idea is confirmed by recent experimental results showing the "rigid" structure of the carbonyl polyhedron and the easy reorientation of the metal core within the ligand cage |24|. However, the method does not give any information on the structure of the metal framework, which is assumed to be almost spherical in shape and characterized by an important "its own" stability.

3.2 Polyhedral Skeleton Electron Pair (PSEP) Model

The PSEP approach |26| is essentially based on the rules developed for the rationalization of the structure of borane compounds |27|. The borane anions $B_nH_n^{2-}$ have deltahedral structure which maximizes the number of nearest neighbours. If all the vertices of the deltahedron are occupied, as in the $B_nH_n^{2-}$ series, the structure is called <u>closo</u> structure. Each B-H bond involves one boron orbital and two electrons thus leaving three orbitals and two electrons to be used in the B_n cage. Two of these three orbitals lie on the "surface" (e.g. p_x or p_y) and one (s, p_z or and sp_z hybrid) is directed towards the center of the polyhedron. This gives 2n "surface" or tangential orbitals, n bonding and n antibonding, one strongly bonding radial and n-1 weakly bonding, non-bonding or antibonding radial orbitals. The total number of bonding orbitals is therefore n+1, the number of cage electrons 2n+2 and the total number of electrons 4n+2. By removing a boron atom from a vertex of a <u>closo</u> structure one obtains a <u>nido</u> structure ($B_{n-1}H_{n+3}$) which obeys the electron formula 4n+4. If two vertex boron atoms are removed, the resulting structure ($B_{n-2}H_{n+4}$) is called <u>arachno</u> and obeys the electronic rule 4n+6 |27|.

The extension of these rules to the case of transition metal clusters is based on the isolobal principle which was introduced |28| in order to rationalize some aspects of the cluster chemistry and some structural analogies found between metal clusters and borane polyhedra. In principle, a wide range of atoms or groups of atoms can substitute the BH fragment of borane. Neutral analogs of BH include $Fe(CO)_3$, $Ni(PR_3)_2$, AlR, Sn, Pb. In fact, all these fragments contribute two electrons and three orbitals to the cluster bonding. In $Fe(CO)_3$ the Fe atom has 9 valence orbitals and 8 electrons. Three empty σ orbitals and three filled π orbitals interact with the CO orbitals; this leaves three orbitals and two electrons available for the bonding with the other metal atoms.

In general, fragments are said to be isolobal if they have an equal number of frontier orbitals characterized by the same symmetry, similar energy and spatial extent |28|. On the basis of these analogies it is possible to assume that isolobal fragments produce isostructural polyhedra clusters. Of course the isolobal principle is a symmetry-based statement about separated fragments which assumes little about the details of their interaction in real clusters. Nevertheless, transition metal analogs of borane are known. They are characterized by a number of valence electrons equal to 14n+2 as a consequence of the filling of the additional five d atomic orbitals with respect to the borane analogues. Also in the case of transition metal polyhedra <u>nido</u> and <u>arachno</u> structures which obey the 14n+4 and 14n+6 rules, respectively, have been synthesized (Table 1).

The PSEP rule represents therefore a very simple but powerful method to rationalize and predict the structure of cluster compounds and has been

Table 1. Examples of electron count in molecular metal clusters

Cluster	Skeletal geometry	electron count			
		obs.[a]	PSEP|26|[b]	TEC|30|	Lauher|31|
$Os_5(CO)_{16}$	trigonal bipyramid[c]	72	72	72	72
$\|Ni_5(CO)_{12}\|^{2-}$	trigonal bipyramid[c]	76	72	76	72
$\|Co_8C(CO)_{18}\|^{2-}$	square antiprism[d]	114	118	114	114
$\|Ni_8C(CO)_{16}\|^{2-}$	square antiprism[d]	118	118	118	114
$\|Os_{10}C(CO)_{24}\|^{2-}$	tetrahedron[c]	134	142(134)[e]	134	140
$\|Rh_{12}Sb(CO)_{27}\|^{3-}$	icosahedron[c]	170	170	170	170

a) Determined by adding the number of valence electrons of the metal centers and of the encapsulated atoms, the number of negative charges and two electrons "donated" by each CO ligand. b) 14n+2=<u>closo</u>, 14n+4=<u>nido</u>, 14n+6= <u>arachno</u>; n is the number of vertices of the polyhedron. c) <u>Closo</u> structure. d) <u>Arachno</u> structure. e) After correction for condensed polyhedra |27d|.

successfully applied to a wide range of systems |26|. Recently, an extension of the original rules has been proposed by Mingos in order to explain the structure of high-nuclearity molecular metal clusters which can be considered as derived by condensed polyhedra |26d|.

Stone has formulated an elegant and general theoretical justification of the PSEP rule |29|. The basic idea is that the MOs of a cluster can be classified according to their nodal structure, so that their energy depends on the nodal characteristics rather than on the detailed form. Stone's Tensor Surface Harmonic Theory |29| constructs a set of appropriate MOs from the solution of the Schrödinger equation for an electron on the surface of a pseudo-spherical deltahedral cluster and, for a transition metal cluster M_n, identifies 2n-1 orbitals which are likely to be vacant, leaving 7n+1 MOs available for cluster electrons in agreement with the 14n+2 rule.

3.3 Topological Electron Counting (TEC) Theory

The PSEP theory has the advantage of easy application to metal clusters derived from triangulated polyhedra, but is hardly applicable to transition metal clusters with unusual polyhedral frameworks which do not correspond to those of boranes. Moreover, examples of metal clusters exhibiting the same geometry but different electron count which do not fit the PSEP model are known (Table 1).

An attempt to overcome these difficulties has been recently done by Teo |30|. He proposed a topological electron count theory (TEC) based on the Euler's theorem which states that for a polyhedron with V vertices, F faces and E edges E = V + F - 2 (1). Assuming that each atom on the surface of the metal framework attains the 18-electron noble-gas configuration and that each edge can be considered as a two-center, two-electron metal-metal bond, than the total electron count for the cluster will be N = 18V - 2E (2) and the occupied Cluster Valence Molecular Orbitals (CVMO) are CVMO = N/2 = 9V - E (3). By substituting for E the expression (1) one obtains CVMO = 8V - F + 2 (4). Indeed, it is well known that the bonding in metal clusters is better described in terms of multi-center bonds (see § 3.6). In order to take into account this important correction Teo proposed that CVMO = 8V - F +2 + X (5), where X is a parameter depending on the shape of a given polyhedron |30a|.

The method succesfully describes polyhedra which follows as well as those that violate the PSEP rule (Table 1). Some exceptions found in the case of Pt and Au clusters have been attributed to the tendency of these atoms to form 16-electron rather than 18-electron complexes |30|.

3.4 Lauher's Approach

According to this method proposed by Lauher |31|, the cluster MOs are determined according to an Extended Hückel (EHMO) calculation on the bare metal cluster and then divided into filled Cluster Valence Molecular Orbitals (CVMO) and High-Lying Antibonding Orbitals (HLAO) by choosing as empirical upper limit for the CVMOs the atomic p level of the isolated metal atom. Only the CVMO are energetically suitable for ligand bonding or for containing metal electrons and the total number of CVMOs is a function of the cluster geometry. In molecular clusters this number of available MOs will determine the stoichiometry, the structure and the chemistry of the compound |31|.

The method has been applied to a series of molecular clusters containing less than 17 metal atoms with good results |31|. For larger clusters it has been shown |32| that any compact closed-packed M_n compound should have 6n+7 CVMOs or a number very close to this, where the factor 6n can be interpreted as originating from the fact that, for a large cluster, each metal atom contributes a maximum of six orbitals (one s and five d) as in the bulk metal.

The method is apparently more grounded on quantum-mechanical principles than the other electron counting rules but suffers a number of limitations (choice of the parameters of the model hamiltonian, dependence of the results upon the assumed bond distances, etc.). Moreover, for very large systems, where the MO spectrum tends to become a continuum, it is rather difficult to separate bonding from non-bonding orbitals.

3.5 Limitations of Qualitative Theories of Bonding

The above described methods have considerable value as support to the design of new synthetic routes for cluster species but do not provide much insight into the real nature of the metal-metal bond in transition metal clusters.

The formal analogy between boranes and transition metal clusters is very fascinating but it should not be overestimated as a true model of bonding in these latter systems. This analogy originates from topological and symmetry arguments, predicting qualitatively similar MO energy spectra for isostructural clusters. However, these arguments do not tell us the details of the composition of these MOs, and in particular do not adequately account for the participation of s, p, and d orbitals to the cluster bonding.

Nevertheless, the simple electron counting rules not only rationalize a large number of known solid state structures, but also agree with the stechiometry of new gas-phase cluster compounds. Recently it has been found |33| that molecular beams of nickel cluster ions reacting with carbon monoxide yield complexes of the type $Ni_n(CO)_k^+$ and $Ni_nC(CO)_k^+$ which correlate extremely well with Lauher's electroun counting rules |31|.

Hopefully, in a near future it will be possible to correlate the electron counting rules, mainly based on topological arguments, with a deeper understanding of the bonding derived from more quantitative theoretical models.

3.6 Quantum-chemical Calculations

Several theoretical studies on molecular metal clusters have been reported since the pioneering work of Brown on $Fe_3(CO)_{12}$ |34|. Most of these studies have been performed using the EHMO, CNDO, Fenske-Hall, and Hartree-Fock Slater (HFS or Xα) theoretical models |35,36| based on the LCAO approach.

These investigations provide a theoretically grounded analysis of the bonding in terms of MO energy levels and, in some cases, of energy considerations. In particular, qualitative correlations have been established between bond orders and experimental bond distances, electronic transitions and observed optical spectra, density of states and photoelectron spectra, and, in few cases, spin distribution and EPR data, HOMO energies and redox potentials |35,36|.

From these studies it appears that, in general, the s and p character in the bonding is not large but essential since the d orbitals are better suited for metal-metal and metal-ligand bonding if s and p orbitals are mixed in. The strength of the metal-metal bond is found to increase along a series with increasing atomic number, due to the larger extension of the d-d overlap in heavier elements.

Although the bonding is often interpreted in terms of HOMO-LUMO interaction of fragments, recently it has been shown |36i| that some cluster properties (e.g. the redox behaviour) are more dependent upon the net cluster core charge rather than simply correlated with the HOMO position. This supports the notion that the energy of individual orbitals does not reflect the molecular properties and that metal clusters behave as delocalized sources or sinks of electrons.

The determination and understanding of the magnetic behaviour of molecular clusters and small particles is one of the aspects which can help to elucidate the electronic structure of these systems |37|. In particular, the different magnetic behaviour of carbonylated and ligand-free clusters can provide valuable information. Most of the isolated cluster complexes are diamagnetic |1| whereas the corresponding naked frameworks usually have high-spin ground states |38|.

We have addressed this problem by means of spin unrestrictred INDO calculations on Ni_n and $Ni_n(CO)_m$ clusters |39|. In agreement with ab initio studies |40|, our approximate calculations indicate that small bare Ni clusters are characterized by high-spin ground states. Apparently, the bonding in these systems originates mainly from 4s-4s interaction since the 3d electrons belong to strongly localized shells. The average configuration of the Ni atoms is $3d^9 4s^1$ and the spin ordering is ferromagnetic as in the bulk metal |39,40|.

The main effect of the progressive carbonylation of the cluster is a partial or total quenching of the cluster magnetism. The Ni_5 bipyramidal trigonal cluster, for instance, is characterized by a quintet ground state. Two unpaired 3d electrons are shared by the three basal Ni atoms and the two other unpaired electrons are localized one on each of the two apical atoms (Fig. 5). In $Ni_5(CO)_9$ the ground state changes from quintet to triplet with the two unpaired electrons localized on the 4s and 3d orbitals of the coordinatively unsaturated Ni atom (Fig. 5). Finally, in the fully carbonylated $Ni_5(CO)_{12}$ system the CO ligands totally annihilate the spin leading to a diamagnetic ground state |39|.

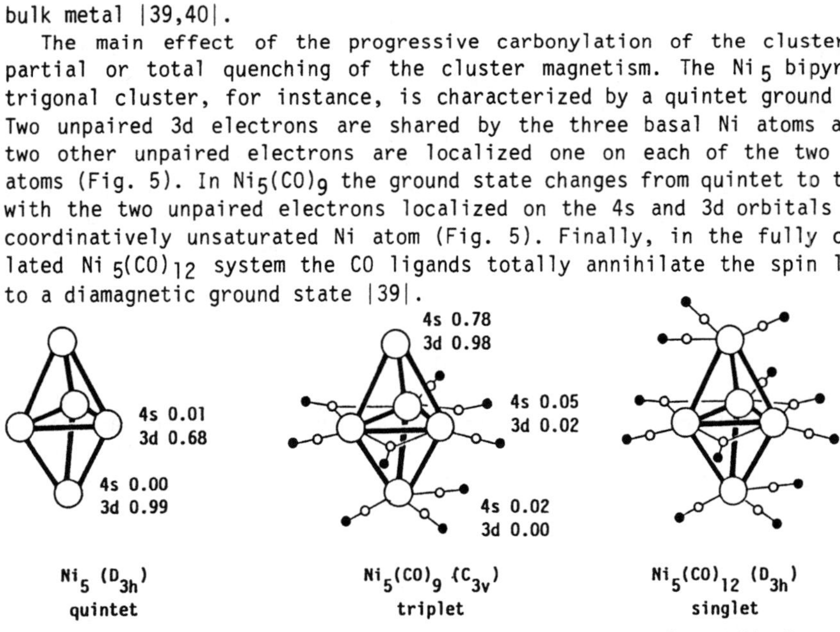

Fig. 5 - Computed spin distribution in bare and carbonylated Ni clusters |39|

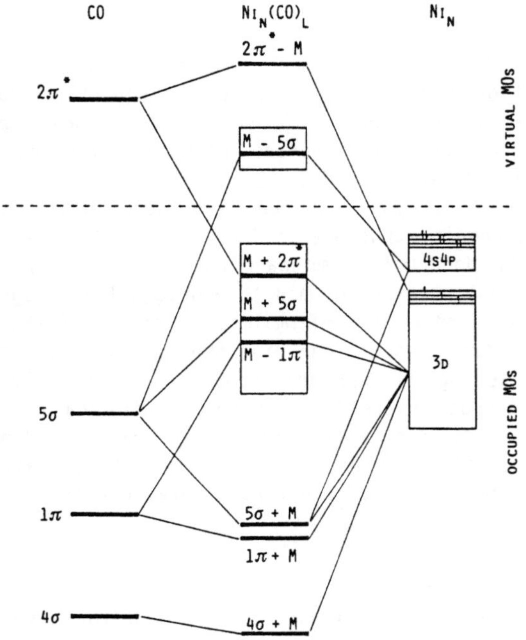

Fig. 6 - Schematic MO diagram for Ni-CO interaction (see text)

Two mechanisms contribute to the quenching of the magnetism of the bare metal cluster (Fig. 6). Firstly, the repulsive interaction with the 5σ MO of CO |41| destabilizes the 4s-derived MOs which are shifted above the "Fermi level" provided that a sufficient number of CO ligands interacts with the metal center (coordinative saturation). Secondly, the π back-donation from the metal orbitals into the $2\pi^*$ MO on CO generates a partial positive charge on the Ni atoms. This causes a 4s ⟶ 3d electron shift associated with a spin pairing which increases the 3d population and favours the d_π-CO back-bonding.

Similarily, according to SCF-Xα calculations |36h|, it has been found that the bare Co_4 cluster is highly magnetic (6 unpaired electrons) while in the carbonylated $(CO)_3Co_4(\mu_2\text{-}CO)_3$ system there are only four unpaired electrons with the spin density localized in the coordinatively unsaturated basal plane. The experimentally known $Co_4(CO)_{12}$ species, which is coordinatively staurated, is diamagnetic. It is worth noting that CO adsorption on Ni clusters and surfaces analogously decreases the magnetic moments of Ni atoms near the adsorption site |42|.

These first results suggest that strong changes occur in the metal-metal interaction as a consequence of the carbonylation. In the case of Ni the most favourable configuration for the formation of metal-metal bonds (high 4s occupancy) opposes the optimum metal-CO interaction which requires the lowest 4s occupancy. This seems to indicate that in carbonylated Ni clusters the strength of the direct metal-metal bonds is smaller than in their bare counterparts. According to this view, the role of the ligands could be that to stabilize a structure with relatively low intrinsic stability |43|.

4. CONCLUSIONS

The similarities between surface science and cluster organometallic chemistry prompted the formulation of the cluster-surface analogy |3|. Many fundamental questions arise as a consequence. Among them, why are small naked metal clusters such good catalysts whereas carbonyl clusters of similar nuclearity are rather inert species |4|?

The theoretical results obtained in the last few years seem to indicate that parallels between the electronic structure of bare metal clusters and molecular metal clusters must be done with some care and that the proposed cluster-surface analogy may be less straigthforward than originally thought.

Only the rigorous theoretical treatment of molecular metal clusters, although extremely difficult at the present, can give the necessary insight to answer basic questions including 1) the relationships between simple topological rules and more rigorous theoretical models, 2) the nature of the metal-metal bond, 3) the s,p, and d contribution to this bonding, 4) the number and spacing of excited states, and, finally, 5) the extent of the similarity between organometallic clusters, bare clusters and metal surfaces.

Acknowledgment

Thanks are due to Profs. P. Fantucci, A. Sironi, and R. Ugo for several interesting and stimulating discussions.

5. LITERATURE

1. a) G. Schmidt, Structure and Bonding 62, 51 (1985). b) P. Chini, G. Longoni, and V. Albano, Adv. Organomet. Chem. 14, 285 (1976). c) P.Chini J. Organomet. Chem. 200, 37 (1980). d) M. Moskovits, Ed., Metal Clusters John-Wiley & Sons, New York 1986.
2. J.M. Basset and R. Ugo, in: Aspects of Homogenous Catalysis, Ugo R., Ed., Reidel, Dordrecht 1977.
3. E.L. Mutetterties, T.N. Rhodin, E. Band, C.F. Brucker, and W. Pretzer, Chem. Rev. 79, 91 (1979).
4. a) B.C. Gates, L. Gutzi, and H. Knözinger, Ed.s, Metal Clusters in Catalysis, Elsevier, Amsterdam 1986. b) E.L. Muetterties and M.J. Krause Angew. Chem. Int. Ed. Eng. 22, 135 (1983).
5. A. Ceriotti, F. Demartin, G. Longoni, M. Manassero, M. Marchiona, G. Piva and M. Sansoni, Angew. Chem. Int. Ed. Eng. 24, 697 (1985).
6. S. Martinengo, G. Ciani and A. Sironi, J. Am. Chem. Soc. 102, 7564 (1980)
7. G.Longoni, M.Manassero and M.Sansoni, J. Am. Chem. Soc. 102, 7974 (1980).
8. G.Doyle, K.A.Eriksen and D.van Engen, J. Am. Chem. Soc. 108, 445 (1986).
9. C.E. Briant, B.R.C. Theobald, J.W. White, L.K. Bell, D.M.P. Mingos, and A.J. Welch, J.C.S. Chem. Comm. 201 (1981).
10. D.M. Washecheck, E.J. Wucherer, L.F. Dahl, A. Ceriotti, G. Longoni, M. Manassero, M. Sansoni and P. Chini, J. Am. Chem. Soc. 101, 6110 (1979).
11. D. Fenske, J. Ohmer, and J. Hachgenei, Angew. Chem. Int. Ed. Eng. 24, 993 (1985).
12. B.K. Teo and K. Keating, J. Am. Chem. Soc. 106, 2224 (1984).
13. J.L. Vidal and J.M. Troup, J. Organomet. Chem. 213, 351 (1981).

14. J.L.Vidal, L.A.Kapicak and J.M.Troup, J. Organomet. Chem. 215, C11 (1981)
15. C. Couture and D.H. Farraw, J.C.S. Chem. Comm. 197 (1985).
16. R. Goddard, P.W. Jolly, C. Krüger, K.P. Schick and G. Wilke, Organometal. 1, 1709 (1982) and references therein.
17. J.W.A.Van d. Velden, P.T.Beurskeus, J.J. Bour, W.P. Bosman, J.H.Noordik, M. Kolenbrander and J.A.K.M. Buskes, Inorg. Chem. 23, 146 (1984).
18. a) V. Albano and S. Martinengo, Nach. Chem. Tech. Lab. 28, 654 (1980). b) J.L. Vidal and J.M. Troup, J. Organomet. Chem. 213, 351 (1981).
19. A. Arrigoni, A. Ceriotti, R. Della Pergola, G. Longoni, M. Manassero, N. Masciocchi and M. Sansoni, Ang. Chem. Int. Ed. Eng. 23, 322 (1984).
20. A. Ceriotti, G. Longoni, M. Manassero, N. Masciocchi, G.Piro, L.Resconi, and M. Sansoni, J.C.S. Chem. Comm. 1402 (1985).
21. M. Tachicawa and E.L. Muetterties, J. Am. Chem. Soc. 102, 4541 (1980).
22. G. Gervasio, R. Rossetti, P.L. Stanghellini, and G. Bor, Inorg. Chem. 23, 2073 (1984).
23. K.M. MacKay, B.K. Nicholson, W.T. Robinson, and A. Sims, J.C.S. Chem. Comm. 1276 (1984).
24. a) B.F.G. Johnson and R.E. Benfield, J.C.S. Dalton 1554 (1978). b) R.E. Benfield and B.F.G. Johnson, ibidem 1743 (1980).
25. R.J. Gillespie, J. Chem. Education 40, 295 (1963).
26. a) R. Mason, K.M. Thomas and D.M.P. Mingos, J. Am. Chem. Soc. 95, 3802 (1973). b) D.M.P. Mingos, Acc. Chem. Res. 17, 311 (1984). c) Chem.Soc. Rev. 15, 31 (1986). d) J.C.S. Chem. Comm. 1352 (1985).
27. a) R.E. Williams, Inorg. Chem. 10, 210 (1971). b) K. Wade, J.C.S. Chem. Comm. 792 (1971). c) D.M.P. Mingos, Nature 236, 99 (1972). d) R.W. Rudolph, Acc. Chem. Res. 9, 446 (1976).
28. a) R. Hoffmann, Science 211, 995 (1981). b) Angew. Chem. Int. Ed. Eng. 21, 711 (1982).
29. a) A.J. Stone, Mol. Phys. 41, 1339 (1980). b) Inorg. Chem. 20, 563 (1981) c) Polyhedron 3, 1299 (1984).
30. a) B.K. Teo, Inorg. Chem. 23, 1251 (1984). b) B.K. Teo, G. Longoni and F.R.K. Chung, ibidem 23, 1257 (1984). c) B.K. Teo, ibidem 24, 1627 (1985). d) ibidem 24, 4209 (1985).
31. a) J.W.Lauher, J.Am.Chem.Soc. 100, 5305 (1978). b) ibidem 101, 2604 (1979)
32. G. Ciani and A. Sironi, J. Organomet. Chem. 197, 233 (1980).
33. P.Fayet, M.J.McGlinchey and L.H.Wöste, J.Am.Chem.Soc. 109, 1733 (1987).
34. D.A. Brown, J. Inorg. Nucl. Chem. 5, 289 (1958).
35. M.K. Manning and W. Trogler, Coord. Chem. Rev. 38, 89 (1981).
36. For some recent work see for instance: a) B.W. Bullett, Chem. Phys. Lett. 115, 450 (1985). b) ibidem 135, 373 (1987). c) P. Fantucci, G. Pacchioni and V. Valenti, Inorg. Chem. 23, 247 (1984). d) R.G. Woolley, Inorg. Chem. 24, 3519 (1985). e) ibidem 24, 3525 (1985). f) J.F. Halet, R. Hoffmann and J.Y. Saillard, Inorg. Chem. 24, 1695 (1985). g) D.J. Underwood, R. Hoffmann, K. Tatsumi, A. Nakamura and Y. Yamamoto, J.Am. Chem. Soc. 107, 5968 (1985). h) G.F. Holland, D.E. Ellis and W.C.Trogler J. Chem. Phys. 83, 3507 (1985). i) J. Am. Chem. Soc. 108, 1884 (1986).
37. a) D.C. Johnson, R.E. Benfield, P.P. Edwards, W.J.H. Nelson and M.B. Vargas, Nature 314, 231 (1985). b) B.K. Teo, F.J. DiSalvo, J.V.Waszczak, G. Longoni and A. Ceriotti, Inorg. Chem. 25, 2262 (1986). c) B.J. Pronk, H.B. Brown, L.J. deJongh, G. Longoni and A. Ceriotti, Solid State Comm. 59, 349 (1986).

38. a) For a review of experimental work see M.D. Morse, Chem. Rev. 86, 1049 (1986). b) For a review of theoretical work see J. Koutecký and P. Fantucci, Chem. Rev. 86, 539 (1986).
39. G. Pacchioni and P. Fantucci, Chem. Phys. Lett. 134, 407 (1987).
40. a) S.P. Walch, J. Chem. Phys. 86, 5082 (1987). b) H. Basch, M.D. Newton and J.W. Moskowitz, J. Chem. Phys. 73, 4492 (1980).
41. a) P.S. Bagus, C.J. Nelin and C.W. Bauschlicher Jr., Phys Rev. B 28, 5423 (1983). b) J. Koutecký, G. Pacchioni and P. Fantucci, Chem. Phys. 99, 87 (1985).
42. a) F. Raatz and D.R. Salahub, Surf. Sci. 176, 219 (1986). b) C.W. Bauschlicher Jr. and C.J. Nelin, Chem. Phys. 108, 275 (1986). d) W. Gopel, Surf. Sci. 85, 400 (1985).
43. G. Pacchioni and J. Koutecký, J. Phys. Chem. 91, 2658 (1987).

Index of Contributors

MIX
Papier aus verantwortungsvollen Quellen
Paper from responsible sources
FSC® C105338

If you have any concerns about our products,
you can contact us on
ProductSafety@springernature.com

In case Publisher is established outside the EU,
the EU authorized representative is:
Springer Nature Customer Service Center GmbH
Europaplatz 3, 69115 Heidelberg, Germany

Printed by Libri Plureos GmbH
in Hamburg, Germany